Differential Equations: From Calculus to Dynamical Systems

Second Edition

AMS/MAA | TEXTBOOKS

VOL 43

Differential Equations: From Calculus to Dynamical Systems

Second Edition

Virginia W. Noonburg

Providence, Rhode Island

Committee on Books
Jennifer J. Quinn, Chair

MAA Textbooks Editorial Board
Stanley E. Seltzer, Editor

Bela Bajnok	Charles R. Hampton	Jeffrey L. Stuart
Matthias Beck	Suzanne Lynne Larson	Ron D. Taylor, Jr.
Heather Ann Dye	John Lorch	Elizabeth Thoren
William Robert Green	Michael J. McAsey	Ruth Vanderpool

2010 *Mathematics Subject Classification.* Primary 34-01, 35-01.

For additional information and updates on this book, visit
www.ams.org/bookpages/text-43

Library of Congress Cataloging-in-Publication Data

Names: Noonburg, V. W. (Virginia Walbran), 1931– author. | Noonburg, V. W. (Virginia Walbran), 1931– Ordinary differential equations.
Title: Differential equations: From calculus to dynamical systems / Virginia W. Noonburg.
Other titles: Ordinary differential equations
Description: Second edition. | Providence, Rhode Island: MAA Press, an imprint of the American Mathematical Society, [2019] | Series: AMS/MAA textbooks; volume 43 | Includes an index.
Identifiers: LCCN 2018025154 | ISBN 9781470444006 (alk. paper)
Subjects: LCSH: Differential equations. | AMS: Ordinary differential equations – Instructional exposition (textbooks, tutorial papers, etc.). msc | Partial differential equations – Instructional exposition (textbooks, tutorial papers, etc.). msc
Classification: LCC QA371 .N58 2019 | DDC 515/.35–dc23
LC record available at https://lccn.loc.gov/2018025154

Softcover ISBN: 978-1-4704-6329-8

Copying and reprinting. Individual readers of this publication, and nonprofit libraries acting for them, are permitted to make fair use of the material, such as to copy select pages for use in teaching or research. Permission is granted to quote brief passages from this publication in reviews, provided the customary acknowledgment of the source is given.

Republication, systematic copying, or multiple reproduction of any material in this publication is permitted only under license from the American Mathematical Society. Requests for permission to reuse portions of AMS publication content are handled by the Copyright Clearance Center. For more information, please visit www.ams.org/publications/pubpermissions.

Send requests for translation rights and licensed reprints to reprint-permission@ams.org.

© 2019 by the American Mathematical Society. All rights reserved.
The American Mathematical Society retains all rights
except those granted to the United States Government.
Printed in the United States of America.

∞ The paper used in this book is acid-free and falls within the guidelines
established to ensure permanence and durability.
Visit the AMS home page at https://www.ams.org/

10 9 8 7 6 5 4 3 2   24 23 22 21 20 19

Contents

Preface ix
    Acknowledgments ix

1 Introduction to Differential Equations 1
  1.1 Basic Terminology 2
      1.1.1 Ordinary vs. Partial Differential Equations 2
      1.1.2 Independent Variables, Dependent Variables, and Parameters 3
      1.1.3 Order of a Differential Equation 3
      1.1.4 What is a Solution? 3
      1.1.5 Systems of Differential Equations 5
  1.2 Families of Solutions, Initial-Value Problems 6
  1.3 Modeling with Differential Equations 11

2 First-order Differential Equations 19
  2.1 Separable First-order Equations 20
      2.1.1 Application 1: Population Growth 23
      2.1.2 Application 2: Newton's Law of Cooling 25
  2.2 Graphical Methods, the Slope Field 28
      2.2.1 Using Graphical Methods to Visualize Solutions 32
  2.3 Linear First-order Differential Equations 36
      2.3.1 Application: Single-compartment Mixing Problem 41
  2.4 Existence and Uniqueness of Solutions 44
  2.5 More Analytic Methods for Nonlinear First-order Equations 51
      2.5.1 Exact Differential Equations 51
      2.5.2 Bernoulli Equations 56
      2.5.3 Using Symmetries of the Slope Field 58
  2.6 Numerical Methods 59
      2.6.1 Euler's Method 60
      2.6.2 Improved Euler Method 64
      2.6.3 Fourth-order Runge-Kutta Method 66
  2.7 Autonomous Equations, the Phase Line 71
      2.7.1 Stability—Sinks, Sources, and Nodes 73
      Bifurcation in Equations with Parameters 74

3 Second-order Differential Equations 81
  3.1 General Theory of Homogeneous Linear Equations 82
  3.2 Homogeneous Linear Equations with Constant Coefficients 88
      3.2.1 Second-order Equation with Constant Coefficients 88

3.2.2 Equations of Order Greater Than Two	93
3.3 The Spring-mass Equation	95
3.3.1 Derivation of the Spring-mass Equation	95
3.3.2 The Unforced Spring-mass System	96
3.4 Nonhomogeneous Linear Equations	102
3.4.1 Method of Undetermined Coefficients	102
3.4.2 Variation of Parameters	109
3.5 The Forced Spring-mass System	114
Beats and Resonance	117
3.6 Linear Second-order Equations with Nonconstant Coefficients	125
3.6.1 The Cauchy-Euler Equation	125
3.6.2 Series Solutions	127
3.7 Autonomous Second-order Differential Equations	135
3.7.1 Numerical Methods	136
3.7.2 Autonomous Equations and the Phase Plane	137

4 Linear Systems of First-order Differential Equations — 145

4.1 Introduction to Systems	146
4.1.1 Writing Differential Equations as a First-order System	146
4.1.2 Linear Systems	147
4.2 Matrix Algebra	150
4.3 Eigenvalues and Eigenvectors	158
4.4 Analytic Solutions of the Linear System $\vec{x}' = \mathbf{A}\vec{x}$	165
4.4.1 Application 1: Mixing Problem with Two Compartments	169
4.4.2 Application 2: Double Spring-mass System	171
4.5 Large Linear Systems; the Matrix Exponential	176
4.5.1 Definition and Properties of the Matrix Exponential	176
4.5.2 Using the Matrix Exponential to Solve a Nonhomogeneous System	178
4.5.3 Application: Mixing Problem with Three Compartments	180

5 Geometry of Autonomous Systems — 183

5.1 The Phase Plane for Autonomous Systems	184
5.2 Geometric Behavior of Linear Autonomous Systems	187
5.2.1 Linear Systems with Real (Distinct, Nonzero) Eigenvalues	188
5.2.2 Linear Systems with Complex Eigenvalues	190
5.2.3 The Trace-determinant Plane	191
5.2.4 The Special Cases	193
5.3 Geometric Behavior of Nonlinear Autonomous Systems	198
5.3.1 Finding the Equilibrium Points	199
5.3.2 Determining the Type of an Equilibrium	200
5.3.3 A Limit Cycle—the Van der Pol Equation	204
5.4 Bifurcations for Systems	207
5.4.1 Bifurcation in a Spring-mass Model	208
5.4.2 Bifurcation of a Predator-prey Model	209
5.4.3 Bifurcation Analysis Applied to a Competing Species Model	211
5.5 Student Projects	215
5.5.1 The Wilson-Cowan Equations	215
5.5.2 A New Predator-prey Equation—Putting It All Together	218

Contents

6 Laplace Transforms — 221
- 6.1 Definition and Some Simple Laplace Transforms — 221
  - 6.1.1 Four Simple Laplace Transforms — 223
  - 6.1.2 Linearity of the Laplace Transform — 224
  - 6.1.3 Transforming the Derivative of $f(t)$ — 225
- 6.2 Solving Equations, the Inverse Laplace Transform — 227
  - 6.2.1 Partial Fraction Expansions — 228
- 6.3 Extending the Table — 232
  - 6.3.1 Inverting a Term with an Irreducible Quadratic Denominator — 233
  - 6.3.2 Solving Linear Systems with Laplace Transforms — 236
- 6.4 The Unit Step Function — 239
- 6.5 Convolution and the Impulse Function — 251
  - 6.5.1 The Convolution Integral — 251
  - 6.5.2 The Impulse Function — 253
  - 6.5.3 Impulse Response of a Linear, Time-invariant System — 256

7 Introduction to Partial Differential Equations — 261
- 7.1 Solving Partial Differential Equations — 261
  - 7.1.1 An Overview of the Method of Separation of Variables — 263
- 7.2 Orthogonal Functions and Trigonometric Fourier Series — 266
  - 7.2.1 Orthogonal Families of Functions — 266
  - 7.2.2 Properties of Fourier Series, Cosine and Sine Series — 269
- 7.3 Boundary-Value Problems: Sturm-Liouville Equations — 274

8 Solving Second-order Partial Differential Equations — 285
- 8.1 Classification of Linear Second-order Partial Differential Equations — 285
- 8.2 The 1-dimensional Heat Equation — 289
  - 8.2.1 Solution of the Heat Equation by Separation of Variables — 291
  - 8.2.2 Other Boundary Conditions for the Heat Equation — 294
- 8.3 The 1-dimensional Wave Equation — 299
  - 8.3.1 Solution of the Wave Equation by Separation of Variables — 300
  - 8.3.2 D'Alembert's Solution of the Wave Equation on an Infinite Interval — 305
- 8.4 Numerical Solution of Parabolic and Hyperbolic Equations — 309
- 8.5 Laplace's Equation — 318
- 8.6 Student Project: Harvested Diffusive Logistic Equation — 324

Appendix — 329
- A Answers to Odd-numbered Exercises — 331
- B Derivative and Integral Formulas — 385
- C Cofactor Method for Determinants — 387
- D Cramer's Rule for Solving Systems of Linear Equations — 389
- E The Wronskian — 391
- F Table of Laplace Transforms — 393
- G Review of Partial Derivatives — 395

Index — 397

Preface

This book is meant for anyone who has successfully completed a course in calculus and who now wants a solid introduction to differential equations. The material contained in the book is the standard material found in a two-semester undergraduate course in differential equations, normally taken by engineering students in their sophomore year. There are two basic differences between this text and the others: first, there is a much greater emphasis on phase plane analysis applied particularly to the study of bifurcations, and second, a concerted effort has been put forth to make the book as readable as possible (think student-centered). This means that in addition to being an up-to-date text for the standard two-semester course in DEs, it is an obvious choice for a course which is being "flipped"; it is also a very appropriate choice for a student who wants to either learn or review differential equations on his or her own.

In the preface to the first edition I emphasized that I would like to see all science students, not only the prospective engineers, take a course in differential equations early in their career. This applies as well to all of the new material in the second edition. The two completely new chapters on partial differential equations contain all of the standard material on second-order linear PDEs, including a section on Fourier series and a section on Sturm-Liouville boundary-value problems; there is also enough material on numerical solutions to make it possible for students to write their own simple programs to solve the type of nonlinear PDEs that appear, for example, in population biology problems.

Acknowledgments. I want to thank all of the people, both at the MAA and the AMS, who have done such a great job in putting this book together. Special thanks are due to my AMS editor, Becky Rivard, whose efforts to make sure that everything was done right are greatly appreciated.

The person most responsible for both the first and second editions of this book being in print is Stephen Kennedy, and it is with deep gratitude that I acknowledge his help and encouragement.

At this point I would like to admit that a large amount of what I know about modern engineering and biology is the result of talking with my two sons, Derek (a computer engineer) and Erik (a population biologist). With much appreciation, therefore, this book is dedicated to them.

1

Introduction to Differential Equations

Differential equations arise from real-world problems and problems in applied mathematics. One of the first things you are taught in calculus is that the derivative of a function is the instantaneous rate of change of the function with respect to its independent variable. When mathematics is applied to real-world problems, it is often the case that finding a relation between a function and its rate of change is easier than finding a formula for the function itself; it is this relation between an unknown function and its derivatives that produces a differential equation.

To give a very simple example, a biologist studying the growth of a population, with size at time t given by the function $P(t)$, might make the very simple, but logical, assumption that a population grows at a rate directly proportional to its size. In mathematical notation, the equation for $P(t)$ could then be written as

$$\frac{dP}{dt} = rP(t),$$

where the constant of proportionality, r, would probably be determined experimentally by biologists working in the field. Equations used for modeling population growth can be much more complicated than this, sometimes involving scores of interacting populations with different properties; however, almost any population model is based on equations similar to this.

In an analogous manner, a physicist might argue that all the forces acting on a particular moving body at time t depend only on its position $x(t)$ and its velocity $x'(t)$. He could then use Newton's second law to express mass times acceleration as $mx''(t)$ and write an equation for $x(t)$ in the form

$$mx''(t) = F(x(t), x'(t)),$$

where F is some function of two variables. One of the best-known equations of this type is the spring-mass equation

$$mx'' + bx' + kx = f(t), \qquad (1.1)$$

in which $x(t)$ is the position at time t of an object of mass m suspended on a spring, and b and k are the damping coefficient and spring constant, respectively. The function f represents an external force acting on the system. Notice that in (1.1), where x is a function of a single variable, we have used the convention of omitting the independent variable t, and have written x, x', and x'' for $x(t)$ and its derivatives.

In both of the examples, the problem has been written in the form of a differential equation, and the solution of the problem lies in finding a function $P(t)$, or $x(t)$, which makes the equation true.

1.1 Basic Terminology

Before beginning to tackle the problem of formulating and solving differential equations, it is necessary to understand some basic terminology. Our first and most fundamental definition is that of a differential equation itself.

Definition 1.1. A ***differential equation*** *is any equation involving an unknown function and one or more of its derivatives.*

The following are examples of differential equations:

1. $P'(t) = rP(t)(1 - P(t)/N) - H$ harvested population growth
2. $\frac{d^2x}{d\tau^2} + 0.9\frac{dx}{d\tau} + 2x = 0$ spring-mass equation
3. $I''(t) + 4I(t) = \sin(\omega t)$ RLC circuit showing "beats"
4. $y''(t) + \mu(y^2(t) - 1)y'(t) + y(t) = 0$ van der Pol equation
5. $\frac{\partial^2}{\partial x^2}u(x,y) + \frac{\partial^2}{\partial y^2}u(x,y) = 0$ Laplace's equation

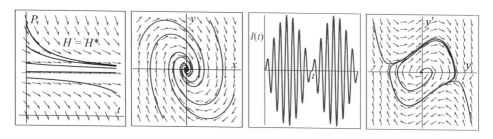

For the first four equations the graphs above illustrate different ways of picturing the solution curves.

1.1.1 Ordinary vs. Partial Differential Equations.
Differential equations fall into two very broad categories, called ordinary differential equations and partial differential equations. If the unknown function in the equation is a function of only one variable, the equation is called an **ordinary differential equation**. In the list of examples, equations 1–4 are ordinary differential equations, with the unknown functions being $P(t)$, $x(\tau)$, $I(t)$, and $y(t)$ respectively. If the unknown function in the equation depends on more than one independent variable, the equation is called a **partial differential equation**, and in this case, the derivatives appearing in the equation will be

1.1 Basic Terminology

partial derivatives. Equation 5 is an example of an important partial differential equation, called Laplace's equation, which arises in several areas of applied mathematics. In equation 5, u is a function of the two independent variables x and y. An introduction to methods for solving partial differential equations is contained in Chapters 7 and 8. One of the basic methods involves reducing the partial differential equation to the solution of two or more ordinary differential equations, so it is important to have a solid grounding in ordinary differential equations first. Note that in the first six chapters of the book, the term differential equation is assumed to mean ordinary differential equation.

1.1.2 Independent Variables, Dependent Variables, and Parameters.

Three different types of quantities can appear in a differential equation. The unknown function, for which the equation is to be solved, is called the **dependent variable**, and when considering ordinary differential equations, the dependent variable is a function of a single **independent variable**. In addition to the independent and dependent variables, a third type of variable, called a **parameter**, may appear in the equation. A parameter is a quantity that remains fixed in any specification of the problem, but can vary from problem to problem. In this book, parameters will usually be real numbers, such as r, N, and H in equation 1, ω in equation 3, and μ in equation 4.

1.1.3 Order of a Differential Equation.

Another important way in which differential equations are classified is in terms of their order.

Definition 1.2. *The **order** of a differential equation is the order of the highest derivative of the unknown function that appears in the equation.*

The differential equation 1 is a first-order equation and the others are all second-order. Even though equation 5 is a partial differential equation, it is still said to be of second order since no derivatives (in this case partial derivatives) of order higher than two appear in the equation.

You may have noticed in the Table of Contents that some of the chapter headings refer to first-order or second-order differential equations. In some sense, first-order equations are thought of as being simpler than second-order equations. By the time you have worked through Chapter 2, you may not want to believe that this is true, and there are special cases where it definitely is not true; however, it is a useful way to distinguish between equations to which different methods of solution apply. In Chapter 4, we will see that solving ordinary differential equations of order greater than one can always be reduced to solving a system of first-order equations.

1.1.4 What is a Solution?

Given a differential equation, exactly what do we mean by a solution? It is first important to realize that we are looking for a function, and therefore it needs to be defined on some interval of its independent variable. Before computers were available, a solution of a differential equation usually referred to an analytic solution; that is, a formula obtained by algebraic methods or other methods of mathematical analysis such as integration and differentiation, from which exact values of the unknown function could be obtained.

Definition 1.3. An **analytic solution** *of a differential equation is a sufficiently differentiable function that, if substituted into the equation, together with the necessary derivatives, makes the equation an identity (a true statement for all values of the independent variable) over some interval of the independent variable.*

It is now possible, however, using sophisticated computer packages, to numerically approximate solutions to a differential equation to any desired degree of accuracy, even if no formula for the solution can be found. You will be introduced to numerical methods in Chapter 2, and many of the equations in later chapters will only be solvable using numerical or graphical methods.

Given an analytic solution, it is usually fairly easy to check whether or not it satisfies the equation. In Examples 1.1.1 and 1.1.2 a formula for the solution is given and you are only asked to verify that it satisfies the given differential equation.

Example 1.1.1. *Show that the function* $p(t) = e^{-2t}$ *is a solution of the differential equation*

$$x'' + 3x' + 2x = 0.$$

Solution. To show that it is a solution, compute the first and second derivatives of $p(t)$:

$$p'(t) = -2e^{-2t}$$
$$p''(t) = 4e^{-2t}.$$

With the three functions $p(t)$, $p'(t)$, and $p''(t)$ substituted into the differential equation in place of x, x', and x'', it becomes

$$(4e^{-2t}) + 3(-2e^{-2t}) + 2(e^{-2t}) = (4 - 6 + 2)(e^{-2t}) = (0)(e^{-2t}) \equiv 0,$$

which is an identity (in the independent variable t) for all real values of t. ∎

When showing that both sides of an equation are identical for all values of the variables, we will use the equivalence sign $\equiv$. This will be used as a convention throughout the book.

For practice, show that the function $q(t) = 3e^{-t}$ is also a solution of the equation $x'' + 3x' + 2x = 0$. It may seem surprising that two completely different functions satisfy this equation, but we will soon see that differential equations can have many solutions, in fact infinitely many. In the above example, the solutions p and q turned out to be functions that are defined for all real values of t. In the next example, things are not quite as simple.

Example 1.1.2. *Show that the function* $\phi(t) = (1 - t^2)^{1/2} \equiv \sqrt{1 - t^2}$ *is a solution of the differential equation* $x' = -t/x$.

Solution. First, notice that $\phi(t)$ is not even defined outside of the interval $-1 \leq t \leq 1$. In the interval $-1 < t < 1$, $\phi(t)$ can be differentiated by the chain rule (for powers of functions):

$$\phi'(t) = (1/2)(1 - t^2)^{-1/2}(-2t) = -t/(1 - t^2)^{1/2}.$$

The right-hand side of the equation $x' = -t/x$, with $\phi(t)$ substituted for x, is

$$-t/\phi(t) = -t/(1 - t^2)^{1/2},$$

which is identically equal to $\phi'(t)$ wherever ϕ and ϕ' are both defined. Therefore, $\phi(t)$ is a solution of the differential equation $x' = -t/x$ on the interval $(-1, 1)$. ∎

1.1 Basic Terminology

You may be wondering if there are any solutions of $x' = -t/x$ that exist outside of the interval $-1 < t < 1$, since the differential equation is certainly defined outside of that interval. This problem will be revisited in Section 2.4 when we study the existence and uniqueness of solutions, and it will be shown that solutions do exist throughout the entire (t, x)-plane.

1.1.5 Systems of Differential Equations.

In Chapter 4 we are going to study systems of differential equations, where two or more dependent variables are related to each other by differential equations. Linked equations of this sort appear in many real-world applications. As an example, ecologists studying the interaction between competing species in a particular ecosystem may find that the growth (think derivative, or rate of change) of each population can depend on the size of some or all of the other populations. To show that a set of formulas for the unknown populations is a solution of a system of this type, it must be shown that the functions, together with their derivatives, make every equation in the system an identity. The following simple example shows how this is done.

Example 1.1.3. *Show that the functions*

$$x(t) = e^{-t}, \quad y(t) = -4e^{-t}$$

form a solution of the system of differential equations

$$\begin{aligned} x'(t) &= 3x + y \\ y'(t) &= -4x - 2y. \end{aligned} \quad (1.2)$$

Solution. The derivatives that we need are $x'(t) = -e^{-t}$ and $y'(t) = -(-4e^{-t}) = 4e^{-t}$. Then substitution into (1.2) gives

$$3x + y = 3(e^{-t}) + (-4e^{-t}) = (3-4)e^{-t} = -e^{-t} \equiv x'(t),$$

$$-4x - 2y = -4(e^{-t}) - 2(-4e^{-t}) = (-4+8)e^{-t} = 4e^{-t} \equiv y'(t);$$

therefore, the given functions for x and y form a solution for the system. ■

Exercises 1.1. *For each equation 1–8 below, determine its order. Name the independent variable, the dependent variable, and any parameters in the equation.*

1. $dy/dt = y^2 - t$

2. $dP/dt = rP(1 - P/k)$

3. $dP/dt = rP(1 - P/k) - \frac{\beta P^2}{\alpha^2 + P^2}$

4. $mx'' + bx' + kx = 2t^5$, *assuming x is a function of t*

5. $x''' + 2x'' + x' + 3x = \sin(\omega t)$, *assuming x is a function of t*

6. $(ty'(t))' = \alpha e^t$

7. $d^2\theta/dt^2 + \sin(\theta) = 4\cos(t)$

8. $y'' + \varepsilon(y^2 - 1)y' + y = 0$, *assuming y is a function of t*

For each equation 9–17 below, show that the given function is a solution. Determine the largest interval or intervals of the independent variable over which the solution is defined, and satisfies the equation.

9. $2x'' + 6x' + 4x = 0$, $x(t) = e^{-2t}$

10. $x'' + 4x = 0$, $x(t) = \sin(2t) + \cos(2t)$

11. $t^2 x'' + 3tx' + x = 0$, $x(t) = 1/t$

12. $t^2 x'' + 3tx' + x = 0$, $x(t) = \ln(t)/t$

13. $P' = rP$, $P(t) = Ce^{rt}$, C any real number

14. $P' = rP(1 - P)$, $P(t) = 1/(1 + Ce^{-rt})$, C any real number

15. $x' = (t + 2)/x$, $x(t) = \sqrt{t^2 + 4t + 1}$

16. $x'' - 2tx' + 6x = 0$, $x(t) = 8t^3 - 12t$

17. $t^2 y'' + ty' + (t^2 - \frac{1}{4})y = 0$, $y(t) = \frac{\sin(t)}{\sqrt{t}}$

In the next four problems, show that the given functions form a solution of the system. Determine the largest interval of the independent variable over which the solution is defined, and satisfies the equations.

18. System: $x' = x - y$, $y' = -4x + y$, solution is $x = e^{-t} - e^{3t}$, $y = 2e^{-t} + 2e^{3t}$.

19. System: $x' = x - y$, $y' = 4x + y$, solution is $x = e^t \cos 2t - \frac{1}{2} e^t \sin 2t$, $y = e^t \cos 2t + 2e^t \sin 2t$.

20. System: $x' = y$, $y' = -10x - 2y$, solution is $x = e^{-t} \sin(3t)$, $y = e^{-t}(3\cos(3t) - \sin(3t))$.

21. System: $x' = x + 3y$, $y' = 4x + 2y$, solution is $x = 3e^{5t} + e^{-2t}$, $y = 4e^{5t} - e^{-2t}$.

1.2 Families of Solutions, Initial-Value Problems

In this section the solutions of some very simple differential equations will be examined in order to give you an understanding of the terms *n-parameter family of solutions* and *general solution* of a differential equation. You will also be shown how to use certain types of information to pick one particular solution out of a set of solutions.

While you do not yet have any formal methods for solving differential equations, there are some very simple equations that can be solved by inspection. One of these is

$$x' = x. \qquad (1.3)$$

This first-order differential equation asks you to find a function $x(t)$ which is equal to its own derivative at every value of t. Any calculus student knows one function that satisfies this property, namely the exponential function $x(t) = e^t$. In fact, one reason mathematicians use e as the basis of their exponential function is that e^t is the only function of the form a^t for which this is true (remember that in general, $\frac{d}{dt}(a^t) = a^t \ln(a)$, and $\ln(a) = 1$ only if $a = e$).

1.2 Families of Solutions, Initial-Value Problems

What may not be immediately clear is that the function

$$x(t) = Ce^t, \tag{1.4}$$

for any constant value of C, is also a solution of (1.3). Check it!

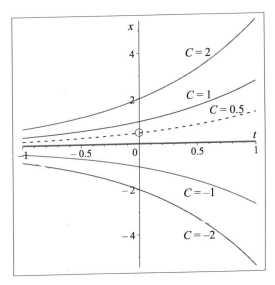

Figure 1.1. Curves in the one-parameter family $x = Ce^t$

This means that we have found an infinite set of solutions (1.4), called a **one-parameter family of solutions**, for the first-order differential equation (1.3). In (1.4) the parameter is denoted by C. We will also be able to show later that this family contains every possible solution of (1.3), and when this is the case we call the one-parameter family (1.4) a **general solution** of the differential equation. The term "general solution" is not standard mathematical notation, but is simply a way of denoting that it contains all of the solutions. In Figure 1.1 we have plotted $x(t) = Ce^t$ for several values of C, and you should be able to convince yourself that these curves are non-intersecting (do not cross), and fill up the (t, x)-plane.

To pick out a single solution curve it is only necessary to specify one point on the curve; that is, one **initial condition** of the form $x(t_0) = x_0$ must be given.

Definition 1.4. *A first-order differential equation with one initial condition specified is called an **initial-value problem**, usually abbreviated as an **IVP**. The solution of an IVP will be called a **particular solution** of the differential equation.*

Example 1.2.1. *Solve the IVP*

$$x' = x, \quad x(0) = \frac{1}{2}.$$

Solution. Since we just found that the general solution of $x' = x$ is $x(t) = Ce^t$, we only need to use the initial condition to determine the value of C. This will pick out one particular curve in the family.

Substituting $t = 0$ and $x(0) = \frac{1}{2}$ into the general solution (1.4),

$$x(0) = Ce^0 = C = \frac{1}{2}.$$

With $C = \frac{1}{2}$, the solution of the IVP is $x(t) = \frac{1}{2}e^t$. This particular solution is the dotted curve shown in Figure 1.1, with the initial point $(0, \frac{1}{2})$ circled. ∎

To see how this changes if the differential equation is of second order, consider the equation

$$x'' = -x. \tag{1.5}$$

There are two trigonometric functions, namely $\sin(t)$ and $\cos(t)$, that have the property that their second derivative is equal to the negative of the function. It is also easy to show that the function

$$x(t) = C_1 \sin(t) + C_2 \cos(t), \tag{1.6}$$

for any constants C_1 and C_2, satisfies (1.5). Check it! This time we have found a **two-parameter family of solutions** for the second-order equation, where the two parameters are denoted by C_1 and C_2. In Chapter 3 we will prove that *every* solution of (1.5) is of this form, and therefore (1.6) is again called a **general solution** of (1.5).

An **initial-value problem** for a second-order differential equation is going to require two initial conditions to determine the two constants C_1 and C_2.

To serve as **initial conditions**, it is necessary that the two conditions be the values of the function and its first derivative, both at the same value of t. If conditions are given at two different values of t, the problem is called a **boundary-value problem** and is, in general, much harder to solve. A very important type of boundary-value problem, called a Sturm-Liouville equation, will be studied in detail in Chapter 7 when we look at methods for solving partial differential equations.

Example 1.2.2. *Solve the IVP*

$$x'' = -x, \quad x(0) = 2, \quad x'(0) = -1.$$

Solution. The general solution was found to be

$$x(t) = C_1 \sin(t) + C_2 \cos(t).$$

We also need to have a formula for $x'(t)$, and this is obtained by differentiating $x(t)$:

$$x'(t) = C_1 \cos(t) - C_2 \sin(t).$$

If we use the two initial conditions, and let $t = 0$ in the formulas for x and x',

$$x(0) = C_1 \sin(0) + C_2 \cos(0) = C_1 \cdot 0 + C_2 \cdot 1 = C_2$$

and

$$x'(0) = C_1 \cos(0) - C_2 \sin(0) = C_1 \cdot 1 - C_2 \cdot 0 = C_1.$$

Therefore, the values of the two constants must be $C_1 = x'(0) = -1$ and $C_2 = x(0) = 2$, and the solution of the IVP is uniquely determined to be

$$x(t) = -\sin(t) + 2\cos(t).$$ ∎

1.2 Families of Solutions, Initial-Value Problems

Notice that we have no obvious way to picture the entire *two-parameter* family of solutions in the (t, x)-plane; much more will be said about this in Section 3.7 of Chapter 3 when the phase plane for a second-order differential equation is defined.

In summary, it will usually be the case that if an algebraic formula for a general solution of an nth order differential equation exists, it will contain n constants, and n initial conditions will be required to find a particular solution. It should be pointed out, however, that for the majority of differential equations, a simple formula for the general solution will not exist.

The equations $x' = x$ and $x'' = -x$ are examples of linear differential equations. This will be defined rigorously in Chapter 2, but for now it will mean that the right-hand side of the differential equation is a linear function of x and its derivatives; that is, a linear first-order equation must be of the form $x' = Ax + B$ where A and B can be functions of t, but not functions of x.

To see what can happen if the differential equation is not linear, let's look for solutions of the nonlinear first-order equation

$$x' = x^2. \tag{1.7}$$

It is not as easy to guess a solution for this equation, but in the first section of Chapter 2 you will be given a method for solving it. The function

$$x(t) = \frac{1}{C - t} \tag{1.8}$$

is a solution of (1.7) for any value of the constant C. Check it!

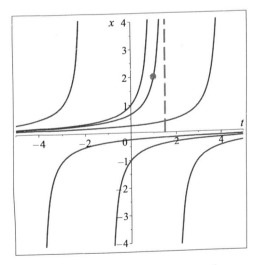

Figure 1.2. The curves $x(t) = \frac{1}{C-t}$

This one-parameter family of curves (1.8) is not quite a general solution; it does not contain the function $x(t) \equiv 0$, which obviously satisfies (1.7). However, the one-parameter family (1.8), together with the zero function does contain all solutions and can be considered a general solution. This family of curves is shown in Figure 1.2, plotted in the (t, x)-plane. It again appears to be a set that fills up the entire plane. For any initial condition $x(0) = x_0$ we can find the corresponding value of C. But there is a definite difference between this family of curves and the family shown in Figure 1.1. Can you spot it?

Every solution except the zero solution becomes infinite at a vertical asymptote. For example, to find the solution satisfying the initial condition $x(1) = 2$ we would set

$$x(1) = \frac{1}{C-1} = 2$$

and solve for C. Thus $C - 1 = \frac{1}{2}$ and $C = \frac{3}{2}$. The particular solution through this initial point is

$$x(t) = \frac{1}{\frac{3}{2} - t}$$

and it can be seen that $\lim_{t \to \frac{3}{2}^-} \left(\frac{1}{\frac{3}{2} - t} \right) = +\infty$. This curve, with its vertical asymptote, is shown in Figure 1.2.

This example is meant to make you suspicious of what can happen when you solve nonlinear equations. There are nice theorems for linear equations that give you exact intervals in which solutions are guaranteed to exist. The comparable theorems for nonlinear equations usually do not guarantee existence of a solution except in some small unspecified interval sufficiently close to the given initial point.

Exercises 1.2. *In problems 1–4, use differentiation to show that the given function is a solution of the equation for all values of the constants.*

1. Equation: $x' = x - 2$, function $x = 2 + Ce^t$.

2. Equation: $\frac{dy}{dx} = 4(x+1)y$, function $y = Ce^{2x^2 + 4x}$.

3. Equation: $\frac{dy}{dt} = y + t$, function $y = Ce^t - 1 - t$.

4. Equation: $x'' + x = 2e^t$, function $x = C_1 \sin(t) + C_2 \cos(t) + e^t$.

Solve the initial value problems 5–10, using the given family of solutions. In each case, first show that the family of solutions satisfies the equation for all values of the constants. Also state the exact interval in which the particular solution of the initial-value problem exists.

5. $x' = 2x$, $x(0) = -1$; family of solutions $x = Ce^{2t}$.

6. $y' = y^2$, $y(0) = -\frac{1}{2}$; family of solutions $y = \frac{1}{C-t}$.

7. $y' = 1 + y^2$, $y(0) = 1$; family of solutions $y = \tan(t + C)$.

8. $\frac{dx}{dt} = 2tx$, $x(0) = 1$; family of solutions $x = Ce^{t^2}$.

9. $x'' + x = t^2$, $x(0) = 0$, $x'(0) = 1$; family of solutions $x(t) = C_1 \sin(t) + C_2 \cos(t) + t^2 - 2$.

10. $x'' + 2x' + x = 0$, $x(0) = 1$, $x'(0) = -1$; family of solutions $x = C_1 e^{-t} + C_2 t e^{-t}$.

11. For the differential equation $x' = x^3$,

   (a) Show that $x(t) = \sqrt{\frac{1}{C-2t}}$ is a one-parameter family of solutions.

1.3 Modeling with Differential Equations

(b) *Find one solution of $x' = x^3$ that cannot be written in this form for any value of C.*

(c) *Find the solution satisfying the initial condition $x(0) = 1$, and sketch its graph.*

(d) *Does the solution in part (c) have a vertical asymptote? If so, for what value of t?*

COMPUTER PROBLEMS. *In Maple, the instructions*

```
dsolve(ODE) or dsolve({ODE,initcond},y(t))
```

can be used to produce the solution of an ordinary differential equation (ODE), if an analytic solution exists. If no initial conditions are specified, the solution will contain one or more parameters. The corresponding instructions in Mathematica are

```
DSolve[ODE,y[t],t] or DSolve[{ODE,initcond},y[t],t].
```

12. *Use a computer to solve the differential equation*

$$\frac{dy}{dt} + 2y(t) = 1. \tag{1.9}$$

The computer instructions are

```
dsolve(y'(t)+2*y(t)=1) in Maple
DSolve[y'[t]+2 y[t]==1,y[t],t] in Mathematica.
```

*In Maple the multiplication operator * cannot be omitted. Write the computer solution and check that it satisfies (1.9). How is the parameter represented in the computer solution?*

13. *Either by hand, or by executing the computer instruction*

   Maple: `dsolve({y'(t)+2*y(t)=1,y(0)=y0},y(t))`
   Mathematica: `DSolve[{y'[t]+2y[t]==1,y(0)==y0},y[t],t]`

   five times, with five different values for $y(0)$, solve equation (1.9) with initial conditions $y(0) = -1, 0, 1, 2,$ and 3. Use an ordinary plotting routine to plot the five solutions on the same set of axes, over the interval $-5 \le t \le 5$. Describe, as precisely as you can, the behavior of this set of curves. Especially note the behavior of $y(t)$ as $t \to \infty$.

14. *Use your own computer algebra system to solve the equation $x'' + x = t^2$. The equation should be entered in the form $x''(t) + x(t) = t^2$. Do you get the solution given in Exercise 9? How are the two parameters represented in the computer solution?*

1.3 Modeling with Differential Equations

To give you an idea of how differential equations arise in real-world problems, and what methods are used to solve them, this section presents examples from five different fields of science. By the time you have worked through Chapter 2 you will have a method for solving all of the first-order equations that appear in these five examples.

Some can be solved analytically by finding an algebraic formula for the solution, but others cannot. In the cases where no analytic solution exists, Maple has been used to draw both a slope field and some numerically computed solution curves. The slope field consists of a collection of short arrows that are drawn tangent to the solution curves. Much more will be said about slope fields in Section 2.2 of Chapter 2.

I PHYSICS

The first problem comes from physics, where we want to model the velocity $v(t)$ of a skydiver falling from a plane. In free fall, Newton's second law can be used to write mass × acceleration = sum of forces. Since acceleration is the derivative of velocity,

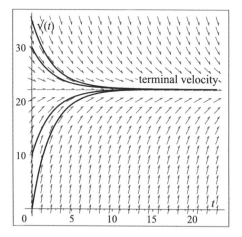

Figure 1.3. Solutions of $v' = 9.8 - 0.6v^{0.9}$

that is, $a(t) = v'(t)$, this results in the differential equation

$$mv'(t) = mg - k(v(t))^p,$$

where v is the velocity in meters/second, m is the mass in kilograms of the falling skydiver, $g = 9.8$ m/s^2 is the downward acceleration due to gravity, k is the coefficient of friction due to air resistance, and p is an exponent usually assumed to be equal to one. In the case $p = 1$, which implies that the friction due to the air resistance is directly proportional to the velocity, the differential equation can be solved analytically, and an exact formula

$$v(t) = \frac{mg}{k} + Ce^{-\frac{k}{m}t}$$

can be found for the velocity at time t. This is a one-parameter family of solutions. It contains a single constant C, which can be determined by specifying a single initial condition of the form $v(t_0) = v_0$. If $p \neq 1$, the problem becomes nonlinear and therefore harder to solve analytically. In this case a numerical solution can be found for any initial value. In Figure 1.3 numerically obtained solutions corresponding to an arbitrarily chosen set of initial values are shown for the equation with parameters $m = 72$, $g = 9.8$, $p = 0.9$, and $k/m = 0.6$. It is clear from the figure that the velocity tends to a constant value as $t \to \infty$; this is called the terminal velocity. When $p = 1$ the exact formula for $v(t)$ shows that the terminal velocity, which is equivalent to $\lim_{t \to \infty} v(t)$, is $\frac{mg}{k}$.

1.3 Modeling with Differential Equations

When we study autonomous equations in Section 2.7 we will see that for any positive p, the terminal velocity is

$$\lim_{t \to \infty} v(t) = \left(\frac{mg}{k}\right)^{1/p}.$$

Does this value agree with the terminal velocity pictured in Figure 1.3?

II MATHEMATICS

A problem that arises in several areas of applied mathematics is that of finding a one-

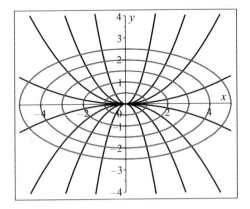

Figure 1.4. Orthogonal curves

parameter family of curves $y = F(x, C)$ that is **orthogonal** to a given family $\bar{y} = f(x, c)$ at each point in the (x, y)-plane. To make the two families of curves orthogonal, at each point in the plane their slopes must be negative reciprocals of each other. This means that at each point (x, y), if the slope of $\bar{y}(x)$ is m, the slope of the curve $y(x)$ must be $-1/m$.

To see how this results in a differential equation, let the given set of curves be the one-parameter family of parabolas $\bar{y} = cx^2$. The slope of a curve in this family is $m = \bar{y}' = \frac{d}{dx}(cx^2) = 2cx$, and using the equation for $\bar{y}$ we have $c = \bar{y}/x^2$, so $m = 2(\bar{y}/x^2)x = 2(\bar{y})/x$. This means that the curve in the orthogonal family will have slope $-x/2y$ at the point (x, y); hence the functions $y(x, C)$ in the orthogonal family must satisfy the differential equation $y' = -x/2y$. You will be shown in Chapter 2 that this is a "separable" differential equation, and you will be given a method for obtaining a formula for the solution. For the equation $y' = -x/2y$, the solution can be written as an implicit function of the form $y^2 + x^2/2 = C$, which can be seen to be a one-parameter family of ellipses. Figure 1.4 shows a set of solution curves from each of the two families. These were simply plotted with a plot routine by using the formulas $\bar{y} = cx^2$ and $y^2 + x^2/2 = C$ with arbitrarily chosen values of the constants c and C.

III ENGINEERING

The current $i(t)$, in amperes, in a simple electrical circuit containing a resistor, an inductor, and a sinusoidal electromotive source can be modeled by the equation $Li'(t) + Ri(t) = p \sin(t)$, where L is the inductance in henrys, R is the resistance in ohms, and

$E(t) = p\sin(t)$ is a periodic electromotive force in volts. This equation turns out to be similar to a simplified version of the spring-mass equation (studied in Chapter 3), and

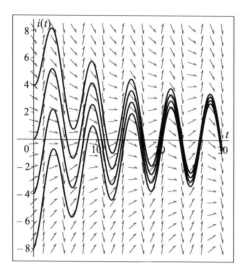

Figure 1.5. Solutions of $i'(t) + 0.1 i(t) = 3\sin(t)$

is linear in the dependent variable i. Because the equation is linear with constant coefficients L and R, it is possible to find a formula for the general solution by an analytic method. For this equation the solution will be found to be

$$i(t) = Ce^{-\frac{R}{L}t} + \frac{p}{L^2 + R^2}(R\sin(t) - L\cos(t)).$$

When we study the uniqueness and existence of solutions in Section 2.4 we will see that this one-parameter family of curves fills the plane, and no two curves in the family can intersect.

For the equation with $L = 1$, $R = 0.1$, and $p = 3$, four solution curves are shown in Figure 1.5, each satisfying a different initial condition at $t = 0$. These curves were drawn by a Maple program that solves differential equations and plots a slope field of tangent vectors to the solution curves. This means that even though an analytic formula exists for the solution, so that you could plot the curves with a simple plot routine, it is possible to check the solutions using the differential equations program. It can be seen, both from the formula for the solution and from the graph itself, that in the limit as $t \to \infty$ the curves $i(t)$ tend to the sinusoidal curve $\frac{p}{L^2+R^2}(R\sin(t) - L\cos(t))$. With the parameter values used for Figure 1.5, this steady state curve is

$$i(t) = \frac{3}{1.01}(0.1\sin(t) - \cos(t)).$$

The function $Ce^{-\frac{R}{L}t}$ in the formula for the solution is the transient, which dies out after the switch is closed. Much more will be said about this in Chapter 3.

IV ECOLOGY

The use of differential equations in biology is currently increasing at a rapid pace. The particular equation $P'(t) = rP(t)(1 - P(t)/N)$, called the logistic growth equation, has

1.3 Modeling with Differential Equations

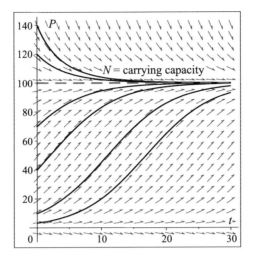

Figure 1.6. Solutions of the growth equation $P'(t) = 0.2P(t)(1 - P(t)/100)$

been around for a long time, but new ways to use it are still being found. The particular version of the equation given here models the growth of a population of size $P(t)$, which is assumed to grow exponentially like e^{rt} when the size of the population is small. The parameter r is called the **intrinsic growth rate** of the population. As the population increases it approaches a limiting value N, called the **carrying capacity** of the population in the ecosystem in which it lives.

This simple version of the equation can be solved analytically, and the formula for the solution is $P(t) = N/(1 - Ce^{-rt})$. The graph in Figure 1.6 shows the characteristic solution curves, all of which tend toward the carrying capacity if the initial condition is positive. For this application, negative initial conditions are not biologically meaningful.

The logistic growth equation has been used by ecologists to yield important information; for example, in the case of fish populations, it has been used to predict when they may be in danger of being depleted. In the remainder of the book, we will have much more to say about this equation and many other population models that are directly related to it.

V NEUROLOGY

Our final example is another application of mathematics to biology, and is currently of interest to research mathematicians as well as to computational neuroscientists. The equation $x' = -x + S(x - \theta + e(t))$ is a simplistic version of an equation representing the activity level of certain nerve cells in the brain. The function $x(t)$ in this equation is assumed to be the percent of nerve cells that are active at time t in a certain region of the brain. The function $e(t)$ represents activity being received from cells outside the region, and θ is a threshold level common to all cells in the region. The function S is called a **response function**, and has shape similar to that of the function $P(t)$ in the previous example (see Figure 1.6). It is used to simulate the effect that a given amount of input to the cells has on the rate of growth of their activity level. The function S is often chosen to be of the form $S(z) = 1/(1 + e^{-Kz})$, which varies from 0 to 1 as z goes

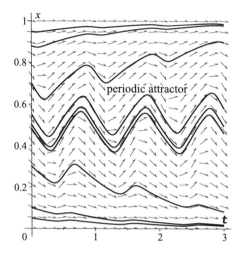

Figure 1.7. Solutions of the equation $x'(t) = -x(t) + 1/(1 + e^{-15(x(t)-0.5-0.3\cos(2\pi t))})$

from $-\infty$ to $+\infty$. This is a highly nonlinear function, making the differential equation difficult to analyze. In fact no exact formula for the solution exists.

Figure 1.7 shows a graphical solution of the differential equation produced by Maple, where the solution curves have been approximately obtained by numerical methods. The input function is $e(t) = -0.3\cos(2\pi t)$, the threshold $\theta = 0.5$, and K is arbitrarily chosen to be 15. With periodic input $e(t)$ to the cells, some very interesting things can happen. From the graph it appears that there is at least one oscillatory solution around the 50% activity level, which means that with an appropriate initial condition the activity level will oscillate with approximately half of the cells active and half inactive at any time $t > 0$. If the initial condition is too small, the activity level will die out and if it is too large it will approach a state where nearly 100% of the cells will be active. In a recent paper (R. Decker and V. W. Noonburg, A periodically forced Wilson-Cowan system with multiple attractors, *SIAM J. Math. Anal.* 2012, Vol. 44, pp. 887–905) a mathematical proof was given to show that there can actually be more than one periodic solution between the very high and very low levels of activity. This means that different initial states might cause different types of oscillatory behavior.

The five examples in this section are meant to show you how widespread the use of differential equations is. In the remainder of the book you will be given methods to solve these problems, and a great deal more. You will also be prepared for a more advanced course in dynamical systems. This is where you will learn what you need to know in order to read the current (and future) scientific literature, both in mathematics and even more importantly in applied fields such as engineering, biology, chemistry, ecology, and any of the many fields in which researchers are beginning to use advanced methods of mathematical analysis.

Exercises 1.3. *To do the exercises for this section you will need to refer back to the given application.*

1.3 Modeling with Differential Equations

- **PHYSICS:** Assuming $p = 1$ and $g = 9.8 m/s^2$, the equation $v' = g - \frac{k}{m}v$ has a solution
$$v(t) = \frac{mg}{k} + Ce^{-\frac{k}{m}t}.$$
A man drops from a high flying plane and falls for 5 seconds before opening his parachute. With the parachute closed, $\frac{k}{m} = 1\,\text{sec}^{-1}$.

  (a) Find the man's velocity when he opens his parachute (use the initial condition $v(0) = 0$ to find the value of the constant C).

  (b) After the chute opens, what must the value of $\frac{k}{m}$ be to get his terminal velocity down to 5 mph? (Use $1\,m/\text{sec} \approx 2.237$ mph.) Assume he has a very long way to fall.

- **MATHEMATICS:** Let $\bar{y}(x) = cx^3$ be a one-parameter family of cubic polynomials.

  (a) Show that the family of curves that is orthogonal at every point of the (x, y)-plane to this family of cubics satisfies the differential equation
  $$y' = -\frac{x}{3y}.$$

  (b) Show that the set of curves $x^2 + 3y^2 = C$ is a one-parameter family satisfying the differential equation $y' = -\frac{x}{3y}$ in part (a).

  (c) Either using a curve-plotting program, or by hand, sketch the curves $y = cx^3$ for $c = \pm 1$, $c = \pm 2$ and the curves $x^2 + 3y^2 = C$ for $C = 1, 2, 3$. If you make the scales on the x and y axes exactly the same, the curves should intersect at right angles.

- **ENGINEERING:** The diagram below shows a circuit containing the elements mentioned in the application; that is, a resistor with $R = 3$ ohms, an inductor with inductance $L = 1$ henry, and a generator producing a periodic electromotive force of $E = 10\sin(t)$ volts.

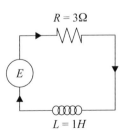

  (a) The current $i(t)$ will approach a periodic steady state value in the limit as $t \to \infty$. Find the equation for the steady state curve.

  (b) What initial value $i(0)$ could you use to produce this curve as the solution to the IVP for $Li'(t) + Ri(t) = E(t)$?

- **ECOLOGY:** A population of ants, initially containing 20 ants, is growing according to the population growth equation

$$P'(t) = 0.5P(t)\left(1 - \frac{P(t)}{N}\right),$$

with time t measured in days.

   (a) If the carrying capacity of the area in which they live is 1000, how many ants will there be one week later?

   (b) If bug spray is used, and it decreases the intrinsic growth rate r from 0.5 to 0.1, how many ants will you have one week later? Assume the carrying capacity remains unchanged at 1000 ants.

- **NEUROLOGY:** The response function $S_K(z) = \frac{1}{1+e^{-Kz}}$ is used to simulate the effect of input on the rate of growth of the activity level of a cell.

   (a) On the interval $-1 \leq z \leq 1$, sketch graphs of $S_K(z)$ for $K = 4$, 10, and 40. Which curve has the fastest rate of growth around $z = 0$? Notice that in every case, S is bounded between 0 and 1.

STUDENT PROJECT. In equations used to model activity of cells in the brain, other response functions $S(z)$ have been used to try to make the resulting differential equation easier to analyze. The main requirement is that a response function must increase monotonically from 0 to 1 as its argument goes from $-\infty$ to ∞.

   (i) Find constants a and b that make

   $$\tilde{S}_K(z) = \frac{\arctan(Kz) + a}{b}$$

   a response function. Show graphically the effect of the parameter K.

   (ii) Find constants c and d that make

   $$\bar{S}_K(z) = \frac{\tanh(Kz) + c}{d}$$

   a response function. Show graphically the effect of the parameter K.

   (iii) Find constants K_1 and K_2 such that $\tilde{S}_{K_1}(z)$ and $\bar{S}_{K_2}(z)$ are close to the function $S_{15}(z) = \frac{1}{1+e^{-15z}}$. Do this graphically and plot the three functions on the same set of axes. Note: there is no *exact* answer to this problem.

   (iv) Can you think of any other function that could be used as a response function? Remember that one goal is to make the function S as simple as possible.

2

First-order Differential Equations

In this chapter, methods will be given for solving first-order differential equations. Remember that first-order means that the first derivative of the unknown function is the highest derivative appearing in the equation. While this implies that the most general first-order differential equation has the form $F(t, x, x') = 0$ for some function F, in this chapter we will assume that the equation can be solved explicitly for x'. This means that our first-order differential equation can always be put in the form:

$$x' = f(t, x), \tag{2.1}$$

where f denotes an arbitrary function of two variables. To see why such an assumption makes sense, suppose the differential equation is

$$(x'(t))^2 + 4x'(t) + 3x(t) = t.$$

It would be messy, but not impossible, to use the quadratic formula to extract two differential equations of the form $x' = f(t, x)$ from this quadratic equation. However, one could also imagine equations where solving for $x'(t)$ is not even possible, and in such a case, some of our methods may not be applicable.

The material in this chapter will cover several analytic methods for solving first-order differential equations, each requiring the function f in (2.1) to have a special form. Two different graphical methods are also described; one for the general equation (2.1), and a more specific method for an autonomous equation where f is a function depending only on x. Numerical methods for first-order equations are introduced, and theoretical issues of existence and uniqueness of solutions are discussed. In the examples you will be presented with some real-world problems in applied mathematics. Acquiring a solid understanding of first-order equations will lay the groundwork for everything that follows.

2.1 Separable First-order Equations

The first analytic method we will consider applies to first-order equations that can be written in the form

$$dx/dt = g(t)h(x); \qquad (2.2)$$

that is, when the function $f(t,x)$ can be factored into a product of a function of t times a function of x. Such a differential equation is called **separable**.

Example 2.1.1. *Determine which of the following first-order differential equations are separable. Hint: try to factor the right-hand side if the equation does not initially appear to be separable.*

(a) $x' = xt + 2x$

(b) $x' = x + \cos(t)$

(c) $x' = xt^2 + t^2 - tx - t$

(d) $x' = x^2 + x + 3$

Solution. Equation (a) can be factored into $x' = x(t+2)$, so it is separable with $g(t) = t+2$ and $h(x) = x$. Equation (b) is not separable as $x + \cos(t)$ cannot be factored into a function of t multiplied by a function of x. With a bit of work, equation (c) can be factored as $x' = (t^2 - t)(x+1)$, so it is separable with $g(t) = (t^2 - t)$ and $h(x) = (x+1)$. Equation (d) can be written as $x' = (1)(x^2 + x + 3)$, with $g(t) = 1$ and $h(x) = x^2 + x + 3$, so it is separable. ∎

Special Cases of $x' = g(t)h(x)$.

(1) If $h(x) = 1$, the separable equation $x' = g(t)$ is just an integration problem, and the solution is

$$x = \int g(t)dt;$$

that is, x is just the **indefinite integral** of the function $g(t)$. Remember that this means that x can be *any* function $G(t)$ such that $G'(t) = g(t)$, and this introduces an arbitrary constant into the solution. As an example, the solution of $x' = t + 1$ is

$$x(t) = \int (t+1)dt = \frac{t^2}{2} + t + C.$$

Even in this simple case the solution is an infinite one-parameter family of functions.

(2) If $g(t) = 1$, the separable equation $x' = h(x)$ is called an **autonomous** first-order differential equation. Unless $h(x)$ is a constant, it is no longer possible to solve the equation by simple integration, and the method given below must be used. Autonomous first-order differential equations are important and will be investigated more thoroughly in Section 2.7. In the above examples, only equation (d) is autonomous. The other three contain functions of t (other than the unknown function $x(t)$) on the right-hand side.

You may have already been shown, in calculus, an easy method for solving separable equations. Consider the following example.

Example 2.1.2. *Solve the differential equation* $\frac{dx}{dt} = -tx^2$.

2.1 Separable First-order Equations

Solution. Split dx/dt into two pieces dx and dt, and do a bit of algebra to write

$$-\frac{dx}{x^2} = t\,dt.$$

Integrate each side with respect to its own variable to obtain

$$\int \left(-\frac{1}{x^2}\right) dx = \int t\,dt \implies \frac{1}{x} = \frac{t^2}{2} + C, \qquad (2.3)$$

where the arbitrary constants on each side have been collected on the right. Solve this equation for x to obtain the one-parameter family of solutions

$$x(t) = \frac{1}{t^2/2 + C}. \qquad \blacksquare$$

You should check that the function $x(t)$ does satisfy the differential equation for any value of the constant C. It appears that this method works, but splitting $\frac{dx}{dt}$ into two pieces is not a mathematically condoned operation; therefore, a justification of the method needs to be given.

If an equation is separable, and $x'(t)$ is written as dx/dt, both sides of the equation $dx/dt = g(t)h(x)$ can be divided by $h(x)$, and the equation becomes

$$\frac{1}{h(x(t))}\frac{dx}{dt} = g(t). \qquad (2.4)$$

The two sides of (2.4) will be identical if, and only if, their indefinite integrals are the same up to an additive constant; that is,

$$\int \frac{1}{h(x(t))}\left(\frac{dx}{dt}\right) dt = \int g(t)dt + C.$$

The method of simple substitution can be applied to the integral on the left. If we substitute $u = x(t)$, then $du = (dx/dt)dt$, and the equation becomes

$$\int [1/h(u)]du = \int g(t)dt + C. \qquad (2.5)$$

Now let $H(u)$ be any function such that $H'(u) = 1/h(u)$ and $G(t)$ any function with $G'(t) = g(t)$. Then (2.5) implies that

$$H(u) + C_1 = G(t) + C_2 \implies H(u) = G(t) + C,$$

where C is the constant $C_2 - C_1$.

Replacing u again by $x(t)$:

$$H(x(t)) = G(t) + C. \qquad (2.6)$$

Check carefully that the expression $H(x) = G(t) + C$ in (2.6) is exactly the same as the solution obtained in (2.3) in the above example. It is an **implicit solution** of (2.2); that is, it defines a relation between the unknown function x and its independent variable t. If it can be solved explicitly for x as a function of t, the result is called an **explicit solution** of the differential equation. As expected, the integration produces an infinite one-parameter family of solutions.

Everything that has been said so far justifies the following step-by-step procedure for solving separable equations.

> **To solve a separable first-order differential equation** $x'(t) = g(t)h(x)$:
>
> - Write the equation in the form $dx/dt = g(t)h(x)$.
>
> - Multiply both sides by dt, divide by $h(x)$, and integrate, to put the equation in the form
> $$\int [1/h(x)]dx = \int g(t)dt.$$
>
> - Find any function $H(x)$ such that $H'(x) = 1/h(x)$ and any function $G(t)$ such that $G'(t) = g(t)$.
>
> - Write the solution as $H(x) = G(t) + C$.
>
> - If possible, solve the equation from the previous step explicitly for x, as a function of t.

The next example shows how this method works.

Example 2.1.3. *Solve the separable differential equation* $x' = t(x+1)$.

Solution. First write the equation in the form
$$dx/dt = t(x+1). \tag{2.7}$$
Separate the variables (including dx and dt) so only the variable x appears on the left and t on the right:
$$\frac{1}{(x+1)}dx = tdt.$$
Integration of each side with respect to its own variable of integration leads to the implicit solution
$$\ln|x+1| = t^2/2 + C,$$
which can be solved explicitly for x by applying the exponential function:
$$e^{\ln|x+1|} = |x+1| = e^{(t^2/2+C)} = e^C e^{(t^2/2)}.$$
If the positive constant e^C is replaced by a nonzero constant A that can be either positive or negative, the absolute value signs can be dropped, to give
$$x + 1 = Ae^{(t^2/2)},$$
where $A = \pm e^C$. Then the explicit solution is
$$x(t) = Ae^{(t^2/2)} - 1. \tag{2.8}$$
∎

The formula for $x(t)$ is the one-parameter family of curves that we expect to get as the solution of a first-order differential equation, and notice that the parameter A was introduced in the step where the equation was integrated.

As we saw in Section 1.2, a **particular solution** to a first-order differential equation is a solution in which there are no arbitrary constants. It will be shown (in Section 2.4) that, in general, to obtain a particular solution of a first-order differential equation it is necessary and sufficient to give one initial condition of the form $x(t_0) = x_0$.

2.1 Separable First-order Equations

Again from Section 1.2, we call a one-parameter family of solutions of a first-order differential equation, containing a single constant of integration, a **general solution** if it contains every solution of the equation. The analytic solution of a separable equation, found by the method just described, will contain all solutions of the equation with the possible exception of **constant solutions**. These are solutions of the form $x(t) \equiv C$ which make the right-hand side of the differential equation $x' = f(t, x)$ identically equal to 0. Note that with $x \equiv C$ the differential equation is satisfied, because the derivative of a constant function is also zero; that is, $\frac{d}{dt}(C) \equiv 0 \equiv f(t, C)$.

Referring back to Example 2.1.3, the function $x \equiv -1$ is the only constant solution of the differential equation $x' = t(x+1)$. In this case it is given by the solution formula (2.8) when the constant $A = 0$; therefore, (2.8) is the general solution of (2.7) if we allow the value $A = 0$.

When solving a separable equation it is wise to find all constant solutions first, since they may be lost when the equation is divided by $h(x)$.

Example 2.1.4. *Solve the initial-value problem* $x' = t/x$, $x(0) = 1$.

Solution. First note that this differential equation has no constant solutions; that is, there are no constant values for x that make $t/x \equiv 0$. Write the equation as $dx/dt = t/x$. Then by multiplying by dt and x, and integrating,

$$x\,dx = t\,dt \implies \int x\,dx = \int t\,dt \implies x^2/2 = t^2/2 + C.$$

The expression $x^2/2 = t^2/2 + C$ is an implicit solution and yields two explicit solutions

$$x(t) = \pm\sqrt{t^2 + 2C}.$$

We can satisfy the initial condition by substituting $t = 0$ into the general solution and setting $x(0) = 1$:

$$x(0) = \pm\sqrt{0 + 2C} = 1.$$

This implies that C must be 1/2 and the sign of the square root must be taken to be positive. Now the unique solution to the initial-value problem, $x(t) = \sqrt{t^2 + 1}$, is completely determined. ∎

The following two applications show how separable differential equations and initial-value problems can arise in real-world situations.

2.1.1 Application 1: Population Growth.

One of the simplest differential equations arises in the study of the growth of biological populations. Consider a population with size $P(t)$ at time t. If it is assumed that the population has a constant birth rate α and constant death rate β per unit of time, then an equation for the rate of growth of the population is

$$dP/dt = \alpha P(t) - \beta P(t) = (\alpha - \beta)P(t) = rP(t), \tag{2.9}$$

where r is called the *net growth rate* of the population. This is a separable differential equation with general solution (Check it!):

$$P(t) = Ke^{rt},$$

where K is the arbitrary constant of integration. The initial value is frequently given as the size of the population at time $t = 0$. Then $P(0) = Ke^{r0} = K$, and the particular solution of this initial-value problem is $P(t) = P(0)e^{rt}$. This means that, t units of time after the initial time, the population will have grown exponentially (or decreased exponentially if $\beta > \alpha$). Populations do not grow exponentially forever, and biologists usually use more complicated equations of growth to take this into account.

One assumption that can be made is that as the population P increases, its growth rate decreases, due to the effects of crowding, intra-species competition, etc. The simplest way to decrease the growth rate as P increases is to assume that the growth rate is linear in P; that is, replace r in (2.9) by $R = r - \gamma P(t)$. Then

$$dP/dt = (r - \gamma P(t))P(t) = rP(t)\left(1 - \frac{\gamma}{r}P(t)\right) = rP(t)(1 - P(t)/N),$$

where we have defined a new constant $N = r/\gamma$. The equation

$$dP/dt = rP(t)(1 - P(t)/N) \tag{2.10}$$

is called the **logistic growth equation**.

Notice that the rate of growth dP/dt goes to 0 as $P(t) \to N$. This limiting value of the population, N, is called the **carrying capacity** of the ecosystem in which the population lives. The parameter r, which now gives the approximate rate of growth when the population is small, is called the **intrinsic growth rate** of P.

The logistic growth equation (2.10) is an autonomous differential equation and therefore is separable, but the expression $dP/[P(1 - P/N)]$ has to be integrated using partial fractions (or with the use of computer algebra). In either case, using our technique for separable equations, we have

$$\frac{dP}{[P(1 - P/N)]} = rdt \implies \int \frac{1}{P(1 - P/N)}dP = \int rdt. \tag{2.11}$$

To compute the integral on the left, we can use partial fractions to write

$$\frac{1}{P(1 - P/N)} \equiv \frac{1}{P} - \frac{1}{P - N}.$$

You should check this last equality carefully (there is a review of partial fraction expansions in Section 2 of Chapter 6).

Integration of (2.11), using the partial fraction expression, now results in

$$\ln |P| - \ln |P - N| = rt + K.$$

To solve for $P(t)$, apply the exponential function to both sides and use the properties of the exponential and logarithmic functions to write

$$e^{\ln |P| - \ln |P - N|} = e^{rt + K} \implies$$

$$\frac{P}{P - N} = K_1 e^{rt}, \text{ where } K_1 = \pm e^K \implies$$

$$P = PK_1 e^{rt} - NK_1 e^{rt} \implies$$

$$P - PK_1 e^{rt} = -NK_1 e^{rt} \implies$$

$$P = \frac{-NK_1 e^{rt}}{1 - K_1 e^{rt}} = \frac{NK_1 e^{rt}}{K_1 e^{rt} - 1} = \frac{N}{1 - Ce^{-rt}} \tag{2.12}$$

where $C = \frac{1}{K_1}$. Note that the differential equation $P' = rP(1 - P/N)$ has two constant solutions $P \equiv 0$ and $P \equiv N$. Using the value $C = 0$ in the solution (2.12) gives $P = N$, but no finite value of C makes this solution identically 0. To have a general solution, we must add the solution $P \equiv 0$ to the formula in (2.12).

2.1 Separable First-order Equations

In Section 2.4 the interval of existence of the solutions of the logistic population equation will be carefully examined and it will be shown that solutions with $P(0)$ between 0 and N exist for all t. Solutions with $P(0) > N$ exist for all $t > 0$ but have a vertical asymptote at a negative value of t. Solutions with $P(0) < 0$ tend to $-\infty$ at a positive value of t, but these are not physically realizable as populations.

2.1.2 Application 2: Newton's Law of Cooling. Newton's Law of Cooling is a well-known law of physics that states that if a small body of temperature T is placed in a room with constant air temperature A, the rate of change of the temperature T is directly proportional to the temperature difference $A - T$. This law can be expressed in the form of a differential equation:

$$T'(t) = k(A - T(t)),$$

where $T(t)$ is the temperature of the small body at time t, A is the surrounding (ambient) air temperature, and k is a positive constant that depends on the physical properties of the small body. The only constant solution of the equation is $T(t) \equiv A$, which says that if the body is initially at the ambient temperature, it will remain there.

The equation can be seen to be separable and can be solved by writing

$$dT/dt = k(A - T) \implies \int \frac{dT}{A - T} = \int k\,dt \implies -\ln|A - T| = kt + C \implies$$

$$|A - T| = e^{-(kt+C)} \implies A - T(t) = \alpha e^{-kt},$$

where $\alpha = \pm e^{-C}$ can be any positive or negative real number. The explicit solution is

$$T(t) = A - \alpha e^{-kt}. \tag{2.13}$$

The constant solution $T \equiv A$ is obtained from the formula by letting α have the value zero. The long-term behavior is very easy to determine here, since if $k > 0$, then $T(t) \to A$ as $t \to \infty$. Thus the temperature of the small body tends to the constant room temperature, which makes good sense physically.

Consider the following very practical example that uses Newton's Law of Cooling.

Example 2.1.5. *A cup of coffee, initially at temperature $T(0) = 210°$, is placed in a room in which the temperature is $70°$. If the temperature of the coffee after five minutes has dropped to $185°$, at what time will the coffee reach a nice drinkable temperature of $160°$?*

Solution. If we assume the cup of coffee cools according to Newton's Law of Cooling, the general solution given by (2.13) with $A = 70$, can be used to write

$$T(t) = 70 - \alpha e^{-kt}.$$

Using the initial condition, we can find the value of α:

$$T(0) = 70 - \alpha e^{-0k} = 70 - \alpha = 210 \implies \alpha = -140.$$

The temperature function can now be written as $T(t) = 70 + 140 e^{-kt}$. To find the value of the parameter k, use the given value $T(5)$:

$$T(5) = 70 + 140 e^{-5k} = 185 \implies e^{-5k} = \frac{115}{140} \implies$$

$$k = -\frac{1}{5}\ln\left(\frac{115}{140}\right) \approx 0.0393.$$

The value for k completely determines the temperature function; that is,
$$T(t) = 70 + 140e^{-0.0393t}$$
for all $t > 0$. Now the answer to the original question can be found by solving the equation $T(\hat{t}) = 160$ for $\hat{t}$. The approximate value for $\hat{t}$ is 11.2 minutes. ∎

In the last example, if the value of the physical parameter k had been known beforehand, only one value of the temperature would have been required to determine the function $T(t)$ exactly. In this problem, the value of the parameter k had to be determined experimentally from the given data, thus necessitating the temperature to be read at two different times. This sort of thing is even more likely to occur in problems that come from nonphysical sciences, where parameters are usually not known physical constants and must be experimentally determined from the data provided.

Exercises 2.1. *In problems 1–4, determine whether the equation is separable.*

1. $x' + 2x = e^{-t}$

2. $x' + 2x = 1$

3. $x' = \dfrac{x+1}{t+1}$

4. $x' = \dfrac{\sin t}{\cos x}$

Put equations 5–14 into the form $x'(t) = g(t)h(x)$, and solve by the method of separation of variables.

5. $x' = \dfrac{x}{t}$

6. $x' = \dfrac{t}{x}$

7. $x' = x + 5$

8. $x' = 3x - 2$

9. $x' = x\cos(t)$

10. $x' = (1+t)(2+x)$

11. $xx' = 1 + 2t + 3t^2$

12. $x' = (t+1)(\cos(x))^2$

13. $x' = t + tx^2$

14. $x' = 2 - tx^2 - t + 2x^2$ (*Hint: factor.*)

In 15–20, solve the initial-value problem.

15. $y' = y + 1,\ y(0) = 2$

16. $y' = ty,\ y(0) = 3$

17. $x' = x\cos(t),\ x(0) = 1$

18. $x' = (1+t)(2+x),\ x(0) = -1$

2.1 Separable First-order Equations

19. $x' = (t+1)(\cos(x))^2$, $x(0) = 1$

20. $P' = 2P(1-P)$, $P(0) = 1/2$

21. (*Newton's Law of Cooling*) A cold Pepsi is taken out of a 40° refrigerator and placed on a picnic table. Five minutes later the Pepsi has warmed up to 50°. If the outside temperature remains constant at 90°, what will be the temperature of the Pepsi after it has been on the table for twenty minutes? What happens to the temperature of the Pepsi over the long term?

22. (*Newton's Law of Cooling*) Disclaimer: The following problem is known not to be a very good physical example of Newton's Law of Cooling, since the thermal conductivity of a corpse is hard to measure; in spite of this, body temperature is often used to estimate time of death.

    At 7 AM one morning detectives find a murder victim in a closed meat locker. The temperature of the victim measures 88°. Assume the meat locker is always kept at 40°, and at the time of death the victim's temperature was 98.6°. When the body is finally removed at 8 AM, its temperature is 86°.

    (a) When did the murder occur?
    (b) How big an error in the time of death would result if the live body temperature was known only to be between 98.2° and 101.4°?

23. (*Orthogonal Curves*) In the exercises for Section 1.3 the family of curves orthogonal to the family $\bar{y} = cx^3$ was shown to satisfy the differential equation $y' = -\frac{x}{3y}$. Use the method for separable equations to solve this equation, and find a formula for the orthogonal family.

24. (*Orthogonal Curves*) Show that the family of curves orthogonal to the family $\bar{y} = ce^x$ satisfies the equation $y' = -\frac{1}{y}$ (you may want to refer back to the model labeled MATHEMATICS in Section 1.3). Solve this separable equation and plot three curves from both families. If you make the scales identical on both axes, the curves should appear perpendicular at their points of intersection.

25. (*Population Growth*) In this problem you are asked to compare two different ways of modeling a population; either by a simple exponential growth equation, or by a logistic growth model.

    Table 2.1. Census

t	year	pop (millions)
0	1800	5.2
0.5	1850	23.2
1.0	1900	76.2
1.5	1950	151.3
2.0	2000	281.4

    One population for which reasonable data is available is the population of the United States. The Census Table above gives census data for the population (in millions) from 1800 to 2000 in 50 year intervals. If the logistic function $P(t) = \frac{N}{1+Ce^{-rt}}$ is fit exactly to the three data points for 1800, 1900, and 2000, the values for the parameters are $N = 331.82$, $C = 62.811$, and $r = 2.93$, and the equation becomes

$$P(t) = \frac{331.82}{1 + 62.811e^{-2.93t}}.$$

Note that t is measured in hundreds of years, with $t = 0$ denoting the year 1800.

(a) *Plot (preferably using a computer) an accurate graph of P(t) and mark the five data points on the graph (three of them should lie* exactly *on the curve).*

(b) *Where does the point* $(t, P(t)) = (1.5, 151.3)$ *lie, relative to the curve? Can you think of a reason why—famine, disease, war?*

(c) *What does the logistic model predict for P(2.1); that is, the population in 2010? (The census data gives 308.7 million.)*

(d) *What does the model predict for the population in 2100? Does this seem reasonable?*

(e) *Now fit a simple exponential model $p(t) = ce^{rt}$ to the same data, using the two points for the years 1900 and 1950 to evaluate the parameters r and c. What does this model predict for the population in 2100? Does this seem more or less reasonable than the result in (d)?*

Note: If you use $t = 0$ for 1900 and $t = 1$ for 1950, then the population in 2100 is p(4).

COMPUTER PROBLEMS. *Use your computer algebra system to solve the equation in each of the odd-numbered exercises 5–13 above. The Maple or Mathematica instructions you need are given at the end of the exercises in Section 1.2. You can use the answers in Appendix A to check your computer results.*

2.2 Graphical Methods, the Slope Field

For any first-order differential equation

$$x' = f(t, x), \tag{2.14}$$

whether or not it can be solved by some analytic method, it is possible to obtain a large amount of graphical information about the general behavior of the solution curves from the differential equation itself. In Section 2.4 you will see that if the function $f(t, x)$ is everywhere continuous in both variables t and x and has a continuous derivative with respect to x, the family of solutions of (2.14) forms a set of nonintersecting curves that fill the entire (t, x)-plane. In this section we will see how to use the slope function $f(t, x)$ to sketch a field of tangent vectors (called a **slope field**) that show graphically how the solution curves (also called **trajectories**) flow through the plane. This can all be done even when it is impossible to find an exact formula for $x(t)$.

Figure 2.1 shows a slope field for the spruce-budworm equation, which has the general form

$$x' = rx\left(1 - \frac{x}{N}\right) - \frac{ax^2}{b^2 + x^2}.$$

This equation models the growth of a population of pests that attack fir trees. It is essentially a logistic growth equation with an added term that models the effect of predation on the pests, primarily by birds.

2.2 Graphical Methods, the Slope Field

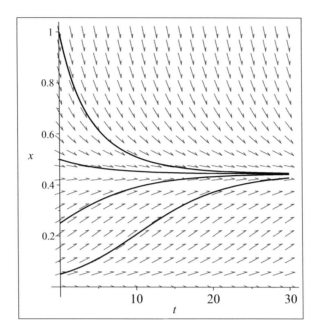

Figure 2.1. Slope field for the equation $x' = 0.2x(1 - x) - 0.3x^2/(1 + x^2)$

You should try to solve this equation with your CAS. It may print out some long unintelligible formula, but basically it is not able to solve the equation in terms of elementary functions, such as polynomials, exponentials, sines and cosines, etc. Notice, however, how easy it is to see how the family of solutions behaves. In Figure 2.1, Maple was used to draw the slope field and four numerically computed solutions (much more will be said about numerical solutions in Section 2.6).

To understand how a slope field is drawn, let $x(t)$ be a solution of (2.14). Then if $x(t)$ passes through some point $(\bar{t}, \bar{x})$ in the plane, the differential equation states that the graph of $x(t)$ at that point must have slope $f(\bar{t}, \bar{x})$. Using just the function $f(t, x)$, a slope field can be drawn by choosing an arbitrary set of points (t_i, x_i), and through each of them drawing a short line with slope $f(t_i, x_i)$; that is, a line that is tangent to the solution curve passing through that point.

Definition 2.1. *A **slope field** for a first-order differential equation $x' = f(t, x)$ is a field of short lines (or arrows) of slope $f(t_i, x_i)$ drawn through each point (t_i, x_i) in some chosen grid of points in the (t, x)-plane.*

The slope lines are often drawn as arrows, all of the same length, pointing in the positive t-direction. They may be drawn with center at the point (t_i, x_i), or alternatively, with the tail of the arrow at (t_i, x_i).

Four of the five models described in Section 1.3 are illustrated by slope fields, and this would be a good time to look back at them to see how much information they provide. The next example will show you exactly how a slope field is constructed.

Example 2.2.1. *Sketch a slope field for the differential equation*

$$x' = x + t.$$

Solution. We will arbitrarily choose a grid of points with integer coordinates in the region $-3 \leq t \leq 3, -3 \leq x \leq 3$ (see Figure 2.2). At each grid point (t_i, x_i), the slope line (which in Figure 2.2 is an arrow centered at the point) will have slope $f(t_i, x_i) = x_i + t_i$. For example, the arrow at $(-1, 2)$ has slope $2 + (-1) = 1$ and the arrow at $(2, -2)$ has slope $(-2) + 2 = 0$. We can put the slopes at all the integer grid points into a table:

Slopes at integer pairs (t, x) for the equation $x' = x + t$

	3	0	1	2	3	4	5	6
	2	−1	0	1	2	3	4	5
	1	−2	−1	0	1	2	3	4
x	0	−3	−2	−1	0	1	2	3
	−1	−4	−3	−2	−1	0	1	2
	−2	−5	−4	−3	−2	−1	0	1
	−3	−6	−5	−4	−3	−2	−1	0
		−3	−2	−1	0	1	2	3
					t			

The slope marks are plotted in Figure 2.2. A solution curve passing through the point $(1, 0)$ has been sketched in as well, by drawing it so that at each point it is tangent to the slope line at that point. In order for this to work, it has to be assumed that the direction of the slope lines changes continuously in both the t and x directions; that is, the function f should at least be a continuous function of both of its variables.

Just by looking at Figure 2.2, certain conclusions can be drawn. It appears that all solutions lying above the line of slope -1 through the point $(0, -1)$ tend to ∞, and solutions lying below that line tend to $-\infty$. What we cannot determine from the slope field is whether the solutions exist for all t, or have a vertical asymptote at a finite value of t.

One might also hazard a guess from looking at the slope field that the straight line $x = -t - 1$ is a solution of the differential equation. To check, let $x(t) = -t - 1$. Then $x'(t) = -1$ and substituting into $x' = x + t$,

$$x(t) + t = (-t - 1) + t = -1 \equiv x'(t),$$

which verifies that it is a solution. ∎

Using a slope field is hardly a precise solution method, but it quickly gives a picture of how the entire family of solutions behaves. Therefore, in a sense, it provides a picture of the *general solution* of the differential equation. The finer the grid, the more information one has to work with.

Figure 2.3 shows a slope field for the logistic growth equation $P' = 0.5P\left(1 - \frac{P}{4}\right)$. We have seen such a picture before, but the point to be made here is that since the equation is autonomous, the slope function P' depends only on P. This means that the slopes along any horizontal line $P = $ constant will *all be the same*, so a slope field for an autonomous equation can be drawn very quickly. Again, it is impossible to tell from the graph whether solutions exist for all t; however, we know from our work with separable equations in Section 2.1 that if $P(0) > N = 4$ or $P(0) < 0$ the corresponding solution has a vertical asymptote.

If it is necessary to sketch a complicated slope field by hand, there is a more efficient way of choosing the grid of points at which the slope lines or arrows are to be

2.2 Graphical Methods, the Slope Field

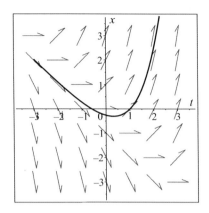 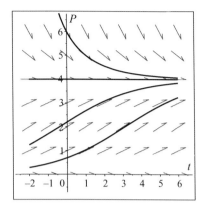

Figure 2.2. Slope field, $x' = x + t$ **Figure 2.3.** Slope field, $P' = \frac{1}{2}P\left(1 - \frac{P}{4}\right)$

drawn. Also, in some cases this method yields information about the long-term behavior of the solutions that is not obvious from a rectangular grid of slopes (as we see in the next example). Consider the right-hand side of a differential equation $x' = f(t, x)$. The equation

$$f(t, x) = m, \quad \text{for } m \text{ any real number,}$$

defines a curve, or set of curves, in the (t, x)-plane along which all the slope vectors must have the same slope m. Such a curve is called an **isocline**, or curve of equal slopes, for the differential equation (2.14). If an isocline for slope m is sketched in the plane, slope lines all of slope m can be drawn along it very quickly.

Example 2.2.2. *Sketch a slope field for $x' = x^2 - t$ by using isoclines.*

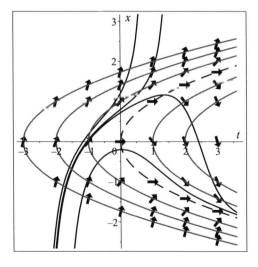

Figure 2.4. Slope field for $x' = x^2 - t$; the parabolic curves are isoclines

Solution. The isoclines are the curves having equation $x^2 - t = m$. They are parabolas, rotated by 90^0, and the isoclines for slope $m = -2, -1, 0, 1, 2,$ and 3 have been sketched in Figure 2.4. Be sure to note that, in general, isoclines are **not** solution curves. Along each of the isoclines, slope lines with appropriate slope have been drawn at equal intervals. It is then possible to sketch some approximate solution curves. There appear to be two types of solutions to this differential equation; those that increase monotonically with t, and those that approach the lower branch of the parabola $x^2 = t$ (and hence ultimately approach $-\infty$). It can be proved analytically that there is a unique solution separating the two families of curves. From Figure 2.4 it can be estimated that the initial condition that separates the different long-term behaviors is around $x(0) = 0.7$. This differential equation cannot be solved analytically in terms of elementary functions, but we will see that it is possible to approximate this special solution, as closely as desired, with the numerical methods described in Section 2.6. ∎

2.2.1 Using Graphical Methods to Visualize Solutions. Drawing slope fields and approximate solution curves is something that computers do very nicely. Almost any computer program that can draw a slope field for a first-order differential equation will also have the facility for drawing approximate solution curves through given initial points. This makes it possible to visualize graphically how the behavior of solutions changes when the parameters in an equation are varied. Be sure to take note of the fact that this can always be done without having an analytic solution of the differential equation.

A simple example of this can be seen in the case of a modified logistic growth equation, one with a harvesting term. This is the differential equation

$$P' = rP\left(1 - \frac{P}{N}\right) - H, \qquad (2.15)$$

described in the previous section, where we have subtracted a constant H, representing the number of individuals being removed from the population (i.e., harvested) per unit of time. The example will demonstrate the power of a graphical method. The equation happens to have an analytic solution, but slope fields can be obtained, and studied, even when the equation cannot be solved analytically.

Example 2.2.3. *Consider a population of deer living in some woods. The deer population is assumed to be growing according to the harvested logistic equation*

$$\frac{dP}{dt} = 0.4P\left(1 - \frac{P}{100}\right) - H, \qquad (2.16)$$

with $P(t)$ equal to the number of deer at time t years after the initial time. We know from previous work that if no deer are removed (that is, $H = 0$) the population will approach the carrying capacity of 100 deer as t increases. The residents living near the wooded area would like to cut down the deer population by allowing hunters to "harvest" H deer each year, and wonder how this would affect the population. Figure 2.5 shows a Maple plot of the slope field with selected solution curves of (2.16) for three different values of the harvesting parameter H. Use the slope fields to describe the effect of increasing H.

Solution. When $H = 0$, we have already seen that the population tends to the carrying capacity $P = 100$ for any positive initial condition. This can be seen in the left-hand graph. Even with an initial population of 2 it predicts a deer population of 100 after about 20 years.

2.2 Graphical Methods, the Slope Field

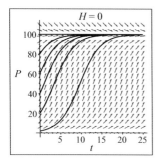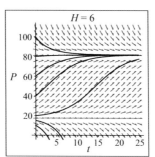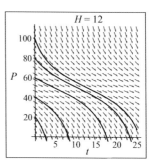

Figure 2.5. Deer population with $H = 0$, $H = 6$, and $H = 12$

If $H = 6$, meaning that 6 deer are removed from the herd each year, it is seen that the limiting population is reduced to about 80. Furthermore, if there are fewer than 20 deer initially, the herd appears to die out.

The right-hand graph with $H = 12$ is significantly different from the other two. It now appears that no matter how many deer are present initially, the herd will ultimately go extinct. This radical change in the behavior of the solutions, as the parameter H crosses a certain value, is called a **bifurcation** of the system modeled by the differential equation. The value of H at which the change occurs is called a **bifurcation value** of the parameter.

We will have a lot more to say about this when we study autonomous first-order equations in Section 2.7. At this point you are just seeing a graphical description of the change; for some equations, including this one, it will turn out to be possible to find the exact bifurcation value of a parameter by an analytic method. ∎

Exercises 2.2. *For the differential equations in 1–5, make a hand sketch of a slope field in the region* $-3 \leq t \leq 3, -3 \leq x \leq 3$, *using an integer grid.*

1. $x' = x + t/2$

2. $x' = -xt/(1 + t^2)$

3. $x' = x(1 - x/2)$

4. $x' = 1 + t/2$

5. $x' = t/x$ (Note: at any point where $x = 0$ the slopes will be vertical.)

For the differential equations in 6–9, sketch a slope field in the region $-3 \leq t \leq 3, -3 \leq x \leq 3$ *using the method of isoclines.*

6. $x' = x + t$

7. $x' = x^3 + t$

8. $x' = x^2$

9. $x' = t^2 - x$

In Exercises 10–12, you are given a differential equation. The corresponding slope field has also been drawn. Sketch enough solution curves in the slope field to be able to describe the behavior of the family of solutions in the entire (t, x)-plane. As part of your description, explain how the long-term behavior depends on the initial condition $x(0)$. Be as specific as possible.

10. $x' = 2x(1 - x/2)$

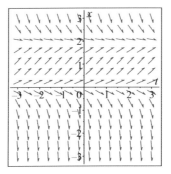

11. $x' + x = t$

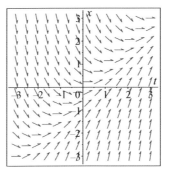

12. $x' = t/x$

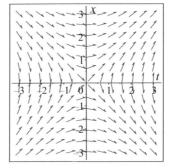

2.2 Graphical Methods, the Slope Field

COMPUTER PROBLEMS. *You will need two new instructions in Maple (or in Mathematica) to produce slope fields and sample solutions for first-order differential equations. In Maple the first instruction,* `with(DEtools):`, *loads the necessary routines. Then the instruction* `DEplot();` *can be used to draw the slope field. The format of the second instruction is*

```
DEplot({deq},[x(t)],t=t0..t1,
[[x(0)=x0],[x(0)=x1],...]],x=a..b);
```

You can also add options at the end of the DEplot instruction. Executing the instruction `?DEplot;` *will show you what is available. It is probably best to specify the option* `stepsize=0.01` *to make sure the solution curves are drawn accurately.*

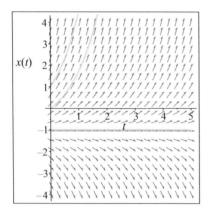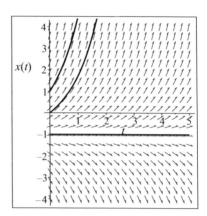

As an example, the two instructions

```
with(DEtools):
DEplot({x'(t)=1+x(t)},[x(t)],t=0..5,
[[x(0)=0],[x(0)=-1],[x(0)=1]],x=-4..4);
```

produced the slope field and three solution curves shown in the left-hand graph above. The graph on the right was done by adding two options; `color=blue` *to make the arrows blue and* `linecolor=black` *to make the curves black.*

The equivalent instructions in Mathematica are `<<VectorFieldPlots`:` *to load the routines, and*

```
VectorFieldPlot[{1,deq}{t,t0,t1},{x,a,b},
Axes->True,Frame->True]
```

to plot the slope field. Again, it would be a good idea to look at the help page for other options. The option `ScaleFunction->(1&)` *will ensure that the arrows are all of equal length, which makes the graph easier to read. The integer 1 in front of the differential equation implies that you are solving a first-order equation.*

You need to get used to these instructions, because they are going to receive heavy use all through the text.
Use your own computer algebra system or graphing calculator to create a detailed slope-field for equations 13–15.

13. $x' = x + t/2$ *(same as 1 above)*

14. $x' = x^2(1-x)$

15. $x' = 1 + t/2$ *(same as 4 above)*

2.3 Linear First-order Differential Equations

Methods for solving linear differential equations have been around for a long time. While it is not correct to say that all linear equations can be solved analytically and all nonlinear equations cannot, it is close to the truth. It is so close, in fact, that before computers were readily available (circa 1955) engineers spent much of their time "linearizing" their problems so they could be solved.

A first-order differential equation is **linear** in $x(t)$ if it can be written in the form
$$a_1(t)x'(t) + a_0(t)x(t) = b(t),$$
where the functions a_0, a_1, and b are arbitrary functions of t. Since we are assuming that our first-order equations can be solved explicitly for $x'(t)$, the function $a_1(t)$ must be nonzero on some interval of t so that we can divide by it and write the equation as
$$x'(t) + \frac{a_0(t)}{a_1(t)}x(t) = \frac{b(t)}{a_1(t)}.$$
This leads to the following definition.

Definition 2.2. *A first-order **linear differential equation in standard form** is an equation that can be written as*
$$x'(t) + p(t)x(t) = q(t), \tag{2.17}$$
*for some functions p and q. If $q(t) \equiv 0$, it is called a **homogeneous linear equation**.*

What linear really means is that the unknown function x and its derivative can only appear in the equation multiplied by functions of t. You cannot have terms like $x^2, x^3, e^x, \ln(x), \sin(x)$, etc. However, the coefficient functions p and q can be arbitrary functions of t.

Example 2.3.1. *For each equation (a)–(d), determine if it is linear with respect to the unknown variables x and x'. If it is linear, state what $p(t)$ and $q(t)$ are. If it is not linear, explain why.*

(a) $x' = -3x + \sin(t)$

(b) $x' = x\sin(t) + t^3$

(c) $4x' + e^x t = 0$

(d) $xx' = 2t$

2.3 Linear First-order Differential Equations

Solution. Equation (a) is linear: write it as $x' + 3x = \sin(t)$, so that $p(t) = 3$ and $q(t) = \sin(t)$. Equation (b) is linear: write it as $x' - x\sin(t) = t^3$, so that $p(t) = -\sin(t)$ and $q(t) = t^3$. Equation (c) is nonlinear: the term e^x is not a linear term in x. Equation (d) is also nonlinear; trying to put it into standard form would result in a term of the form $\frac{1}{x}$ which is not a linear function of x. ∎

Observe that if a linear first-order equation is homogeneous, then it is also separable; that is, an equation of the form $x' + p(t)x = 0$ can also be written in the form $\frac{dx}{dt} = -p(t)x \equiv g(t)h(x)$, with $h(x) = x$ and $g(t) \equiv -p(t)$. This means that our method for solving separable equations can be applied to show that

$$\frac{dx}{dt} = -p(t)x \implies \int \frac{dx}{x} = \int (-p(t))dt \implies \ln|x| = -\int p(t)dt;$$

therefore, the **general solution of the homogeneous equation** can be written as

$$x(t) = e^{-\int p(t)dt}. \tag{2.18}$$

We will usually choose the simplest function $P(t)$ such that $P'(t) = p(t)$, and write the solution in the form

$$x(t) = \alpha e^{-P(t)}. \tag{2.19}$$

Example 2.3.2. *Solve the homogeneous linear equation $x' + \cos(t)x = 0$.*

Solution. In this case, $p(t) = \cos(t)$ and $\int p(t)dt = \sin(t) + C$. If we take $P(t) = \sin(t)$, the one-parameter family of solutions can be written as

$$x(t) = \alpha e^{-\sin(t)}.$$

To show that the solution is correct for any value of α, compute the derivative of x:

$$x'(t) = \alpha e^{-\sin(t)} \frac{d}{dt}(-\sin(t)) = -\alpha \cos(t)e^{-\sin(t)}.$$

Now substitution of x and x' into the equation results in the identity

$$x' + \cos(t)x = -\alpha \cos(t)e^{-\sin(t)} + \cos(t)\alpha e^{-\sin(t)} \equiv 0,$$

as expected. ∎

Solving a nonhomogeneous linear equation is not quite as simple, as the next example illustrates.

Example 2.3.3. *Solve the equation*

$$x' + 2x = te^{-2t}. \tag{2.20}$$

Solution. This time we have no way of integrating the left-hand side of the equation, since we do not know what the function $x(t)$ is. The trick here is to multiply the entire equation by a *positive* function that will make the left side easy to integrate. Note that this will not change the set of solutions. In the case of (2.20) we are going to multiply by the function e^{2t}, and it will be explained below how we picked this particular multiplier.

Making sure to multiply both sides of (2.20) by e^{2t}, we get

$$e^{2t}x' + e^{2t}2x = e^{2t}(te^{-2t}) = t. \tag{2.21}$$

The left side of the equation can be seen to be the exact derivative of the product of the multiplier e^{2t} and the unknown function x; that is,

$$e^{2t}x' + e^{2t}2x \equiv \frac{d}{dt}\left(e^{2t}x\right).$$

This is true for *any* function $x(t)$, since it is just a particular case of the product rule for differentiation.

If we now write (2.21) in the form $\frac{d}{dt}\left(e^{2t}x(t)\right) = t$ and integrate, we get the implicit solution

$$\frac{d}{dt}\left(e^{2t}x(t)\right) = t \implies \int \frac{d}{dt}\left(e^{2t}x(t)\right)dt = \int t\,dt \implies e^{2t}x(t) = \frac{t^2}{2} + C.$$

Solving for x results in the explicit solution

$$x(t) = e^{-2t}\left(\frac{t^2}{2}\right) + Ce^{-2t}.$$

Check it! ∎

The only question that remains, in the case of an arbitrary linear differential equation, is how to find a function $\mu(t)$ such that multiplying the equation by μ makes it integrable. Such a function μ is called an **integrating factor** for the differential equation.

The left side of (2.17), after multiplication by a function $\mu(t)$, is

$$\mu[x' + px] = \mu x' + (\mu p)x. \tag{2.22}$$

If $\mu p \equiv \mu'$, the right-hand side of (2.22) is exactly equal to the derivative of the product μx; that is, $\mu x' + \mu' x \equiv \frac{d}{dt}(\mu x)$ by the product rule for differentiation, and this holds for any functions μ and x.

We already know how to find a function μ that satisfies $\mu p \equiv \mu'$. This is just a homogeneous linear equation $\mu' - p(t)\mu = 0$, and we have already shown that its solution is

$$\mu(t) = e^{\int p(t)dt}. \tag{2.23}$$

Be very careful of plus and minus signs here. We had to write the condition $\mu' = \mu p$ as $\mu' - p\mu = 0$ to put it into standard form. Then the general solution of the homogeneous equation (2.18) gave us another minus sign.

Using the integrating factor μ in (2.23), it is now possible to solve the equation

$$\mu[x' + px] = \mu q$$

by integrating both sides with respect to t:

$$\int \mu[x' + px]dt = \int \mu q\,dt \implies \int \frac{d}{dt}(\mu x)dt = \int \mu q\,dt$$

$$\implies \mu x = \int \mu q\,dt + C.$$

Dividing the final expression by the nonzero function μ gives the explicit solution

$$x(t) = \frac{1}{\mu(t)}\left[\int \mu(t)q(t)dt + C\right]. \tag{2.24}$$

2.3 Linear First-order Differential Equations

Although the equation (2.24) is important theoretically as the **general solution of the first-order linear differential equation**, you do not need to memorize it to solve a simple linear equation. To do that, just follow the steps used in the derivation. This can be written in the form of a 5-step procedure.

To solve a linear first-order differential equation:

- Put the equation in standard form,
$$x'(t) + p(t)x(t) = q(t).$$
Be sure to divide through by the coefficient of $x'(t)$ if it is not already equal to 1.

- Find the simplest possible antiderivative $P(t)$ of the coefficient function $p(t)$.

- Let $\mu(t) = e^{P(t)}$, and multiply both sides of the equation in step one by μ (do not forget to multiply $q(t)$ by μ).

- Integrate both sides of the resulting equation with respect to t (be sure to add a constant of integration on one side). If you have done the first three steps correctly, the integral of the left-hand side is just $\mu(t)x(t)$.

- Divide both sides of the equation by $\mu(t)$ to obtain the explicit solution $x(t)$.

The easiest way to see how all of this works is by example.

Example 2.3.4. *Solve the equation* $x' = x + te^t$.

Solution. This is easily seen to be a linear equation with standard form
$$x' - x = te^t. \tag{2.25}$$

The coefficient $p(t)$ of x is -1, and $\int p(t)dt = \int (-1)dt = -t + C$ implies that we can use the integrating factor $\mu(t) = e^{-t}$. Now multiply both sides of the equation by μ:
$$e^{-t}x' - e^{-t}x = e^{-t}(te^t) = t.$$

Using the product rule for differentiation, the left side is the derivative of $\mu x \equiv e^{-t}x$; therefore, the equation can be written as
$$\frac{d}{dt}(e^{-t}x) = t.$$

It is now possible to integrate both sides of the equation with respect to t:
$$\int \frac{d}{dt}(e^{-t}x)dt = \int t\,dt \implies e^{-t}x = t^2/2 + C$$

and solve explicitly for $x(t)$:
$$x(t) = e^t(t^2/2 + C).$$

This is the general solution of $x' = x + te^t$, and you should check it by differentiating $x(t)$ and substituting x and x' into the differential equation. ∎

It can be seen that the solution of a first-order linear differential equation always contains a single constant that can be used to satisfy one initial condition. The next example shows how this is done.

Example 2.3.5. *Solve the initial-value problem*

$$tx' + x = \cos(t), \quad x(1) = 2. \tag{2.26}$$

Also, determine the behavior of the solution as $t \to \infty$, both by looking at the formula for the solution and by using a slope field.

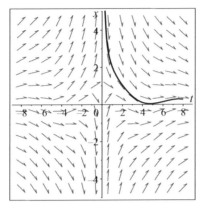

Figure 2.6. $tx' + x = \cos(t)$

Solution. To put (2.26) into standard form, it must be divided by t:

$$x' + (1/t)x = \cos(t)/t,$$

where it must now be assumed that $t \neq 0$. The coefficient of x in the standard form of the equation is $p(t) = 1/t$ and a simple antiderivative is $P(t) = \ln(t)$; therefore, the integrating factor we will use is $\mu(t) = e^{\ln(t)} = t$. Remember, we can always choose the simplest possible antiderivative of p. If the differential equation is multiplied by $\mu(t) = t$ and integrated,

$$\int [tx'(t) + x(t)] dt \equiv \int \frac{d}{dt}[tx(t)] dt = \int \cos(t) dt.$$

This implies that $tx(t) = \sin(t) + C$, and solving explicitly for x,

$$x(t) = [\sin(t) + C]/t, \quad t \neq 0.$$

To satisfy the initial condition $x(1) = 2$, substitute $t = 1$ and $x = 2$ into the equation:

$$x(1) = \sin(1) + C = 2,$$

and solve for C. Then $C = 2 - \sin(1)$, and the solution to the initial-value problem is $x(t) = [\sin(t) + 2 - \sin(1)]/t$. From the exact solution we can see that $x \to 0$ as $t \to \infty$. At $t = 0$, $x(t)$ has a vertical asymptote. A slope field for this equation is shown in Figure 2.6, with the solution through the initial point $(1, 2)$ drawn in. From the slope field it appears that all solutions with initial conditions given at $t > 0$ approach zero as $t \to \infty$. The analytic solution can be used to verify this. ∎

2.3 Linear First-order Differential Equations

2.3.1 Application: Single-compartment Mixing Problem. We will end this section on linear equations with a very important application that comes up in several applied areas. We will see that they can be as diverse as pollution control and medical dosing. In this application we consider a problem called a **one-compartment mixing problem**. At this point we are only able to solve a mixing problem involving a single compartment in which mixing takes place. By the end of Chapter 4 we will be able to treat mixing problems with any number of compartments.

Basically, the problem consists of finding a formula for the amount of some "pollutant" in a container, into which the pollutant is entering at a fixed rate and also flowing out at a fixed rate. The general rule used to model this situation is conservation of mass:

rate of change of pollutant per unit time = rate in − rate out.

If we denote by $x(t)$ the amount of pollutant in the container at time t, its rate of change per unit time is given by $\frac{dx}{dt}$; therefore, the problem leads to a differential equation for the function $x(t)$.

Example 2.3.6. *Consider a fish tank that initially contains 150 liters of water with 20 grams of salt (pollutant) dissolved in it. The salt concentration in the tank needs to be increased from 20/150 grams per liter to 1 gram per liter to accommodate a new species of fish. Water containing 3 grams of salt per liter is allowed to run into the tank at a rate of 2 liters per minute. The thoroughly stirred mixture in the tank is also allowed to drain out at the same rate of 2 liters per minute. Find a differential equation for the amount of salt $x(t)$ in the tank at time t. Use the initial condition $x(0) = 20$ to find the time it will take to increase the salt concentration in the tank to 1 gram per liter.*

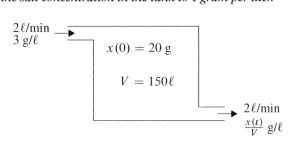

Solution. We are assuming that the mixture in the tank is thoroughly stirred, so that the salt concentration $x(t)/\text{vol}$ is instantaneously the same throughout the tank. Let $x(t)$ denote the number of grams of salt in the tank at time t. Then the rate at which salt enters or leaves the tank is measured in units of grams/minute, so

$$\frac{dx}{dt} = \text{rate in} - \text{rate out} = (2\ell/\min)(3g/\ell) - (2\ell/\min)\left(\frac{x(t)}{\text{vol}} g/\ell\right)$$

and the differential equation we are looking for is

$$\frac{dx}{dt} = 6 - 2\frac{x(t)}{150} \equiv 6 - \frac{x(t)}{75}. \tag{2.27}$$

This differential equation is both separable and linear, and we will arbitrarily choose to solve it by the method for separable equations:

$$\frac{dx}{dt} = \frac{450 - x}{75} \implies \int \frac{dx}{450 - x} = \int \frac{dt}{75} \implies -\ln(|450 - x|) = \frac{t}{75} + C.$$

Exponentiating both sides,

$$\frac{1}{450-x} = Ke^{t/75} \implies x = 450 - (1/K)e^{-t/75}, \quad \text{where } K = \pm e^C.$$

Using the initial condition $x(0) = 20$ to find the value of $1/K$,

$$x(0) = 450 - 1/K = 20 \implies 1/K = 430$$

and the solution of the IVP is

$$x(t) = 450 - 430e^{-t/75}. \tag{2.28}$$

To find the time when the concentration of salt in the tank reaches 1 g/ℓ, set

$$1 = x(t)/\text{vol} = (450 - 430e^{-t/75})/150$$

and solve for t. This gives $t \approx 27$ minutes as the time it takes to increase the concentration to the required value. ∎

It can be seen from (2.28) that as $t \to \infty$, the solution $x(t)$ approaches 450 as we would expect. This is the value of x for which the concentration $x(t)/\text{vol}$ in the tank is the same as the concentration of the solution constantly flowing in.

If the flow rates in and out of the container are not the same, the volume will change over time, and the differential equation will no longer be autonomous, and therefore no longer separable. In the next example it will be shown that we have the tools to obtain an analytic solution in this case.

Example 2.3.7. *In the previous example it took about 27 minutes for the solution to reach the desired concentration. One way to speed up the process would be to let the salt water flow in at a faster rate. Assume that the input flow is increased to 2.5ℓ/min, but that the output flow cannot be increased. If the maximum amount of water the container can hold is 160 liters, will the salt concentration reach the desired level before the container overflows?*

Solution. The differential equation now becomes

$$x'(t) = (2.5\ \ell/\text{min})(3\ \text{g}/\ell) - (2\ \ell/\text{min})\left(\frac{x(t)}{\text{vol}}\ \text{g}/\ell\right).$$

Since the rate of water flowing in is 2.5ℓ/min and the rate out is still 2ℓ/min, the volume of fluid in the tank will increase by 0.5 liter each minute; therefore, the volume at time t is $V(t) = 150 + \frac{1}{2}t$.

The differential equation $x' = 7.5 - \frac{2x}{150+\frac{1}{2}t}$ is not separable, but it is linear and its standard form is

$$x' + \left(\frac{4}{300+t}\right)x = 7.5.$$

2.3 Linear First-order Differential Equations

The integrating factor is $e^{\int \frac{4}{300+t} dt}$, and choosing the simplest antiderivative we can take $\mu(t) = e^{4(\ln(300+t))} = (300+t)^4$. Multiplying by μ and integrating gives

$$(300+t)^4 x' + 4(300+t)^3 x = 7.5(300+t)^4$$

$$\implies \int \left((300+t)^4 x' + 4\left((300+t)^3 x\right)\right) dt$$

$$\equiv \int \frac{d}{dt}\left((300+t)^4 x\right) dt = \int 7.5(300+t)^4 dt;$$

therefore,

$$(300+t)^4 x = \frac{7.5(300+t)^5}{5} + C$$

and

$$x(t) = 1.5(300+t) + \frac{C}{(300+t)^4}.$$

Using the initial condition $x(0) = 20$,

$$20 = 1.5(300) + \frac{C}{(300)^4}$$

and the value of C is $-430(300)^4$. To make the concentration in the tank equal $1 g/\ell$, set

$$1 = x(t)/V(t) = \frac{x(t)}{150 + t/2} = \left(\frac{2}{300+t}\right) x(t) = 3.0 - \frac{860(300)^4}{(300+t)^5},$$

and solve for t. This gives $t \approx 22.4$ minutes, but $V(22.4) = 150 + 0.5(22.4) = 161.2$ liters; therefore, the tank will overflow just before the solution reaches the desired concentration. ∎

Exercises 2.3. *In Exercises 1–6, state whether the equation is linear, separable, both, or neither.*

1. $x' = tx$
2. $x' = \cos(t)x$
3. $x' = t^2 \sin(x)$
4. $x' = t + x$
5. $y' = 1/(1 + e^{-y})$
6. $y' = e^t y + e^t$

In Exercises 7–10, use the method described in this section to find an analytic solution to the equation and describe what happens to the solution as $t \to \infty$.

7. $x' + 2x = e^{-2t} \sin(t)$
8. $x' = 2x + 1$ *This equation is also separable.*
9. $tx' - 2x = 1 + t$
10. $tx' = 2x + t^2$

In Exercises 11–14, solve the initial-value problems.

11. $x' = -x + e^{2t}, \quad x(0) = 1$

12. $x' + 2x = e^{-2t}\cos(t), \quad x(0) = -1$

13. $tx' + x = 3t^2 - t, \quad x(1) = 0$

14. $x' + 2tx = 3t, \quad x(0) = 4$

15. In Section 1.3, the velocity $v(t)$ of a free-falling body was shown to satisfy the differential equation
$$mv' = mg - kv^p,$$
where m is the mass of the body in kilograms, g is the gravitational constant, and p is a real constant.

    (a) For what value(s) of p is the equation linear in $v(t)$?
    (b) Solve the equation with $p = 1$, $\frac{k}{m} = 0.6$, and $g = 9.8 m/sec^2$.
    (c) What is the value of the terminal velocity; that is, $\lim_{t \to \infty} v(t)$? Convert the velocity from meters per second to miles per hour.
    (d) Does the terminal velocity depend on the initial velocity $v(0)$? Explain.

16. In Exercises 1.3 (under ENGINEERING), the RL circuit problem
$$i'(t) + 3i(t) = 10\sin(t)$$
was given, and you were asked to find the value of $i(0)$ that makes the solution be $i(t) = 3\sin(t) - \cos(t)$. Use the method described in this section to solve the equation, and determine the correct value of $i(0)$. (The following integral formula may help: $\int e^{at}\sin(bt)dt = e^{at}\left(\frac{a\sin(bt) - b\cos(bt)}{a^2 + b^2}\right)$.) Use your CAS to draw a slope field for the equation and plot the solution using the initial condition you found.

The next two problems refer to the mixing problem in subsection 2.3.1.

17. Continuing Examples 2.3.6 and 2.3.7, assume now that instead of increasing the input flow as in Example 2.3.7, we decrease the input flow to $1.5\ell/min$ (the output flow is not changed). Will the salt concentration reach the desired level before the container empties?

18. Continuing Example 2.3.7, find the maximum input flow rate such that the tank will just reach its capacity of 160 liters when the salt concentration reaches 1 gram per liter. How many minutes does it take? Your answer should be exact to the nearest second. (This one is hard!)

2.4 Existence and Uniqueness of Solutions

In this section we will consider the question of whether a given initial-value problem
$$x' = f(t, x), \quad x(t_0) = x_0 \qquad (2.29)$$
has a solution passing through a given initial point (t_0, x_0), and, if it does, whether there can be more than one such solution through that point. These are the questions of existence and uniqueness, and they become very important when one is solving applied problems.

There is a basic theorem, proved in advanced courses on differential equations, which can be applied to the initial-value problem (2.29).

2.4 Existence and Uniqueness of Solutions

Theorem 2.1. (Existence and Uniqueness): *Given the differential equation $x' = f(t, x)$, if f is defined and continuous everywhere inside a rectangle $\mathbf{R} = \{(t, x) \mid a \leq t \leq b, c \leq x \leq d\}$ in the (t, x)-plane, containing the point (t_0, x_0) in its interior, then there exists a solution $x(t)$ passing through the point (t_0, x_0), and this solution is continuous on an interval $t_0 - \varepsilon < t < t_0 + \varepsilon$ for some $\varepsilon > 0$. If $\frac{\partial f}{\partial x}$ is continuous in $\mathbf{R}$, there is exactly one such solution; that is, the solution is unique.*

Comment 1. Even if the function f is differentiable, the theorem does not imply that the solution exists for all t. For example, it may have a vertical asymptote close to $t = t_0$.

Comment 2. To prove the uniqueness part of Theorem 2.1 it is not necessary for the function f to be differentiable. It only needs to satisfy what is called a **Lipschitz condition**.

Definition 2.3. *A function $f(t, x)$ defined on a bounded rectangle $\mathbf{R}$ is said to satisfy a **Lipschitz condition** on $\mathbf{R}$ if for any two points (t, x_1), (t, x_2) in $\mathbf{R}$ there exists a positive constant M such that*

$$|f(t, x_2) - f(t, x_1)| \leq M|x_2 - x_1|.$$

Note that if the derivative $\frac{\partial f}{\partial x}$ does exist and is continuous in $\mathbf{R}$, then we can use the maximum value of $|\frac{\partial f}{\partial x}|$ in $\mathbf{R}$ as a Lipschitz constant M, since the Mean Value Theorem implies that the slope of the secant $\frac{f(t,x_2)-f(t,x_1)}{x_2-x_1}$ is equal to $\frac{\partial f}{\partial x}$ at some intermediate point between x_1 and x_2. One of the exercises at the end of this section will ask you to find a Lipschitz constant for a nondifferentiable slope function. This may convince you that if f is differentiable it is a much easier condition to use.

The proof of Theorem 2.1 requires advanced analysis, but the idea is straightforward. First notice that any solution $x(t)$ of the IVP (2.29) is also a solution of the integral equation

$$x(t) = x_0 + \int_{t_0}^{t} f(s, x(s))ds. \qquad (2.30)$$

To see this, simply differentiate both sides of (2.30) using the Fundamental Theorem of Calculus on the right. Therefore, solving (2.29) is equivalent to solving (2.30).

To solve (2.30), start by letting $x_0(t)$ be the constant $x_0(t) \equiv x_0$. Then create a sequence of functions $\{x_n(t)\}_{n=1}^{\infty}$ as follows:

$$x_1(t) = x_0 + \int_{t_0}^{t} f(s, x_0(s))ds,$$

$$x_2(t) = x_0 + \int_{t_0}^{t} f(s, x_1(s))ds,$$

$$\ldots, x_{n+1}(t) = x_0 + \int_{t_0}^{t} f(s, x_n(s))ds, \ldots.$$

In order to show that $x(t) = \lim_{n \to \infty} x_n(t)$ is the solution of our IVP we would need to be able to claim that

$$\lim_{n \to \infty} \int_{t_0}^t f(s, x_n(s))ds = \int_{t_0}^t f(s, \lim_{n \to \infty} x_n(s))ds.$$

This statement requires a theorem called the monotone convergence theorem for integrals, and this is where the hard analysis comes in.

The following example illustrates how such a successive approximation scheme might work. Be warned that this is a very simple equation, and in general the iteration procedure does not give a constructive method for producing a solution.

Example 2.4.1. *Show that the procedure described above leads to the correct solution of the IVP*

$$x' = x, \; x(0) = 1.$$

Solution. This IVP is equivalent to the integral equation $x(t) = 1 + \int_0^t x(s)ds$. Let $x_0(t) \equiv 1$. Then the next three functions in the sequence are:

$$x_1(t) = 1 + \int_0^t x_0(s)ds = 1 + \int_0^t 1 ds = 1 + t;$$

$$x_2(t) = 1 + \int_0^t x_1(s)ds = 1 + \int_0^t (1+s)ds = 1 + t + t^2/2;$$

$$x_3(t) = 1 + \int_0^t x_2(s)ds = 1 + \int_0^t (1 + s + s^2/2)ds = 1 + t + t^2/2 + t^3/(3 \cdot 2);$$

and it can be seen that $x_n(t)$ is the nth Taylor polynomial for the function e^t. As $n \to \infty$ we know that this approaches the function e^t which *is* the solution of the IVP. ■

Theorem 2.1 does, however, give us a very useful piece of information. In some of the slope fields pictured in Section 2.2, it was noted that solutions appear to form a space-filling set of curves. We are now able to test that assertion, using the following lemma.

Lemma 2.1. *Let $x' = f(t, x)$ be a first-order differential equation with f and $\frac{\partial f}{\partial x}$ both continuous for all values of t and x in some region* **R** *in the plane. Then inside the region* **R** *the solution curves of the differential equation will form a nonintersecting space-filling family of curves.*

Proof. Theorem 2.1 guarantees that there exists a unique solution through any point in **R** since f and $\frac{\partial f}{\partial x}$ are both continuous; this means that every point in the region has a solution curve passing through it. Furthermore, two solutions cannot intersect anywhere inside **R**, since if they did, there would be a point with two or more solutions through it and this would contradict uniqueness guaranteed by the theorem. □

We first apply these results to the logistic growth equation.

Example 2.4.2. *What does Theorem 2.1 tell us about solutions of the equation*

$$x' = x(1 - x)? \tag{2.31}$$

2.4 Existence and Uniqueness of Solutions

Solution. The functions $f(t,x) = x(1-x) = x-x^2$ and $\frac{\partial f}{\partial x} = 1-2x$ are both continuous everywhere in the (t,x)-plane; therefore, there exists a unique solution through any point (t_0, x_0). Lemma 2.1 tells us that the set of solutions fills the entire (t,x)-plane and that two different solution curves never intersect.

Since $x(t) \equiv 0$ and $x(t) \equiv 1$ are both constant solutions of the differential equation, the theorem also implies that any solution with $0 < x(0) < 1$ must remain bounded between 0 and 1 for all t for which it exists. This is easily seen, since a solution starting inside the strip $0 < x < 1$ cannot have a point of intersection with either of the bounding solutions, hence it can never exit the strip.

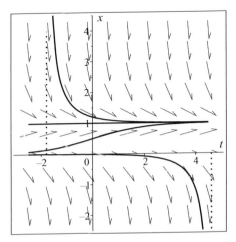

Figure 2.7. Solutions of $x' = x(1-x)$

Be very careful however not to assume that every solution must exist for all values of t. Theorem 2.1 does not tell us how big the t-interval of existence is for a solution of $x' = x(1-x)$. We solved this equation in Section 2.1.1 (it is separable). The general solution of $x' = x(1-x)$ is $x(t) = \frac{1}{1+Ce^{-t}}$, where $C = \frac{1}{x(0)} - 1$ if $x(0) \neq 0$. If $0 < x(0) < 1$ the constant C is positive, and therefore the denominator of the solution is never equal to zero. This means that the solution exists for all t. (This could also have been determined from the fact that solutions starting in this strip can never leave.) If $x(0) > 1$ or $x(0) < 0$, the denominator of $x(t)$ will have a zero at $t = \ln(-C) = \ln(1 - \frac{1}{x(0)})$ and $x(t)$ will have a vertical asymptote there. Figure 2.7 shows the solutions through $x(0) = -0.01, 0.5, 1.0$, and 1.2. The vertical asymptote for the solution with $x(0) = -0.01$ is at $t = \ln(101) \approx 4.6$, and for the solution with $x(0) = 1.2$ the asymptote is at $t = \ln(1/6) \approx -1.8$. ∎

A second example will show what happens when the equation is not well-behaved everywhere.

Example 2.4.3. *Does the IVP $tx' + x = \cos(t)$, $x(1) = 2$, have a unique solution?*

Solution. The function $f(t,x) = (-x + \cos(t))/t$ is continuous everywhere except where $t = 0$. As shown in Figure 2.8, a rectangle that contains no points of discontinuity of f can be drawn around the initial point $(t_0, x_0) = (1, 2)$. This implies that

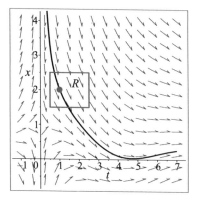

Figure 2.8. Solutions of $tx' + x = \cos(t)$

there exists at least one solution of the IVP. To check its uniqueness, it is necessary to compute
$$\frac{\partial f}{\partial x} = \frac{\partial}{\partial x}\left(-\frac{1}{t}x + \frac{\cos(t)}{t}\right) = -\frac{1}{t}.$$
This function is also continuous wherever $t \neq 0$; therefore, the IVP has a unique solution (shown in Figure 2.8). We found this solution in Example 2.3.5 to be $x(t) = (\sin(t) + 2 - \sin(1))/t$, so the solution of our IVP turns out to only be defined for $t > 0$, and has a vertical asymptote at $t = 0$. ∎

The next example will show what can happen when the slope function does not satisfy all of the hypotheses of Theorem 2.1. Figure 2.9 shows a slope field for the

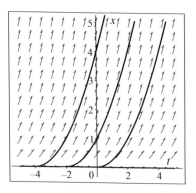

Figure 2.9. Slope field for the equation $x' = \sqrt{x}$

equation $x' = \sqrt{x}$. No arrows appear in the lower half of the plane. This is because the slope function $\sqrt{x}$ is not defined for negative x. It can also be seen that this is an autonomous differential equation, and slopes along any horizontal line (for constant x) will all be the same.

Example 2.4.4. *Does the IVP $x' = \sqrt{x}$, $x(0) = 0$ have a solution? Is it unique?*

2.4 Existence and Uniqueness of Solutions

Solution. The function $f(t, x) = \sqrt{x}$ is defined at $x = 0$, but we cannot enclose the initial point $(0, 0)$ *inside* a rectangle containing only points with $x \geq 0$. When $x < 0$, $\sqrt{x}$ is not even defined; therefore, f does not satisfy the continuity hypothesis at the initial point $(0,0)$, and Theorem 2.1 gives us *no* information about solutions of this IVP.

The differential equation $x' = \sqrt{x}$ is separable and can be solved by integration, as follows:

$$\frac{dx}{dt} = \sqrt{x} \Longrightarrow \frac{dx}{\sqrt{x}} = dt \Longrightarrow \int \frac{dx}{\sqrt{x}} \equiv \int x^{-1/2} dx = \int dt$$

$$\Longrightarrow 2x^{1/2} = t + C \Longrightarrow x(t) = \left(\frac{t+C}{2}\right)^2.$$

To satisfy the initial condition $x(0) = (\frac{0+C}{2})^2 = 0$, we must have $C = 0$; therefore, $x(t) = \frac{t^2}{4}$ is a solution of the IVP. However, $x(t) \equiv 0$ is also a solution; therefore, in this case the solution of the IVP is definitely *not* unique. This should not be surprising, since Theorem 2.1 requires the function f to have a continuous derivative with respect to x, and $\frac{\partial f}{\partial x} = \frac{1}{2\sqrt{x}}$ is not even defined at $x = 0$. ∎

Back in Section 1.1 we showed in Example 1.1.2 that one solution of the differential equation $x' = -t/x$ is $x(t) = \sqrt{1-t^2}$. The question was raised as to whether the equation has any solutions that exist outside of the interval $-1 \leq t \leq 1$. We are now able to answer that question.

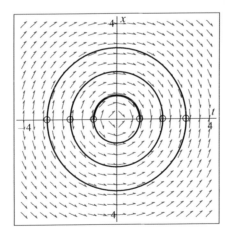

Figure 2.10. Solutions of $x' = -t/x$

The slope function $f(t, x) = -t/x$ and its derivative $\frac{\partial f}{\partial x} = t/x^2$ are both continuous everywhere except on the t-axis, where $x = 0$. This means that the upper half-plane, and also the lower half-plane, will be filled with nonintersecting solution curves.

The differential equation $x' = -t/x$ is separable and can be solved by writing

$$\frac{dx}{dt} = -\frac{t}{x} \Longrightarrow \int x\, dx = \int (-t)\, dt \Longrightarrow \frac{x^2}{2} = -\frac{t^2}{2} + C \Longrightarrow x^2 + t^2 = 2C \equiv \alpha^2.$$

This implicit solution produces a family of circles in the (t, x)-plane, but each circle gives us the solutions $x(t) = \pm\sqrt{\alpha^2 - t^2}$ of two initial-value problems; one satisfying $x(0) = \alpha$ and the other satisfying $x(0) = -\alpha$. What this means is that every point on the t-axis is a point at which two solution curves meet. Any point not on the t-axis will have a unique solution passing through it. This is illustrated in the slope field shown in Figure 2.10. Note very carefully the direction of the vectors in the slope field. What is happening to the slope of a solution curve as x approaches zero?

Exercises 2.4. *Exercises 1–6 test your understanding of Theorem 2.1 and Lemma 2.1.*

1. Use Theorem 2.1 to show that the differential equation $x' = \dfrac{x^2}{1+t^2}$ has a unique solution through every initial point (t_0, x_0). Can solution curves ever intersect?

2. Use Theorem 2.1 to prove that the solution of an initial-value problem for the equation $x' = \dfrac{x}{1+t^2}$, with $x(0) > 0$, can never become negative. Hint: First find a constant solution of the differential equation for some constant C.

3. Does the equation $x' = x^2 - t$ have a unique solution through every initial point (t_0, x_0)? Can solution curves ever intersect for this differential equation? If so, where?

4. Does Theorem 2.1 imply that the solution of $x' = x^2 - t$, $x(0) = 1.0$, is defined for all t? What can you say about the size of the interval for which it exists?

5. Draw a slope field for the equation $x' = x^{2/3}$. Can solution curves ever intersect? If so, where? (Note that this is a separable equation.)

6. Consider the differential equation $x' = t/x$.

   (a) Use Theorem 2.1 to prove that there is a unique solution through the initial point $x(1) = 1/2$.
   (b) Show that for $t > 0$, $x(t) = t$ is a solution of $x' = t/x$.
   (c) Use (b) to show that the solution in (a) is defined for all $t > 1$.
   (d) Solve the initial-value problem in (a) analytically (see Example 2.1.4). Find the exact t-interval on which the solution is defined. Sketch the solution in the slope field in Figure 2.11.

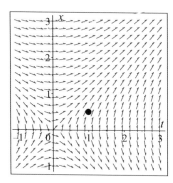

Figure 2.11. Slope field for $x' = t/x$

2.5 More Analytic Methods for Nonlinear First-order Equations

Student Project. *The Student Project at the end of Chapter 1 compared several functions S(z) which varied from 0 to 1 as z went from $-\infty$ to $+\infty$. These functions, called response functions, are used to model the growth of activity in a collection of nerve cells. One of the simplest functions of this type is the piecewise linear function*

$$S(z) = \begin{cases} 0 & \text{if } z < -1/2, \\ z + 1/2 & \text{if } -1/2 \leq z \leq 1/2, \\ 1 & \text{if } z > 1/2. \end{cases} \quad (2.32)$$

A simple neural model using the function S is the equation

$$x'(t) = f(t,x) = -x(t) + S(20(x(t) - 0.5 - 0.4\cos(2\pi t))).$$

To study the behavior of solutions of this equation in the square $B = \{(t,x) \mid 0 \leq t \leq 1, 0 \leq x \leq 1\}$ we need to know if there exist unique solutions through any given initial point in B.

1. *Show that the derivative $\frac{\partial f}{\partial x}$ is NOT a continuous function in B.*

2. *Show that f does satisfy a Lipschitz condition in B and find a Lipschitz constant M.*

3. *Plot five solutions of $x' = f(t,x)$ in B. If you are using Maple, the function S(z) can be defined by the command:* `S:=z->piecewise(z<-0.5,0,z<0.5,z+0.5,1.0)`. *Plot solutions through the initial points $(t_0, x_0) = (0, 0.1), (0, 0.3), (0, 0.5), (0, 0.7), (0, 0.9)$.*

2.5 More Analytic Methods for Nonlinear First-order Equations

Two additional methods for solving first-order differential equations are described in this section. Together with the previous methods we have looked at, our list is still far from exhaustive. If there is any analytic method for solving a given differential equation, your computer algebra system will probably know how to apply it. Discovering and programming methods for finding analytic solutions of differential equations is one of the things that keeps the programmers for Maple, *Mathematica*, etc. busy; however, understanding how these methods work will enable you to be an intelligent user of these programs.

2.5.1 Exact Differential Equations.
In this section we use the variables x and y instead of t and x since this is the way exact equations are usually written in textbooks, as functions of two space variables. However, everything remains true if x is replaced by t and y is replaced by x.

Let F be a differentiable function of two independent variables x and y. The expression

$$dF = \frac{\partial F}{\partial x}dx + \frac{\partial F}{\partial y}dy$$

is called the *total differential* of the function F. From calculus it is known that if dF is identically equal to zero in some region of the (x,y)-plane, then the function $F(x,y)$ must be a constant in that region. This can be used to find analytic solutions for a certain type of first-order differential equation.

Given a first-order differential equation written in the form

$$g(x,y)dx + h(x,y)dy = 0, \qquad (2.33)$$

suppose we can show that there exists a differentiable function $F(x,y)$ such that $\frac{\partial F}{\partial x} = g(x,y)$ and $\frac{\partial F}{\partial y} = h(x,y)$. (This is not going to be true in general.) If it is true, (2.33) states that the total differential of the function F is zero, and therefore we have found an implicit solution of (2.33), namely $F(x,y) = C$.

Example 2.5.1. *Solve the differential equation*

$$y' = -x/y. \qquad (2.34)$$

This is the same equation, namely $x' = -t/x$, that we looked at the end of Section 2.4, with the variables renamed. A slope field for (2.34) is pictured in Figure 2.10.

Solution. We first write (2.34) in the form $dy/dx = -x/y$ and then expand it, by assuming that the differentials dx and dy can be separated, as

$$-xdx = ydy \quad \text{or} \quad xdx + ydy = 0.$$

The function $F(x,y) = \frac{x^2+y^2}{2}$ satisfies $\frac{\partial F}{\partial x} = x$ and $\frac{\partial F}{\partial y} = y$. (Check it!) This means that F is a function such that $\frac{\partial F}{\partial x}dx + \frac{\partial F}{\partial y}dy \equiv 0$; therefore, $F(x,y) = \frac{x^2+y^2}{2} \equiv C$ is an implicit solution of (2.34). ∎

This is exactly the same implicit solution we found for $x' = -t/x$ using the method for separable equations.

Definition 2.4. *A first-order differential equation $y' = f(x,y)$ is called **exact** if it can be written in the form*

$$M(x,y)dx + N(x,y)dy = 0 \qquad (2.35)$$

where $M(x,y) = \frac{\partial F}{\partial x}$ and $N(x,y) = \frac{\partial F}{\partial y}$ for some differentiable function $F(x,y)$.

In the calculus of several variables it is shown that if F is a twice continuously differentiable function of two variables, then the second-order mixed partial derivatives

$$\frac{\partial^2 F}{\partial x \partial y} \quad \text{and} \quad \frac{\partial^2 F}{\partial y \partial x}$$

are equal. If $M = \frac{\partial F}{\partial x}$ and $N = \frac{\partial F}{\partial y}$, then

$$\frac{\partial M}{\partial y} = \frac{\partial}{\partial y}\left(\frac{\partial F}{\partial x}\right) \equiv \frac{\partial}{\partial x}\left(\frac{\partial F}{\partial y}\right) = \frac{\partial N}{\partial x}.$$

It is also true, but slightly more difficult to prove, that if $\frac{\partial M}{\partial y} = \frac{\partial N}{\partial x}$, then there exists a function $F(x,y)$ with $\frac{\partial F}{\partial x} = M$ and $\frac{\partial F}{\partial y} = N$; therefore, a simple way to test (2.35) for exactness is to check that the functions M and N satisfy $\frac{\partial M}{\partial y} = \frac{\partial N}{\partial x}$.

Example 2.5.2. *Determine whether or not the following differential equations are exact.*
 (i) $(x^2 + y)dx + (x - \sin(y))dy = 0$

2.5 More Analytic Methods for Nonlinear First-order Equations

(ii) $(x^2 y)dx + (x^3/3 + 4y^2 + 1)dy = 0$

(iii) $(x+y)dx - (x-y)dy = 0$

Solution. In the first equation, $M = x^2 + y$ and $N = x - \sin(y)$. The partial derivatives are

$$\frac{\partial M}{\partial y} = \frac{\partial}{\partial y}(x^2 + y) = 1 \quad \text{and} \quad \frac{\partial N}{\partial x} = \frac{\partial}{\partial x}(x - \sin(y)) = 1;$$

therefore, the equation is exact.

In the second equation, $M = x^2 y$ and $N = (x^3/3 + 4y^2 + 1)$. The partial derivatives are

$$\frac{\partial M}{\partial y} = \frac{\partial}{\partial y}(x^2 y) = x^2 \quad \text{and} \quad \frac{\partial N}{\partial x} = \frac{\partial}{\partial x}(x^3/3 + 4y^2 + 1) = x^2;$$

therefore, the equation is exact.

In the third equation, $M = x + y$ and $N = -(x - y)$. The partial derivatives are

$$\frac{\partial M}{\partial y} = \frac{\partial}{\partial y}(x + y) = 1 \quad \text{and} \quad \frac{\partial N}{\partial x} = \frac{\partial}{\partial x}(-x + y) = -1.$$

Since the partial derivatives are not the same, equation (iii) is not exact. ∎

To solve an exact equation $M(x,y)dx + N(x,y)dy = 0$: Make sure that $\frac{\partial M}{\partial y} = \frac{\partial N}{\partial x}$, and then use the following three steps.

(1) Set $\frac{\partial F}{\partial x} = M(x,y)$ and integrate once with respect to x to get

$$F(x,y) = \int M(x,y)dx + Q(y).$$

Note that $Q(y)$ represents the "constant" of integration with respect to x, and the integration is done as though x is the only variable, and y is a parameter.

(2) Differentiate the function F found in step (1), partially with respect to y, and set the result equal to $N(x,y)$:

$$\frac{\partial F}{\partial y} = \frac{\partial}{\partial y}\left(\int M(x,y)dx\right) + Q'(y) \equiv N(x,y).$$

(3) If you have done steps (1) and (2) correctly, the equation resulting from step (2) will define $Q'(y)$ as a function of y only. Antidifferentiate $Q'(y)$ to obtain $Q(y)$. The function F from step (1), with this value of $Q(y)$, will provide an implicit solution $F(x,y) = C$ of the given exact differential equation.

The examples below demonstrate the use of this method.

Example 2.5.3. *Solve the differential equation* $(x^2 + y)dx + (x - \sin(y))dy = 0$.

Solution. We showed in Example 2.5.2 that $\frac{\partial M}{\partial y} = \frac{\partial N}{\partial x} = 1$; therefore, this equation is exact. As suggested in step (1), we set $\frac{\partial F}{\partial x} = M = x^2 + y$. Then

$$F = \int (x^2 + y)dx = x^3/3 + yx + Q(y).$$

When integrating partially with respect to x, an arbitrary function $Q(y)$ acts as the constant of integration since $\frac{\partial}{\partial x}(Q(y)) \equiv 0$.

Step (2) says to differentiate this version of F partially with respect to y and set the result equal to N:

$$\frac{\partial F}{\partial y} = \frac{\partial}{\partial y}\left(x^3/3 + yx + Q(y)\right) = 0 + x + Q'(y) \equiv N = x - \sin(y).$$

Therefore, the function Q must satisfy $Q'(y) = -\sin(y)$, and integration gives $Q(y) = \cos(y)$. It is not necessary to add a constant to $Q(y)$ since the function F will be set equal to an arbitrary constant.

Substituting $Q(y)$ into $F = x^3/3 + yx + Q(y)$ we have the complete function $F(x, y) = x^3/3 + yx + \cos(y)$; therefore, an implicit solution of the differential equation is given by

$$x^3/3 + yx + \cos(y) = C.$$

In this case it is not possible to solve for $y(x)$ explicitly.

Alternatively, we could have started by setting $\frac{\partial F}{\partial y} = N = x - \sin(y)$ and integrated with respect to y to obtain $F = \int (x - \sin(y))\,dy = xy + \cos(y) + P(x)$. Then $\frac{\partial F}{\partial x} = y + P'(x) \equiv M = x^2 + y$ implies $P'(x) = x^2$ and $P(x) = x^3/3$. This results in the same function $F(x, y) = xy + \cos(y) + x^3/3$. Sometimes one of the two methods turns out to be much easier than the other. ∎

Example 2.5.4. *Solve the initial-value problem*

$$y' = f(x, y) = \frac{-2xy}{1 + x^2 + 3y^2}, \quad y(0) = 1. \tag{2.36}$$

Solution. The differential equation (2.36) is neither separable nor linear. It is easy to see that $f(x, y)$ is defined and continuous for all x and y, and so is its partial derivative with respect to y; therefore, the Existence and Uniqueness Theorem tells us that there is a unique solution through any initial point and solutions cannot intersect in the (x, y)-plane. Note that $y \equiv 0$ is a constant solution of (2.36). A slope field for this equation is shown in Figure 2.12.

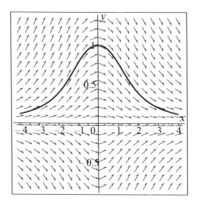

Figure 2.12. Slope field for $y' = \frac{-2xy}{1+x^2+3y^2}$

2.5 More Analytic Methods for Nonlinear First-order Equations

To find an algebraic solution of the IVP, we first set

$$\frac{dy}{dx} = \frac{-2xy}{1+x^2+3y^2},$$

and then write the equation in the form

$$M(x,y)dx + N(x,y)dy = (2xy)dx + (1+x^2+3y^2)dy = 0.$$

Noting that $\frac{\partial M}{\partial y} = 2x = \frac{\partial N}{\partial x}$, we see that the method for solving exact equations can be used.

Now, following the three-step method described above,

(1) Let $F(x,y) = \int M(x,y)dx + Q(y) = \int (2xy)dx + Q(y) = x^2y + Q(y)$.

(2) Differentiating the function found in (1) with respect to y,

$$\frac{\partial F}{\partial y} = x^2 + Q'(y) \equiv N(x,y) = 1 + x^2 + 3y^2$$

and the equation for $Q'(y) \equiv \frac{dQ}{dy}$ is $Q'(y) = 1 + 3y^2$.

(3) Integrating $Q'(y) = 1 + 3y^2$ gives $Q(y) = y + y^3$.

We have now found that $F(x,y)$ is the function $x^2y + y + y^3$ and therefore

$$x^2y + y + y^3 = C \tag{2.37}$$

is an implicit solution of (2.36). In this example, we can use the initial condition $y(0) = 1$ to find the constant C. Letting $x = 0$ and $y = 1$ in (2.37) gives $(0)(1) + 1 + (1)^3 = 2 = C$, and the unique solution to (2.36) is $x^2y + y + y^3 = 2$.

We can go one step further and find an explicit formula for the solution y. There is a formula for roots of cubic polynomials (see the CRC tables) which is messy, but can sometimes be used to produce a reasonable result. In this case it provides the following explicit solution:

$$y(x) = \left(\sqrt{1 + \left(\frac{x^2+1}{3}\right)^3} + 1\right)^{1/3} - \left(\sqrt{1 + \left(\frac{x^2+1}{3}\right)^3} - 1\right)^{1/3}. \tag{2.38}$$

This solution curve is shown in the slope field in Figure 2.12. At this point, it would be interesting to see what your computer algebra system gives as a solution to (2.36). It may not be written in exactly the same form as (2.38), but you can check that it is the same function by graphing both functions on the same set of axes. ■

The method for solving exact equations can be extended by also allowing for multiplication by an integrating factor. This technique can be used to make an equation of the form $P(x,y)dx + Q(x,y)dy = 0$ into an exact equation. An excellent discussion of this method can be found in the book by Polking, Boggess, and Arnold[1].

---

[1] J. Polking, A. Boggess, and D. Arnold, *Differential Equations with Boundary Value Problems*, 2nd ed., Pearson, Prentice Hall, 2005.

2.5.2 Bernoulli Equations.
An equation of the form
$$x' = p(t)x + q(t)x^n, \quad n \neq 0, 1 \tag{2.39}$$
is called a **Bernoulli equation**. If $n = 0$ or $n = 1$ the equation is easily seen to be linear, and our method for solving linear equations can be used. For any other value of n, the substitution of a new dependent variable $v(t) = (x(t))^{1-n}$ turns the equation into a linear equation in v. To see this, differentiate $v(t)$ by the chain rule:
$$v' = \frac{d}{dt}(x^{1-n}) = (1-n)x^{-n}x'.$$
Multiplying equation (2.39) by x^{-n},
$$x^{-n}x' = p(t)x^{1-n} + q(t) = p(t)v + q(t),$$
and therefore
$$v' = (1-n)(p(t)v + q(t)) = (1-n)p(t)v + (1-n)q(t), \tag{2.40}$$
which is a linear differential equation for v.

Example 2.5.5. *As a first example we will show that the logistic growth equation*
$$P' = rP(1 - P/N) \tag{2.41}$$
is a Bernoulli equation for any values of the parameters r and N.

Solution. We can rewrite (2.41) in the form
$$P' = rP - \frac{r}{N}P^2$$
and this has the form of a Bernoulli equation with $n = 2$, $p(t) = r$, and $q(t) = -r/N$. Letting $v = P^{1-n} = P^{-1} = 1/P$ and using (2.40), the equation for v is
$$v' = (1-2)p(t)v + (1-2)q(t) = -rv + \frac{r}{N}.$$
The linear equation $v' + rv = \frac{r}{N}$ can be solved by multiplying by the integrating factor $\mu = e^{rt}$, which gives
$$e^{rt}v' + re^{rt}v = \frac{r}{N}e^{rt} \implies \frac{d}{dt}(e^{rt}v) = \frac{r}{N}e^{rt},$$
and integrating,
$$e^{rt}v = \frac{r}{N} \cdot \frac{e^{rt}}{r} + C.$$
If both sides are multiplied by e^{-rt},
$$v = 1/N + Ce^{-rt}.$$
To find P, we substitute back into $v = 1/P$ to write
$$P = 1/v = \frac{1}{1/N + Ce^{-rt}} = \frac{N}{1 + NCe^{-rt}},$$
which is equivalent to the solution we found in Section 2.1 by separation of variables. If we let $C_1 = -NC$ the two forms are the same. ■

Another Bernoulli equation is considered in the next example.

2.5 More Analytic Methods for Nonlinear First-order Equations

Example 2.5.6. *Solve the initial-value problem*

$$y' = -2y + e^t y^3, \quad y(0) = 1. \tag{2.42}$$

Solution. This is a Bernoulli equation with $n = 3$, $p(t) = -2$, and $q(t) = e^t$. If we make the substitution $v = y^{1-n} = y^{-2}$, then the equation for v is

$$v' = (1-3) \cdot (-2) \cdot v + (1-3) \cdot e^t = 4v - 2e^t.$$

Solving for v by our method for linear equations, with $\mu = e^{-4t}$,

$$e^{-4t}v' - 4e^{-4t}v = -2e^t \cdot e^{-4t} = -2e^{-3t} \implies \int \frac{d}{dt}(e^{-4t}v)dt = \int(-2e^{-3t})dt$$

$$\implies e^{-4t}v = (-2e^{-3t})/(-3) + C \implies v(t) = \frac{2}{3}e^t + Ce^{4t}.$$

To find y, note that $v = y^{-2}$ implies that $y = \pm\sqrt{1/v}$. Therefore,

$$y(t) = \pm\left(\frac{2}{3}e^t + Ce^{4t}\right)^{-\frac{1}{2}}. \tag{2.43}$$

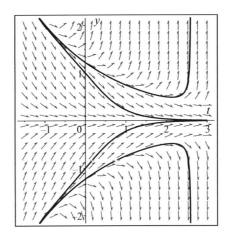

Figure 2.13. Slope field for $y' = -2y + e^t y^3$

To satisfy the initial condition $y(0) = 1$, set $\pm(\frac{2}{3} + C)^{-\frac{1}{2}} = 1$. We must use the plus sign, and then $C = \frac{1}{3}$. The unique solution of the initial value problem (2.42) is

$$y(t) = 1/\sqrt{\frac{2}{3}e^t + \frac{1}{3}e^{4t}} = \sqrt{\frac{3}{e^t(2+e^{3t})}},$$

and it is defined for all real t. ∎

It is important in Example 2.5.6 to note that $y = 0$ is a constant solution of the differential equation (2.42), but it is not given by the "general" formula (2.43). This means that when using Bernoulli's Method it is necessary to check for constant solutions first.

The Uniqueness and Existence Theorem can be used to show that (2.42) has unique solutions through every point of the (t, y)-plane. It is interesting to speculate whether the positive solutions will tend to 0 or to $+\infty$ as $t \to \infty$. One might suppose that the term $e^t y^3$ in the slope function would drive solutions to $+\infty$. However, the slope field shown in Figure 2.13, as well as the algebraic solution, suggests that there is a positive value of $y(0)$ below which all positive solutions tend to 0 and above which they tend to $+\infty$, possibly at a vertical asymptote. Exercise 21. below asks you to find this value of $y(0)$.

2.5.3 Using Symmetries of the Slope Field.
In Figure 2.13 a certain type of symmetry is evident in the slope field for the equation in Example 2.5.6. If we write the equation as $y' = f(t, y) = -2y + e^t y^3$, we can see analytically that the slope function f satisfies

$$f(t, -y) = -2(-y) + e^t(-y)^3 = -(-2y + e^t y^3) = -f(t, y);$$

that is, the slope function is symmetric about the t-axis.

Any first-order differential equation $y' = f(t, y)$ with f satisfying $f(t, -y) \equiv -f(t, y)$ has the property that if $y(t)$ is any solution of the equation, then $-y(t)$ is also a solution. To prove this, assume $y' = f(t, y)$, $f(t, -y) \equiv -f(t, y)$, and let $w(t) = -y(t)$. Then

$$w' = -y' = -f(t, y) = -f(t, -w) \equiv -(-f(t, w)) = f(t, w),$$

where the next-to-last equality uses the symmetry property of the slope field. Having shown that $w' = f(t, w)$, we can conclude that the function $w \equiv -y$ is a solution of the same differential equation satisfied by y. This symmetry of the solution curves about the t-axis can be clearly seen in Figure 2.13.

Exercises 2.5. *Determine whether or not equations 1–6 are exact.*

1. $(x + y)dx + xdy = 0$
2. $(2x + y)dx + (x - y)dy = 0$
3. $\sin(y)dx + x\cos(y)dy = 0$
4. $ye^x dx + xe^y dy = 0$
5. $2xydx + (x^2 + y^2)dy = 0$
6. $(x^2 + xy + 3y^2)dx + (y^2 + xy + 3x^2)dy = 0$

Find a general solution for equations 7–10.

7. $(x + y)dx + (x + 1)dy = 0$
8. $(2 + y)dx + (x - 3)dy = 0$
9. $(y + \sin(y))dx + (1 + x + x\cos(y))dy = 0$
10. $(xy^2 + 2y)dx + (x^2 y + 2x + 1)dy = 0$

For equations 11–14, show that the equation is exact, and solve the IVP:

11. $(1 + xy)dx + \frac{1}{2}x^2 dy = 0$, $y(1) = 1$
12. $(x + y)dx + (x + 1)dy = 0$, $y(0) = 2$

13. $\sin(y)dx + (x\cos(y))dy = 0$, $y(1) = \frac{\pi}{2}$

14. $ye^x dx + (2 + e^x)dy = 0$, $y(0) = 1$

Solve the following Bernoulli equations.

15. $P' = 2P(1 - P/4)$

16. $P' = P(2 - 5P)$

17. $y' = -y + e^t y^2$, $y(0) = 1$

18. $y' = 2y + e^{3t} y^2$, $y(0) = -1$

19. $y' + y = ty^3$

20. $y' + 2y = -4ty^3$

21. For the Bernoulli equation $y' = -2y + e^t y^3$ in Example 2.5.6, find the positive solution that separates solutions tending to zero from those having a vertical asymptote at a positive value of t. What is the initial value $y(0)$ for this solution?

22. In the PHYSICS example in Chapter 1.3, the equation for the velocity of a skydiver was given as
$$v'(t) = g - \frac{k}{m}(v(t))^p.$$
Assume now that $p = 2$, so the term representing the friction due to the air is proportional to the square of the velocity.

(a) Define the constants $\alpha = \frac{k}{m}$ and $b = \sqrt{\frac{mg}{k}}$. Show that the function $x(t) = v(t) + b$ satisfies the Bernoulli equation
$$x'(t) = (2b\alpha)x(t) - \alpha(x(t))^2.$$

(b) Solve this Bernoulli equation for $x(t)$, and show that the general solution is
$$v(t) \equiv x(t) - b = \frac{b(1 - 2bCe^{-2b\alpha t})}{1 + 2bCe^{-2b\alpha t}}.$$

(c) Find the terminal velocity $\bar{v} = \lim_{t\to\infty}(v(t))$. Is it equal to $\left(\frac{mg}{k}\right)^{\frac{1}{p}}$ as stated in Chapter 1.3?

2.6 Numerical Methods

In this section you will be shown how solutions to initial-value problems
$$x'(t) = f(t, x), \quad x(t_0) = x_0, \tag{2.44}$$
can be numerically approximated. The idea is not to make you an expert in numerical analysis, but to provide you with enough information so you will know when an approximate solution is accurate to the number of significant digits required by your work.

A numerical method for solving (2.44) produces a set of discrete points
$$(t_0, x_0), (t_1, x_1), \ldots, (t_N, x_N)$$

where $t_j = t_0 + j\Delta t$ and x_j is an approximation to the value of the solution at t_j; that is, $x_j \approx x(t_j)$. Once this list of points is obtained, it is usually plotted by connecting the points by straight-line segments; or, if you are using a sophisticated computer package, a curve fitting routine may be used to fit a smooth curve through the points.

For the three methods described here, it will be seen that the accuracy of the method depends directly on the step size Δt. For a given method, the smaller the value of Δt, the more accurate the approximation as long as the word size of the computer is large enough to avoid round-off error. However, it will also be shown that the error depends significantly on the method being used to compute the approximation.

If the slope function f satisfies the conditions of the Existence and Uniqueness Theorem 2.1, it is always possible to compute a numerical approximation to the solution. In theory, if f can be differentiated enough times, $x(t)$ could be approximated by the first $N + 1$ terms in its Taylor series at $t = t_0$; that is, by

$$x(t_0 + h) = x(t_0) + x'(t_0)h + \frac{x''(t_0)h^2}{2!} + \cdots + \frac{x^{(N)}(t_0)h^N}{N!} + E_N(h), \qquad (2.45)$$

where the error $E_N(h)$ in this approximation is known from calculus to be

$$E_N(h) = \frac{x^{(N+1)}(\xi)h^{N+1}}{(N+1)!} \qquad (2.46)$$

for ξ some value of t between t_0 and $t_0 + h$.

In general, this is not a good procedure to use because, as most calculus students already know, it may take a very large number of terms in the series to approximate $x(t)$ accurately for t very far from t_0. Furthermore, for each different differential equation the formula $f(t, x)$ for x' changes and the derivatives $x'', x''', \ldots$ would all have to be recalculated. These can be very complicated calculations; for example, the chain rule gives $x''(t) = \frac{d}{dt}(x'(t)) = \frac{d}{dt}(f(t, x(t))) = \frac{\partial f}{\partial t} + \frac{\partial f}{\partial x}x'(t)$, and higher derivatives become successively more complex. The numerical methods described in this section avoid these problems by starting at $t = t_0$, with a *small* Δt. The value $x(t_0 + \Delta t)$ is approximated by a *small* number of terms in the Taylor series, and then the process is repeated at $t_1 = t_0 + \Delta t, t_2 = t_0 + 2\Delta t, \ldots$. There will be a small error at each step, due to truncating the series, and these errors can accumulate.

2.6.1 Euler's Method. Euler's Method is one of the oldest and simplest numerical methods for obtaining approximate solutions to the initial value problem (2.44). It uses only the first two terms of the Taylor series; that is, on each interval $[t_j, t_{j+1}]$ it approximates $x(t)$ by the *tangent line approximation* at the left end of the interval. This approximation is

$$x(t_j + \Delta t) \approx x(t_j) + x'(t_j)\Delta t \equiv x(t_j) + f\left(t_j, x(t_j)\right)\Delta t.$$

Notice that *no derivatives* of f have to be evaluated, only the function f itself.

2.6 Numerical Methods

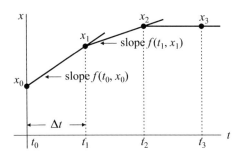

Algorithm for Euler's Method
Given $x' = f(t, x)$, $x(t_0) = x_0$, to find approximate values of $x(t)$ on the interval $t_0 \leq t \leq t_{max}$:
(1) Choose a small stepsize $\Delta t = \frac{t_{max} - t_0}{N}$, with N a positive integer.
(2) For $j = 0, 1, \ldots, N - 1$ compute

$$x_{j+1} = x_j + f(t_j, x_j)\Delta t$$
$$t_{j+1} = t_j + \Delta t.$$

(3) Plot the points (t_j, x_j), $j = 0, 1, \ldots, N$. If straight lines are drawn between consecutive points, this results in a piecewise linear approximation to the solution $x(t)$ on the interval $[t_0, t_{max}]$.

Example 2.6.1. *Approximate the solution of* $x' = t - x$, $x(0) = 1$, *on the interval* $[0, 2]$.

Solution. As a first try we arbitrarily let $\Delta t = \frac{2-0}{N}$ with N chosen to be 4. Then $\Delta t = 0.5$. It helps to make a table, as shown below. The values $t_j = t_0 + j\Delta t$ are all determined once Δt is chosen.

j	t_j	x_j	$f(t_j, x_j) = t_j - x_j$	$x_{j+1} = x_j + f(t_j, x_j)\Delta t$
0	0	1.0	-1.0	$1.0 + (-1.0)(0.5) = 0.5$
1	0.5	0.5	0	$0.5 + 0(0.5) = 0.5$
2	1.0	0.5	0.5	$0.5 + 0.5(0.5) = 0.75$
3	1.5	0.75	0.75	$0.75 + (0.75)(0.5) = 1.125$
4	2.0	1.125		

∎

Comments.

- The value of x_{j+1} at the end of each row is used as the value of x_j in the following row. The initial conditions give the values of t_0 and x_0 in the first line.

- The function $f(t_j, x_j)$ is the slope function and depends on the differential equation being solved.

Using the analytic method for solving linear differential equations, the exact solution of $x' = t - x$, $x(0) = 1$ is $x(t) = t - 1 + 2e^{-t}$. Check it! This gives the exact value $x(2) = 2 - 1 + 2e^{-2} \approx 1.27067$. The absolute error in our numerically computed value of $x(2)$ in the abovetable is $|1.27067 - 1.125| \approx 0.14567$.

If the calculations are redone with $N = 8$, that is, with $\Delta t = 0.25$, the corresponding table is

j	t_j	x_j	$f(t_j, x_j) = t_j - x_j$	$x_{j+1} = x_j + f(t_j, x_j)\Delta t$
0	0	1.0	−1.0	0.75
1	0.25	0.75	−0.50	0.625
2	0.50	0.625	−0.125	0.59375
3	0.75	0.59375	0.15625	0.6328125
4	1.00	0.6328125	0.3671875	0.72460938
5	1.25	0.72460938	0.52539063	0.85595703
6	1.50	0.85595703	0.64404297	1.01696777
7	1.75	1.01696777	0.73303223	1.20022583
8	2.00	1.20022583		

The error in the approximate value for $x(2)$ is now $|1.27067 - 1.200226| \approx 0.0704$. By cutting the step size Δt in half the error has been cut approximately in half, from 0.146 to 0.070. Figure 2.14 shows the two approximate solutions plotted with the exact solution.

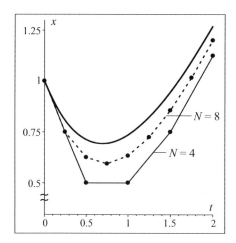

Figure 2.14. Euler approximations to $x' = t - x$, $x(0) = 1$, with $N = 4, 8$

Error and the Order of a Numerical Method. Two different types of error can occur when solving a differential equation numerically. The first is called **round-off error**, and is due to rounding or chopping the result after each step in the procedure, in order to store it in a computer with fixed word size. For example, if $x(0) = \frac{1}{3}$, and the computer running the program has a word size of 4 decimal digits, then it would probably store $x(0)$ as 0.3333, already introducing an error greater than 0.00003 into the

2.6 Numerical Methods

calculation. Fortunately, modern computers can store a lot more digits than this so it is usually not a problem, unless the procedure requires a huge number of steps.

The type of error you really need to be concerned about is caused by the approximations made by whatever method you are using. This is called **truncation error**, and in the case of Euler's Method it is caused by dropping all but the first two terms in the Taylor series for $x(t + \Delta t)$. From (2.46), it is seen that the maximum error made by using the approximation

$$x(t + \Delta t) \approx x(t) + x'(t)\Delta t = x(t) + f(t, x(t))\Delta t$$

is given by $E_1(\Delta t) = x''(\xi)(\Delta t)^2/2$ where ξ is some value of t between t and $t + \Delta t$. If we know that $|x''(t)| \leq M$ on the entire interval $[t_0, t_{max}]$, then we can say that the error E made in a single step (called the **local truncation error**) is bounded by

$$E \leq \frac{M}{2}(\Delta t)^2.$$

Numerical analysts refer to this as an **error of order 2 in** Δt, and usually write it as $\mathcal{O}((\Delta t)^2)$.

Definition 2.5. *An error of order n, denoted by $\mathcal{O}((\Delta t)^n)$, is an error which approaches a constant times $(\Delta t)^n$ as $\Delta t \to 0$.*

Since Euler's Method takes N steps to go from t_0 to t_{max}, and $N = \frac{t_{max}-t_0}{\Delta t}$, it might seem that the total error in the calculation would be bounded by N times the local error E. The product NE is a constant times Δt, hence is $\mathcal{O}(\Delta t)$. Unfortunately this is not quite correct. After the first step we are using the approximation x_1 instead of the true value of $x(t_1)$. When we write $x_2 = x_1 + \Delta t f(t_1, x_1)$ we not only have the truncation error to worry about, but also the error in x_1, and these errors can propagate.

In fact it can be shown that the total error increases exponentially over the total number of steps. An excellent discussion of this is given in the book by Hubbard and West[2], in which it is shown that the total error over the N steps is bounded by

$$E(t_{max}) \leq \frac{C}{K}\left(e^{K|t_{max}-t_0|} - 1\right)\Delta t$$

for some constant C, where K can be taken to be the maximum value of $|\frac{\partial f}{\partial x}|$ on a rectangle containing the entire solution. For a fixed length interval $|t_{max} - t_0|$, this will still be just a constant times Δt, but if the interval of integration is increased, the error can grow exponentially. This should warn you that using Euler's Method over very long time intervals may produce large errors even if Δt is very small.

Based on the preceding discussion, we say that Euler's Method is a *first-order* numerical method. What this means, practically, is that if an equation is solved using Euler's Method with a given Δt, and then again with $\frac{\Delta t}{2}$, the error in the second calculation should be about $\frac{1}{2}$ of the error in the first.

---

[2] J. H. Hubbard and B. H. West, *Differential Equations, A Dynamical Systems Approach, Part I*, Springer Verlag, 1990.

An nth-order numerical method is one whose error is $\mathcal{O}\left((\Delta t)^n\right)$; that is, the error is equal to a constant times $(\Delta t)^n$, in the limit as $\Delta t \to 0$. If an equation is solved by an nth-order method with a given Δt, and then again with $\frac{\Delta t}{2}$, the error in the second calculation should be about $(\frac{1}{2})^n$ times the error in the first. This is what makes higher-order methods so accurate for small values of Δt.

We will see that higher-order methods are needed in real-world applications, but Euler's Method is by far the easiest one to understand.

2.6.2 Improved Euler Method. A second-order numerical method called the Improved Euler Method, or Heun's Method, uses two values of the slope function in each interval $[t_j, t_{j+1}]$. The slope $m_0 = f(t_j, x_j)$ at the left end of the interval is computed first. Then the Euler approximation $\bar{x}_{j+1} = x_j + m_0 \Delta t$ is used to calculate an approximation m_1 to the slope at the right-hand end of the t-interval. The average slope $m = \frac{m_0 + m_1}{2}$ is used to compute $x_{j+1} = x_j + m \Delta t$ which is the improved approximation of $x(t_{j+1})$. See Figure 2.15.

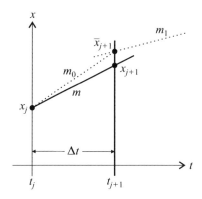

Figure 2.15. Improved Euler Method

This results in the formula

$$x_{j+1} = x_j + \frac{\Delta t}{2}(m_0 + m_1) = x_j + \frac{\Delta t}{2}\left[f(t_j, x_j) + f\left(t_j + \Delta t, x_j + \Delta t f(t_j, x_j)\right)\right].$$

With some difficulty, it can be shown that this formula for x_{j+1} agrees with the Taylor series for $x(t_j + \Delta t)$ in its first *three* terms. This means that the truncation error at each step is $\mathcal{O}\left((\Delta t)^3\right)$; therefore, using the same argument as before, the accumulated error in the Improved Euler Method, over the interval $[t_0, t_{max}]$, is $\mathcal{O}\left((\Delta t)^2\right)$. Hence, this is a **second-order method**, and cutting the step size in half reduces the error by approximately $\frac{1}{4} = (\frac{1}{2})^2$.

2.6 Numerical Methods

Algorithm for the Improved Euler Method

Replace Step (2) in the Euler Method by the following:

(2′) For $j = 0, 1, \ldots, N - 1$ compute

$$m_0 = f(t_j, x_j)$$
$$\tilde{x}_{j+1} = x_j + m_0 \Delta t$$
$$t_{j+1} = t_j + \Delta t$$
$$m_1 = f(t_{j+1}, \tilde{x}_{j+1})$$
$$m = \frac{m_0 + m_1}{2}$$
$$x_{j+1} = x_j + m\Delta t.$$

Example 2.6.2. *Use the Improved Euler Method, with $\Delta t = 0.5$, to approximate the solution of $x' = t - x$, $x(0) = 1$, on the interval $[0, 2]$.*

j	t_j	x_j	$m_0 =$ $t_j - x_j$	$\tilde{x}_{j+1} =$ $x_j + m_0\Delta t$	$m_1 =$ $t_{j+1} - \tilde{x}_{j+1}$	$m = \frac{m_0+m_1}{2}$	x_{j+1}
0	0	1.0	−1.0	0.5	0	−0.5	0.75
1	0.5	0.75	−0.25	0.625	0.375	0.0625	0.78125
2	1.0	0.78125	0.21875	0.89063	0.60938	0.41406	0.98828
3	1.5	0.98828	0.51172	1.24414	0.75586	0.63379	1.30518
4	2.0	1.30518					

Solution. The absolute error in $x(2)$ is approximately $|1.30518 - 1.27067| = 0.03451$, which is even less than the error we found using Euler's Method with twice as many steps. ∎

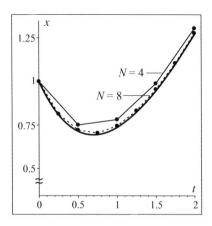

Figure 2.16. Improved Euler approximations to $x' = t - x$, $x(0) = 1$, with $N = 4, 8$

With $N = 8$, the Improved Euler Method gives an approximation of 1.27756 for $x(2)$, which is in error by approximately 0.00689. Figure 2.16 shows the two approximate solutions with $N = 4$ and $N = 8$. Comparing these with Figure 2.14 shows the distinct improvement, and also the fact that the solutions now converge to the exact solution from above.

By halving the step size, the error has been reduced to less than a quarter of its size, which is what we would expect with a second-order method. The Improved Euler Method requires two evaluations of $f(t, x)$ for each step, but again no derivatives of f have to be calculated.

2.6.3 Fourth-order Runge-Kutta Method.

A lot of work was done in the first half of the twentieth century to develop methods that are far more accurate. Most computer algebra systems use what are called Runge-Kutta Methods, developed by two German mathematicians. One of the major motivations behind these methods was to find a way to avoid having to differentiate the slope function. The fourth-order Runge-Kutta Method, for example, uses four evaluations of the slope function on each t-interval to produce an approximation which agrees with the first five terms of the Taylor series, and therefore has an accumulated error $\mathcal{O}\left((\Delta t)^4\right)$ on the interval $[t_0, t_{max}]$.

There are many different Runge-Kutta formulas for any given order, but the algorithm below is the one that has gained acceptance because of its simplicity. Although it is tedious to compute by hand, it is easy to program for a calculator or computer. If a differential equation needs to be solved within a Maple or *Mathematica* procedure, for example, the algorithm given below can be used exactly as written.

Algorithm for the fourth-order Runge-Kutta Method
Replace Step (2) in Euler's Method by the following:
(2*) For $j = 0, 1, \ldots, N-1$ compute
$$m_1 = f(t_j, x_j)$$
$$m_2 = f(t_j + \frac{\Delta t}{2}, x_j + m_1 \frac{\Delta t}{2})$$
$$m_3 = f(t_j + \frac{\Delta t}{2}, x_j + m_2 \frac{\Delta t}{2})$$
$$m_4 = f(t_j + \Delta t, x_j + m_3 \Delta t)$$
$$x_{j+1} = x_j + \frac{\Delta t}{6}(m_1 + 2m_2 + 2m_3 + m_4)$$
$$t_{j+1} = t_j + \Delta t.$$

For the IVP $x' = t - x$, $x(0) = 1$, four steps of the fourth-order Runge-Kutta Method with $\Delta t = 0.5$ result in a value of $x(2) \approx 1.27110$. The absolute error in this value is $|1.27067 - 1.27110| = 0.00043$, which is much smaller than the error in either Euler's Method or the Improved Euler Method, as expected. If Δt is halved, the error in $x(2)$ will be approximately $\frac{1}{16} = (\frac{1}{2})^4$ times its original size. This means that halving the step size Δt will result in an answer with at least one more significant decimal digit. The following table shows the values of $x(2)$, using the fourth-order Runge-Kutta method with $\Delta t = 0.25, 0.125, 0.0625$, and 0.03125.

2.6 Numerical Methods

Δt	$x(2)$
0.25	1.2706 9228
0.125	1.2706 7178
0.0625	1.2706 7063
0.03125	1.2706 7057

When solving real-world problems, it is often the case that no exact solution is possible; it then becomes critical to be able to estimate the accuracy of a numerical solution. Based on the preceding discussion, if one is using the fourth-order Runge-Kutta Method, it is reasonable to compare two solutions having step sizes Δt and $\Delta t/2$, and use the position of the digit in which they differ to estimate how many significant digits are correct.

Example 2.6.3. *In Example 2.2.2 we drew a slope field for the differential equation $x' = x^2 - t$, and remarked that no simple analytic formula for the solution exists. The slope field and four numerically computed trajectories for this equation are seen in Figure 2.17. It appears that there may be a unique solution that separates solutions that tend to infinity as t increases from those that enter the region inside the parabola $x(t) = \pm\sqrt{t}$ and then tend asymptotically to $-\sqrt{t}$ as $t \to \infty$. Notice that once a solution crosses the upper branch of the parabola, the slope function $f(t, x) = x^2 - t$ becomes negative, implying that the solution must go down.*

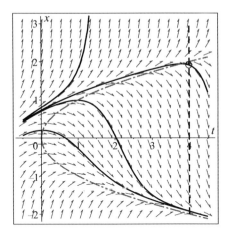

Figure 2.17. Solutions of $x' = x^2 - t$, $x(0) = 0.2, 0.7, 0.72901, 0.8$

One way to obtain an estimate for the initial value $\hat{x}_0$ for the special solution separating the two families is to pick initial conditions $(K, \sqrt{K})$ far out on the parabola (that is, with K large) and then integrate the solution back from $t = K$ to $t = 0$. Most numerical differential equation solvers allow you to enter a negative value for Δt, or alternatively to specify a range of t from 0 to K and enter the initial condition as $x(K) = \sqrt{K}$. This was done with a very accurate Maple routine, and resulted in the following values:

Each of the initial values in the table must be slightly below $\hat{x}_0$ since these three solution curves have already met the parabola at $x(K) = \sqrt{K}$, and therefore are destined to go down as t increases. It is easy to test the accuracy of this initial value by using initial

K	x(0)
4	0.7290109
9	0.7290112
16	0.7290112

conditions $x(0) = 0.7290112 \pm \varepsilon$ to see if the solutions ultimately go in different directions. This would be a good way to test the accuracy of your differential equation solver. In most applications an accuracy of 7 decimal places is much more than is needed.

You should now feel fairly confident in your ability to estimate the accuracy of the solution of a given IVP, even when the equation cannot be solved analytically. The table below shows the value of $x(4)$ obtained by using the simple fourth-order Runge-Kutta method described above to solve the IVP

$$x' = x^2 - t, \quad x(0) = 0.72901.$$

To include the possibility of round-off error, two calculations were made, one using 8 significant digits in the calculations (comparable to having a word size of eight decimal digits), and one made with 16 digits. This is possible to do in Maple by using a variable called Digits. The table below shows the computed values of $x(4)$ for step sizes $\Delta t = 0.125, 0.125/2, 0.125/2^2, \ldots, 0.125/2^{10}$, for the two word sizes.

Δt	$x(4)$, Digits=8	$x(4)$, Digits= 16
0.125	2.0645 996	2.0649 8090
0.0625	1.9151 634	1.9164 0489
0.03125	1.9036 732	**1.9**072 4317
0.015625	1.9099 448	**1.9066** 6497
0.0078125	1.9131 016	**1.9066** 2862
0.00390625	1.9417 954	**1.9066** 2634

In the right-hand column the values of $x(4)$ are gaining one more significant digit each time the step size Δt is divided by two; and $x(4) = 1.9066$ can be assumed accurate to 5 significant digits. This is what we would expect to happen with our fourth-order method, and as long as the number of steps needed, relative to the significant digits used in the calculation is reasonable, this is what you will see.

What you need to realize is that if you are working on a machine that keeps only a small number of significant digits in its calculations, then if Δt is very small, or a huge number of steps are needed, bad things can happen. In the center column, with an 8-digit word size, the value of $x(4)$ never really converged to its true value. If Δt was made even smaller, the answer would soon become "garbage," just due to the round-off error at each step.

Conclusion. For reasonable problems where only three or four significant digits are needed in the answer, a fourth-order method is fine; and all you need to do is run it with a reasonably small step size, and then compare the answer to what you get with half that step size. You may need to repeat this until you see the final value agree to the number of significant digits you require. If an agreement does not occur, then you may need to increase the word size of the computer or calculator you are using, or you may need to

2.6 Numerical Methods

find a different type of equation solver that is geared to your particular problem. There are many more sophisticated methods for numerically solving differential equations. To find out about these you should consult a book on numerical analysis.

Exercises 2.6. *Problems 1 and 2 ask you to compute tables for the Euler Method and Modified Euler Method by hand, for the IVP $x' = t - x$, $x(0) = 1$. To make these a reasonable length, you are going to find values of $x(1)$, instead of $x(2)$ (as in the Examples). The exact solution is $x(t) = t - 1 + 2e^{-t}$, which gives $x(1) \approx 0.735759$.*

1. For the IVP $x' = t - x$, $x(0) = 1$,

    (a) *Do 8 steps of Euler's Method, with $\Delta t = 0.125$, to find an approximation to $x(1)$.*

    (b) *Complete the following table (values are from Example 2.6.1):*

Δt	$x(1)$	error in $x(1)$
0.5	0.50000	0.235759
0.25	0.632813	0.102946
0.125	?	?

    (c) *Is the error in $x(1)$ what you would expect from a first-order method? Explain.*

2. For the IVP $x' = t - x$, $x(0) = 1$,

    (a) *Do 4 steps of the Modified Euler Method, with $\Delta t = 0.25$, to find an approximation to $x(1)$. Then do 8 steps with $\Delta t = 0.125$.*

    (b) *Complete the following table (values are from Example 2.6.2):*

Δt	$x(1)$	error in $x(1)$
0.5	0.781250	0.045491
0.25	?	?
0.125	?	?

    (c) *Are the errors in $x(1)$ what you would expect from a second-order method? Explain.*

3. *Use whatever technology you have available to solve (numerically) the IVP*

    $$x' = t - x, \quad x(0) = 1.$$

    Find out how to set the parameters and step size to obtain a value of $x(2)$ accurate to 6 decimal places. Explain exactly what you had to do.

4. *Use the same settings as in Exercise 3 to solve the IVP*

    $$x' = x^2 - t, \quad x(0) = 0.729011.$$

    Decrease the step size at least once to make sure your answer is exact to 5 decimal places.

    (a) *Does the solution intersect the parabola $t = x^2$?*

(b) *If it does, at what value of t?*

(c) *Does this seem reasonable? Explain. (Refer to Figure 2.17.)*

Problem 5.b demonstrates an inherent danger in using approximate solutions.

5. (a) *Compute the Euler Method approximation to the solution of $x' = x^2$, $x(0) = 1$, on the interval $[0, 1.2]$ with $\Delta t = 0.2$.*

   (b) *Solve the differential equation exactly (it is separable), and explain what you find. What is happening in (a)?*

6. *Use your numerical equation solver to find the value of $x(0.8)$ for the solution of $x' = x^2$, $x(0) = 1$. Compare this with the exact solution obtained in problem 5.b.*

7. *Using a computer algebra system, such as Maple, see what it gives for $x(1)$ and $x(1.2)$ for the IVP $x' = x^2$, $x(0) = 1$. Explain what happens.*

For each problem 8–13, use an appropriate analytic method to solve the IVP. Then use your own numerical equation solver to solve the problem, and compare the value of the numerical solution at $t = 5$ with the analytic solution at $t = 5$. Determine the exact error in your numerical result.

8. $x' = \frac{t+2}{x}$, $x(0) = 2$

9. $x' + \frac{1}{5}x = e^{-0.2t}\sin(t)$, $x(0) = -1$

10. $x' = \frac{1-t/2}{2+x}$, $x(0) = -1$

11. $tx' = 2x + t^2$, $x(1) = 1$

12. $\sin(y)dt + t\cos(y)dy = 0$, $y(1) = \pi/2$ *(Exact equation)*

13. $y' = -y + e^t y^2$, $y(0) = 1$ *(Bernoulli equation)*

COMPUTER PROBLEMS. In Section 1.2, the Maple instruction

```
dsolve({ODE,initcond},y(t))
```

was introduced to produce the exact solution of a solvable differential equation. If the equation is not solvable, this same instruction can be used to produce a numerical solution. The example below will produce a numerical solution of the IVP $y' = t/y$, $y(0) = 1$, which has the exact solution $y(t) = \sqrt{t^2 + 1}$, but will be used to illustrate the method.

In Maple, enter the instruction

```
sol:=dsolve({y'(t)=t/y(t),y(0)=1},type=numeric);
```

This returns a procedure for computing numerical values of the solution.

```
sol(1.5);
``` produces the output [t=1.5,y(t)=1.80277548...]

To pick off the value of $y(1.5)$, use

```
ans:=op(2,op(2,sol(1.5)));
``` This will be just the number 1.80277548....

## 2.7 Autonomous Equations, the Phase Line

The corresponding instructions in *Mathematica* are

    approxsol = NDSolve[{y'[t]==t/y[t],y[0]==1},y[t],{t,0,2}]

ya[t_]=y[t]/.First[approxsol] returns a list ya containing approximate values of the solution on the interval [0, 2]. To obtain a value at a specific point, ya[1.5] returns the value 1.80278, found by interpolation using the list of approximate values of $y$ on the interval [0, 2].

## 2.7 Autonomous Equations, the Phase Line

A first-order differential equation $x' = f(t, x)$ is called **autonomous** if the slope function $f$ depends only on $x$, and not explicitly on $t$. In other words, the rate of change of $x$ does not depend on time, but only on the current state of the dependent variable $x$. The equations $x' = x^2$ and $P' = rP(1-P)$ are autonomous; the equations $x' = x + t$ and $x' = x^2 - t$ are not. An autonomous first-order equation can be written in the form

$$x' \equiv \frac{dx}{dt} = f(x). \tag{2.47}$$

An autonomous first-order equation is always separable.

Solutions of autonomous first-order equations have very limited types of behavior. Suppose that $f(r) = 0$ for some real number $r$. Then the constant function $x(t) \equiv r$ is a solution of (2.47), since both $f(x)$ and $x'$ are identically zero for all values of $t$.

**Definition 2.6.** *A constant function $x(t) \equiv r$, such that $f(r) = 0$, is called an **equilibrium solution** of* (2.47).

Since the slopes depend only on $x$, a slope field (as defined in Section 2.2) for an autonomous equation $x' = f(x)$ is completely determined once slopes along any vertical line $t = \bar{t}$ are plotted. In fact, if $f(x)$ is defined and continuous for all $x$, the behavior of solutions of the equation can be determined from the slope lines along the $x$-axis. This leads to the construction of what is called a **phase line** for the differential equation (2.47).

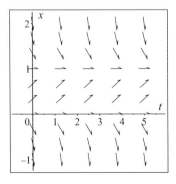

Slope field for $x' = 2x(1-x)$

phase line

> **To draw a phase line for the equation** $x' = f(x)$:
>
> - Find all real numbers $r_1 < r_2 < \cdots$ such that $f(r_i) = 0$, and label these values on a vertical $x$-axis. These points represent the equilibrium solutions. We assume first that the function $f$ is equal to zero at only finitely many values of $x$; if not, a phase line can always be drawn using a finite interval on the $x$-axis.
>
> - For each interval $(-\infty, r_1), (r_1, r_2), \ldots$ pick any value $x = \bar{x}$ in the interval and determine whether $f(\bar{x})$ is positive or negative. Draw an arrow on the axis, in the given interval, pointing up if $f(\bar{x})$ is positive and down if $f(\bar{x})$ is negative. Note that if $f$ is a continuous function, it will have constant sign between any two zeros.

A solution with initial value satisfying $r_i < x(0) < r_{i+1}$ is monotonically increasing if the arrow points up or monotonically decreasing if the arrow points down. If $f(x)$ satisfies the conditions of the Existence and Uniqueness Theorem 2.1, then the solution must remain bounded between the two equilibrium values, since they are solutions and solutions cannot intersect.

**Example 2.7.1.** *Draw a phase line for the differential equation*

$$x' = f(x) = 0.5(x-1)(x+2).$$

*Solution.* This equation has two equilibrium solutions: $x_1 \equiv 1$ and $x_2 \equiv -2$. These are shown plotted on a phase line in Figure 2.18. When $x = 0$, $f(x) = 0.5(-1)(2) = -1 < 0$ so an arrow is drawn in the downward direction in the interval $(-2, 1)$. In the intervals $(-\infty, -2)$ and $(1, \infty)$, $f(x) > 0$ so the arrows point up. Now consider a solution with initial condition less than $-2$, for example $x(0) = -3$. Its slope will always be positive, but it can never cross the equilibrium solution $x(t) \equiv -2$ (do you see why?); therefore, it must increase monotonically and approach $-2$ as a horizontal asymptote, as $t \to \infty$. Similarly, a solution with $-2 < x(0) < 1$ must be monotonically decreasing and bounded between $-2$ and $1$ for all $t$; therefore, it must approach $-2$ asymptotically as $t \to \infty$. If $x(0) > 1$, the solution is monotonically increasing. Whether it exists for all $t$ or has a vertical asymptote at some positive value of $t$ cannot be determined geometrically.

In Figure 2.19 solutions have been drawn for certain initial values at $t_0 = 0$, but the picture would look exactly the same if the same initial conditions were specified at any value of $t_0$. ∎

The phase line contains almost all of the information needed to construct the graphs of solutions shown in Figure 2.19. It does not contain information on how fast the curves approach their asymptotes, or where the curves have inflection points, however. This information, which does depend on $t$, is lost in going to the phase line representation, but note that we did not need to solve the differential equation analytically in order to draw the phase line.

Solutions below the lowest equilibrium solution $r_1$ and above the highest equilibrium solution $r_N$ may either be defined for all $t$ or become infinite at a finite value of $t$. The logistic equation that was graphed in Figure 2.7 is an example of an autonomous equation. In that example, the solutions below $x \equiv 0$ and above $x \equiv 1$ were both

## 2.7 Autonomous Equations, the Phase Line 73

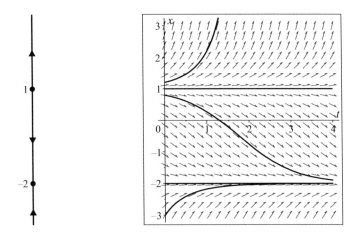

Figure 2.18. Phase line for
$x' = 0.5(x - 1)(x + 2)$

Figure 2.19. Solutions of
$x' = 0.5(x - 1)(x + 2)$

shown to have vertical asymptotes, but in order to do this we had to be able to obtain an analytic solution of the equation.

If the slope function $f$ has infinitely many zeros, for example $f(x) = \sin(x)$, which is zero at $x = 0, \pm\pi, \pm 2\pi, \ldots$, the phase line will only be able to show the behavior around a finite number of these.

**2.7.1 Stability—Sinks, Sources, and Nodes.** In Figure 2.19 it can be seen that if solutions start initially close enough to the equilibrium solution $x \equiv -2$, they will tend toward it as $t \to \infty$. An equilibrium solution of this type (with arrows pointing toward it from both sides on the phase line) is called a **sink** and is said to be a *stable equilibrium*. On the other hand, solutions starting close to $x = 1$ all tend to move away from this solution as $t$ increases. An equilibrium solution of this type (with arrows pointing away from it on both sides) is called a **source**. It is said to be an *unstable equilibrium*. If the arrows on the phase line point toward an equilibrium on one side and away from it on the other side, the equilibrium is called a **node**. It is *semi-stable* in the sense that if a solution starts on one side of the equilibrium it will tend toward it, and on the other side it will tend away as $t \to \infty$.

**Example 2.7.2.** *Draw a phase line for the equation*

$$x' = x(x + 1)^2(x - 3) = f(x)$$

*and label each equilibrium point as a sink, source, or node.*

*Solution.* The equilibrium solutions are the zeros of $f(x)$, namely $x = 0, -1$, and 3. These are shown plotted on the phase line in between Figures 2.20 and 2.21. To determine the direction of the arrows, it helps to draw a graph of the slope function $f(x) = x(x + 1)^2(x - 3)$. This is shown in Figure 2.20. Be careful not to confuse the graph of $f(x)$ with graphs of the solution curves $x(t)$ shown in Figure 2.21. The graph of $f(x)$ is only used to determine whether the arrow between two equilibria points up or down. It can be seen that $f(x)$ is positive between all pairs of equilibrium points except 0 and 3; therefore, all of the arrows point up except the arrow between 0 and 3.

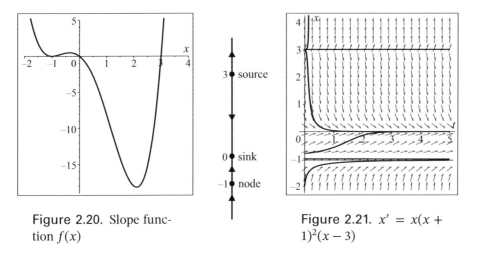

Figure 2.20. Slope function $f(x)$

Figure 2.21. $x' = x(x + 1)^2(x - 3)$

Once the arrows are drawn, it is easy to see that $-1$ is a node, $0$ is a sink, and $3$ is a source. Some solutions of this equation are shown in Figure 2.21.

**Bifurcation in Equations with Parameters.** If an autonomous differential equation contains a parameter $\alpha$, so that

$$x' = f(x, \alpha),$$

then the phase line will change as $\alpha$ varies. Sometimes no significant change will occur, but there can be values of $\alpha$ at which a small change in the parameter can produce a large change in the qualitative structure of the vector field.

**Definition 2.7.** *If a small smooth change to the value of a parameter causes a sudden qualitative change in the behavior of the solutions of a differential equation, we say that a **bifurcation** has occurred. The value of the parameter at which the change occurs is called a **bifurcation value** of the parameter.*

According to Wikipedia, the name "bifurcation" was first introduced in a paper by Henri Poincaré in 1885, a fairly recent event in mathematical terms. For a simple first-order autonomous equation, the only way a bifurcation can occur is if the number or type of one or more of the equilibrium solutions changes.

**Example 2.7.3.** *Show that the differential equation*

$$x' = f(x, \alpha) = \alpha x - x^3$$

*bifurcates when the parameter $\alpha$ passes through the value $0$.*

*Solution.* For any $\alpha \leq 0$, $f(x, \alpha) = x(\alpha - x^2) = 0$ has only the solution $x = 0$, but for $\alpha > 0$ there are three equilibrium points $x = 0, \pm\sqrt{\alpha}$. Since the number of equilibria changes as $\alpha$ passes through the value $0$, the number $0$ is a bifurcation value of $\alpha$. The direction of the arrows on the phase lines can be determined from the two graphs in Figure 2.22 showing the slope functions $f(x, \alpha)$ for $\alpha = -1$ and $\alpha = +1$, and a characteristic phase line for each case. ∎

## 2.7 Autonomous Equations, the Phase Line

If a collection of phase lines is sketched for values close to the bifurcation value, we get what is called a **bifurcation diagram**. By convention, stable equilibria (sinks) are connected by solid curves and unstable equilibria (sources) by dotted curves. These curves show the values of $x$ at which the equilibria occur for values of $\alpha$ between those for which phase lines are drawn. A bifurcation diagram for the equation in Example 2.7.3 is shown in Figure 2.23.

If you have ever seen a pitchfork, it should be clear from the picture in Figure 2.23 why this type of bifurcation is referred to as a **pitchfork bifurcation**. It occurs when a sink turns into a source surrounded by two new sinks.

### 2.7.1.1 Bifurcation of the Harvested Logistic Equation.
In Section 2.2 an example was given (Example 2.2.3) of a harvested logistic growth problem for a herd of deer. Figure 2.5 showed three slope fields for the equation

$$P' = 0.4P\left(1 - \frac{P}{100}\right) - H \tag{2.48}$$

for three values of the harvesting parameter $H$. It was suggested by the slope fields that something strange happened to the behavior of the solution curves between the values $H = 6$ and $H = 12$. It is now possible to completely describe this phenomenon in terms of a bifurcation caused by the change in the parameter $H$.

For any fixed values of $r$, $N$, and $H$, the equilibrium solutions of the harvested logistic equation

$$P' = f(P, H) = rP\left(1 - \frac{P}{N}\right) - H$$

are the constant values of $P$ where $f(P, H) = 0$; that is, they are the zeros of the quadratic polynomial

$$f(P, H) = -\frac{r}{N}P^2 + rP - H = -\frac{r}{N}\left(P^2 - NP + \frac{NH}{r}\right).$$

Using the quadratic formula, the zeros of $f(P, H)$ are at

$$P = \frac{N \pm \sqrt{N^2 - \frac{4NH}{r}}}{2} = \frac{N}{2} \pm \frac{N}{2}\sqrt{1 - \left(\frac{4}{rN}\right)H}.$$

Figure 2.24 shows the parabolic slope function $f(P, H)$ for several values of $H$. When $H = 0$, there are two equilibria, at $P = 0$ and $P = N$. The vertex of the parabola is at

$$f\left(\frac{N}{2}, 0\right) = r\frac{N}{2}\left(1 - \frac{(N/2)}{N}\right) = \frac{rN}{4}.$$

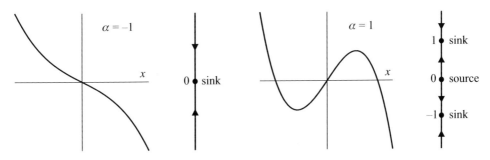

Figure 2.22. $f(x)$ and phase line for $x' = f(x) = \alpha x - x^3$, $\alpha = -1, 1$

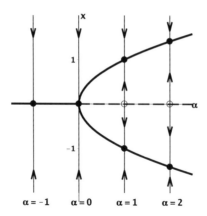

Figure 2.23. Bifurcation diagram for $x' = \alpha x - x^3$

As $H$ increases, the parabola moves down due to the term $-H$, and the two roots get closer together. At the bifurcation value $H^* = \frac{rN}{4}$ there is a single root at $P = \frac{N}{2}$. For $H > \frac{rN}{4}$, there are no roots. Using the graph we can see how the arrows should go on phase lines for values of $H < \frac{rN}{4}, H = H^* = \frac{rN}{4}$, and $H > \frac{rN}{4}$. A characteristic phase line for each case is also shown in Figure 2.24.

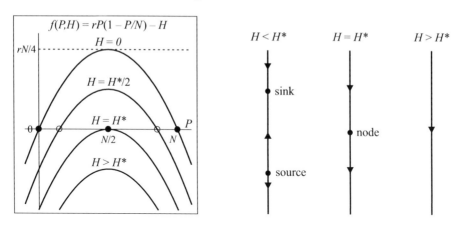

Figure 2.24. Phase lines for the harvested logistic equation $P' = rP(1 - P/N) - H$

For the problem in Example 2.2.3, the value of the carrying capacity was $N = 100$ and the intrinsic growth rate was $r = 0.4$. This gives a bifurcation value for the harvesting parameter $H^* = \frac{rN}{4} = (0.4)(100)/4 = 10$. You should now be able to explain exactly what is happening to the solutions shown in the slope fields in Figure 2.25.

In the left-hand graph $H = 9 < H^*$, and there are two equilibrium solutions; a stable solution with $P(0) \approx 65.8$ and an unstable solution with $P(0) \approx 34.2$. These values are given by the formula $P = \frac{N}{2} \pm \frac{N}{2}\sqrt{1 - \left(\frac{4}{rN}\right)H}$ for the roots of the quadratic. If the size of the herd is initially 66 or greater, it will decrease, but not go below the

## 2.7 Autonomous Equations, the Phase Line

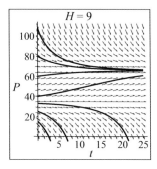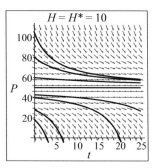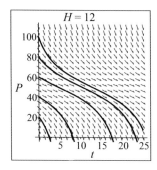

Figure 2.25. Deer population with $H = 9$, $H = H^* = 10$, and $H = 12$

value 65.8. If it is initially between 34 and 65 deer, it will increase to 65. If the herd initially has fewer than 34 deer, it will gradually go extinct.

In the middle graph, ten deer are being removed each year, and the equilibrium point is a node. If there are more than 50 deer initially, the herd will gradually decrease and stabilize at a size of 50, but if there are initially fewer than 50 deer, it will go extinct over time. In the right-hand graph we can see that if more than 10 deer are removed each year, the herd will go extinct no matter how large it is initially. The bifurcation value of the harvesting parameter is called the **critical harvesting rate**. The harvested logistic equation first appeared in a 1975 paper by Fred Brauer and David Sánchez[3], and has been written about by several authors since then.

**Exercises 2.7.** *In problems 1–6, draw a phase line for the autonomous differential equation and label each equilibrium point as a sink, source, or node.*

1. $x' = x(1 - x/4)$

2. $x' = x^2 - 1$

3. $x' = x^2$

4. $x' = (x^2 - 1)(x + 2)^2$

5. *(spruce-budworm growth equation)* $dP/dt = P(1 - P/5) - \dfrac{0.7P^2}{(0.05)^2 + P^2}$

6. $x' = \sin(x)$  *Hint: plot $f(x) = \sin(x)$ to determine the direction of the arrows.*

7. *The logistic growth equation $x' = rx\left(1 - \dfrac{x}{N}\right)$, with $N$ and $r$ strictly positive constants, is an autonomous equation. Find all equilibrium solutions, draw a phase line, and label each equilibrium with its type. If $x(0) > 0$, what must happen to the population as $t \to \infty$?*

8. *Some populations are assumed to grow according to an equation of the form*
$$y' = ry(y - A)(B - y),$$

---

[3]F. Brauer and D. A. Sánchez, Constant rate population harvesting: equilibrium and stability, *Theor. Population Biology* **8** (1975), 12–30

where $r, A,$ and $B$ are all positive constants, and $B > A$. Find all equilibrium solutions, draw a phase line, and label each equilibrium with its type. Determine what happens to the population as $t \to \infty$ if

(a) $0 < y(0) < A$,
(b) $A < y(0) < B$,
(c) $y(0) > B$.

9. In Section 1.3 the equation for the velocity of a free falling skydiver was assumed to be the autonomous equation
$$v'(t) = g - \frac{k}{m}(v(t))^p,$$
where $p, g, k,$ and $m$ are all strictly positive constants.

(a) Find all equilibrium solutions, draw a phase line, and label each equilibrium with its type.
(b) What value must $v(t)$ approach as $t \to \infty$? (Remember that this is what we called the terminal velocity.)

10. The RL-circuit equation $Li'(t) + Ri(t) = E(t)$ was described in Chapter 1, Section 3. The inductance $L$ and resistance $R$ are positive constants. If the voltage $E(t)$ is also a constant, this is an autonomous equation. Draw a phase line for this equation and determine the limiting value of the current $i(t)$ as $t \to \infty$.

11. A population of fish in a lake satisfies the growth equation:
$$dx/dt = f(x, h) = 0.5x(4 - x) - h,$$
where $x(t)$ is thousands of fish in the lake at time $t$ (in years), and $h$ is thousands of fish harvested per year.

(a) If the harvesting term $h$ is zero, how many fish will the lake support (i.e., what is its carrying capacity)?
(b) If the harvesting term is constant at $h = 1$, find all equilibrium solutions and draw a phase line. Label each equilibrium as a sink, source, or node.
(c) If $h = 1$ and the initial condition is $x(0) = 0.5$, what happens to the solution as $t \to \infty$? Explain this in terms that a biologist might use.
(d) Sketch phase lines for $h = 0, 0.5, 1.0, 1.5, 2.0, 2.5$ and place them side by side as in Figure 2.23 to form a bifurcation diagram.
(e) What is the bifurcation point $h = h^*$ for this problem?

**STUDENT PROJECT: Single Neuron Equation.** Biologists are currently applying mathematical modeling to a wide range of problems. One such problem involves the function of nerve cells (neurons) in the brain. These cells generate electrical impulses that move along their axons, affecting the activity of neighboring cells. Even for a single neuron, a complete model using what is currently known about its firing mechanism may take several equations. Since the human brain contains on the order of $10^{10}$ neurons, an exact model is not feasible. In this application you will be looking at a very simplified model (see Figure 2.26) for an isolated population of neurons all having similar properties.

## 2.7 Autonomous Equations, the Phase Line

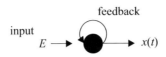

Figure 2.26. Input and output of the neuron population

Let $x(t)$ denote the percent of neurons firing at time $t$, normalized to be between 0 (low activity) and 1 (high activity). A simple model representing the change of activity level in the population is given by the differential equation

$$\frac{dx}{dt} = -x + S_a(x + E - \theta) \qquad (2.49)$$

where $E$ is the level of input activity coming from cells external to the population, $\theta$ is a common threshold level for cells in the set, and $S_a$ is a response function that models the change in activity level due to a given input. We will use a standard "sigmoidal" response function

$$S_a(z) = \frac{1}{1 + e^{-az}}.$$

The nonlinear function $S_a$ can be seen to increase monotonically from 0 to 1 as $z$ increases from $-\infty$ to $\infty$. It is called a sigmoidal function because it has a sort of stylized S-shape. You may remember that solutions of the logistic growth equation had this same shape.

(1) Find a formula for the derivative of $S_a(z)$, and show that it satisfies the identity

$$S'_a(z) \equiv aS_a(z)(1 - S_a(z)).$$

(2) Draw a graph of $S_a(z)$ for $a = 3, 10$, and $20$. Where is the slope a maximum? Is it the same in each case? Explain how the graph of $S_a(z - \theta)$ differs from the graph of $S_a(z)$.

The differential equation (2.49) for $x$ is autonomous and therefore it can be analyzed by drawing a phase line. Assume $a = 10$, and the incoming activity $E$ is constant at $E = 0.2$. Equation (2.49) now becomes

$$\frac{dx}{dt} = -x + \frac{1}{1 + e^{-10(x+0.2-\theta)}}. \qquad (2.50)$$

Because the value of $S_a$ is always between 0 and 1, if $x > 1$ the slope $x' = -x + S_a(x + 0.2 - \theta)$ is negative and if $x < 0$ it is positive. This means that any equilibrium solutions, that is, values of $x$ where $\frac{dx}{dt} \equiv 0$, must lie between 0 and 1. It also implies that the arrows on the phase line are always pointing down above $x = 1$ and up below $x = 0$.

Figure 2.27 shows graphs of $y = x$ (the dashed line) and the response function $y = 1/(1 + e^{-10(x+0.2-\theta)})$ for $\theta = 0.4, 0.7$, and $1.0$. It can be seen that for small $\theta$ there will be one equilibrium solution near $x = 1$ and for large $\theta$ there will be one equilibrium solution near $x = 0$. This seems reasonable since a high threshold means it takes a lot of input to produce a great amount of activity. For $\theta$ in a middle range, however, there can be three equilibrium solutions.

(3) Draw phase lines for equation (2.50) with $\theta = 0.4, 0.5, \ldots, 0.9, 1.0$. Label each equilibrium point as a sink, source, or node. You will need a numerical equation

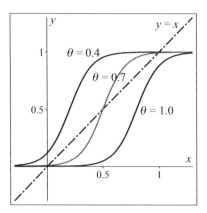

**Figure 2.27.** $y = 1/(1 + e^{-10(x+0.2-\theta)})$ for $\theta = 0.4, 0.7, 1.0$

solver to find the equilibrium values; that is, the values of $x$ where $x = 1/(1 + e^{-10(x+0.2-\theta)})$. As a check, the three equilibria for $\theta = 0.7$ are $x_1 \approx 0.007188$, $x_2 = 0.5$, and $x_3 \approx 0.992812$.

(4) Make a bifurcation diagram, as in Figure 2.23, by putting the seven phase lines from (3) in a row and joining the equilibrium points.

(5) From the bifurcation diagram, estimate the two bifurcation values of $\theta$ where the number of equilibrium points changes from one to three, and then from three back to one.

(6) Find the two bifurcation values of $\theta$ analytically. You will need to solve two simultaneous equations obtained by using the fact that at a bifurcation value of $\theta$ the curves $y = 1/(1 + e^{-10(x+0.2-\theta)})$ and $y = x$ have a point of tangency. At this point the $y$-values of the curves are equal and their slopes are also equal.

(7) Use your computer algebra system to draw a slope field for (2.49) with $a = 10$, $E = 0.2$, and $\theta = 0.7$. Let $t$ vary from 0 to 20. Use initial values $x(0) = 0.1, 0.3, 0.5, 0.7$, and 0.9, and describe what happens to the activity as $t \to \infty$.

(8) Redo problem (7) with periodic input $E = E(t) = 0.2(1 + \sin(t))$. With this time-varying input, the equation is no longer autonomous. Explain carefully how the activity differs from that described in (7). Do you think there is a periodic solution separating the two types of solutions in the periodic case? This is an interesting problem, and it might be a good time to look at the paper "Qualitative tools for studying periodic solutions and bifurcations as applied to the periodically harvested logistic equation", by Diego Benardete, V. W. Noonburg, and B. Pollina, *Amer. Math. Monthly*, vol. 115, 202–219 (2008). It discusses, in a very readable way, how one goes about determining the answer to such a question.

# 3

# Second-order Differential Equations

Recalling our definitions from Chapter 1, a **second-order differential equation** is any equation that can be written in the form $G(t, x, x', x'') = 0$ for some function $G$, and just as we did in Chapter 2 for the first-order equation, we will assume it can be solved for the highest derivative. This means that our most general second-order equation will be assumed to be of the form

$$x'' = F(t, x, x'), \tag{3.1}$$

for some function $F$.

The differential equation (3.1) is called **linear** if it can be put into the **standard form**

$$x'' + p(t)x' + q(t)x = f(t), \tag{3.2}$$

where $p, q$, and $f$ can be arbitrary functions of $t$. Equation (3.2) is called a **homogeneous linear equation** if $f(t) \equiv 0$; that is, if the equation can be written as

$$x'' + p(t)x' + q(t)x = 0. \tag{3.3}$$

The majority of second-order differential equations are not solvable by analytic methods. Even (3.3), with nonconstant coefficients, can be very difficult to solve; although many famous mathematicians (as well as physicists, engineers, astronomers, etc.) have spent a good part of their lives working on these equations. The following are examples of some important second-order equations named after the scientists who studied them:

- Bessel's equation: $t^2 x'' + tx' + (t^2 - \nu^2)x = 0$
- Legendre's equation: $(1 - t^2)x'' - 2tx' + \lambda(\lambda + 1)x = 0$
- van der Pol's equation: $x'' + \varepsilon(x^2 - 1)x' + x = 0$

The first two equations are linear and homogeneous, with variable coefficients; however, the third equation is nonlinear because of the $x^2 x'$ term. Some sample solutions of these equations are shown in Figure 3.1.

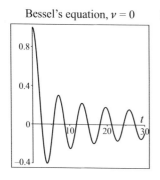
Bessel's equation, $\nu = 0$

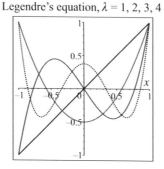
Legendre's equation, $\lambda = 1, 2, 3, 4$

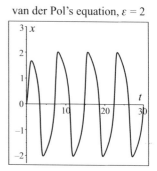
van der Pol's equation, $\varepsilon = 2$

Figure 3.1. Solutions of three special second-order equations

If the equation (3.2) has constant coefficients, that is, if it is of the form

$$ax'' + bx' + cx = f(t) \tag{3.4}$$

for some constants $a \neq 0$, $b$, and $c$, and if $f(t)$ is any continuous function, the methods in this chapter will provide an analytic solution.

One reason for the importance of second-order equations is that if $x(t)$ represents the position of an object at time $t$, then $x'(t)$ is its instantaneous velocity and $x''(t)$ is its acceleration; thus, real-world problems from the physical sciences often lead to equations of this type. A very well-known and important example of (3.4) is the spring-mass equation. Analytic formulas will be given for solutions of the spring-mass equation, and the behavior of the solutions in terms of real-world problems will be studied in this chapter. Methods for solving (3.4) with discontinuous right-hand sides $f(t)$, such as step functions and impulse functions, will be given in Chapter 6, using the method of Laplace transforms.

For the majority of *nonlinear* second-order differential equations it will be necessary to again resort to numerical and graphical methods. In Section 3.7, the phase line, defined for autonomous first-order equations, will be generalized to a "phase plane" for autonomous second-order equations. This will provide a graphical method for studying the behavior of solutions of autonomous equations of the form $x'' = F(x, x')$.

## 3.1 General Theory of Homogeneous Linear Equations

All the sections in this chapter, except the last, apply only to *linear* second-order equations; that is, equations of the form

$$x'' + p(t)x' + q(t)x = f(t). \tag{3.5}$$

It is also going to be possible, using Theorem 3.1 below, to split the solution of (3.5) into two parts. If we can find a general solution $x_h$ of the **associated homogeneous equation**

$$x'' + p(t)x' + q(t)x = 0, \tag{3.6}$$

## 3.1 General Theory of Homogeneous Linear Equations

the general solution of (3.5) can be written as the sum of $x_h$ plus a single solution $x_p$ of (3.5). The solution $x_p$ is usually referred to as a **particular solution**.

**Theorem 3.1.** *If $x_h$ is a general solution of (3.6) and $x_p$ is any solution of (3.5), the sum $x = x_h + x_p$ is a general solution of (3.5).*

The proof of this theorem is straightforward and will be left to the exercises.

In this section we will therefore concentrate on the tools needed to find a general solution of the homogeneous equation (3.6). We have seen that nonlinear differential equations can rarely be solved analytically, but homogeneous linear differential equations with constant coefficients are always solvable, and even some with nonconstant coefficients are tractable. This is basically due to the fact that sums of solutions and constant multiples of solutions of (3.6) are also solutions. If you have had a course in linear algebra this just amounts to recognizing that you are working with *linear* operators. Since we are not assuming you have had such a course, we will provide a simple proof of this statement in the following lemma. The proof of the lemma does, however, assume that you were shown in calculus that the ordinary derivative satisfies the two properties

$$\frac{d}{dt}(f(t) + g(t)) = \frac{d}{dt}(f(t)) + \frac{d}{dt}(g(t)) \quad \text{and} \quad \frac{d}{dt}(Cf(t)) = C\frac{d}{dt}(f(t)).$$

**Lemma 3.1.** *If $x_1(t)$ and $x_2(t)$ are solutions of (3.6), then $Ax_1 + Bx_2$ is also a solution for any values of the constants A and B.*

*Proof.* We are given that $x_1$ and $x_2$ both satisfy (3.6), and this means that

$$x_1'' + px_1' + qx_1 \equiv 0 \quad \text{and} \quad x_2'' + px_2' + qx_2 \equiv 0.$$

To show that $x = Ax_1 + Bx_2$ is a solution, we use the given formulas for derivatives of sums and constant multiples of functions to write

$$x'' + px' + qx = (Ax_1 + Bx_2)'' + p(Ax_1 + Bx_2)' + q(Ax_1 + Bx_2)$$
$$= (Ax_1'' + Bx_2'') + p(Ax_1' + Bx_2') + q(Ax_1 + Bx_2).$$

Factoring gives

$$x'' + px' + qx = A(x_1'' + px_1' + qx_1) + B(x_2'' + px_2' + qx_2) \equiv 0 + 0 \equiv 0,$$

and this tells us that $x$ is also a solution of the differential equation (3.6). □

Note, for future reference, that the proof can be easily extended to show that a linear combination of any finite number of solutions is again a solution. Also note that the proof does not work for a nonhomogeneous equation or a nonlinear equation. Do you see why?

For second-order linear homogeneous equations this means that if we can find two distinct solutions $x_1(t)$ and $x_2(t)$, not constant multiples of each other, we will already have a two-parameter family of solutions $x(t) = Ax_1(t) + Bx_2(t)$. The next example, which you may have already seen in a calculus course, exhibits one case where you can easily find a pair of solutions $x_1$ and $x_2$.

**Example 3.1.1.** *Find all constants r such that the exponential function $x(t) = e^{rt}$ satisfies the second-order differential equation*

$$x'' + 3x' + 2x = 0. \tag{3.7}$$

*Solution.* Substituting the function $x$ and its two derivatives, $x' = re^{rt}$ and $x'' = r^2 e^{rt}$, into the equation results in the following condition on $r$:

$$(r^2 e^{rt}) + 3(re^{rt}) + 2(e^{rt}) = e^{rt}(r^2 + 3r + 2) \equiv 0.$$

An exponential function is never equal to zero, so this condition is satisfied if, and only if, $r$ is one of the roots, $-1$ or $-2$, of the quadratic polynomial $r^2 + 3r + 2$.

This produces two solutions for (3.7), namely $x_1(t) = e^{-t}$ and $x_2(t) = e^{-2t}$, and you should convince yourself that they are not constant multiples of each other. ∎

Using Lemma 3.1, a linear combination of the two solutions in Example 3.1.1 gives us a two-parameter family of solutions, which is what we expect for a second-order equation; for practice you can check, by substituting $x$ and its derivatives into the equation, that $x(t) = Ae^{-t} + Be^{-2t}$ does in fact satisfy (3.7) for any constants $A$ and $B$.

**Question.** Can we now conclude that we have the **general solution** of (3.7); that is, can every solution of (3.7) be written as $x(t) = Ae^{-t} + Be^{-2t}$ for some values of the constants $A$ and $B$?

To answer this question we first need an existence and uniqueness theorem. The theorem stated below is proved in more advanced texts on differential equations. It applies only to linear equations, and, as we previously pointed out, this means that it can specify an exact interval on the $t$-axis for which the solution is guaranteed to exist. Note that the differential equation it applies to must be linear but not necessarily homogeneous.

**Theorem 3.2. (Existence and Uniqueness Theorem)** *For the linear equation*

$$x'' + p(t)x' + q(t)x = f(t), \qquad (3.8)$$

*given two initial conditions of the form $x(t_0) = x_0$ and $x'(t_0) = v_0$, if the functions $p$, $q$, and $f$ are all continuous in some interval $t_1 < t < t_2$ containing $t_0$, then there exists a unique solution $x(t)$ that is continuous in the entire interval $(t_1, t_2)$.*

**Example 3.1.2.** *For each linear equation below, assume initial conditions are given at $t = 0$. Find the largest interval in which Theorem 3.2 guarantees the existence and continuity of the solution.*

(a) $x'' + 2x' + 6x = e^t \sin(t)$

(b) $x'' + \tan(t)x' + \frac{1}{1-t}x = 0$

(c) $(1 - t^2)x'' - 2tx' + 6x = 0$ *(Legendre equation of order 2)*

*Solution.* In (a) the coefficient functions $p(t) = 2$, $q(t) = 6$, and $f(t) = e^t \sin(t)$ are continuous for all $t$, so a unique solution will exist and be continuous on the whole real axis.

In (b) the coefficient $\tan(t)$ is continuous on $(-\frac{\pi}{2}, \frac{\pi}{2})$ but has vertical asymptotes at each end of the interval, and $\frac{1}{1-t}$ is continuous on $(-\infty, 1)$ but not at $t = 1$; therefore, a solution with initial condition given at $t = 0$ is guaranteed to exist and be continuous at least on the interval $(-\frac{\pi}{2}, 1)$.

## 3.1 General Theory of Homogeneous Linear Equations

In (c) the equation must first be put into standard form by dividing by $(1 - t^2)$:

$$x'' - \frac{2t}{1-t^2}x' + \frac{6}{1-t^2}x = 0.$$

Now we see that $p(t)$ and $q(t)$ both have vertical asymptotes at $t = \pm 1$. Therefore the theorem guarantees continuity of the solution on the open interval $(-1, 1)$. ∎

From Theorem 3.2 it is clear that to check a linear combination of two solutions, $x(t) = C_1 x_1(t) + C_2 x_2(t)$, to see if it is a general solution of the homogeneous equation (3.6), we must show that $x(t)$ can be made to satisfy arbitrary initial conditions of the form $x(t_0) = x_0$ and $x'(t_0) = v_0$ at an arbitrary time $t = t_0$. That means that we must be able to solve the following system of linear equations for $C_1$ and $C_2$:

$$x(t_0) = C_1 x_1(t_0) + C_2 x_2(t_0) = x_0 \qquad (3.9)$$
$$x'(t_0) = C_1 x_1'(t_0) + C_2 x_2'(t_0) = v_0.$$

If (3.6) has nonconstant coefficients $p$ and $q$, we will agree that this only needs to be done on the interval around $t_0$ in which $p$ and $q$ are both continuous.

From linear algebra, it is known that this system of linear equations has a unique solution $C_1$, $C_2$, for arbitrary values of $x_0$ and $v_0$, if and only if the determinant of the matrix of coefficients of $C_1$ and $C_2$ is unequal to zero. This leads to the following definition.

**Definition 3.1.** *If $x_1(t)$ and $x_2(t)$ are solutions of the second-order linear homogeneous equation $x'' + p(t)x' + q(t)x = 0$, the determinant*

$$W(x_1, x_2)(t) \equiv \det \begin{pmatrix} x_1(t) & x_2(t) \\ x_1'(t) & x_2'(t) \end{pmatrix} \equiv x_1(t) x_2'(t) - x_1'(t) x_2(t)$$

*is called the **Wronskian** of the functions $x_1$ and $x_2$.*

The condition that needs to be satisfied is given by the following theorem.

**Theorem 3.3.** *If $x_1$ and $x_2$ are any two solutions of $x'' + px' + qx = 0$, and their Wronskian $x_1 x_2' - x_1' x_2$ is unequal to zero for all values of $t$, then $x = C_1 x_1 + C_2 x_2$ is a **general solution** of $x'' + px' + qx = 0$.*

The next lemma shows that it is only necessary to check the value of the Wronskian at a single value of $t$.

**Lemma 3.2.** *If $x_1$ and $x_2$ are any two solutions of the equation $x'' + px' + qx = 0$, then $W(x_1, x_2)$ is an exponential function and is either zero for all $t$ or unequal to zero for all $t$.*

The proofs of Theorem 3.3 and Lemma 3.2 are contained in Appendix E.

Since the Wronskian tells us whether or not a pair of solutions can be used to form a general solution, we will use the following definition:

**Definition 3.2.** *A pair of solutions $\{x_1, x_2\}$ of (3.6) satisfying $W(x_1, x_2) \neq 0$, for all values of $t$ in an interval in which the coefficient functions are continuous, will be called a **fundamental solution set** in that interval.*

Using the Wronskian we are now able to show that the solution we found in Example 3.1.1 is a general solution for the equation.

**Example 3.1.3.** *Show that the function $x(t) = Ae^{-t} + Be^{-2t}$ is a general solution of the equation $x'' + 3x' + 2x = 0$; that is, show that the two solutions $x_1(t) = e^{-t}$ and $x_2(t) = e^{-2t}$ form a fundamental solution set for the equation.*

*Solution.* We already showed in Example 3.1.1 that $x_1(t) = e^{-t}$ and $x_2(t) = e^{-2t}$ are solutions of the equation, and using Lemma 3.1 we also know that $x(t) = Ae^{-t} + Be^{-2t}$ is a solution for any constants $A$ and $B$. All that remains is to compute the Wronskian of $x_1$ and $x_2$:

$$W(x_1, x_2) = x_1 x_2' - x_2 x_1' = e^{-t}(-2e^{-2t}) - e^{-2t}(-e^{-t}) = -e^{-3t}.$$

Since $W(x_1, x_2)$ is an exponential function, it is unequal to zero for all $t$, and that is all we need to conclude that $x(t)$ is a general solution of the differential equation. ∎

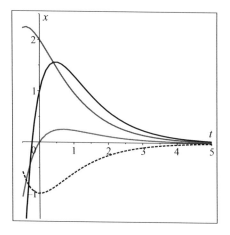

Figure 3.2. Solutions of $x'' + 3x' + 2x = 0$

Figure 3.2, obtained using DEplot in Maple, shows four solutions of $x'' + 3x' + 2x = 0$. The four sets of initial conditions, all given at $t = 0$, are

$$(x(0), x'(0)) = (-1, 0), (0, 1), (1, 3), \text{ and } (2, -1).$$

You should take careful note of two very important differences between the graph in Figure 3.2 and the graphs we saw in Chapter 2.

(1) Even though the same computer program was used, no slope field has been drawn in the $(t, x)$-plane.

(2) Solutions can, and obviously do, intersect in the $(t, x)$-plane.

Remember that in Chapter 2 a slope field was drawn by using the first-order equation $x' = f(t, x)$ to compute the slopes at each point of a grid in the $(t, x)$-plane, but a second-order equation $x'' = f(t, x, x')$ only gives us a formula for the second derivative, and without having an analytic solution we have no way of knowing what the value of $x'$ is at a particular point $(t, x)$. In Section 3.7 we will look very carefully at this problem, and we will show that for *autonomous* second-order equations, that is, equations of the form $x'' = F(x, x')$, there is a way to portray the entire family of solutions in a single 2-dimensional plot in which solution curves do not intersect. This is one of

## 3.1 General Theory of Homogeneous Linear Equations

the things that motivated mathematicians to look into geometric methods for studying higher-order differential equations.

**Exercises 3.1.** *Determine whether each of the following equations is linear or nonlinear in the dependent variable x. If it is linear, state whether or not it is homogeneous.*

1. $x'' + t^2 x' + \sin(t)x = 0$
2. $x'' + t^2 x' + t \sin(x) = 0$
3. $t^2 x'' + tx' + x + t = 0$
4. $t^2 x'' + tx' + 3x = 0$
5. $x'' + x^2 = 0$
6. $x'' + t^2 = 0$

*For each linear equation below, assume initial conditions are given at $t = 0$. Find the largest interval in which Theorem 3.2 guarantees continuity of the solution.*

7. $3x'' + 2x' + x = 0$
8. $x'' + 10x' + 4x = e^t$
9. $x'' + \sec(t)x' + \dfrac{1}{2+t}x = 0$
10. $x'' + \sqrt{2-t^2}\,x' + \dfrac{1}{t-3}x = \sec(t)$
11. $(t^2 - 1)x'' + tx = 4$
12. $(t^2 + t - 6)x'' + e^{-t}x' + \sin(t)x = (t-2)t$

*For equations 13–16 two functions $x_1$ and $x_2$ are given. Show that $x_1$ and $x_2$ are both solutions of the equation, and compute their Wronskian. Does $\{x_1, x_2\}$ form a fundamental solution set, and if so, on what interval of t?*

13. $x'' + x = 0$, $\quad x_1 = \cos(t)$, $\quad x_2 = \sin(t)$
14. $x'' + 3x' + 2x = 0$, $\quad x_1 = e^{-t}$, $\quad x_2 = e^{-2t}$
15. $t^2 x'' - 2tx' + 2x = 0$, $t > 0$, $\quad x_1 = t$, $\quad x_2 = t^2$
16. $x'' + 4x = 0$, $\quad x_1 = \cos(2t)$, $\quad x_2 = (\cos(t))^2 - \dfrac{1}{2}$

17. In Appendix E it is shown that the Wronskian of two solutions $x_1$ and $x_2$ of the equation $x'' + p(t)x' + q(t)x = 0$ satisfies the first order differential equation $\dfrac{dW}{dt} = -p(t)W$, where $p(t)$ is the coefficient of $x'$ in the equation. This means that $W(t) = Ce^{-\int p(t)\,dt}$.

   (a) Show that this holds for the solutions $x_1$ and $x_2$ in Problem 13.
   (b) Show that this holds for the solutions $x_1$ and $x_2$ in Problem 14.
   (c) Show that this holds for the solutions $x_1$ and $x_2$ in Problem 15. Hint: be sure to put the equation into standard form first.

18. *Use the following steps to write a proof of Theorem 3.1. We want to show that if $x_p$ is any solution of the equation*

$$x'' + px' + qx = f(t) \tag{3.10}$$

*and $x_h$ is a general solution of the associated homogeneous equation*

$$x'' + px' + qx = 0, \tag{3.11}$$

*then $x = x_h + x_p$ is a general solution of (3.10).*

(a) *First show that if $x_h$ satisfies (3.11) and $x_p$ satisfies (3.10), then $x = x_h + x_p$ is a solution of (3.10). Use the proof of Lemma 3.1 as a model.*

(b) *In a similar manner, show that if $x_{p_1}$ and $x_{p_2}$ are two solutions of (3.10), then their difference $x = x_{p_1} - x_{p_2}$ is a solution of (3.11).*

*You should now be able to conclude the proof of the theorem.*

## 3.2 Homogeneous Linear Equations with Constant Coefficients

We are now going to use the tools from the previous section to show that a general solution $x_h$ of the homogeneous linear constant coefficient equation can always be found. This will be carried out in detail for the second-order equation $ax'' + bx' + cx = 0$, and at the end of this section it will be shown that the method extends to homogeneous linear constant coefficient equations of arbitrary order. In the next section the formulas will be used to solve the spring-mass equation, a very important equation in physics and engineering.

### 3.2.1 Second-order Equation with Constant Coefficients.
Using exponential functions, a general solution of the **second-order homogeneous linear constant coefficient equation**

$$ax''(t) + bx'(t) + cx(t) = 0 \tag{3.12}$$

will be found for any values of the constants $a \neq 0, b,$ and $c$.

As we saw in Example 3.1.1 in the previous section, an exponential function $x(t) = e^{rt}$ is a reasonable guess for a solution of (3.12). Substituting $x(t) = e^{rt}, x'(t) = re^{rt}$, and $x''(t) = r^2 e^{rt}$ into (3.12) shows that $e^{rt}$ will be a solution if, and only if,

$$a(r^2 e^{rt}) + b(re^{rt}) + ce^{rt} = e^{rt}(ar^2 + br + c) \equiv 0;$$

that is, (3.12) has a solution $x(t) = e^{rt}$ if, and only if, $r$ is a root of the quadratic polynomial

$$ar^2 + br + c.$$

The polynomial $P(r) = ar^2 + br + c$ is called the **characteristic polynomial** of the differential equation (3.12).

Recalling the quadratic formula, the roots of $P(r)$ are

$$r = \frac{-b \pm \sqrt{b^2 - 4ac}}{2a}, \tag{3.13}$$

## 3.2 Homogeneous Linear Equations with Constant Coefficients

and the value of the **discriminant** $K = b^2 - 4ac$ determines whether the two roots are real and distinct, real and equal, or complex. For the second-order equation these are the only three possibilities.

In each of the three cases we need to obtain two solutions $x_1$ and $x_2$ of (3.12) and show, by computing their Wronskian, that they form a fundamental solution set for (3.12); that is, we need to show that $W(x_1, x_2) \neq 0$.

**Case 1 ($K > 0$):** The two real roots $r_1 \neq r_2$ given by (3.13) both satisfy $P(r) = 0$, so we know that the two exponential functions $x_1(t) = e^{r_1 t}$ and $x_2(t) = e^{r_2 t}$ are solutions of (3.12). To show that they form a fundamental solution set, compute the Wronskian:

$$W(x_1, x_2) = x_1 x_2' - x_1' x_2 = e^{r_1 t} r_2 e^{r_2 t} - r_1 e^{r_1 t} e^{r_2 t} = (r_2 - r_1) e^{(r_1 + r_2)t}.$$

Since $r_1 \neq r_2$, and an exponential function is never zero, $W(x_1, x_2)$ is unequal to zero for all values of $t$, and the general solution of the equation can be written in the form $x = C_1 x_1 + C_2 x_2$.

**Case 2 ($K = 0$):** Let the single real root given by (3.13) be denoted by $\bar{r}$. We know that $P(\bar{r}) = a\bar{r}^2 + b\bar{r} + c = 0$; therefore, $x_1(t) = e^{\bar{r}t}$ is a solution of (3.12). It turns out that in this case we can show that $x_2(t) = te^{\bar{r}t}$ is also a solution. Using the fact that $K = 0$, it can be seen from the quadratic formula that $\bar{r} = -b/2a$. This means that $2a\bar{r} + b = 0$. Substituting $x_2(t) = te^{\bar{r}t}$ into the differential equation (check the derivatives),

$$a(te^{\bar{r}t})'' + b(te^{\bar{r}t})' + c(te^{\bar{r}t}) = a(t\bar{r}^2 e^{\bar{r}t} + 2\bar{r}e^{\bar{r}t}) + b(t\bar{r}e^{\bar{r}t} + e^{\bar{r}t}) + cte^{\bar{r}t}$$

$$= te^{\bar{r}t} \underbrace{(a\bar{r}^2 + b\bar{r} + c)}_{0} + e^{\bar{r}t} \underbrace{(2a\bar{r} + b)}_{0} \equiv 0;$$

therefore, $x_2 = te^{\bar{r}t}$ is also a solution. To show $x_1$ and $x_2$ form a fundamental solution set, compute the Wronskian:

$$W(e^{\bar{r}t}, te^{\bar{r}t}) = e^{\bar{r}t}\left(\frac{d}{dt}(te^{\bar{r}t})\right) - \left(\frac{d}{dt}(e^{\bar{r}t})\right)(te^{\bar{r}t}) = e^{\bar{r}t}(t\bar{r}e^{\bar{r}t} + e^{\bar{r}t}) - \bar{r}e^{\bar{r}t}te^{\bar{r}t} = e^{2\bar{r}t}.$$

Therefore, $W$ is unequal to 0 for all values of $t$.

**Case 3 ($K < 0$):** In this case, the two roots of the characteristic polynomial will be complex conjugates $\alpha \pm \beta\iota$. This means that the two complex functions $z_1 = e^{(\alpha+\beta\iota)t}$ and $z_2 = e^{(\alpha-\beta\iota)t}$ are solutions of (3.12). To find real solutions, we use Euler's Formula $e^{\theta\iota} = \cos(\theta) + \iota\sin(\theta)$ to write

$$z_1 = e^{\alpha t}(\cos(\beta t) + \iota\sin(\beta t)), \quad z_2 = e^{\alpha t}(\cos(\beta t) - \iota\sin(\beta t)).$$

Since linear combinations of solutions are solutions, we have that

$$x_1 = \frac{z_1 + z_2}{2} = e^{\alpha t}(\cos(\beta t)), \quad x_2 = \frac{z_1 - z_2}{2\iota} = e^{\alpha t}(\sin(\beta t))$$

are both solutions. To show that they form a fundamental solution set, the reader is asked to compute the Wronskian $W(x_1, x_2)$ and show that $W(x_1, x_2) = \beta e^{2\alpha t}$. Since we would not be in the complex root case if $\beta = 0$, $W(x_1, x_2)$ is not equal to zero, and the solutions $x_1$ and $x_2$ form a fundamental solution set. We have now shown that the formulas in **Table 3.1** give a general solution for each case.

The following three examples will demonstrate how easy it is to solve homogeneous linear equations with constant coefficients, as well as initial-value problems for

Table 3.1. General Solution of $ax'' + bx' + cx = 0$

| Discriminant $K = b^2 - 4ac$ | Roots of P(r) $P(r) = ar^2 + br + c$ | General Solution |
|---|---|---|
| $K > 0$ | distinct real roots $r_1 \neq r_2$ | $x(t) = C_1 e^{r_1 t} + C_2 e^{r_2 t}$ |
| $K = 0$ | single real root $\bar{r}$ | $x(t) = C_1 e^{\bar{r} t} + C_2 t e^{\bar{r} t}$ |
| $K < 0$ | complex roots $\alpha \pm \beta \iota$ | $x(t) = C_1 e^{\alpha t} \cos(\beta t) + C_2 e^{\alpha t} \sin(\beta t)$ |

these equations, using the formulas in the table. In every case, both terms in the solution are multiplied by exponential functions. If the exponents are both negative, it can be seen that the solutions will all tend to zero as $t \to \infty$. This is usually true in the case of problems involving real-world physical systems, and we will see many more examples in the sections on the spring-mass equation.

**Example 3.2.1.** *Find the general solution of the equation $x'' + 4x' + 4x = 0$.*

*Solution.* The characteristic polynomial is $P(r) = r^2 + 4r + 4 = (r + 2)^2$, and it has a double root $\bar{r} = -2$. From Case 2 in Table 3.1 we see that the general solution is
$$x(t) = C_1 e^{-2t} + C_2 t e^{-2t}.$$
Even without determining the constants $C_1$ and $C_2$, we can see that $x(t)$ will tend to zero as $t \to \infty$. It is clear that $\lim_{t \to \infty} e^{-2t} = 0$, and L'Hôpital's Rule can be used to show that $\lim_{t \to \infty} t e^{-2t}$ is also zero. ■

**Example 3.2.2.** *Solve the IVP*
$$2x'' + 7x' + 3x = 0, \quad x(0) = 1, \quad x'(0) = 2.$$

*Solution.* The characteristic polynomial is $P(r) = 2r^2 + 7r + 3$, and its roots are
$$r = \frac{-7 \pm \sqrt{49 - 24}}{4} = \frac{-7}{4} \pm \frac{5}{4} = -3, -\frac{1}{2}.$$

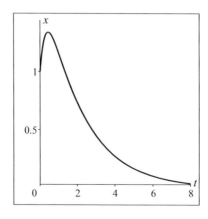

Figure 3.3. Solution of $2x'' + 7x' + 3x = 0, x(0) = 1, x'(0) = 2$

## 3.2 Homogeneous Linear Equations with Constant Coefficients

With $r_1 = -3$ and $r_2 = -\frac{1}{2}$, the general solution given in line 1 of Table 3.1 is

$$x(t) = C_1 e^{-3t} + C_2 e^{-\frac{1}{2}t}.$$

To satisfy the initial conditions we need to compute $x'(t) = -3C_1 e^{-3t} - \frac{1}{2}C_2 e^{-\frac{1}{2}t}$. Solving the two equations $x(0) = C_1 + C_2 = 1$, $x'(0) = -3C_1 - \frac{1}{2}C_2 = 2$ for $C_1$ and $C_2$ gives $C_1 = -1$, $C_2 = 2$; therefore, the solution of the IVP is

$$x(t) = -e^{-3t} + 2e^{-\frac{1}{2}t}.$$

A graph of $x(t)$ is shown in Figure 3.3, and it is clear that as $t \to \infty$ the solution tends exponentially to zero. ∎

At this point, you should think carefully about what the solution would look like if either one or both real roots of the quadratic equation are positive.

In Case 3, where the roots of the characteristic polynomial are complex, the graph of the general solution $x(t) = e^{\alpha t}(C_1 \cos(\beta t) + C_2 \sin(\beta t))$ can be hard to visualize, even after the constants $C_1$ and $C_2$ are determined. To make it simpler to understand the behavior of this solution, we can use a well-known formula from trigonometry to write the linear combination $C_1 \cos(\beta t) + C_2 \sin(\beta t)$ as a single sine function.

**Writing $C_1 \cos(\beta t) + C_2 \sin(\beta t)$ in the form $R \sin(\beta t + \phi)$.** We want to find values for the amplitude $R$ and phase angle $\phi$ such that

$$R \sin(\beta t + \phi) \equiv C_1 \cos(\beta t) + C_2 \sin(\beta t). \tag{3.14}$$

Using the formula for the sine of a sum of two angles, $\sin(A + B) = \sin(A)\cos(B) + \cos(A)\sin(B)$, we can write

$$R \sin(\beta t + \phi) = R(\sin(\beta t)\cos(\phi) + \cos(\beta t)\sin(\phi)). \tag{3.15}$$

If the right-hand side of (3.15) is rearranged in the form

$$R \sin(\phi)\cos(\beta t) + R \cos(\phi)\sin(\beta t),$$

it can be seen that (3.14) will be an identity if

$$R \sin(\phi) = C_1, \quad R \cos(\phi) = C_2.$$

Now $R$ can be found in terms of $C_1$ and $C_2$ by writing

$$C_1^2 + C_2^2 = R^2(\sin(\phi))^2 + R^2(\cos(\phi))^2 = R^2[(\sin(\phi))^2 + (\cos(\phi))^2] = R^2.$$

Similarly, the value of $\phi$ can be found by writing $\frac{C_1}{C_2} = \frac{R\sin(\phi)}{R\cos(\phi)} = \tan(\phi)$. This leads to the two identities

$$R = \sqrt{C_1^2 + C_2^2} \quad \text{and} \quad \phi = \tan^{-1}(C_1/C_2).$$

If $C_2 < 0$, then you need to use the formula $\phi = \tan^{-1}(C_1/C_2) + \pi$ to put $\phi$ into the quadrant where $\cos(\phi) < 0$.

In the next example this technique will be used to make it easier to understand the long-term behavior of the solution when the roots of the characteristic polynomial are complex.

**Example 3.2.3.** *Solve the IVP $x'' + 2x' + 5x = 0$, $x(0) = 1$, $x'(0) = 3$, and sketch a graph of the solution on the interval $0 \leq t \leq 7$.*

*Solution.* The characteristic polynomial is $P(r) = r^2 + 2r + 5$ and the roots are

$$r = \frac{-2 \pm \sqrt{4 - 4(5)}}{2} = \frac{-2 \pm \sqrt{-16}}{2} = -1 \pm 2\iota.$$

Using the formula for Case 3 in Table 3.1,

$$x(t) = e^{-t}(C_1 \cos(2t) + C_2 \sin(2t)).$$

The product rule for differentiation gives

$$x'(t) = -e^{-t}(C_1 \cos(2t) + C_2 \sin(2t)) + e^{-t}(-2C_1 \sin(2t) + 2C_2 \cos(2t)).$$

Applying the initial conditions at $t = 0$, $x(0) = C_1 = 1$ and $x'(0) = -C_1 + 2C_2 = 3$. The solution of these simultaneous equations is $C_1 = 1$, $C_2 = 2$ and therefore, the solution of the IVP is

$$x(t) = e^{-t}(\cos(2t) + 2\sin(2t)).$$

Using the formula derived above, with $R = \sqrt{C_1^2 + C_2^2} = \sqrt{5}$ and $\phi = \tan^{-1}(C_1/C_2) = \tan^{-1}(1/2)$, $x$ can be written in the form

$$x(t) = \sqrt{5} e^{-t} \sin(2t + \tan^{-1}(1/2)).$$

The function $\sin(2t + \phi)$ oscillates between $\pm 1$ with period $2\pi/2$, so it is clear that the solution is a damped sine wave; that is, a sine wave bounded by the two decaying exponential curves $\pm \sqrt{5} e^{-t}$, and therefore oscillates infinitely often, with period $\pi$, and approaches zero as $t \to \infty$.

A graph of this function is shown in Figure 3.4. It is the sign of the real part $\alpha$ of the complex root that determines whether the solution goes to zero or becomes infinitely large as $t \to \infty$, and the imaginary part $\beta$ that determines the period of oscillation. ∎

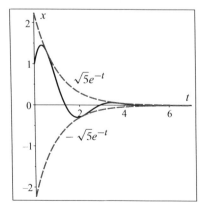

Figure 3.4. Solution of $x'' + 2x' + 5x = 0$, $x(0) = 1$, $x'(0) = 3$

## 3.2 Homogeneous Linear Equations with Constant Coefficients

**3.2.2 Equations of Order Greater Than Two.** The method used for $n = 2$ is easily extended to a linear homogeneous constant coefficient equation of order $n > 2$. To solve the equation

$$a_n x^{(n)} + a_{n-1} x^{(n-1)} + \cdots + a_1 x' + a_0 x = 0, \tag{3.16}$$

where the $a_i$ are all real constants, we again assume an exponential solution $x(t) = e^{rt}$. If this function and its derivatives are substituted into (3.16), it can be seen that $r$ must be a root of the $n$th degree polynomial

$$P(r) = a_n r^n + a_{n-1} r^{n-1} + \cdots + a_1 r + a_0, \tag{3.17}$$

again called the **characteristic polynomial** of the differential equation (3.16). If the coefficients $a_i$ are all real, it is shown in an advanced algebra course that the roots of $P(r)$ are either real, or complex conjugate pairs of the form $\alpha \pm \beta\iota$.

For each real root $\bar{r}$, find the largest positive integer $k$ such that $(r - \bar{r})^k$ divides $P(r)$. The integer $k$ is called the **multiplicity of the root** $\bar{r}$, and the solution $x_h$ of (3.16) must contain the $k$ terms

$$c_1 e^{\bar{r}t} + c_2 t e^{\bar{r}t} + \cdots + c_k t^{k-1} e^{\bar{r}t}.$$

Similarly, for each pair of complex roots $\alpha \pm \beta\iota$, find the largest positive integer $j$ such that the irreducible quadratic term

$$(r - (\alpha + \beta\iota))(r - (\alpha - \beta\iota)) = r^2 - 2\alpha r + (\alpha^2 + \beta^2),$$

raised to the power $j$, divides $P(r)$. Then $j$ is called the multiplicity of the roots $\alpha \pm \beta\iota$, and the solution $x_h$ must contain the $j$ terms

$$e^{\alpha t}(A_1 \cos(\beta t) + B_1 \sin(\beta t)) + t e^{\alpha t}(A_2 \cos(\beta t) + B_2 \sin(\beta t)) + \cdots$$
$$+ t^{j-1} e^{\alpha t}(A_j \cos(\beta t) + B_j \sin(\beta t)).$$

The resulting function $x_h$ can be shown to contain exactly $n$ constants, which can be used to satisfy $n$ arbitrary initial conditions. The proof of this involves an $n$-dimensional version of the Wronskian.

The following example will give you an idea of what is involved. It should be pointed out that finding the roots of an $n$th degree polynomial is a hard problem, even for a computer, and is definitely the hardest part of solving (3.16). We will be using the well-known result from algebra that $\alpha$ is a root of a polynomial $P(r)$ if, and only if, the factor $(r - \alpha)$ divides $P(r)$.

**Example 3.2.4.** *Find the general solution of the differential equation*

$$x'''' + 6x''' + 17x'' + 20x' + 8x = 0. \tag{3.18}$$

*Solution.* The characteristic polynomial is

$$P(r) = r^4 + 6r^3 + 17r^2 + 20r + 8$$

and it is easy to show that $r = -1$ is a root of $P(r)$ by showing that $P(-1) = 1 - 6 + 17 - 20 + 8 = 0$. Dividing $P(r)$ by the factor $(r - (-1)) = (r + 1)$,

$$P(r) = (r + 1)(r^3 + 5r^2 + 12r + 8) = (r + 1)Q(r).$$

Since $Q(-1) = -1 + 5 - 12 + 8 = 0$, we can divide $Q(r)$ by $(r + 1)$ and write

$$P(r) = (r + 1)^2 (r^2 + 4r + 8).$$

The quadratic term does not have $r+1$ as a factor; therefore, $r = -1$ is a real root of $P(r)$ of multiplicity 2. The roots of the quadratic $r^2 + 4r + 8$ are $\alpha \pm \beta \imath = -2 \pm 2\imath$; therefore, the general solution of the differential equation (3.18) can be written as

$$x_p(t) = c_1 e^{-t} + c_2 t e^{-t} + e^{-2t}(A_1 \cos(2t) + B_1 \sin(2t)).$$

This formula contains four constants that can be used to make the solution satisfy four arbitrary initial conditions of the form $x(t_0) = x_0$, $x'(t_0) = y_0$, $x''(t_0) = w_0$, and $x'''(t_0) = z_0$ at any value of $t = t_0$. ∎

**Exercises 3.2.** *In problems 1–8, find the general solution.*

1. $x'' + 5x' + 6x = 0$

2. $2x'' + 7x' + 3x = 0$

3. $x'' + 6x' + 9x = 0$

4. $x'' - 9x = 0$

5. $x'' + 4x' + 5x = 0$

6. $x'' + x' + x = 0$

7. $x''' + x'' + 4x' + 4x = 0$

8. $x''' + 2x'' + x' + 2x = 0$

*For the initial value problems 9 and 10, find the solution and graph it on the interval $0 \leq t \leq 10$.*

9. $x'' + 4x' + 4x = 0$, $x(0) = 2$, $x'(0) = -1$

10. $x'' + 6x' + 7x = 0$, $x(0) = 0$, $x'(0) = 1$

*For problems 11 and 12, write the solution in the form $x(t) = Re^{\alpha t} \sin(\omega t + \phi)$ and then graph it on the interval $0 \leq t \leq 10$.*

11. $x'' + 2x' + 10x = 0$, $x(0) = 1$, $x'(0) = 0$

12. $x'' + 6x' + 13x = 0$, $x(0) = 1$, $x'(0) = 1$

13. Find formulas for $R$ and $\theta$ (in terms of $C_1$ and $C_2$) to make
$$C_1 \sin(bt) + C_2 \cos(bt) \equiv R \cos(bt + \theta).$$
Hint: use the formula $\cos(A + B) = \cos(A)\cos(B) - \sin(A)\sin(B)$.

*Solve each of the initial value problems below, and write the solution in the form $x(t) = Re^{\alpha t} \cos(bt + \theta)$. Sketch a graph.*

14. $x'' + 6x' + 13x = 0$, $x(0) = 1$, $x'(0) = 1$

15. $x'' + 2x' + 10x = 0$, $x(0) = 1$, $x'(0) = 0$

## 3.3 The Spring-mass Equation

One of the primary applications of constant coefficient second-order linear equations is to harmonic oscillators, and the particular example we will consider here is the simple spring-mass system. In this section we will derive the equation and show that, at this point, we are able to completely analyze such a system if no external forces are acting on it.

**3.3.1 Derivation of the Spring-mass Equation.** Assume that a massive object is suspended on a spring as shown in Figure 3.5. Let $x(t)$ denote the vertical position of the object at time $t$. The **equilibrium position**, where the object comes to rest after hanging it on the spring, will be denoted by $x = 0$. The upward direction is arbitrarily chosen to be the positive $x$-direction. Assuming the motion of the object is always in the vertical direction, the instantaneous velocity of the mass at time $t$ is $x'(t)$, and $x''(t)$ is its acceleration. Newton's Second Law: "mass times acceleration equals sum of the forces" can then be used to write

$$mx''(t) = F_s + F_d + f(t), \tag{3.19}$$

where $m$ is the mass of the suspended object, $F_s$ is the restoring force due to the spring, $F_d$ is the force due to damping in the system, and $f(t)$ represents any external forces acting on the system.

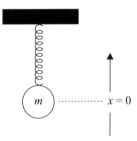

Figure 3.5. A simple spring-mass system

For a **linear spring**, **Hooke's Law** states that the restoring force of the spring is proportional to its displacement; therefore, $F_s = -kx(t)$, where the constant of proportionality $k$ is called the **spring constant**. The restoring force acts in the direction opposite to the displacement. The damping force $F_d$ is usually assumed to be proportional to the velocity, and acts in the direction opposite to the direction of motion; therefore, $F_d = -bx'(t)$, where $b$ is called the **damping constant**. Any external forces acting on the system are lumped into the term $f(t)$.

Now (3.19) can be written as

$$mx''(t) = -kx(t) - bx'(t) + f(t),$$

and adding $kx(t) + bx'(t)$ to both sides, we have

$$mx''(t) + bx'(t) + kx(t) = f(t). \tag{3.20}$$

Equation (3.20) is called the **spring-mass equation**, and we can see that it is a linear second-order equation with constant coefficients.

Each term in (3.20) is in units of force, and if we use the SI system, with distance in meters (m), mass in kilograms (kg) and time in seconds (s), the force is in Newtons (N), where

$$1 \text{ Newton} = \text{force required to accelerate 1 kg of mass by 1 m/s}^2 \equiv \frac{1 \text{ kg} \cdot \text{m}}{\text{s}^2}.$$

The damping constant $b$ is then measured in kg/s $= \frac{\text{N}}{\text{m/s}} = \frac{\text{N} \cdot \text{s}}{\text{m}}$ and the spring constant $k$ is in kg/s$^2$ = N/m.

If $f(t) \equiv 0$, (3.20) is called an **unforced** spring-mass system, otherwise it is called a **forced** system.

**3.3.2 The Unforced Spring-mass System.** The equation for an **unforced spring-mass system** is just the homogeneous linear equation with constant coefficients

$$mx'' + bx' + kx = 0,$$

and we already have the necessary formulas for its solution in Table 3.1.

**System with no damping.** The simplest case to consider is an unforced system where no damping is present. This is not a realistic case since there is always some damping in a physical system, but in the limit as $b \to 0$, the behavior predicted by the model approaches the behavior of solutions of the equation

$$mx'' + kx = 0. \tag{3.21}$$

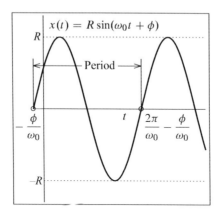

**Figure 3.6.** The function $R \sin(\omega_0 t + \phi)$

For (3.21) the roots of the characteristic polynomial $P(r) = mr^2 + k$ are $r = 0 \pm \iota\sqrt{\frac{k}{m}}$ and we know from Table 3.1 that the solution has the form of a simple sine function

$$x(t) = e^{0t}(C_1 \cos(\omega_0 t) + C_2 \sin(\omega_0 t)) = R \sin(\omega_0 t + \phi) \tag{3.22}$$

shown in Figure 3.6, where we have defined $\omega_0 = \sqrt{\frac{k}{m}}$. The values of the amplitude $R$ and phase angle $\phi$ were shown previously to be $R = \sqrt{C_1^2 + C_2^2}$ and $\phi = \tan^{-1}(C_1/C_2)$, with $\pi$ added to $\phi$ if $C_2$ is negative.

## 3.3 The Spring-mass Equation

This is called **simple harmonic motion** with period $\frac{2\pi}{\omega_0}$ seconds, and frequency equal to $\frac{\omega_0}{2\pi}$ cycles per second (hertz). The value $\omega_0 = \sqrt{\frac{k}{m}}$ is called the **natural frequency** of the system.

**System with damping.** If there is damping in the unforced system, we are looking at the homogeneous constant coefficient equation

$$mx'' + bx' + kx = 0 \qquad (3.23)$$

with $b \neq 0$, and Table 3.1 gives us an exact formula for the solution. Since the characteristic polynomial is

$$P(r) = mr^2 + br + k,$$

there are three possible types of solutions, depending on the discriminant $K = b^2 - 4mk$. In the case of a spring-mass equation, the three cases determined by the value of $K$ are given special names:

- $b^2 - 4mk > 0$, the system is called **overdamped**
- $b^2 - 4mk = 0$, the system is called **critically damped**
- $b^2 - 4mk < 0$, the system is called **underdamped**

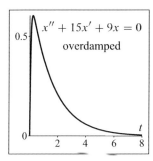
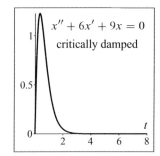
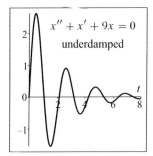

Figure 3.7. Solutions of $x'' + bx' + 9x = 0$ with three different values of $b$

It is very important, at this point, to observe that if the constants $m$, $b$, and $k$ are all positive (which is the usual case for a physical system), the roots of the characteristic polynomial will all have real part less than zero. To see this, write the roots as

$$r = \frac{-b \pm \sqrt{b^2 - 4mk}}{2m}.$$

If the system is overdamped, $b^2 - 4mk$ is positive, but less than $b^2$, so the square root has absolute value less than $b$. Adding $-b$ produces a negative value for both real roots. In the critically damped case, the root is just $-\frac{b}{2m}$ which is negative, and in the underdamped case the real part of the two complex roots is also $-\frac{b}{2m}$. What this means is that in all three cases, if the damping is strictly greater than zero, the solution given by Table 3.1 will tend to zero as $t \to \infty$. This is what one would expect for an unforced physical system with damping.

We have already seen what solutions look like in the underdamped case back in Figure 3.4, and the right-hand graph in Figure 3.7 shows another example. Underdamped solutions oscillate infinitely often, but tend to zero exponentially as $t \to \infty$. In the overdamped case, seen in the left-hand graph in Figure 3.7, $x(t)$ is a linear combination of exponential functions with negative exponents and it tends to zero as $t \to \infty$, without oscillating. The graph of the critically damped equation is very similar to that of the overdamped equation. It is simply the dividing point between solutions that oscillate and those that do not. Depending on the values of the constants $C_1$ and $C_2$, it can be shown in the overdamped and critically damped cases that the mass can pass through its equilibrium point at most once and then tend monotonically to zero. The following example illustrates this behavior in an overdamped system.

**Example 3.3.1.** *A 1 kg object is suspended on a spring with spring constant 4 N/m. The system is submerged in water, causing it to have a large damping constant $b = 5$ N s/m. Find the equation for $x(t)$, and solve the IVP if*

(a) *The object is lifted up 1 m and let go.*

(b) *The object is lifted up 1 m and given a downward velocity of 5 m/s.*

*Before solving for $x(t)$, using your common sense about the physical problem, in which case, (a) or (b), do you think the object is more likely to go below its equilibrium position $x = 0$? Why?*

*Solution.* The equation for $x$,
$$x'' + 5x' + 4x = 0,$$
is overdamped ($b^2 - 4mk = 25 - 16 > 0$) and the characteristic polynomial $P(r) = r^2 + 5r + 4 = (r+1)(r+4)$ has unequal real roots $r = -1, -4$. The general solution of the homogeneous equation is
$$x(t) = C_1 e^{-t} + C_2 e^{-4t}.$$

- In case (a) the initial conditions are $x(0) = 1$ and $x'(0) = 0$ (since the object is let go with an initial velocity of zero). The derivative of $x(t)$ is $x'(t) = -C_1 e^{-t} - 4C_2 e^{-4t}$; therefore, $C_1$ and $C_2$ satisfy the system of equations
$$x(0) = C_1 + C_2 = 1, \quad x'(0) = -C_1 - 4C_2 = 0.$$
Check that these have solution $C_1 = \frac{4}{3}, C_2 = -\frac{1}{3}$; therefore, the solution in (a) is
$$x_a(t) = \frac{4}{3} e^{-t} - \frac{1}{3} e^{-4t}.$$

- In case (b) the initial conditions are $x(0) = 1$ and $x'(0) = -5$, and solving for the constants,
$$x(0) = C_1 + C_2 = 1, \quad x'(0) = -C_1 - 4C_2 = -5,$$
gives $C_1 = -\frac{1}{3}, C_2 = \frac{4}{3}$; therefore
$$x_b(t) = -\frac{1}{3} e^{-t} + \frac{4}{3} e^{-4t}.$$

## 3.3 The Spring-mass Equation

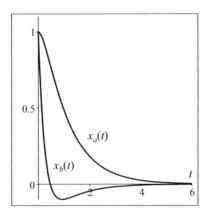

**Figure 3.8.** Solutions of $x'' + 5x' + 4x = 0$, $x(0) = 1$; $x'_a(0) = 0$ and $x'_b(0) = -5$

The graphs of $x_a$ and $x_b$ are shown in Figure 3.8. In which case did the object go below its equilibrium position? Is this the case you chose at the start of the example? ∎

The next example involves both setting up and analyzing a model of a simple physical system.

**Example 3.3.2.** *A mechanic raises a car on a lift as shown in Figure 3.9. The mechanism for lowering the lift acts like a spring-mass system where the mass m is the mass of the lift (200 kg) plus the mass of the car. The spring constant of the system is $k = 500$ N/m.*

*The equilibrium position of the mass is at $x = 0$, and when a car is being worked on, the center of gravity of the mass is at $x = 2.5$ m. There is a well under the lift that lets the car go a small distance below its equilibrium position $x = 0$, but for safety reasons it is necessary that it not go below $x = -0.3$ m.*

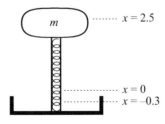

**Figure 3.9.** Schematic of the car lift

*If a car of mass 800 kg is raised on the lift, find the value of the damping constant b that will allow the car to come down most quickly without letting it go below $x = -0.3$ m. The smaller the damping, the more quickly the lift will come down. Do you see why?*

*Solution.* The spring-mass equation that needs to be solved is

$$(200 + 800)x'' + bx' + 500x = 0, \tag{3.24}$$

and the conditions at the initial time (assumed to be $t = 0$) when the lift starts to drop are
$$x(0) = 2.5\text{m}, \quad x'(0) = 0.$$
If (3.24) is divided by 1000 and a new constant $B = b/1000$ is defined, the equation becomes
$$x'' + Bx' + \frac{1}{2}x = 0. \tag{3.25}$$
The characteristic polynomial $P(r) = r^2 + Br + \frac{1}{2}$ has roots
$$r = \frac{-B \pm \sqrt{B^2 - 4(1)(\frac{1}{2})}}{2} = -\frac{B}{2} \pm \frac{\sqrt{B^2 - 2}}{2}.$$
To find the smallest allowable value for $B$, we need to find a value for which $x(t)$ does not go below $x = -0.3$, but we can see that the system can be slightly underdamped, so we need a system with complex roots; that is, with $B^2 < 2$. This means we can write
$$r = -\frac{B}{2} \pm \frac{\sqrt{2 - B^2}}{2}\, \iota.$$
Make sure you see why this is true. Remember that the aim is to make $b$, and therefore $B$, as small as possible.

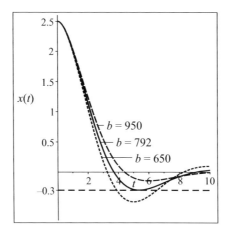

**Figure 3.10.** Solutions of $1000x'' + bx' + 500x = 0$ with $b = 650$, 792, and 950

Using line 3 in Table 3.1, the solution of (3.25) is
$$x(t) = C_1 e^{-\frac{B}{2}t} \cos\left(\frac{\sqrt{2 - B^2}}{2}t\right) + C_2 e^{-\frac{B}{2}t} \sin\left(\frac{\sqrt{2 - B^2}}{2}t\right),$$
and the derivative is
$$x'(t) = C_1 e^{-\frac{B}{2}t}\left(-\frac{B}{2}\cos\left(\frac{\sqrt{2-B^2}}{2}t\right) - \frac{\sqrt{2-B^2}}{2}\sin\left(\frac{\sqrt{2-B^2}}{2}t\right)\right)$$
$$+ C_2 e^{-\frac{B}{2}t}\left(-\frac{B}{2}\sin\left(\frac{\sqrt{2-B^2}}{2}t\right) + \frac{\sqrt{2-B^2}}{2}\cos\left(\frac{\sqrt{2-B^2}}{2}t\right)\right).$$

## 3.3 The Spring-mass Equation

Using these equations and the fact that $\sin(0) = 0$ and $\cos(0) = 1$, it can be seen that $x(0) = C_1$ and $x'(0) = -\frac{B}{2}C_1 + \frac{\sqrt{2-B^2}}{2}C_2$. With the initial conditions $x(0) = 2.5$, $x'(0) = 0$, the constants must satisfy

$$C_1 = 2.5, \quad -\frac{B}{2}C_1 + \frac{\sqrt{2-B^2}}{2}C_2 = 0,$$

and, therefore,

$$C_1 = 2.5, \quad C_2 = \frac{2}{\sqrt{2-B^2}}\left(\frac{B}{2} \cdot C_1\right) = \frac{2.5B}{\sqrt{2-B^2}}.$$

Both $x$ and $x'$ can be written as products of an exponential function and a sine function with period $P = \frac{2\pi}{\frac{\sqrt{2-B^2}}{2}} = \frac{4\pi}{\sqrt{2-B^2}}$. Since $x' = 0$ at $t = 0$, its next zero, where $x$ takes its minimum value, will be at $t = P/2 = \frac{2\pi}{\sqrt{2-B^2}}$. This means that if

$$x\left(\frac{2\pi}{\sqrt{2-B^2}}\right) \geq -0.3,$$

the value of $x$ will never go below $-0.3$. All that needs to be done is to solve

$$x\left(\frac{2\pi}{\sqrt{2-B^2}}\right) = \exp\left(-\frac{B}{2}\frac{2\pi}{\sqrt{2-B^2}}\right)\left(2.5\cos(\pi) + \frac{2.5B}{\sqrt{2-B^2}}\sin(\pi)\right)$$

$$= -2.5e^{-\frac{B\pi}{\sqrt{2-B^2}}} = -0.3.$$

This gives the approximate value $B \approx 0.7911$ and therefore the damping constant should be $b = 792\frac{Ns}{m}$. The graph in Figure 3.10 shows the position function $x(t)$ for equation (3.24) for three values of the damping constant $b$. Note carefully that as $b$ decreases the lift descends more quickly and goes further below the equilibrium position. ∎

**Exercises 3.3.** *For each of the spring-mass equations 1–6, determine whether the system is undamped, underdamped, critically damped, or overdamped and whether it is forced or unforced.*

1. $2x'' + 4x' + 6x = 0$
2. $x'' + 5x = 0$
3. $3x'' + x' + \frac{1}{2}x = \sin(t)$
4. $x'' + 4x' + 4x = 1$
5. $2x'' + 10x' + x = e^t$
6. $100x'' + 10x' + 0.02x = 0$

*For problems 7–10, set up the spring-mass equation. Determine whether it is undamped, under or overdamped. Solve the IVP and draw a graph of the solution on the interval $0 < t < 12$. If the system is underdamped, convert the solution to the form $Re^{\alpha t}\sin(\beta t + \phi)$ and check that you get exactly the same graph.*

7. *A 1 kg mass is suspended on a spring with spring constant $k = 10$ N/m. There is no damping in the system. The mass is lifted up two meters and let go.*

8. *A 10 kg mass is suspended on a spring with spring constant $k = 20$ N/m. The damping coefficient $b = 30$ kg/s. The mass is initially at equilibrium and is given an initial velocity of 2 m/s in the downward direction.*

9. *A 1 kg mass suspended from a spring with spring constant 5 N/m is pulled down 1 meter and given an initial velocity of 1 m/s in the downward direction. The damping coefficient of the system is 2 kg/s.*

10. *A 2 kg mass is suspended on a spring with spring constant $k = 2$ N/m. The damping coefficient is $b = 4$ kg/s. The mass is lifted up 1 meter and let go.*

11. *Redo the problem in Example 3.3.2. This time, assume that the damping constant is fixed at $b = 500$ kg/s and cannot be changed. Find the spring constant $k$ that will let the car come down most quickly without going below $-0.3$ meters. Are you maximizing or minimizing the value of $k$?*

12. *Redo the problem in Example 3.3.2, assuming the mass of the car is changed to 1300kg and the maximum well depth is limited to 0.5m (instead of 0.3m). Again find the damping constant $b$ that brings the car down most quickly. Draw a graph of $x(t)$, and compare the time of descent with the result in Example 3.3.2.*

## 3.4 Nonhomogeneous Linear Equations

In this section we will describe two methods for finding a particular solution $x_p$ of a nonhomogeneous equation. The simplest method is called the method of undetermined coefficients and is essentially a guess and test method. It will work for an equation of the form $ax'' + bx' + cx = f(t)$, where the coefficients $a$, $b$, and $c$ are constants and the function $f$ contains only exponential functions, sines and cosines, or polynomials. A second method for finding particular solutions, called variation of parameters, is also given. It is more general in that it applies whenever it is possible to find a fundamental solution set for the associated homogeneous equation, and the right-hand side of the equation, $f(t)$, is an integrable function. However, the method of variation of parameters often requires some very complicated integration.

### 3.4.1 Method of Undetermined Coefficients. 
Given a linear constant coefficient differential equation of any order $n$,

$$a_n x^{(n)} + a_{n-1} x^{(n-1)} + \cdots + a_1 x' + a_0 x = f(t). \tag{3.26}$$

We showed in Section 3.1 that if a general solution $x_h$ of the associated homogeneous equation

$$a_n x^{(n)} + a_{n-1} x^{(n-1)} + \cdots + a_1 x' + a_0 x = 0 \tag{3.27}$$

can be found, then we only need to find a single solution $x_p$ of (3.26), called a **particular solution**, and the general solution of (3.26) will be $x = x_h + x_p$. We will now describe one of the simplest methods for finding $x_p$, called the **method of undetermined coefficients**. We will concentrate on equations of second order, but keep in

## 3.4 Nonhomogeneous Linear Equations

mind that this method can be applied to equations of the form (3.26) of arbitrary order $n$.

The method of undetermined coefficients only works if the coefficients $a_i$ in equation (3.26) are all constants, and the function $f(t)$ is an exponential function, a linear combination of sines and cosines, a polynomial, or possibly an algebraic combination of such functions.

The method is very easy to use, and consists of guessing an appropriate function $x_p$ containing undetermined constants. The function $x_p$ and its derivatives are substituted for $x$ and its derivatives in (3.26), and the values of the constants can then be determined by solving a system of ordinary linear equations.

Before formally defining the method, consider the following very simple example.

**Example 3.4.1.** *Find a particular solution $x_p$ of the equation*

$$x'' + 4x' + 4x = 2e^{-3t}. \tag{3.28}$$

*Solution.* Since the function $f(t) = 2e^{-3t}$ is an exponential function, it seems reasonable to try the particular solution

$$x_p = Ae^{-3t}.$$

If $x_p$, $x_p'$, and $x_p''$ are substituted into (3.28), the left side is

$$x_p'' + 4x_p' + 4x_p = (9Ae^{-3t}) + 4(-3Ae^{-3t}) + 4Ae^{-3t} = (9A - 12A + 4A)e^{-3t} = Ae^{-3t}.$$

This will be identical to the right-hand side $f(t) = 2e^{-3t}$ if, and only if, $A = 2$. In fact, you should check that the function $x_p(t) = 2e^{-3t}$ does satisfy (3.28). ∎

There are two important things to notice here:

(1) For this method to work, all of the derivatives of the assumed solution $x_p$ must have the same form as $x_p$. We will see that this is true for exponential functions, linear combinations of sines and cosines, and polynomials. It is not true for functions like $\tan(t)$, $\ln(t)$, or $t^{\frac{1}{2}}$, for example; for these we will have to use a different method.

(2) If the function you choose for $x_p$ happens to be a solution of the associated homogeneous equation, then you have a problem, since the left-hand side of (3.26) will be identically zero, and cannot be made to equal $f(t)$. There is a way to solve this problem, and it will be explained at the end of this subsection.

In Table 3.2 we summarize the forms that $x_p$ should take, in the three simplest cases that can be solved by the method of undetermined coefficients. We have already seen in Example 3.4.1 how the method works for $f(t) = ae^{bt}$, and the next two examples will illustrate the method for polynomial and trigonometric right-hand sides.

In each of the three cases in Table 3.2, the function $x_p$ can be seen to be general enough so that its derivatives of any order can be expressed in the same form as $x_p$. Also notice that in lines 1 and 3 the value of $b$ in $x_p(t)$ must be the same as the value of $b$ in $f(t)$; that is, $b$ is not an undetermined coefficient.

**Example 3.4.2.** *Find the solution of the IVP*

$$x'' + 3x' + 2x = 1 - 2t^2, \quad x(0) = 0, \quad x'(0) = -4. \tag{3.29}$$

*First find the homogeneous solution* $x_h$. The characteristic polynomial is $P(r) = r^2 + 3r + 2 = (r+1)(r+2)$, with real, unequal roots $r_1 = -1$ and $r_2 = -2$. Using the first line in Table 3.1,

$$x_h(t) = C_1 e^{-t} + C_2 e^{-2t}.$$

**Warning.** *It is very tempting at this point to try to determine $C_1$ and $C_2$ from the initial conditions; but $x_h(t)$ is not the solution of the equation, and you need to wait until you have $x(t) = x_h(t) + x_p(t)$.*

*Solution.* Using Table 3.2, with $f(t) = 1 - 2t^2$, we see that with $f$ a polynomial of degree 2, the form to be used for $x_p$ is

$$x_p(t) = A + Bt + Ct^2.$$

Note that even though $f(t)$ has no term in $t^1$, we have to include this term in $x_p$.

This is the time to check carefully that the function we are assuming for $x_p$ is not a solution of the homogeneous equation, and in this case it clearly is not, since $x_h$ is a sum of exponential functions.

The derivatives of $x_p$ are $x_p' = B + 2Ct$ and $x_p'' = 2C$. Substituting into the differential equation it can be seen that the constants $A, B$, and $C$ must be chosen to make the left side,

$$x_p'' + 3x_p' + 2x_p = (2C) + 3(B + 2Ct) + 2(A + Bt + Ct^2),$$

identical to $f(t) = 1 - 2t^2$. It is best to write both sides of (3.29) as linear combinations of the three functions $t^0, t^1$, and $t^2$, and it is then clear that the constants $A, B$, and $C$ must make

$$(2C + 3B + 2A)t^0 + (6C + 2B)t^1 + (2C)t^2 \equiv 1t^0 + 0t^1 + (-2)t^2.$$

This identity will hold for all values of $t$ if, and only if, the coefficient of each power of $t$ is the same on both sides. This leads to a system of three equations in three unknowns:

$$2C + 3B + 2A = 1, \quad 6C + 2B = 0, \quad 2C = -2$$

that are easily solved by starting with the right-most equation. The solution is $C = -1$, $B = 3$, and $A = -3$; therefore,

$$x_p(t) = A + Bt + Ct^2 = -3 + 3t - t^2.$$

We can now write the general solution of (3.29) as

$$x(t) = x_h(t) + x_p(t) = C_1 e^{-t} + C_2 e^{-2t} - 3 + 3t - t^2.$$

Table 3.2. Method of Undetermined Coefficients

| Function $f(t)$ | Form of $x_p^*$ | Sample $f(t)$ | $x_p$ used |
|---|---|---|---|
| Exponential $ae^{bt}$ | $Ae^{bt}$ | $5e^{-2t}$ | $Ae^{-2t}$ |
| Polynomial (degree $n$) | $A_0 + A_1 t + \cdots + A_n t^n$ | $1 + 2t - 7t^3$ | $A + Bt + Ct^2 + Dt^3$ |
| $\alpha \sin(bt) + \beta \cos(bt)$ | $A \sin(bt) + B \cos(bt)$ | $-2 \cos(5t)$ | $A \cos(5t) + B \sin(5t)$ |

---

* If $x_p$ is a solution of the associated homogeneous equation, then use $t^k x_p$ where $k$ is the smallest positive integer such that $t^k x_p$ is not a homogeneous solution.

## 3.4 Nonhomogeneous Linear Equations

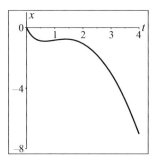

Figure 3.11. Solution of the IVP in (3.29)

This is the function that needs to satisfy the initial conditions $x(0) = 0, x'(0) = -4$. Differentiating $x(t)$ gives

$$x'(t) = -C_1 e^{-t} - 2C_2 e^{-2t} + 3 - 2t.$$

Substituting $t = 0$ into $x$ and $x'$, we see that $C_1$ and $C_2$ must be chosen to make $x(0) = C_1 + C_2 - 3 = 0$ and $x'(0) = -C_1 - 2C_2 + 3 = -4$. Solving for $C_1$ and $C_2$ gives $C_1 = -1$, $C_2 = 4$; therefore the solution of the IVP is

$$x(t) = -e^{-t} + 4e^{-2t} - 3 + 3t - t^2,$$

and you can check that it satisfies both the differential equation and the two initial conditions. ∎

It is clear from Figure 3.11 that we can no longer assume that solutions of nonhomogeneous equations are always going to tend to zero even if the homogeneous solution is a sum of negative exponential functions.

In the next example we consider the case where $f(t)$ is a sine function.

**Example 3.4.3.** *Find the solution of the initial value problem*

$$x'' + x' + x = 3\sin(2t), \quad x(0) = 1, \quad x'(0) = 0. \tag{3.30}$$

*Solution.* The characteristic polynomial of the homogeneous equation is $P(r) = r^2 + r + 1$, and it has complex roots $r = -\frac{1}{2} \pm \frac{\sqrt{3}}{2}\iota$; therefore, the homogeneous solution is

$$x_h(t) = C_1 e^{-t/2} \sin\left(\frac{\sqrt{3}}{2}t\right) + C_2 e^{-t/2} \cos\left(\frac{\sqrt{3}}{2}t\right).$$

From line 3 in Table 3.2 we see that with $f(t) = 3\sin(2t)$, the suggested form for $x_p$ is $x_p = A\sin(2t) + B\cos(2t)$. Notice that even though $f(t)$ does not contain a cosine term, the derivative of $x_p$ will, so we must allow for it by including the term $B\cos(2t)$. The derivatives of $x_p$ are $x'_p = 2A\cos(2t) - 2B\sin(2t)$ and $x''_p = -4A\sin(2t) - 4B\cos(2t)$. It is clear that $x_p$ is not a solution of the homogeneous equation, so we need to find constants $A$ and $B$ that make $x''_p + x'_p + x_p \equiv 3\sin(2t) + 0\cos(2t)$. Substituting $x_p$ and its derivatives into the differential equation, the left side of (3.30) can be written as

$$(-4A\sin(2t) - 4B\cos(2t)) + (2A\cos(2t) - 2B\sin(2t)) + (A\sin(2t) + B\cos(2t)),$$

and collecting the sine terms and cosine terms on both sides, $x_p$ is a solution if, and only if,

$$(-4A - 2B + A)\sin(2t) + (-4B + 2A + B)\cos(2t) \equiv f(t) = 3\sin(2t) + 0\cos(2t).$$

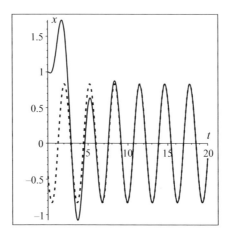

Figure 3.12. The solid curve is the solution $x = x_h + x_p$ of (3.30) and the dotted curve is the particular solution $x_p$

To make this an identity in $t$, the coefficients of $\sin(2t)$ and $\cos(2t)$ must be the same on both sides; therefore, $A$ and $B$ must satisfy the simultaneous equations

$$-3A - 2B = 3, \quad 2A - 3B = 0,$$

and these have solution $A = -\frac{9}{13}$, $B = -\frac{6}{13}$. The general solution of the differential equation is

$$x(t) = x_h(t) + x_p(t)$$
$$= C_1 e^{-t/2} \sin\left(\frac{\sqrt{3}}{2}t\right) + C_2 e^{-t/2} \cos\left(\frac{\sqrt{3}}{2}t\right) - \frac{9}{13}\sin(2t) - \frac{6}{13}\cos(2t).$$

We could also write $x_p$ as a single sine function $R\sin(2t + \phi)$ where

$$R = \sqrt{A^2 + B^2} = \sqrt{\left(-\frac{9}{13}\right)^2 + \left(-\frac{6}{13}\right)^2} \approx 0.83205$$

and

$$\phi = \tan^{-1}(B/A) + \pi = \tan^{-1}(2/3) + \pi \approx 3.7296.$$

The constant $\pi$ is added to the arctangent since $A \equiv R\cos(\phi) = -\frac{9}{13}$ is negative. The solution of (3.30) can now be written as

$$x(t) = C_1 e^{-t/2} \sin\left(\frac{\sqrt{3}}{2}t\right) + C_2 e^{-t/2} \cos\left(\frac{\sqrt{3}}{2}t\right) + 0.83205 \sin(2t + 3.7296).$$

## 3.4 Nonhomogeneous Linear Equations

To satisfy the initial conditions we need the derivative

$$x'(t) = C_1 e^{-t/2}\left(-\frac{1}{2}\sin(\frac{\sqrt{3}}{2}t) + \frac{\sqrt{3}}{2}\cos(\frac{\sqrt{3}}{2}t)\right)$$

$$+ C_2 e^{-t/2}\left(-\frac{1}{2}\cos(\frac{\sqrt{3}}{2}t) - \frac{\sqrt{3}}{2}\sin(\frac{\sqrt{3}}{2}t)\right) + 2(0.83205)\cos(2t + 3.7296).$$

Then

$$x(0) = C_2 + 0.83205\sin(3.7296) = 1,$$

$$x'(0) = \frac{\sqrt{3}}{2}C_1 - \frac{1}{2}C_2 + 1.6641\cos(3.7296) = 0.$$

Solving for $C_1$ and $C_2$,

$$C_1 = 2.44263, \quad C_2 = 1.46154$$

and $R = \sqrt{C_1^2 + C_2^2} \approx 2.8465$, $\phi = \tan^{-1}\left(\frac{C_2}{C_1}\right) \approx 0.53920$. The solution to the initial value problem can now be written as

$$x(t) = 2.8465 e^{-t/2}\sin\left(\frac{\sqrt{3}}{2}t + 0.53920\right) + 0.83205\sin(2t + 3.7296).$$

Figure 3.12 shows a graph of this function. The homogeneous solution decays to zero as $t \to \infty$, so in the long run the solution becomes essentially equal to the particular solution $x_p$. We will look more carefully at equations of this type when we study the forced spring-mass equation in Section 3.5. ∎

**Additional Rule for the Method of Undetermined Coefficients.** In applying the method of undetermined coefficients to a differential equation, if the value of $x_p$ obtained from Table 3.2 satisfies the associated homogeneous equation, then $x_p$ must be multiplied by $t$. You need to keep multiplying by $t$ until you find the smallest positive integer $k$ such that $t^k x_p$ is no longer a solution to the homogeneous equation. The following example demonstrates how this works.

**Example 3.4.4.** *Suppose the IVP in Example 3.4.1 is slightly changed, and you are asked to find the solution to the IVP*

$$x'' + 4x' + 4x = 2e^{-2t}, \quad x(0) = 1, \quad x'(0) = 1, \tag{3.31}$$

where $f(t)$ has been changed from $2e^{-3t}$ to $2e^{-2t}$. The characteristic polynomial $P(r) = r^2 + 4r + 4 \equiv (r+2)^2$ has the double root $r = -2$, which means that the homogeneous solution is

$$x_h(t) = C_1 e^{-2t} + C_2 t e^{-2t}.$$

From Table 3.2 we see that with $f(t) = 2e^{-2t}$ the suggested form of the particular solution is $x_p = Ae^{-2t}$. This, however, is a solution of the associated homogeneous equation (it is equal to $x_h$ with $C_1 = A$ and $C_2 = 0$), and this implies that we should use $x_p = Ate^{-2t}$. This is again a homogeneous solution ($x_h$ with $C_1 = 0$ and $C_2 = A$); therefore, the rule implies that we need to use

$$x_p = At^2 e^{-2t}.$$

Using the product rule for differentiation, the derivatives of $x_p$ can be shown to be

$$x_p'(t) = 2Ate^{-2t} - 2At^2 e^{-2t} = At\left(2e^{-2t} - 2te^{-2t}\right)$$

and
$$x_p''(t) = At\left(-6e^{-2t} + 4te^{-2t}\right) + A\left(2e^{-2t} - 2te^{-2t}\right).$$
If we substitute these into (3.31) and collect the terms involving $e^{-2t}, te^{-2t}$, and $t^2 e^{-2t}$ we see that
$$\begin{aligned} x_p'' + 4x_p' + 4x_p &= At(-6e^{-2t} + 4te^{-2t}) + A(2e^{-2t} - 2te^{-2t}) \\ &\quad + 4At(2e^{-2t} - 2te^{-2t}) + 4(At^2 e^{-2t}) \\ &= At^2 e^{-2t}(4 - 8 + 4) + Ate^{-2t}(-6 - 2 + 8) + Ae^{-2t}(2). \end{aligned}$$
The coefficients of the terms in $Ate^{-2t}$ and $At^2 e^{-2t}$ are both zero, so the left side of (3.31) can be made to equal $f(t)$ by letting $A = 1$. The full solution of the differential equation is
$$x(t) = x_h(t) + x_p(t) = C_1 e^{-2t} + C_2 t e^{-2t} + t^2 e^{-2t}.$$

The derivative is
$$x'(t) = -2e^{-2t}(C_1 + C_2 t + t^2) + e^{-2t}(C_2 + 2t).$$
At $t = 0$, $x(0) = C_1$ and $x'(0) = -2C_1 + C_2$; therefore, $C_1 = 1$, $C_2 = 3$ and the solution of the IVP can be seen to be
$$x(t) = e^{-2t} + 3te^{-2t} + t^2 e^{-2t}. \qquad \blacksquare$$

It should be clear that if we had used $x_p = Ae^{-2t}$ in the previous example, we would have been faced with having to satisfy the equation $0(e^{-2t}) = f(t) = 2e^{-2t}$, which is impossible since $f(t)$ is not identically zero.

**Right-hand sides that are sums of terms in Table 3.2.** The method of undetermined coefficients can also be easily applied to nonhomogeneous linear equations with right-hand sides equal to sums of terms of the form given in Table 3.2. This can be done using the following lemma.

**Lemma 3.3.** *If $x_a$ is a solution of $x'' + px' + qx = f(t)$ and $x_b$ is a solution of $x'' + px' + qx = g(t)$, then the sum $x = x_a + x_b$ is a solution of $x'' + px' + qx = f(t) + g(t)$.*

*Proof.* With $x = x_a + x_b$, we can write
$$\begin{aligned} x'' + px' + qx &= (x_a + x_b)'' + p(x_a + x_b)' + q(x_a + x_b) \\ &= (x_a'' + px_a' + qx_a) + (x_b'' + px_b' + qx_b) = f(t) + g(t). \qquad \square \end{aligned}$$

As an example, consider the following problem.

**Example 3.4.5.** *Find the general solution of the differential equation*
$$x'' + 3x' - 10x = 14e^{2t} + 100t.$$

*Solution.* The roots of the characteristic polynomial $P(r) = r^2 + 3r - 10 = (r-2)(r+5)$ are $r = 2, -5$; therefore,
$$x_h(t) = C_1 e^{2t} + C_2 e^{-5t}.$$
We can write $x_p = x_a + x_b$ where $x_a$ is a solution of $x'' + 3x' - 10x = 14e^{2t}$ and $x_b$ is a solution of $x'' + 3x' - 10x = 100t$.

## 3.4 Nonhomogeneous Linear Equations

From Table 3.2, $x_a$ must be taken to be $x_a(t) = Ate^{2t}$, since $Ae^{2t}$ is a solution of the homogeneous equation. Substituting $x_a$ and its derivatives $x'_a = 2Ate^{2t} + Ae^{2t}$ and $x''_a = 4Ate^{2t} + 4Ae^{2t}$ into the differential equation, we have

$$x''_a + 3x'_a - 10x_a = (4Ate^{2t} + 4Ae^{2t}) + 3(2Ate^{2t} + Ae^{2t}) - 10(Ate^{2t})$$
$$= te^{2t}(4A + 6A - 10A) + e^{2t}(4A + 3A) \equiv 14e^{2t}.$$

The term in $te^{2t}$ is identically zero, so we can let $7A = 14$, and the solution is $x_a(t) = 2te^{2t}$.

To find $x_b$ we can assume the form $x_b = A + Bt$. Then $x'_b = B$ and $x''_b = 0$. Substituting these into the differential equation, $x_b$ must satisfy

$$x''_b + 3x'_b - 10x_b = 0 + 3(B) - 10(A + Bt) = (3B - 10A) + (-10B)t \equiv 0 + 100t,$$

and this requires that $3B - 10A = 0$ and $-10B = 100$; therefore, $B = -10$ and $A = -3$. Now we can write the solution of the problem in the form

$$x(t) = x_h + x_a + x_b = C_1 e^{2t} + C_2 e^{-5t} + 2te^{2t} - 3 - 10t.$$

Check that this function satisfies the differential equation for any values of the constants $C_1$ and $C_2$. ∎

### 3.4.2 Variation of Parameters.

We have seen that finding a particular solution $x_p$ by using undetermined coefficients only works when the forcing function $f(t)$ is a combination of polynomials, exponential functions, sines, and cosines. The reason for this is that when we assume $x_p$ is of a particular form, if differentiating it introduces new types of functions, we will not be able to make the two sides of the differential equation identical. For example, if $f(t) = \tan(t)$, and we assume $x_p = A\tan(t)$, the derivative of $x_p$ is $A(\sec(t))^2$. Suppose we then let $x_p = A\tan(t) + B(\sec(t))^2$. Now $x'_p$ will contain another new function. There is no way to express all of the derivatives of $\tan(t)$ as a linear combination of a finite set of functions.

The method of **variation of parameters** is another method for finding a particular solution $x_p$. If the forcing function $f$ is integrable, this method will be shown to work for any equation of the form

$$x'' + p(t)x' + q(t)x = f(t) \tag{3.32}$$

for which a general solution to the homogeneous equation can be found. We can always find a homogeneous solution if $p$ and $q$ are both constants, but it is only in rare cases that we will be able to do this if $p$ and $q$ are functions of $t$. However, the method is definitely more general in that $f(t)$ is not limited to a small set of functions.

To use the method of variation of parameters, assume that we are given a general solution of the associated homogeneous equation, say $x_h = C_1 x_1 + C_2 x_2$; that is, $\{x_1, x_2\}$ is a fundamental solution set for (3.32), when $f(t) \equiv 0$. The particular solution $x_p$ is then assumed to be a linear combination of $x_1$ and $x_2$ of the form

$$x_p = v_1 x_1 + v_2 x_2. \tag{3.33}$$

The functions $v_1$ and $v_2$ must be nonconstant functions of $t$; otherwise, $x_p$ would satisfy the homogeneous equation.

Since we have two unknown functions $v_1$ and $v_2$ to determine, we can specify two conditions that they must satisfy. The derivative of $x_p$ by the product rule is

$$x'_p = v_1 x'_1 + v_2 x'_2 + v'_1 x_1 + v'_2 x_2,$$

and in order to simplify $x'_p$ and $x''_p$, the first condition we impose on $v_1$ and $v_2$ is that
$$v'_1 x_1 + v'_2 x_2 \equiv 0. \tag{3.34}$$
With (3.34) satisfied,
$$x'_p = v_1 x'_1 + v_2 x'_2$$
and therefore
$$x''_p = v_1 x''_1 + v_2 x''_2 + v'_1 x'_1 + v'_2 x'_2.$$
Substituting $x_p, x'_p$ and $x''_p$ into (3.32) requires that
$$(v_1 x''_1 + v_2 x''_2 + v'_1 x'_1 + v'_2 x'_2) + p(v_1 x'_1 + v_2 x'_2) + q(v_1 x_1 + v_2 x_2) = f(t),$$
and the terms can be regrouped as
$$v_1 \underbrace{(x''_1 + p x'_1 + q x_1)}_{\equiv 0} + v_2 \underbrace{(x''_2 + p x'_2 + q x_2)}_{\equiv 0} + v'_1 x'_1 + v'_2 x'_2 = f(t).$$
Using the fact that $x_1$ and $x_2$ satisfy the homogeneous equation, we see that our second condition on the functions $v_1$ and $v_2$ must be
$$v'_1 x'_1 + v'_2 x'_2 = f(t). \tag{3.35}$$
We now have two equations for the derivatives $v'_1$ and $v'_2$, and if Cramer's Rule is used to solve the simultaneous equations (see Appendix D for Cramer's Rule)
$$v'_1 x_1 + v'_2 x_2 = 0$$
$$v'_1 x'_1 + v'_2 x'_2 = f(t)$$
we obtain the following formulas for $v'_1$ and $v'_2$:
$$v'_1 = \frac{\begin{vmatrix} 0 & x_2 \\ f & x'_2 \end{vmatrix}}{\begin{vmatrix} x_1 & x_2 \\ x'_1 & x'_2 \end{vmatrix}} = \frac{-x_2 f}{x_1 x'_2 - x'_1 x_2}, \quad v'_2 = \frac{\begin{vmatrix} x_1 & 0 \\ x'_1 & f \end{vmatrix}}{\begin{vmatrix} x_1 & x_2 \\ x'_1 & x'_2 \end{vmatrix}} = \frac{x_1 f}{x_1 x'_2 - x'_1 x_2}.$$

The determinant in the denominator of both terms is the Wronskian of $x_1$ and $x_2$, and since we are assuming that $\{x_1, x_2\}$ is a fundamental solution set for the homogeneous equation we know that $W(x_1, x_2)$ is not equal to zero. We therefore have formulas for $v'_1$ and $v'_2$, which can be integrated to give
$$v_1 = -\int \left( \frac{x_2 f}{W(x_1, x_2)} \right) dt, \quad v_2 = \int \left( \frac{x_1 f}{W(x_1, x_2)} \right) dt. \tag{3.36}$$
This method is straightforward, but the integrals in (3.36) are usually not easy to evaluate. We first give a relatively simple example:

**Example 3.4.6.** *Solve the differential equation*
$$x'' + x = \tan(t). \tag{3.37}$$

*Solution.* The associated homogeneous equation $x'' + x = 0$ has characteristic polynomial $r^2 + 1 = 0$ with roots $r = \pm \iota$, so its general solution is
$$x_h = C_1 \cos(t) + C_2 \sin(t).$$
If we let $x_1 = \cos(t)$ and $x_2 = \sin(t)$ the Wronskian is
$$W(x_1, x_2) = x_1 x'_2 - x'_1 x_2 = \cos(t) \cdot \cos(t) - (-\sin(t)) \cdot \sin(t) \equiv 1;$$

## 3.4 Nonhomogeneous Linear Equations

therefore, according to (3.36) $v_1$ and $v_2$ are given by

$$v_1 = -\int \sin(t)\tan(t)dt = -\int \frac{\sin^2(t)}{\cos(t)}dt$$

and

$$v_2 = \int \cos(t)\tan(t)dt = \int \sin(t)dt.$$

Integration gives

$$v_1 = \sin(t) - \ln|\sec(t) + \tan(t)|,$$
$$v_2 = -\cos(t);$$

therefore, a particular solution of (3.37) is

$$\begin{aligned}x_p &= x_1 v_1 + x_2 v_2 \\ &= \cos(t)[\sin(t) - \ln|\sec(t) + \tan(t)|] + \sin(t)(-\cos(t)) \\ &= -\cos(t)\ln|\sec(t) + \tan(t)|.\end{aligned}$$

It is not necessary to add constants to the integrals for $v_1$ and $v_2$, since that would just make

$$x_p = (x_1(v_1 + C_1) + x_2(v_2 + C_2)) = x_1 v_1 + x_2 v_2 + C_1 x_1 + C_2 x_2,$$

which is just $x_p$ with the homogeneous solution already added to it.

The general solution of (3.37) can now be written as

$$x(t) = x_h + x_p = C_1 \cos(t) + C_2 \sin(t) - \cos(t)\ln|\sec(t) + \tan(t)|. \quad \blacksquare$$

If initial conditions are given at $t = 0$, the solution of $x'' + x = \tan(t)$ becomes unbounded as $t \to \pm\frac{\pi}{2}$. This is not surprising because the forcing function $\tan(t)$ can be seen to have vertical asymptotes at $\pm\frac{\pi}{2}$, and the Existence and Uniqueness Theorem 3.2 only assures a continuous solution on the interval $(-\frac{\pi}{2}, \frac{\pi}{2})$.

The function $f(t) = \tan(t)$ is not a forcing function that occurs very often in real-world problems. A more reasonable forcing function for a problem arising in physics, or even in biology, might be

$$f(t) = \frac{1}{1 + e^{-t}}.$$

This function $f(t)$ could represent the output of a transistor which increases linearly for small $t$, but saturates at a value of 1 as $t \to \infty$. An example of a problem involving this function is given below.

**Example 3.4.7.** Consider the second-order differential equation

$$x'' + 3x' + 2x = f(t) = \frac{1}{1 + e^{-t}}. \tag{3.38}$$

Since $f(t) \to 1$ and the homogeneous solution tends to zero as $t \to \infty$ it is reasonable to assume that $\lim_{t\to\infty} x(t) = 0.5$. Suppose an engineer is asked to answer the following question: "If $x(0) = 1.0$, will $x$ always go below 0.5 for some positive value of $t$, or are there values of $x'(0)$ for which it could approach 0.5 from above as $t \to \infty$?" Figure 3.13 shows numerical solutions of (3.38) with $x(0) = 1$ and several values of $x'(0)$, and while this gives a general impression of how solutions behave, it would not be easy to answer the specific question being asked.

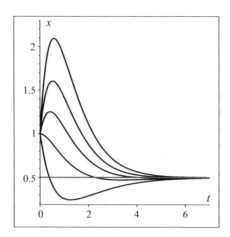

Figure 3.13. Solutions of (3.38) with $x(0) = 1$ and $x'(0) = -1, 0, 1.5, 3, 5$

We will therefore try the method of variation of parameters to find an analytic solution of (3.38). The general solution of the homogeneous equation is $x_h = C_1 e^{-2t} + C_2 e^{-t}$, so we will let $x_1 = e^{-2t}$ and $x_2 = e^{-t}$. The Wronskian is

$$W(e^{-2t}, e^{-t}) = e^{-2t}(-e^{-t}) - (e^{-t})(-2e^{-2t}) = e^{-3t};$$

therefore, letting $x_p = v_1 x_1 + v_2 x_2$, we need to compute the integrals

$$v_1 = -\int \left(\frac{x_2 f}{W(x_1, x_2)}\right) = -\int \left(\frac{e^{-t}}{e^{-3t}} \cdot \frac{1}{1 + e^{-t}}\right) dt = -\int \frac{e^{2t}}{1 + e^{-t}} dt$$

and

$$v_2 = \int \left(\frac{x_1 f}{W(x_1, x_2)}\right) = \int \left(\frac{e^{-2t}}{e^{-3t}} \cdot \frac{1}{1 + e^{-t}}\right) dt = \int \frac{e^t}{1 + e^{-t}} dt.$$

Both of these integrals can be evaluated by substituting $u = e^{-t}$ and then using partial fractions. You should check carefully the following calculations.

Substituting $u = e^{-t}$, the integral for $v_1$ becomes

$$v_1 = \int \frac{du}{u^3(1 + u)} = \int \left(\frac{1}{u} - \frac{1}{u^2} + \frac{1}{u^3} - \frac{1}{u + 1}\right) du$$

$$= \ln u + \frac{1}{u} - \frac{1}{2u^2} - \ln(1 + u) = -t + e^t - e^{2t}/2 - \ln(1 + e^{-t})$$

and the integral for $v_2$ is

$$v_2 = -\int \frac{du}{u^2(1 + u)} = \int \left(\frac{1}{u} - \frac{1}{u^2} - \frac{1}{u + 1}\right) du$$

$$= \ln u + \frac{1}{u} - \ln(1 + u) = -t + e^t - \ln(1 + e^{-t}).$$

Writing the particular solution as $x_p = x_1 v_1 + x_2 v_2$, the general solution of (3.38) is

$$x(t) = C_1 e^{-2t} + C_2 e^{-t} + e^{-2t}\left(e^t - t - \frac{e^{2t}}{2} - \ln(1 + e^{-t})\right)$$
$$+ e^{-t}\left(e^t - t - \ln(1 + e^{-t})\right). \tag{3.39}$$

## 3.4 Nonhomogeneous Linear Equations

*The constants $C_1$ and $C_2$ depend on the initial conditions, but if the terms of (3.39) are regrouped and $x(t)$ is written as*

$$x(t) = \frac{1}{2} + e^{-t}((C_2 + 1) - t - \ln(1 + e^{-t})) + e^{-2t}(C_1 - t - \ln(1 + e^{-t})) \quad (3.40)$$

*it can be clearly seen that for all $t > \max\{C_1, C_2 + 1\}$, the function $x(t)$ is less than 0.5. Thus, for any finite initial conditions, $x$ always becomes less than 0.5 and ultimately approaches 0.5 from below as $t \to \infty$.* □

One object of Example 3.4.7 is to reinforce the idea that sometimes a numerical solution is all you need, but other times trying to find an analytic solution may be worth the effort.

**Exercises 3.4.** *For equations 1–6 determine the form to be used for $x_p$ when using the method of undetermined coefficients.*

1. $x'' + 3x' + 2x = 6e^{2t}$
2. $x'' + 5x' + 6x = 5e^{-t}$
3. $x'' + 4x = 3\cos(2t)$
4. $x'' + 2x = 5\sin(2t)$
5. $x'' + x = t^3$
6. $x'' + x' + x = 1 + t^4$

*For equations 7–10, use the method of undetermined coefficients to find a particular solution $x_p$.*

7. $x'' + 3x' + 2x = 6e^{2t}$
8. $x'' + 5x' + 6x = 5e^{-t}$
9. $x'' + 4x' + 4x = 2 + t^2$
10. $x'' + 4x' + 5x = t^3$

*For equations 11–14, find the general solution $x_h + x_p$.*

11. $x'' + x = \cos(2t)$
12. $x'' + 4x = 2\sin(t)$
13. $x'' + 5x' + 4x = t + e^t$
14. $2x'' + 5x' + 2x = 2e^{-2t} - t$

*For problems 15–18, solve the initial value problem.*

15. $x'' - 5x' + 4x = t$, $x(0) = 1$, $x'(0) = 0$
16. $x'' + 3x' + 2x = 1 + t^2$, $x(0) = 0$, $x'(0) = 1$
17. $x'' + x = 4e^{-t}$, $x(0) = 1$, $x'(0) = 0$
18. $x'' + 4x' + 5x = 10$, $x(0) = 0$, $x'(0) = 0$

*For problems 19–20, solve the initial value problem. Draw a graph of the solution and describe the behavior of x(t) as t → ∞.*

19. $x'' + 4x' + 5x = \cos(t), \quad x(0) = 0, \quad x'(0) = 0$

20. $2x'' + 5x' + 2x = 10\sin(t), \quad x(0) = -2, \quad x'(0) = 1$

*In problems 21–24 use variation of parameters to find a particular solution $x_p$.*

21. $x'' + 4x' + 3x = 2e^t, \quad x_1(t) = e^{-t}, \quad x_2(t) = e^{-3t}$

22. $x'' + 2x' + x = 4e^{-t}, \quad x_1(t) = e^{-t}, \quad x_2(t) = te^{-t}$

23. $x'' + x' = \frac{1}{1+e^{-t}}, \quad x_1(t) = e^{-t}, \quad x_2(t) = 1$

    (Compare the $x_p$ you obtain to the particular solution of this equation with $f(t) \equiv 1$.)

24. $x'' + x = \cot(t), \quad x_1(t) = \cos(t), \quad x_2(t) = \sin(t)$

*In Section 3.6 formulas will be given for the solution of an equation called a Cauchy-Euler equation. It has the form*

$$at^2 x'' + btx' + cx = 0,$$

*where a, b, and c are real constants. It will be shown that solutions are of the form $t^r$, where r is a root of the quadratic polynomial $P(r) = ar^2 + (b - a)r + c$. In problems 25–27, use this to find a fundamental solution set $\{x_1, x_2\}$, and then use variation of parameters to find a particular solution $x_p$. Be sure to divide the whole equation by $t^2$ to put it into standard form before deciding what the function $f(t)$ is.*

25. $t^2 x'' - 2tx' + 2x = 3t$

26. $t^2 x'' + 2tx' - 6x = t^2$

27. $t^2 x'' - 2x = 2t$

## 3.5 The Forced Spring-mass System

In this section we will apply our techniques for solving nonhomogeneous equations to the solution of problems involving spring-mass equations with external forcing. Remember, from Section 3.3, that if there is external forcing, the equation for the spring-mass system becomes the nonhomogeneous equation

$$mx'' + bx' + kx = f(t), \qquad (3.41)$$

which we now know how to solve, at least for continuous functions $f(t)$. The general solution is

$$x(t) = x_h(t) + x_p(t),$$

and if the coefficients $m$, $b$, and $k$ are all positive and $b > 0$ we know that the homogeneous solution $x_h(t)$ will approach zero as $t \to \infty$. This means that in the long run, the solution of (3.41) will approach the particular solution $x_p(t)$. For this reason, in the case of a damped forced spring-mass system, the function $x_h$ is referred to as the **transient part** of the solution, and $x_p$ is called the **steady-state solution**.

The first example is fairly simple, and will give you an idea of what to expect.

## 3.5 The Forced Spring-mass System

**Example 3.5.1.** *Consider a spring-mass system with spring constant $k = 6$ N/m and damping coefficient $b = 5$ kg/sec. The 1 kg mass is lifted up one meter and given a downward velocity of 8 m/sec. If there is no forcing on the system, the position $x(t)$ of the mass will satisfy the homogeneous equation*

$$x'' + 5x' + 6x = 0,$$

*and you can easily show that the solution of this equation with initial values $x(0) = 1$, $x'(0) = -8$ is*

$$x_h(t) = -5e^{-2t} + 6e^{-3t}.$$

*Suppose there is an external force acting on the system. It will arbitrarily be assumed to be a periodic force, so that the resulting equation is given by*

$$x'' + 5x' + 6x = 4\sin(t). \tag{3.42}$$

*The position function $x(t)$ will now be of the form $x_h + x_p$, where $x_h$ is the transient solution and $x_p$ is the steady-state solution. You know from Table 3.2 that $x_p$ can be written in the form $A\sin(t) + B\cos(t)$, and you know exactly how to find the constants $A$ and $B$. Since the programmers for Maple and Mathematica also know how to do this, it may be time to let them do their job. With initial conditions $x(0) = 1$, $x'(0) = -8$, Maple returns the following solution of (3.42):*

$$x(t) = -\frac{21}{5}e^{-2t} + \frac{28}{5}e^{-3t} - \frac{2}{5}\cos(t) + \frac{2}{5}\sin(t).$$

*It is clear that as $t \to \infty$, the exponential terms die out and the solution approaches the steady-state solution $x_p(t) = -\frac{2}{5}\cos(t) + \frac{2}{5}\sin(t)$. The middle graph in Figure 3.14 shows this solution.*

*Now suppose the forcing is oscillating periodically but is also decaying exponentially and the equation is*

$$x'' + 5x' + 6x = 4e^{-0.2t}\sin(t). \tag{3.43}$$

*It should now be clear that as $t \to \infty$, the full solution of the equation will also decay to zero. Every term in the solution will be multiplied by a negative exponential function. In the exercises you will be asked to show how to use the method of undetermined coefficients to handle a case where $f(t)$ is a product of a sine or cosine and an exponential function.* ∎

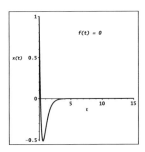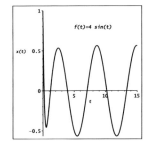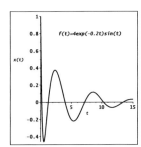

Figure 3.14. Solutions of $x'' + 5x' + 6x = f(t)$, with three different types of forcing

The three graphs in Figure 3.14 show the three solutions described in the above example, and we are now ready to consider a slightly more interesting problem.

**Example 3.5.2.** *The vertical motion of a car driving along a bumpy road (think of driving along old train tracks) is modeled by the equation*

$$2x'' + bx' + 3x = 4\sin(t/2). \tag{3.44}$$

*How large would the damping coefficient b have to be so that the long-term oscillatory up-and-down motion of the car would have a vertical amplitude less than 0.2 m?*

*Solution.* This is asking for the steady-state solution to be a sine function with amplitude less than 0.2. It isn't necessary in this case to find the homogeneous solution (i.e., the transient part of the solution), since for any value of $b > 0$ it dies out as $t \to \infty$. From Table 3.2 it can be seen that we need to write the particular solution as

$$x_p(t) = A\sin(t/2) + B\cos(t/2),$$

and substituting $x_p$ and its derivatives into (3.44), we need to find $A$ and $B$ to make the left-hand side of the equation $2x_p'' + bx_p' + 3x_p$ equal to the right-hand side $f(t)$. This requires that

$$2\left(-\frac{1}{4}A\sin(t/2) - \frac{1}{4}B\cos(t/2)\right) + b\left(\frac{1}{2}A\cos(t/2) - \frac{1}{2}B\sin(t/2)\right)$$
$$+ 3(A\sin(t/2) + B\cos(t/2))$$

be made identical to the right-hand side

$$f(t) \equiv 4\sin(t/2) + 0\cos(t/2).$$

Collecting the coefficients of the sine and cosine on each side of the equation, we must have

$$\sin(t/2)\left(-\frac{1}{2}A - \frac{1}{2}Bb + 3A\right) + \cos(t/2)\left(-\frac{1}{2}B + \frac{1}{2}Ab + 3B\right)$$
$$\equiv 4\sin(t/2) + 0\cos(t/2).$$

The resulting equations for $A$ and $B$ are

$$\frac{5}{2}A - \frac{1}{2}Bb = 4, \quad \frac{5}{2}B + \frac{1}{2}Ab = 0.$$

Using Cramer's Rule to solve the linear system $\begin{pmatrix} \frac{5}{2} & -\frac{b}{2} \\ \frac{b}{2} & \frac{5}{2} \end{pmatrix} \begin{pmatrix} A \\ B \end{pmatrix} = \begin{pmatrix} 4 \\ 0 \end{pmatrix}$ gives

$$A = \frac{\begin{vmatrix} 4 & -\frac{b}{2} \\ 0 & \frac{5}{2} \end{vmatrix}}{\begin{vmatrix} \frac{5}{2} & -\frac{b}{2} \\ \frac{b}{2} & \frac{5}{2} \end{vmatrix}} = \frac{10}{\frac{25}{4} + \frac{b^2}{4}} = \frac{40}{25 + b^2}$$

and

$$B = \frac{\begin{vmatrix} \frac{5}{2} & 4 \\ \frac{b}{2} & 0 \end{vmatrix}}{\begin{vmatrix} \frac{5}{2} & -\frac{b}{2} \\ \frac{b}{2} & \frac{5}{2} \end{vmatrix}} = \frac{-2b}{\frac{25}{4} + \frac{b^2}{4}} = \frac{-8b}{25 + b^2}.$$

## 3.5 The Forced Spring-mass System

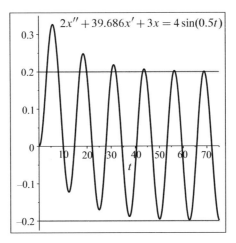

Figure 3.15. Solution of the forced equation (3.44) with $b = 39.686$

Remember that we can write $x_p$ in the form of a single sine function, with amplitude $\sqrt{A^2 + B^2}$. To make $\sqrt{A^2 + B^2} \leq 0.2$, we solve

$$A^2 + B^2 = \left(\frac{40}{25 + b^2}\right)^2 + \left(\frac{-8b}{25 + b^2}\right)^2 = 0.04.$$

This can be written as a quadratic equation in $b^2$:

$$b^4 - 1550b^2 - 39375 = 0,$$

and using the quadratic formula, the solution is $b \approx 39.686$.

A graph of the solution of the given equation, with $b = 39.686$, is shown in Figure 3.15. ∎

**Beats and Resonance.** We will now look briefly at two well-known phenomena that can occur in a sinusoidally forced undamped spring-mass system. These are referred to as **beats** and **resonance**. They are both encountered when an undamped system is forced by a sinusoidal forcing function having frequency very close to the natural frequency of the unforced system. Resonance occurs when these two frequencies are identical, and beats occurs when they are very close, but not the same.

**Pure Resonance.** Consider an undamped spring-mass system $mx'' + kx = \rho \sin(\omega t)$ that is being forced by a sinusoidal forcing function. In Section 3.3 we defined the **natural frequency** of an undamped unforced system to be $\frac{\omega_0}{2\pi}$, where $\omega_0 = \sqrt{k/m}$; therefore we will divide our equation by $m$ and write it in the form

$$x'' + \frac{k}{m}x = x'' + \omega_0^2 x = f(t) = R \sin \omega t, \quad \text{where } R = \frac{\rho}{m}. \qquad (3.45)$$

The characteristic polynomial $P(r) = r^2 + \omega_0^2$ has pure imaginary roots $r = \pm \omega_0 \iota$, and the homogeneous solution is

$$x_h(t) = C_1 \sin(\omega_0 t) + C_2 \cos(\omega_0 t).$$

If the forcing function has the same frequency, that is, $f(t) = R\sin(\omega_0 t)$, then Table 3.2 implies that the function to be used for $x_p$ is

$$x_p(t) = t(A\sin(\omega_0 t) + B\cos(\omega_0 t)).$$

It is clear that this function continues to grow with $t$ as $t \to \infty$. To see what can happen, consider the following example.

**Example 3.5.3.** *Solve the IVP*

$$x'' + 4x = \sin(2t), \quad x(0) = 0, \quad x'(0) = 0, \qquad (3.46)$$

*and draw a graph of the solution on the interval $0 \leq t \leq 20$.*

*Solution.* From the discussion above, we know that $x_h = C_1 \sin(2t) + C_2 \cos(2t)$, and the form to be used for $x_p$ is $x_p(t) = t(A\sin(2t) + B\cos(2t))$. The derivatives of $x_p$ are

$$x'_p(t) = A\sin(2t) + B\cos(2t) + t(2A\cos(2t) - 2B\sin(2t)),$$

$$x''_p(t) = 4(A\cos(2t) - B\sin(2t)) + t(-4A\sin(2t) - 4B\cos(2t)).$$

Substituting into the differential equation (3.46), the terms multiplied by $t$ cancel, and we need to have

$$x''_p + 4x_p = 4(A\cos(2t) - B\sin(2t)) \equiv f(t) = 0 \cdot \cos(2t) + 1 \cdot \sin(2t).$$

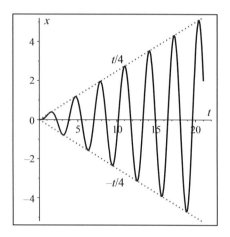

Figure 3.16. Solution of $x'' + 4x = \sin(2t)$, $x(0) = 0$, $x'(0) = 0$, showing pure resonance

This will be an identity if $A = 0$ and $B = -\frac{1}{4}$; therefore, the particular solution is seen to be

$$x_p(t) = -\frac{1}{4}t\cos(2t).$$

The general solution of the differential equation is

$$x(t) = C_1 \sin(2t) + C_2 \cos(2t) - \frac{1}{4}t\cos(2t),$$

and the derivative is

$$x'(t) = 2C_1 \cos(2t) - 2C_2 \sin(2t) - \frac{1}{4}\cos(2t) + \frac{1}{2}t\sin(2t).$$

## 3.5 The Forced Spring-mass System

Setting $x(0) = C_2 = 0$ and $x'(0) = 2C_1 - \frac{1}{4} = 0$ we have $C_1 = \frac{1}{8}$ and

$$x(t) = \frac{1}{8}\sin(2t) - \frac{1}{4}t\cos(2t).$$

A graph of this function is shown in Figure 3.16. The homogeneous solution does not die out in this case, but after a while the particular solution is so large that it is all that is visible. ∎

It is clear that $x$ oscillates with increasing amplitude as $t \to \infty$, and a solution of this sort is said to be unstable. Notice, however, that with even a small amount of damping in the system, the size of the oscillation can be controlled. We will have more to say about this when we define resonance for a damped spring-mass system.

**Beats.** Beats is a phenomenon which can occur when the frequency of the forcing function is very close to the natural frequency of the system. An easy-to-understand physical example is the sound one hears when two tuning forks of almost the same frequency are struck simultaneously. The sound seems to increase and decrease periodically, hence the name "beats". To see why this happens, we will solve the IVP

$$x'' + \omega_0^2 x = f(t) = R\sin\omega t, \quad x(0) = x'(0) = 0 \quad (3.47)$$

assuming $|\omega - \omega_0|$ is small, but not equal to zero.

We know that $x_h(t) = C_1 \sin(\omega_0 t) + C_2 \cos(\omega_0 t)$, and since $\omega \neq \omega_0$, the particular solution simplifies to

$$x_p = A\sin(\omega t).$$

It is not even necessary to include the term $B\cos(\omega t)$, since with zero damping $x'_p$ does not appear in the equation. This means that only sine terms will be involved in the calculation.

If $x_p$ and its second derivative $x''_p(t) = -\omega^2 A \sin(\omega t)$ are substituted into (3.47) for $x$ and $x''$, it is necessary to satisfy the equation

$$-\omega^2 A \sin(\omega t) + \omega_0^2 A \sin(\omega t) = A(\omega_0^2 - \omega^2)\sin(\omega t) \equiv R\sin(\omega t);$$

therefore, the value of the constant $A$ is $\frac{R}{\omega_0^2 - \omega^2}$, and the particular solution is

$$x_p(t) = \frac{R}{\omega_0^2 - \omega^2}\sin(\omega t).$$

Adding $x_p$ to $x_h$,

$$x(t) = x_h + x_p = C_1 \sin(\omega_0 t) + C_2 \cos(\omega_0 t) + \frac{R}{\omega_0^2 - \omega^2}\sin(\omega t)$$

and

$$x'(t) = \omega_0 C_1 \cos(\omega_0 t) - \omega_0 C_2 \sin(\omega_0 t) + \frac{\omega R}{\omega_0^2 - \omega^2}\cos(\omega t).$$

Setting $x(0) = 0$ gives $C_2 = 0$, and $x'(0) = \omega_0 C_1 + \frac{\omega R}{\omega_0^2 - \omega^2} = 0$ implies that

$$C_1 = -\frac{\omega R}{\omega_0(\omega_0^2 - \omega^2)}.$$

The function $x(t)$ can now be written as

$$x(t) = \frac{R}{\omega_0^2 - \omega^2}\sin(\omega t) - \frac{\omega R}{\omega_0(\omega_0^2 - \omega^2)}\sin(\omega_0 t) = \frac{R}{\omega_0^2 - \omega^2}\left(\sin(\omega t) - \frac{\omega}{\omega_0}\sin(\omega_0 t)\right).$$

To see what the function x looks like, consider the particular IVP

$$x'' + 4x = \sin(1.9t), \quad x(0) = 0, \quad x'(0) = 0.$$

Using the above formula, the solution of this IVP is

$$x(t) = \frac{1.0}{4 - 1.9^2}\left(\sin(1.9t) - \frac{1.9}{2}\sin(2t)\right),$$

and a graph of $x$, showing the characteristic pattern of beats, is seen in Figure 3.17.

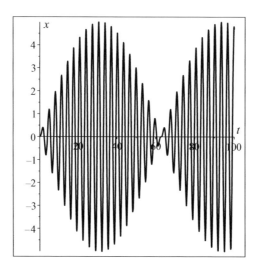

**Figure 3.17.** Solution of $x'' + 4x = \sin(1.9t)$, $x(0) = 0$, $x'(0) = 0$, showing beats

To understand mathematically why the beats occur, notice that if we let $K = \frac{R}{\omega_0^2 - \omega^2}$, our formula for $x(t)$ is essentially equal to

$$x(t) = K\big(\sin(\omega t) - \sin(\omega_0 t)\big),$$

where we have eliminated the multiplier $\frac{\omega}{\omega_0}$ of the term $\sin(\omega_0 t)$. If $\omega$ and $\omega_0$ are close, this multiplier is very close to one.

We can use the formula

$$\sin(A) - \sin(B) = 2\cos\left(\frac{A+B}{2}\right)\sin\left(\frac{A-B}{2}\right).$$

With $A = \omega t$ and $B = \omega_0 t$, it gives us

$$x(t) \approx K\big(\sin(\omega t) - \sin(\omega_0 t)\big) = 2K\cos\left(\frac{\omega + \omega_0}{2}t\right)\sin\left(\frac{\omega - \omega_0}{2}t\right).$$

Since $\omega + \omega_0$ is large relative to the difference $\omega - \omega_0$, it can be seen that the cosine term oscillates much more rapidly than the sine term, so that what you see is a high frequency cosine term with its amplitude modulated by the slowly oscillating sine function.

## 3.5 The Forced Spring-mass System

**Resonance in a Damped Spring-mass System.** We have seen that in the "nonrealizable case" of an undamped spring-mass system it is possible to drive the solution to infinity by forcing it with a sine function of exactly the right frequency, namely, the natural frequency of the undamped system.

It was previously pointed out that real physical systems will always have some damping, so the more practical question becomes "can we maximize the amplitude of the steady-state function $x_p(t)$ by forcing an underdamped system by a sine function of a particular frequency?" This can be an important question in the case of some real physical models.

Consider the equation

$$mx'' + bx' + kx = F\sin(\omega t). \tag{3.48}$$

If we consider just the steady state solution

$$x_p(t) = A\sin(\omega t) + B\cos(\omega t),$$

is there a value of $\omega$ that maximizes the amplitude $R = \sqrt{A^2 + B^2}$ of $x_p$? If such a value of $\omega$ exists, it will certainly be a function of the physical parameters $m, b$, and $k$ of the system. To find out, we will determine $A$ and $B$ by substituting $x_p, x_p'$, and $x_p''$ into (3.48). This results in the equation

$$m\big(-\omega^2 A\sin(\omega t) - \omega^2 B\cos(\omega t)\big) + b\big(\omega A\cos(\omega t) - \omega B\sin(\omega t)\big)$$
$$+ k\big(A\sin(\omega t) + B\cos(\omega t)\big) \equiv F\sin(\omega t) + 0\cos(\omega t).$$

Equating the coefficients of the sine and the cosine functions on both sides of the equation requires that

$$(k - m\omega^2)A - b\omega B = F \quad \text{and} \quad b\omega A + (k - m\omega^2)B = 0.$$

Solving these two equations using Cramer's Rule,

$$A = \frac{\begin{vmatrix} F & -b\omega \\ 0 & k - m\omega^2 \end{vmatrix}}{\begin{vmatrix} k - m\omega^2 & -b\omega \\ b\omega & k - m\omega^2 \end{vmatrix}} = \frac{F(k - m\omega^2)}{(k - m\omega^2)^2 + (b\omega)^2},$$

$$B = \frac{\begin{vmatrix} k - m\omega^2 & F \\ b\omega & 0 \end{vmatrix}}{\begin{vmatrix} k - m\omega^2 & -b\omega \\ b\omega & k - m\omega^2 \end{vmatrix}} = \frac{-Fb\omega}{(k - m\omega^2)^2 + (b\omega)^2}.$$

The denominator of $A$ and $B$ is always positive if $b$ and $\omega$ are both greater than zero.

Now the steady state solution can be written as

$$x_p(t) = A\sin(\omega t) + B\cos(\omega t)$$
$$= \frac{F}{(k - m\omega^2)^2 + (b\omega)^2}\big[(k - m\omega^2)\sin(\omega t) - b\omega\cos(\omega t)\big].$$

If the two terms inside the brackets are written as a single sine function $R\sin(\omega t + \phi)$, the amplitude will be

$$R = \sqrt{(k - m\omega^2)^2 + (b\omega)^2}.$$

Then $x_p(t) = \left(\frac{F}{R^2}\right) R \sin(\omega t + \phi) = \frac{F}{R} \sin(\omega t + \phi)$, and it can be seen that the amplitude of the steady-state solution $x_p$ is equal to the amplitude $F$ of the forcing function multiplied by the function

$$G(\omega) = \frac{1}{R} = \frac{1}{\sqrt{(k - m\omega^2)^2 + (b\omega)^2}}.$$

$G$ is called the **gain function**, and our question now becomes "does the gain function have a maximum value for some value of $\omega$?" To find out, set $G'(\omega) = 0$ and solve for $\omega$:

$$G'(\omega) = \frac{d}{d\omega}\left((k - m\omega^2)^2 + (b\omega)^2\right)^{-\frac{1}{2}} = -\frac{1}{2}\left(\frac{2(k - m\omega^2)(-2m\omega) + 2b^2\omega}{[(k - m\omega^2)^2 + (-b\omega)^2]^{\frac{3}{2}}}\right).$$

The only way $G'(\omega)$ can be zero is if the numerator is zero, so we need

$$(k - m\omega^2)(-2m\omega) + b^2\omega = \omega\big[-2m(k - m\omega^2) + b^2\big] = 0.$$

Setting the term in brackets equal to zero gives

$$\omega^2 = \frac{2mk - b^2}{2m^2} = \frac{k}{m} - \frac{b^2}{2m^2};$$

therefore, if $b^2 < 2mk$, the gain function has a maximum at $\omega = \sqrt{\frac{k}{m} - \frac{b^2}{2m^2}}$, otherwise the maximum occurs at $\omega = 0$ and $G$ decreases for all $\omega > 0$. To have a maximum, we see that not only must the equation be underdamped, but the square of the damping must be less than half of the critical damping value $b^2 = 4mk$. The value of $\omega$ where the maximum gain occurs, if it exists, is called the **resonant frequency** of the system. If the damping constant $b$ is small, then the resonant frequency is close to the natural frequency $\omega_0 = \sqrt{\frac{k}{m}}$.

**Example 3.5.4.** *An engineer wishes to maximize the amplitude of the steady-state current $I(t)$ in an RLC-circuit which satisfies the equation*

$$100 I''(t) + 50 I'(t) + 40 I(t) = \sin(\omega t). \tag{3.49}$$

*What value should $\omega$ have?*

*Solution.* This equation is identical to a forced spring-mass equation with $m = 100$, $b = 50$, and $k = 40$.

Since $b^2 = 2500$ is less than $2mk = 8000$, the steady-state solution $I_p(t)$ will have maximum amplitude if $\omega$ has the value that maximizes the gain function, that is,

$$\omega = \bar{\omega} = \sqrt{\frac{k}{m} - \frac{b^2}{2m^2}} \approx 0.52441.$$

This value can be seen to be close, but not exactly equal, to the resonant frequency $\omega_0 = \sqrt{0.4} \approx 0.63$ of the undamped system. A plot of the gain function $G(\omega)$ is shown in Figure 3.18. The right-hand graph in Figure 3.18 shows three solutions of (3.49) for values $\omega = 0.2$, $\bar{\omega}$, and $1.0$. The period $\frac{2\pi}{\omega}$ of the solutions decreases with increasing $\omega$, but the amplitude of the solution is clearly a maximum when $\omega$ has the value $\bar{\omega}$. ∎

## 3.5 The Forced Spring-mass System

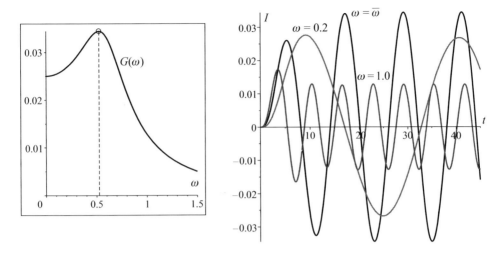

Figure 3.18. Gain function and solution curves of (3.49)

Even though the transient part of the solution goes to zero as $t \to \infty$, a large transient in a system such as an electrical circuit can cause problems. For this reason engineers need a way to know how long it will take for the transient to die out. If the equation is underdamped and written as

$$mx'' + bx' + kx = f(t),$$

we know that the transient part of the solution has the form

$$x_h(t) = e^{-\frac{b}{2m}t}\bigl(C_1 \cos(\omega t) + C_2 \sin(\omega t)\bigr).$$

The decay rate of the transient is determined by the exponential factor $e^{-\frac{b}{2m}t}$. Letting $c \equiv \frac{b}{2m}$, the **time constant** $T$ of the system is defined as

$$T = \frac{1}{c}.$$

After a time interval equal to $T$, the transient has decayed to $e^{-cT} = e^{-1}$ times its initial value. A general rule of thumb for engineers is to use a time of $4T$, which means a decay of $e^{-4} \approx 0.0183$, to be the time at which the transient becomes negligible. In the circuit in Example 3.5.4, $c = b/2m = 1/4 \longrightarrow T = 4$. This means that the transient should die out in approximately $4T = 16$ seconds. The graph in Figure 3.19 shows the current in the circuit with $\omega = 0.52441$, and with initial conditions $I(0) = 0.5$, $I'(0) = 0$. It can be clearly seen that it takes about 16 seconds for the effect of the transient to disappear.

**Exercises 3.5.** *In problems 1–4 you will need to do some work by hand, but you will also need a computer algebra system that can solve and plot solutions for second-order differential equations.*

1. *Solve the IVP*

$$x'' + 4x = 1 + e^{-t}, \quad x(0) = 0, \quad x'(0) = 1,$$

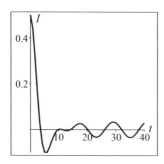

Figure 3.19. Solution of $100I'' + 50I' + 40I = \sin(0.52441t)$

and draw a graph of the solution for $0 < t < 20$. Notice that $x_p$ has the form $A + Be^{-t}$, so in the long run it approaches the constant value $A$. Does your solution $x(t)$ approach a constant as $t \to \infty$? If not, why not?

2. Use your CAS to plot graphs of the solutions $x(t)$ of the initial-value problems:

   (a) $2x'' + x = 3\cos(0.7t)$, $x(0) = 1$, $x'(0) = 0$,
   (b) $4x'' + 9x = \sin(1.5t)$, $x(0) = 1$, $x'(0) = 0$.

   In which case are you seeing "pure resonance", and in which case "beats"? Explain.

3. A forced spring-mass system satisfies the IVP

   $$x'' + x' + Kx = \sin(t), \quad x(0) = x'(0) = 0.$$

   Find the maximum value of the spring constant $K$ such that the steady-state displacement $x_p(t)$ never goes above $\frac{1}{2}$ meter. Find the value of $K$ by hand, and then use a computer d.e. solver to plot your solution.

4. For the spring-mass system modeled by the equation

   $$4x'' + 4x' + 3x = \sin(\omega t),$$

   what value $\omega = \omega^*$ will produce steady-state output with maximum amplitude? Find the correct value by hand, and then use the computer to draw graphs of $x(t)$ with $\omega = \omega^*$ and then with $\omega = \frac{\omega^*}{2}$. You can use the initial values $x(0) = x'(0) = 0$.

5. An RLC electrical circuit is diagrammed below. It contains a resistor, an inductor, and a capacitor in series, together with a generator to supply current. If $R$ is the resistance in ohms, $L$ is the inductance in henrys, and $C$ is the capacitance in farads, the equation for the current $I(t)$ (in amperes) is

   $$LI''(t) + RI'(t) + \frac{1}{C}I(t) = \frac{d}{dt}(E(t)),$$

### 3.6 Linear Second-order Equations with Nonconstant Coefficients

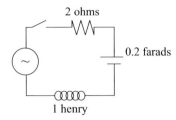

where $E(t)$ is the electromotive force (in volts) supplied by the generator. Find the current $I(t)$ in the circuit if $I(0) = 10$, $I'(0) = 0$, and $E(t) = 2\sin(3t)$. Graph your solution and mark the time $4T$ on the graph to show where the transient dies out.

6. For the circuit problem in Exercise 5 find the value of $\omega$ which will maximize the amplitude of the output. Redo the problem using this value of $\omega$ in the forcing function, and again graph the solution on $0 < t < 30$.

7. Determine the form of the function $x_p$ to use to solve the equation

$$x'' + 5x' + 6x = 4e^{-0.2t}\sin(t)$$

by the method of undetermined coefficients. (Hint: Remember that $x_p$ must be a linear combination, with arbitrary constant coefficients, of all functions that result from differentiating terms of the form $e^{bt}\cos(t)$ and $e^{bt}\sin(t)$. It only needs to contain four terms.) Use your formula to find $x_p$.

## 3.6 Linear Second-order Equations with Nonconstant Coefficients

In this section we will briefly consider two different ways to approach linear equations with nonconstant coefficients. We will only consider homogeneous equations, but remember that if we can find a fundamental solution set for a homogeneous equation $x'' + p(t)x' + q(t)x = 0$, we only need one solution of the full equation $x'' + p(t)x' + q(t)x = f(t)$ to have its general solution as well.

It was pointed out at the beginning of Chapter 3 that solving second-order linear equations with nonconstant coefficients can be difficult. We will produce exactly one equation of this type, the Cauchy-Euler equation, for which an analytic solution always exists. The other method of solution is by using infinite series, and some examples of this technique will also be given.

### 3.6.1 The Cauchy-Euler Equation.
There is one linear homogeneous differential equation with nonconstant coefficients for which the general solution can always be found. It is called a **Cauchy-Euler equation** and has the form

$$at^2 x'' + btx' + cx = 0, \qquad (3.50)$$

where $a$, $b$, and $c$ are real numbers, with $a \neq 0$.

The standard form of this equation is $x'' + \dfrac{b}{at}x' + \dfrac{c}{at^2}x = 0$, so that at $t = 0$ (3.50) does not even satisfy the hypotheses of the Existence and Uniqueness Theorem 3.2; therefore, at $t = 0$ bad things may happen to the solution.

Table 3.3. General Solution of $at^2x'' + btx' + cx = 0$, $t \neq 0$

| Discriminant $K = (b-a)^2 - 4ac$ | Roots of Q(r) $Q(r) = ar^2 + (b-a)r + c$ | General Solution |
|---|---|---|
| $K > 0$ | two real roots $r_1 \neq r_2$ | $x(t) = C_1 t^{r_1} + C_2 t^{r_2}$ |
| $K = 0$ | single real root $\bar{r}$ | $x(t) = C_1 t^{\bar{r}} + C_2 \ln(t) t^{\bar{r}}$ |
| $K < 0$ | complex roots $\alpha \pm \beta \iota$ | $x(t) = C_1 t^\alpha \cos(\beta \ln(t)) + C_2 t^\alpha \sin(\beta \ln(t))$ |

This equation comes up when solving certain types of partial differential equations, but our reason for including it here is that it can be solved by making a change of independent variable to turn it into an equation we already know how to solve. This is a very useful technique in mathematics, and this is a good place to see how it works.

Define a new independent variable $s = \ln(t)$. This means that $t = e^s$ and also that $\frac{ds}{dt} = \frac{1}{t} = e^{-s}$. Then if we let $Y(s) \equiv x(t)$, $Y$ will be a function of $s$, and the chain rule can be used to compute

$$x'(t) = \frac{d}{dt}(x(t)) = \frac{d}{dt}(Y(s(t))) = \frac{dY}{ds}\frac{ds}{dt} = Y'(s)e^{-s}.$$

Similarly,

$$x''(t) = \frac{d}{dt}(x'(t)) = \frac{d}{dt}(Y'(s)e^{-s}) = \frac{d}{ds}(Y'(s)e^{-s})\frac{ds}{dt},$$

and using the product rule for differentiation,

$$x''(t) = (Y''(s)e^{-s} - e^{-s}Y'(s))e^{-s} = (Y''(s) - Y'(s))e^{-2s}.$$

If these values of $x$, $x'$, and $x''$ are substituted into (3.50), with $t$ also replaced by $e^s$, the equation becomes

$$at^2x'' + btx' + cx = ae^{2s}(e^{-2s}(Y''(s) - Y'(s))) + be^s(e^{-s}Y'(s)) + cY(s)$$
$$= aY''(s) + (b-a)Y'(s) + cY(s) = 0.$$

The equation in terms of $Y(s)$ is the constant coefficient equation we already know how to solve, but be careful to note that the characteristic polynomial $Q(r) = ar^2 + (b-a)r + c$ is not quite the same one we found for the constant coefficient case. Letting $Y(s) = e^{rs}$, where $r$ is a root of $Q(r)$, and remembering that $e^{rs} = e^{r\ln(t)} \equiv t^r$, we can use our previous results in Table 3.1 to write formulas for the general solution of (3.50).

In each case we have written $x(t)$ in terms of $Y(s)$, and then let $s = \ln t$. For example, in Case 1 (real, unequal roots), $x(t) = Y(s) = C_1 e^{r_1 s} + C_2 e^{r_2 s} = C_1 e^{r_1 \ln(t)} + C_2 e^{r_2 \ln(t)} = C_1 t^{r_1} + C_2 t^{r_2}$. The other two cases are treated similarly.

It is clear from Table 3.3 that the behavior of $x(t)$ near $t = 0$ can be very strange indeed. The example below will show what can happen when the roots of the characteristic polynomial are complex. For $0 \leq t \leq 1$, the function $\ln(t)$ takes on every value between 0 and $-\infty$; therefore, the functions $\sin(\beta \ln(t))$ and $\cos(\beta \ln(t))$ oscillate infinitely often as $t$ approaches zero from the right.

## 3.6 Linear Second-order Equations with Nonconstant Coefficients

**Example 3.6.1.** *Solve the IVP*

$$t^2 x'' + tx' + 16x = 0, \quad x(1) = 1, \quad x'(1) = 0.$$

*Solution.* Initial conditions can no longer be specified at $t = 0$. With $a = 1, b = 1$, and $c = 16$, the roots of the characteristic polynomial $Q(r) = r^2 + (1-1)r + 16 = r^2 + 16$ are $r = \pm 4\iota$, and from the formula in Table 3.3, for the case with complex roots, with $\alpha = 0, \ \beta = 4$, we have

$$x(t) = C_1 \cos(4\ln(t)) + C_2 \sin(4\ln(t)).$$

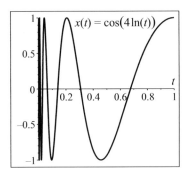

**Figure 3.20.** Solution of $t^2 x'' + tx' + 16x = 0$ with $x(1) = 1$ and $x'(1) = 0$

Differentiation, using the chain rule, gives

$$x'(t) = C_1\left(-\sin(4\ln(t)) \cdot \frac{d}{dt}(4\ln(t))\right) + C_2\left(\cos(4\ln(t)) \cdot \frac{d}{dt}(4\ln(t))\right)$$

$$= -\frac{4}{t} C_1 \sin(4\ln(t)) + \frac{4}{t} C_2 \cos(4\ln(t)).$$

Since $\ln(1) = 0$, we see that the coefficients must satisfy $x(1) = C_1 \cos(0) + C_2 \sin(0) = C_1 = 1$ and $x'(1) = -4C_1 \sin(0) + 4C_2 \cos(0) = 4C_2 = 0$. Therefore the solution of the IVP is

$$x(t) = \cos(4\ln(t)).$$

A graph of this function is shown in Figure 3.20. As $t \to 0$ from the right, the solution oscillates infinitely often. Do you see what happens to the solution as $t \to \infty$? ■

**3.6.2 Series Solutions.** In this subsection you will be given a brief introduction to series solutions of linear second-order equations with nonconstant coefficients. Before computers were available this was the standard "analytic" method for attacking problems of this kind, and many of the famous equations, such as Bessel's equation $t^2 x'' + tx' + (t^2 - v^2)x = 0$, have series named after them.

It is now as easy for a computer to produce a value of the series solution of Bessel's equation as it is to ask it to compute values of $\sin(t)$ or $\cos(t)$. But remember, $\sin(t)$ and $\cos(t)$ are solutions of the second order equation $x'' + x = 0$, and when a computer produces a numeric value of $\sin(t)$ or $\cos(t)$, it usually does so by summing a series.

In Section 3.1 we saw that a linear second-order differential equation

$$x'' + p(t)x' + q(t)x = f(t)$$

can be solved by first considering the associated homogeneous equation
$$x'' + p(t)x' + q(t)x = 0 \tag{3.51}$$
and then adding a particular solution $x_p$. The method of variation of parameters can be used to find $x_p$ whenever

(1) $f(t)$ is an integrable function and

(2) a fundamental solution set $\{x_1, x_2\}$ of the homogeneous equation can be found.

We showed that it is always possible to find a fundamental solution set in the constant coefficient case, but our method of assuming an exponential solution $x(t) = e^{rt}$ does not work if $p$ and $q$ are nonconstant functions of $t$. If $p$ and $q$ are "nice" functions, however, it may be possible to find Taylor series for functions $x_1$ and $x_2$ that do form a fundamental solution set.

If you have not looked at Taylor series for a while, you may want to consult your calculus book. First, we need a definition that may not have been given in the calculus course.

**Definition 3.3.** *A function $p(t)$ is said to be **analytic** at a point $t = t_0$ if it can be expanded in a Taylor series*

$$p(t) = \sum_{n=0}^{\infty} \frac{p^{(n)}(t_0)}{n!}(t-t_0)^n = p(t_0) + \frac{p'(t_0)(t-t_0)}{1!} + \frac{p''(t_0)(t-t_0)^2}{2!} + \cdots$$

*that converges in some interval about $t = t_0$.*

We will assume here that the coefficient functions $p$ and $q$ in (3.51) are both analytic for all values of $t$. If a function is analytic at a point, all its derivatives have to exist at that point; therefore, a function that is analytic everywhere is certainly differentiable everywhere. Since (3.51) is a linear equation, the Existence and Uniqueness Theorem 3.2, stated in Section 3.1, then guarantees that a general solution for (3.51) exists and is continuous for all $t$.

The basic idea is to assume a series solution for $x(t)$. For simplicity we will use the Maclaurin series; that is, the Taylor series about $t_0 = 0$ that has the form

$$x(t) = \sum_{n=0}^{\infty} a_n t^n, \quad a_n = \frac{x^{(n)}(0)}{n!}, \tag{3.52}$$

and substitute it and its derivatives into (3.51). If the coefficients $a_n$ can be found, we will have a solution to the equation. We are assuming that our initial conditions are given at $t_0 = 0$, but if this is not the case, a simple change of the independent variable can be used to make it so.

Sometimes it is only possible to find a **recursion formula** for the coefficient $a_n$ in terms of coefficients $a_0, a_1, \ldots, a_{n-1}$ with lower indices, but sometimes a formula for $a_n$ can be found, and in rare cases it may even be possible to recognize the resulting series as the Maclaurin series for a known function of $t$.

Remember from calculus that if a convergent power series $\sum_{n=0}^{\infty} a_n t^n$ for $x(t)$ exists at $t$, then the derivatives of the function $x$ at $t$ can be found by differentiating the power series term by term; that is,

$$x'(t) = \sum_{n=0}^{\infty} n a_n t^{n-1}, \quad x''(t) = \sum_{n=0}^{\infty} n(n-1) a_n t^{n-2}. \tag{3.53}$$

## 3.6 Linear Second-order Equations with Nonconstant Coefficients

The example below shows what is involved in finding series solutions for a linear differential equation with nonconstant coefficients.

**Example 3.6.2.** *Solve the IVP*

$$x'' + tx' + x = 0, \quad x(0) = 1, \quad x'(0) = -1. \tag{3.54}$$

*Solution.* This equation can be thought of as a spring-mass equation with nonconstant damping coefficient $b = t$, with the damping starting at zero and increasing with $t$. Use your physical intuition to make a guess as to what you think the solution should look like over the long term. For what range of $t$ is the equation underdamped? When is it overdamped?

If the series (3.52) and (3.53) for $x$ and its derivatives are substituted into the differential equation, we must make the following expression identically equal to zero:

$$x'' + tx' + x = \sum_{n=0}^{\infty} n(n-1)a_n t^{n-2} + t\sum_{n=0}^{\infty} na_n t^{n-1} + \sum_{n=0}^{\infty} a_n t^n. \tag{3.55}$$

The second series can be written as

$$t\sum_{n=0}^{\infty} na_n t^{n-1} \equiv \sum_{n=0}^{\infty} na_n t^n$$

since each term in the series can be multiplied by $t$. To write (3.55) as a single series we use the property that if $\sum A_n t^n$ and $\sum B_n t^n$ both converge in some interval, then their sum is $\sum A_n t^n + \sum B_n t^n = \sum (A_n + B_n)t^n$, and it is also a convergent series in the same interval.

It is first necessary to have all three sums in (3.55) written in terms of coefficients of the same power of $t$. To do this, we need to make a change of index in the first sum; that is, we need to rewrite

$$\sum_{n=0}^{\infty} n(n-1)a_n t^{n-2} \equiv \sum_{m=?}^{\infty} C_m t^m,$$

where the index of summation $n$ is changed but the two series still have identical terms.

In this case, this can be accomplished by letting $m = n - 2$, so that the power of $t$ in the left-hand sum is just the index $m$. Then we replace every occurrence of $n$ in the left-hand sum by $m + 2$. First note that when $n = 0$ or $n = 1$, the corresponding term $n(n-1)a_n t^{n-2}$ in the sum is zero, so the index can start at $n = 2$ without losing any terms. This is equivalent to starting the index $m = n - 2$ at zero. Thus the series for $x''$ can be written as

$$\sum_{n=2}^{\infty} n(n-1)a_n t^{n-2} = \sum_{m=0}^{\infty} (m+2)(m+2-1)a_{m+2} t^{m+2-2} = \sum_{m=0}^{\infty} (m+2)(m+1)a_{m+2} t^m.$$

You should write out the first few terms in each of these series and check that the sums on the left and right contain exactly the same terms.

The differential equation (3.54) now requires us to make

$$\sum_{m=0}^{\infty} (m+2)(m+1)a_{m+2} t^m + \sum_{m=0}^{\infty} ma_m t^m + \sum_{m=0}^{\infty} a_m t^m$$

identically equal to zero. The index in the second and third sums has just been given a new name; that is, $n$ is replaced everywhere by $m$. Make sure you see that this does not change the terms in the series; it is exactly the same as renaming the variable of integration in an integral.

Adding the three sums, by adding their coefficients, the equation becomes

$$\sum_{m=0}^{\infty} \left[(m+2)(m+1)a_{m+2} + ma_m + a_m\right] t^m \equiv 0.$$

It is known mathematically that a power series is identically equal to zero if and only if the coefficient of each power of $t$ is zero. This gives us what is called a recursion relation for the coefficients. For each integer $m = 0, 1, \ldots$, we set the coefficient

$$(m+2)(m+1)a_{m+2} + ma_m + a_m = 0$$

and solve for the $a_i$ with largest index:

$$a_{m+2} = \frac{-(m+1)a_m}{(m+2)(m+1)}.$$

Simplifying this result gives us our recursion formula

$$a_{m+2} = -\frac{a_m}{m+2}, \quad m = 0, 1, \ldots. \tag{3.56}$$

The final goal is to try to write each of the coefficients in terms of $a_0$ and $a_1$. In the Maclaurin series for $x(t)$ the initial conditions $x(0)$ and $x'(0)$ are, respectively, the values of $a_0$ and $a_1$. Using (3.56), for $m$ even,

$$a_2 = -\frac{a_0}{2}, \quad a_4 = -\frac{a_2}{4} = -\frac{1}{4}\left(-\frac{a_0}{2}\right) = \frac{a_0}{2 \cdot 4}, \quad a_6 = -\frac{a_4}{6} = -\frac{a_0}{2 \cdot 4 \cdot 6}, \ldots$$

and in general

$$a_{2m} = (-1)^m \frac{a_0}{2 \cdot 4 \cdot 6 \cdots (2m)}, \quad m = 1, 2, \ldots.$$

Similarly, for $m$ odd,

$$a_3 = -\frac{a_1}{3}, \quad a_5 = -\frac{a_3}{5} = \frac{a_1}{3 \cdot 5}, \quad a_7 = -\frac{a_5}{7} = -\frac{a_1}{3 \cdot 5 \cdot 7}, \ldots$$

and this gives the formula

$$a_{2m+1} = (-1)^m \frac{a_1}{1 \cdot 3 \cdot 5 \cdots (2m+1)}, \quad m = 0, 1, \ldots.$$

If the solution $x(t)$ is written in the form

$$x(t) = \sum_{n=0}^{\infty} a_n t^n = a_0 + a_1 t + a_2 t^2 + a_3 t^3 + a_4 t^4 + a_5 t^5 + \cdots$$

$$= a_0 + a_1 t - \frac{a_0}{2} t^2 - \frac{a_1}{3} t^3 + \frac{a_0}{2 \cdot 4} t^4 + \frac{a_1}{3 \cdot 5} t^5 + \cdots,$$

and the terms involving $a_0$ and $a_1$ are grouped separately, then for any values of $a_0$ and $a_1$,

$$x(t) = a_0 \left(1 - \frac{1}{2}t^2 + \frac{1}{2 \cdot 4}t^4 - \cdots\right) + a_1 \left(t - \frac{1}{1 \cdot 3}t^3 + \frac{1}{1 \cdot 3 \cdot 5}t^5 - \cdots\right)$$

$$\equiv a_0 x_1(t) + a_1 x_2(t),$$

where

$$x_1(t) = 1 + \sum_{n=1}^{\infty} \frac{(-1)^n t^{2n}}{2 \cdot 4 \cdot 6 \cdots (2n)}, \quad x_2(t) = \sum_{n=0}^{\infty} \frac{(-1)^n t^{2n+1}}{1 \cdot 3 \cdot 5 \cdots (2n+1)}.$$

## 3.6 Linear Second-order Equations with Nonconstant Coefficients

Since all this holds for any values of $a_0$ and $a_1$, including the value zero, we see that $x$ is a linear combination of two solutions of the homogeneous equation. In fact, $x_1$ and $x_2$ form a fundamental solution set. To see this, note that since $x_1(0) = 1$ and $x_2(0) = 0$, they cannot be constant multiples of each other, and it is shown in Appendix E that the Wronskian of two solutions $x_1$ and $x_2$ is identically zero if, and only if, the two functions are constant multiples of each other. Both of these series can easily be shown to converge for all $t$; they are alternating series, and for any fixed value of $t$ the $n$th term approaches zero as $n \to \infty$.

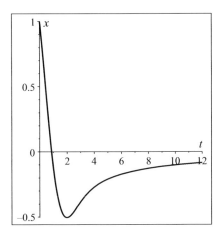

Figure 3.21. Solution of $x'' + tx' + x = 0, \quad x(0) = 1, \quad x'(0) = -1$

The solution to the initial-value problem, with $a_0 = x(0) = 1$ and $a_1 = x'(0) = -1$, is

$$x(t) = x(0)x_1(t) + x'(0)x_2(t) = x_1(t) - x_2(t).$$

Figure 3.21 shows a graph of $x(t)$ for $0 \le t \le 12$. Does this look the way you expected it to look? ■

---

**USEFUL ADVICE:** Whenever a second-order differential equation can be thought of as an example of a spring-mass system it pays to use your intuition about the behavior of the physical system to decide whether the solution you obtain is reasonable!

---

In Example 3.6.2 the damping constant $b = t$ is tending to infinity with $t$, and as the value of the damping increases the vertical motion should slow down. Do you see any way to determine the limiting value of $x(t)$ as $t \to \infty$? This is not a simple problem.

Remember that we have two possible ways to graph a solution. In this case we have a formula, but it is in terms of infinite series. If you try to sum terms in these series for large values of $t$, you will find that you get very large positive and negative terms that essentially cancel before the series converges to its final value (which is small). The graph in Figure 3.21 was produced by a numerical solver, which will be described in the next section. As a check, the series were summed for $0 \le t \le 5$ and these values were compared to the values on the graph to show that they gave the same result.

If you enjoy working with series, you might convince yourself that

$$x_1(t) = 1 + \sum_{n=1}^{\infty} \frac{(-1)^n t^{2n}}{2 \cdot 4 \cdot 6 \cdots (2n)} \equiv \sum_{n=0}^{\infty} \frac{(-1)^n t^{2n}}{2^n n!} \equiv \sum_{n=0}^{\infty} \frac{(-t^2/2)^n}{n!}, \quad \text{(note that } 0! = 1\text{)},$$

and this is just the Maclaurin series for the function $x_1(t) = e^{-\frac{t^2}{2}}$. In the exercises you will be asked to show that this function is a solution of the differential equation $x'' + tx' + x = 0$.

On the other hand, the function $x_2(t)$ cannot be written in terms of elementary functions; it is one example of a class of "special" functions related to the **error function**, where the error function $erf(t)$ has the formula

$$erf(t) = \frac{2}{\sqrt{\pi}} \int_0^t e^{-s^2} ds.$$

As a point of interest, the function $erf(t)$ is another example of a function that has a sigmoidal shape, very much like the logistic growth function. It is sometimes used as a "response" function in modeling the activity of brain cells, as in the Student Project at the end of Chapter 2.

**Series Solutions at a Singular Point.** We now have a method for finding a series solution for a second-order linear ODE

$$x'' + p(t)x' + q(t)x = 0, \tag{3.57}$$

if $p(t)$ and $q(t)$ are both analytic functions. A point where one or both of these functions is not analytic is referred to as a **singular point** for the equation. By making a change of independent variable, we can assume that the singularity occurs at $t = 0$. In this case it may still be possible to obtain a series solution if both $tp(t)$ and $t^2 q(t)$ are analytic functions at $t = 0$. If these conditions are satisfied, the point $t = 0$ is called a **regular singular point** and the method for finding a series solution is called the Method of Frobenius.

The given conditions on $p$ and $q$ can be written in the form

$$p(t) = \frac{1}{t} \sum_{n=0}^{\infty} p_n t^n, \quad q(t) = \frac{1}{t^2} \sum_{n=0}^{\infty} q_n t^n, \tag{3.58}$$

where $\sum_{n=0}^{\infty} p_n t^n$ and $\sum_{n=0}^{\infty} q_n t^n$ are both convergent power series in some neighborhood of $t = 0$.

Using the forms (3.58) for $p$ and $q$, the equation

$$t^2(x'' + p(t)x' + q(t)x) = t^2 x'' + t(p_0 + p_1 t + \cdots)x' + (q_0 + q_1 t + \cdots)x = 0$$

looks very much like the Cauchy-Euler ODE

$$t^2 x'' + p_0 t x' + q_0 x = 0 \tag{3.59}$$

near $t = 0$. The associated characteristic equation $Q(r)$ for (3.59) (see Table 3.3), also called the **indicial equation**, is

$$r^2 + (p_0 - 1)r + q_0 = 0, \tag{3.60}$$

## 3.6 Linear Second-order Equations with Nonconstant Coefficients

and Frobenius used this to show that at a regular singular point there always exists at least one series solution of (3.57) of the form

$$x(t) = t^{r_1} \sum_{n=0}^{\infty} a_n t^n$$

where $r_1$ is the larger root of (3.60). Note that this solution will not be defined at $t = 0$ if $r_1$ is negative or complex. His method also gives a formula for a second solution which may, or may not, be analytic at $t = 0$.

In Chapter 7, we are going to need a series solution for **Bessel's equation of order zero**:

$$tx''(t) + x'(t) + tx(t) = 0. \tag{3.61}$$

If this equation is divided by $t$ to put it into standard form, it can be seen that the coefficient $p(t)$ has a singularity at $t = 0$. To see if zero is a regular singular point, write

$$\frac{1}{t}(tx'' + x' + tx) = x'' + \frac{1}{t}(1 + 0t + 0t^2 + \cdots)x' + \frac{1}{t^2}(0 + 0t + 1t^2 + 0t^3 + \cdots)x = 0,$$

and note that $p_0 = 1$ and $q_0 = 0$. The indicial equation

$$r^2 + (1-1)r + 0 = r^2 = 0$$

has a double root $r = 0$, so there is one ordinary series solution $x(t) = t^0 \sum_{n=0}^{\infty} a_n t^n$.

If we let $x(t) = \sum_{n=0}^{\infty} a_n t^n$ and insert the series for $x, x'$, and $x''$ into (3.61), then

$$t \sum_{n=2}^{\infty} n(n-1)a_n t^{n-2} + \sum_{n=1}^{\infty} na_n t^{n-1} + t \sum_{n=0}^{\infty} a_n t^n \equiv 0,$$

and multiplying the series in the first and third terms by $t$,

$$\sum_{n=2}^{\infty} n(n-1)a_n t^{n-1} + \sum_{n=1}^{\infty} na_n t^{n-1} + \sum_{n=0}^{\infty} a_n t^{n+1} \equiv 0.$$

If the change of index $n + 1 = m - 1$ is made in the third sum, and $m = n$ in the first two, then

$$\sum_{m=2}^{\infty} m(m-1)a_m t^{m-1} + \sum_{m=1}^{\infty} ma_m t^{m-1} + \sum_{m=2}^{\infty} a_{m-2} t^{m-1} \equiv 0.$$

The coefficient $a_0 \equiv x(0)$ can be seen to be arbitrary, and the only term in the series for index $m = 1$ implies that $a_1 \equiv x'(0) = 0$.

For $m \geq 2$,

$$m(m-1)a_m + ma_m + a_{m-2} = 0 \longrightarrow m^2 a_m = -a_{m-2}.$$

Since $a_1 = 0$, all coefficients $a_m$ with $m$ odd must be zero. For $m$ even,

$$a_m = -\frac{a_{m-2}}{m^2},$$

and therefore

$$a_2 = -\frac{a_0}{2^2}, \quad a_4 = -\frac{a_2}{4^2} = -\frac{1}{4^2}\left(-\frac{a_0}{2^2}\right) = \frac{a_0}{2^2 4^2}, \quad \ldots$$

The series for $x(t)$ can now be written as

$$x(t) = a_0 \left(1 - \frac{t^2}{2^2} + \frac{t^4}{2^2 4^2} - \frac{t^6}{2^2 4^2 6^2} \cdots \right).$$

Letting the arbitrary constant $a_0$ be one, the resulting function $x(t)$ is called $J_0(t)$, the **Bessel function of order zero of the first kind**, and it has the form

$$J_0(t) = \sum_{n=0}^{\infty} \frac{(-1)^n t^{2n}}{2^{2n}(n!)^2}.$$

Check the formula for the $n$th term in this series carefully, remembering that $0! = 1$ by definition.

**Exercises 3.6.** 1. Match each Cauchy-Euler equation below to the graph of its solution (in Figures A–C). You should be able to do this just by determining the roots of the characteristic polynomial and looking at the behavior of the solution, either as $t \to \infty$ or as $t \to 0$. The initial conditions in each case are $x(1) = 1$, $x'(1) = 0$.

(i) $t^2 x'' + 5tx' + 3x = 0$

(ii) $t^2 x'' + tx' + 9x = 0$

(iii) $t^2 x'' - 3tx' + 3x = 0$

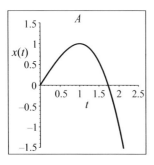
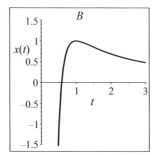
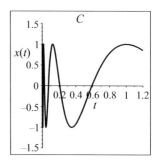

For the following Cauchy-Euler equations, find two solutions of the homogeneous equation and then use variation of parameters to find $x_p$. Before solving for $x_p$ you need to divide the equation by $t^2$ to have the correct forcing function $f(t)$.

2. $t^2 x'' - 2tx' + 2x = 3t$

3. $t^2 x'' - 2x = 2t$

4. $t^2 x'' + \frac{1}{4}x = t^{\frac{5}{2}}$

Solve the initial-value problems 5–7 and sketch your solution on $0 < t \leq 3$.

5. $x'' - \frac{2}{t}x' + \frac{2}{t^2}x = \frac{3}{t}$, $x(1) = 1$, $x'(1) = 0$

6. $x'' - \frac{2}{t^2}x = \frac{2}{t}$, $x(1) = 0$, $x'(1) = 1$

7. $x'' + \frac{1}{4t^2}x = \sqrt{t}$, $x(1) = 0$, $x'(1) = 0$

What is the limit of $x(t)$ as $t$ approaches zero?

8. Use a computer algebra system to solve the IVP

$$t^2 x''(t) + 3tx'(t) + 10x(t) = 1 \quad x(1) = 1, \quad x'(1) = 0.$$

Find $\lim_{t \to \infty} x(t)$ and also describe what happens to the solution as $t \to 0$.

*The following problems involve series solutions.*

9. Show that $x(t) = e^{-\frac{t^2}{2}}$ is a solution of the initial-value problem
$$x'' + tx' + x = 0, \quad x(0) = 1, \quad x'(0) = 0.$$

10. Assume a series solution $x(t) = \sum_{n=0}^{\infty} a_n t^n$ for the equation
$$x'' + x = 0.$$

    Show that the recursion formula for the coefficients is
    $$a_{m+2} = \frac{-a_m}{(m+1)(m+2)}, \quad m = 0, 1, 2, \ldots$$

    and that this leads to the general solution
    $$x(t) = a_0 \underbrace{\left(1 - \frac{t^2}{2!} + \frac{t^4}{4!} + \cdots\right)}_{\text{series for cos}} + a_1 \underbrace{\left(t - \frac{t^3}{3!} + \frac{t^5}{5!} + \cdots\right)}_{\text{series for sin}} \equiv a_0 \cos(t) + a_1 \sin(t).$$

11. The Bessel equation of order N is
$$t^2 x'' + tx' + (t^2 - N^2)x = 0. \tag{3.62}$$

    (i) Show that for any real number N, the indicial equation of (3.62) is $r^2 - N^2 = 0$. Hint: Divide (3.62) by $t^2$ to find the coefficient functions $p(t)$ and $q(t)$, and then use
    $$p_0 = \lim_{t \to 0} t p(t) \quad \text{and} \quad q_0 = \lim_{t \to 0} t^2 q(t)$$
    to find $p_0$ and $q_0$.

    (ii) For $N = 1$, the solution of (3.62) satisfying $x(0) = 1$ is called $J_1(t)$, the **Bessel function of order one of the first kind**. Find the coefficients $a_n$ in the series $J_1(t) = t^1 \sum_{n=0}^{\infty} a_n t^n$.

    (iii) Using the first 11 terms in the series, plot $J_1(t) \approx \sum_{n=0}^{10} a_n t^{n+1}$ on $0 \le t \le 10$.

    (iv) Find a computer algebra system that has Bessel functions available, and plot $J_1(t)$ on $0 \le t \le 10$. In Maple, this can be done by executing the instruction "plot(BesselJ(1,t),t=0..10)".

## 3.7 Autonomous Second-order Differential Equations

We will not spend a lot of time on autonomous second-order equations because in the next chapter we are going to show that any nth order autonomous equation can be written as a system of first-order equations, and Chapter 5 will go a lot further with graphical methods for analyzing systems of equations.

This section will be used primarily to introduce you to the concept of a phase plane, and in order to do that we first need to show that the numerical methods used to solve first-order differential equations can easily be extended to higher-order equations and systems of equations.

### 3.7.1 Numerical Methods.

You have seen that graphs are helpful in understanding the behavior of solutions of differential equations, and there are two ways to produce these graphs. One way is to find a formula for the solution of the equation in terms of what are commonly referred to as the "elementary functions". If such a formula can be found, we say we have found an **analytic solution** of the equation. In this case, a graph of the function $x(t)$ can be plotted using any plot routine available on your computer. This can always be done, for example, for a homogeneous constant-coefficient linear differential equation of any order.

If the differential equation cannot be solved analytically, it is usually still possible to obtain a very accurate graph of the solution. In Section 2.6 we showed how to find **numerical solutions** of first-order differential equations, and these methods are easily extended to second- and higher-order equations.

Consider first an **autonomous second-order equation**

$$x'' = F(x, x'). \tag{3.63}$$

For this equation the following existence and uniqueness theorem holds.

**Theorem 3.4. (Existence and Uniqueness, second-order Nonlinear)**: *Given the initial-value problem*

$$x'' = F(x, x'), \quad x(t_0) = x_0, \quad x'(t_0) = y_0,$$

*if $\frac{\partial F}{\partial x}$ and $\frac{\partial F}{\partial x'}$ are both continuous in a region $R = \{a < x < b, c < x' < d\}$ in the $(x, x')$-plane containing the point $(x_0, y_0)$, then there exists a unique solution of the differential equation through the point $(x_0, y_0)$. The solution is continuous in an interval $(t_0 - \varepsilon, t_0 + \varepsilon)$ for some $\varepsilon > 0$.*

This is another case where the possible nonlinearity of the equation makes it impossible to say how big the $t$-interval will be, so in looking for a numerical approximation to the solution it is necessary to be on the alert for vertical asymptotes. This problem was already encountered in Section 2.6, in the case of numerical approximations to nonlinear first-order equations.

The simplest numerical method, described in Section 2.6 for first-order equations, was Euler's Method. It consisted of partitioning the time interval $[t_0, t_{\max}]$ into $N$ subintervals of size $\Delta t = \frac{t_{\max} - t_0}{N}$, and in each subinterval $(t_j, t_{j+1})$ of length $\Delta t$ approximating $x(t_j + \Delta t)$ by $x(t_j) + x'(t_j)\Delta t$. This is just the tangent line approximation, and we found that we ended up at $t = t_{\max}$ with an error in $x(t_{\max})$ which was on the order of $\Delta t$.

To extend this method to the autonomous second-order equation we write (3.63) in terms of two first-order equations by defining a new variable $y \equiv x'$. Then our second-order equation (3.63) is equivalent to the pair of first-order equations

$$x' = y, \quad y' = F(x, y),$$

and Euler's Method is done exactly as before, except that at each step both $x$ and $y = x'$ must be incremented by a term equal to $\Delta t$ times the slope at the left-hand end of the interval.

## 3.7 Autonomous Second-order Differential Equations

> **Algorithm for Euler's Method**
> Given $x' = y$, $y' = F(x, y)$, $x(t_0) = x_0$, $y(t_0) = x'(t_0) = y_0$, to find approximate values of $x(t), y(t)$ on the interval $[t_0, t_{max}]$:
> (1) Choose a small step size $\Delta t = \frac{t_{max} - t_0}{N}$, with $N$ a positive integer.
> (2) For $j = 0, 1, \ldots, N-1$ compute
> $$x_{j+1} = x_j + y_j \Delta t$$
> $$y_{j+1} = y_j + F(x_j, y_j)\Delta t$$
> $$t_{j+1} = t_j + \Delta t.$$
> (3) Plot the points $(t_j, x_j)$, $j = 0, 1, \ldots, N$. If straight lines are drawn between consecutive points, the result is a piecewise linear approximation to the solution $x(t)$ on the interval $[t_0, t_{max}]$.

We could also plot a piecewise linear approximation to $x'(t)$ by using the values $(t_j, y_j), j = 0, 1, \ldots, N$.

Since both functions $x$ and $x'$ are approximated at each step by the first two terms in their Taylor series, their values at $t = t_{max}$ will have an error $\mathcal{O}(\Delta t)$.

Remember that Euler's Method is not practical if one is trying to solve differential equations and obtain 5 or 6 digit accuracy. Most computer algebra systems use very sophisticated methods that are explained in courses in numerical analysis. However, if you need to write your own differential equation solver for some problem, probably the simplest method that gives you reasonable accuracy is a fourth-order Runge-Kutta method. The algorithm given in Section 2.6 can be easily extended to an algorithm for a second-order equation.

Remember that at each step you are adding an increment to $x(t)$ and an increment to $y(t)$. These increments should be good estimates of the slope of $x$ times $\Delta t$, and the slope of $y$ times $\Delta t$, respectively. At any value of $t$, the slope of $x$ is given by the value of $y \equiv x'$ at $t$, and the slope of $y$ is given by $y' = x'' = F(x, y)$. The Runge-Kutta method uses an average of four estimates of the slope for each variable, between $t$ and $t + \Delta t$.

It was pointed out in Section 2.6 that fourth-order Runge-Kutta is equivalent to using the first five terms in the Taylor series, and the errors in $x(t_{max})$ and $x'(t_{max}) \equiv y(t_{max})$ are both on the order of $(\Delta t)^4$; therefore, if the step size $\Delta t$ is halved, the error in $x$ and $x'$ at $t_{max}$ should be cut by about $\frac{1}{16} = \left(\frac{1}{2}\right)^4$. This means that you can do two calculations, one with $\Delta t$ and a second one with $\frac{1}{2}\Delta t$, and compare the end result to see where they differ. This will give you a reasonable way to check on the number of correct significant digits in your result.

### 3.7.2 Autonomous Equations and the Phase Plane.
For an autonomous second-order differential equation

$$x'' = F(x, x') \tag{3.64}$$

it is possible to extend the concepts of slope field and phase line defined in Sections 2.2 and 2.7 for first-order differential equations.

The numerical methods for solving (3.64), described above, produce lists of points of the form $(t_k, x(t_k), x'(t_k))$ at each value $t_k$ on a partition of the $t$-interval $[t_0, t_{max}]$. This information can be plotted in a 2-dimensional graph in three different ways: a graph of $x(t)$ versus $t$, a graph of $x'(t)$ versus $t$, or a parametric graph of $x'(t)$ versus $x(t)$. The first two graphs are called **time plots**, and we are quite familiar with these; however, the third type of graph is the one we are interested in here.

As an example, consider the underdamped spring-mass equation

$$2x'' + 2x' + 5x = 0, \tag{3.65}$$

which can also be written in the form

$$x'' = F(x, x') = -\frac{5}{2}x - x'. \tag{3.66}$$

Since there is no forcing function, the equation can be seen to be autonomous. For this simple equation the characteristic polynomial method gives a general solution of the form

$$x(t) = e^{-t/2}\left(C_1 \sin\left(\frac{3}{2}t\right) + C_2 \cos\left(\frac{3}{2}t\right)\right). \tag{3.67}$$

Arbitrarily choosing the initial conditions to be $x(0) = 3$ and $x'(0) = 0$, the constants $C_1$ and $C_2$ can be found; in this case $C_1 = 1$, $C_2 = 3$. This gives the following analytic formulas for $x$ and $x'$:

$$x(t) = e^{-t/2}\left(\sin\left(\frac{3}{2}t\right) + 3\cos\left(\frac{3}{2}t\right)\right), \quad x'(t) = -5e^{-t/2}\sin\left(\frac{3}{2}t\right).$$

The list $(t_k, x(t_k), x'(t_k))$, $k = 0, 1, \ldots, N$ produced by a numerical solution should be a piecewise linear approximation to the parametric space curve

$$\mathcal{C}(t) = (t, x(t), x'(t))$$
$$= \left(t, e^{-t/2}\left(\sin\left(\frac{3}{2}t\right) + 3\cos\left(\frac{3}{2}t\right)\right), -5e^{-t/2}\sin\left(\frac{3}{2}t\right)\right), \quad 0 \le t \le t_{\max}.$$

This 3-dimensional curve can be plotted in Maple, and the result is shown in Figure 3.22.

The three types of graphs referred to above can be obtained by projecting the space curve $\mathcal{C}$ on three of the 2-dimensional faces of the cube in Figure 3.22; that is, in each case the cube is turned so that exactly one of the directions is hidden. If this is done, the three projections are as shown in Figure 3.23.

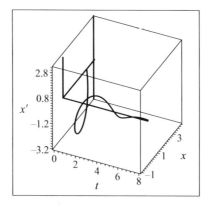

Figure 3.22. Space curve $\mathcal{C}(t) = (t, x(t), x'(t))$

## 3.7 Autonomous Second-order Differential Equations

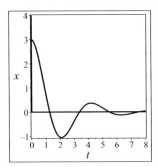
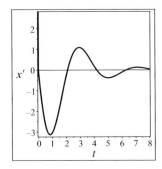
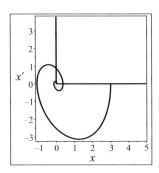

**Figure 3.23.** Three views of $\mathcal{C}(t)$: $(t, x)$-plane, $(t, x')$-plane, and $(x, x')$-plane

The $(t, x)$ and $(t, x')$ graphs are exactly what we expect for $x$ and its derivative, where $x$ is the solution of an underdamped mass spring equation. But why should we be interested in the graph of $x'$ versus $x$?

It turns out that for an autonomous second-order equation this is the only graph in which solution curves are guaranteed not to intersect; that is, if the slope function $F(x, x')$ satisfies the hypotheses of the Existence and Uniqueness Theorem 3.4 in the entire $(x, x')$-plane, then given any initial conditions $(x(0), x'(0)) = (x_0, x'_0)$ there will be a unique solution through the point $(x_0, x'_0)$, so solution curves cannot cross each other (this would violate the uniqueness part of Theorem 3.4).

**Definition 3.4.** *For an autonomous second-order differential equation $x'' = F(x, x')$, the $(x, x')$-plane is called the **phase plane**.*

It is in the phase plane that the behavior of all the solutions of the differential equation can be easily seen; that is, the phase plane is a picture of the general solution of the equation just as the slope field was for a first-order equation $x' = f(t, x)$.

There is a second important reason for using the phase plane. Solutions in this plane are plotted as 2-dimensional parametric curves $(x(t), x'(t))$, $t_0 \leq t \leq t_{\max}$, and from calculus we know how to find tangent vectors to these parametric curves at any point in the plane. Remember that the tangent vector to the parametric curve $(x(t), y(t))$, $a \leq t \leq b$, at a particular value of $t$ has the formula

$$\vec{T}(t) = x'(t)\vec{\imath} + y'(t)\vec{\jmath}.$$

Without even solving the equation $x'' = F(x, x')$, given any point $(\tilde{x}, \tilde{x}')$ in the phase plane, we can compute the tangent vector at that point as

$$\vec{T}(\tilde{x}, \tilde{x}') = \tilde{x}'\vec{\imath} + (\tilde{x}')'\vec{\jmath} = \tilde{x}'\vec{\imath} + F(\tilde{x}, \tilde{x}')\vec{\jmath}.$$

This makes it possible to sketch a vector field in the phase plane, showing exactly how solution curves move through the plane; and it can be done without solving the equation.

Almost any computer program for solving differential equations will produce a phase plane complete with a vector field for an equation of the form $x'' = F(x, x')$. You may first need to write the second-order equation as a pair of first-order differential equations in the form

$$x' = y, y' = F(x, y).$$

**Definition 3.5.** *For an autonomous second-order differential equation, a set of tangent vectors drawn at a grid of points in the phase plane is called a* **vector field** *for the differential equation.*

**Definition 3.6.** *A phase plane with a vector field and a set of solution curves drawn on it, showing how solutions look in various regions of the plane, is called a* **phase portrait**.

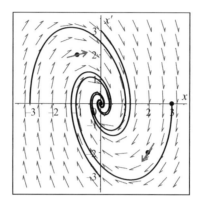

Figure 3.24. Phase portrait for $2x'' + 2x' + 5x = 0$

Figure 3.24 shows a phase portrait for the underdamped mass-spring equation (3.66), with slope function $F(x, x') = -\frac{5}{2}x - x'$. As a check on the direction of the arrows in the vector field, we will compute the tangent vectors $\vec{T}(\tilde{x}, \tilde{x}') = \tilde{x}'\vec{i} + F(\tilde{x}, \tilde{x}')\vec{j}$ at two points $(\tilde{x}, \tilde{x}') = (2, -2)$ and $(\tilde{x}, \tilde{x}') = (-1, 2)$:

$$\vec{T}(2, -2) = -2\vec{i} + F(2, -2)\vec{j} = -2\vec{i} + \left(-\frac{5}{2}(2) + 2\right)\vec{j} = -2\vec{i} - 3\vec{j}$$

and

$$\vec{T}(-1, 2) = 2\vec{i} + F(-1, 2)\vec{j} = 2\vec{i} + \left(\frac{5}{2} - 2\right)\vec{j} = 2\vec{i} + \frac{1}{2}\vec{j}.$$

These two vectors are sketched on the phase portrait to show that they are in the direction given by the computer-generated vector field at those points.

In the case of a second-order equation, since the $\vec{i}$ component of the tangent vector is $x'_k$, the vectors in the upper half of the $(x, x')$-plane all point to the right, and those in the lower half point to the left (this results from the fact that $x$ is increasing when $x'$ is positive and decreasing when it is negative). You need to be warned that this may no longer be true in Chapter 5, when we look at phase planes for an arbitrary system of two first-order equations $x' = f(x, y)$ and $y' = g(x, y)$ with $y$ not necessarily being the derivative of $x$.

To see the connection between a phase plane for an autonomous second-order equation and the phase line for an autonomous first-order differential equation, note that for the solutions plotted in the phase plane we have again lost the dependence on time. Just as for the phase line, we cannot tell when maxima and minima of $x$ or $x'$ occur, although it is easy to see how large each of these can get on any given solution curve. For example, in Figure 3.24, it can be seen that if one moves along the solution curve (also referred to as a **trajectory**) with initial conditions $(x, x') = (3, 0)$,

## 3.7 Autonomous Second-order Differential Equations

the value of $x$ first decreases to a value just below $-1$ and then begins to increase to approximately 0.5 before spiralling in toward the origin. Simultaneously, $x'$ goes from zero to a minimum of approximately $-3$, then back up to approximately 1.3 before spiralling in toward the origin. The spirals in the phase plane are due to the damped oscillations of $x(t)$ and $x'(t)$ and the fact that both functions approach zero as $t \to \infty$. Much more will be said about this in Chapter 5.

To see how a phase portrait can be used to visualize the behavior of solutions, consider the following physical example.

**Example 3.7.1.** *A simple pendulum consists of a bob of mass m suspended by a rigid weightless wire of length L.*

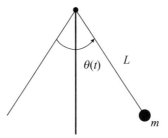

*The pendulum is free to swing back and forth in the plane. Using vector calculus and Newton's second law the following equation can be derived for the angle $\theta(t)$ between the wire and the vertical axis at time t:*

$$mL\theta''(t) + cL\theta'(t) + mg\sin(\theta(t)) = 0,$$

*where c is the damping coefficient and g is the gravitational constant.*

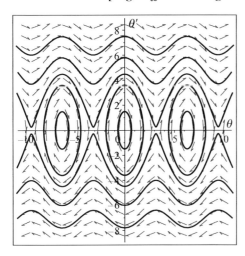
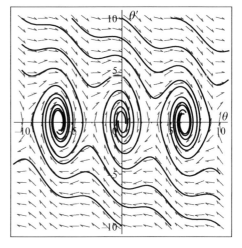

Figure 3.25. Pendulum equation $\theta'' + c\theta' + 5\sin(\theta) = 0$, with $c = 0$ on the left (undamped), and $c = 0.5$ on the right

*The equation is second-order, autonomous, but nonlinear due to the term $mg\sin(\theta(t))$. Phase portraits for the pendulum equation*

$$\theta'' + c\theta' + 5\sin(\theta) = 0$$

are shown in Figure 3.25. In the left-hand graph $c = 0$, so there is no damping in the system, and it is clear that if the initial velocity $\theta'(0)$ is not too large, the bob oscillates back and forth about an equilibrium point at $\theta = 2n\pi, \theta' = 0$. Note that "equilibrium point" means a point in the phase plane where both the velocity $\theta'$ and acceleration $\theta''$ are equal to zero. If the initial velocity is large enough, the bob can go all the way around and, since there is no damping, it will repeat this forever.

In the right-hand portrait, you can see the effect of damping. If the initial velocity is not too large, the bob cycles back and forth around an equilibrium nearest to its starting position and the motion dies out over time. If the initial velocity is a bit larger, the bob goes all the way around and heads for the next equilibrium, with an angle $2\pi$ radians larger or smaller, depending on whether the initial velocity is positive or negative. With an even larger initial velocity it makes two cycles and then oscillates around an angle $4\pi$ radians larger or smaller, and so on. ∎

**Exercises 3.7.** *Write equations 1–4 in the form $x'' = F(x, x')$, and compute the tangent vectors*

$$\vec{T}(x, x') = x'\vec{i} + F(x, x')\vec{j}$$

*at the three given points $(x, x')$. Sketch the tangent vectors in the given phase plane. Note: to draw the tangent vector $\vec{T} = A\vec{i} + B\vec{j}$ at the point $(x, x')$, draw a short line from the point $(x, x')$ toward the point $(x + A, x' + B)$.*

1. $x'' + x'(1 - x^2) + x = 0$, at the points $(1.5, 2), (-1, 1)$, and $(-2, -2)$. (Use the left graph in Figure 3.26.)

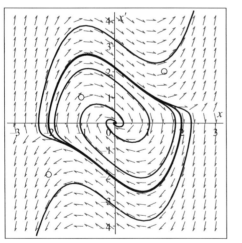 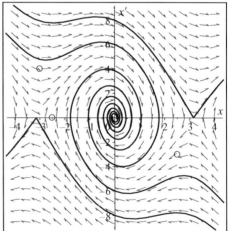

Figure 3.26. Graph for problems 1 and 2.

2. $x'' + 0.8x' + 10\sin(x) = 0$, at the points $(-2.5, 0), (2.5, -3)$, and $(-3, 4)$. (Use the right graph in Figure 3.26.)

3. $x'' + 3x' + (x + x^3) = 0$, at the points $(-2, 3), (0, 1)$, and $(2, -3)$. (Use the left graph in Figure 3.27.)

4. $x'' + 3x' + (x - x^3) = 0$, at the points $(2.3, 1), (-1.5, -.8)$, and $(2, 2.5)$. (Use the right graph in Figure 3.27.)

## 3.7 Autonomous Second-order Differential Equations

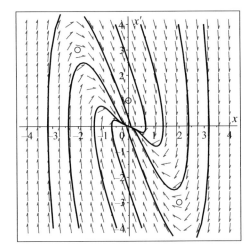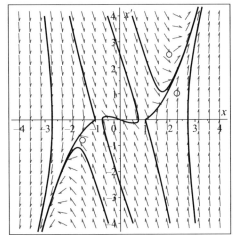

Figure 3.27. Graphs for problems 3 and 4

5. The equation in Exercise 3 models a spring-mass system with a "hard" spring. This term is used to indicate that the restoring force of the spring is proportional to $x + x^3$, rather than $x$. Using the phase portrait, describe as precisely as you can what happens to the mass if it starts at $x = 0$ with velocity $x' = 2.5$. What happens if it starts at $x = 0$ with velocity $x' = 4$?

6. The equation in Exercise 4 is a spring-mass model with a "soft" spring. A "soft" spring has a restoring force proportional to $x - x^3$, rather than $x$. Using the phase portrait, describe as precisely as you can what happens to the mass if it starts at $x = 0$ with velocity $x' = 2.5$. What happens if it starts at $x = 0$ with velocity $x' = 4$?

In problems 7–10, use your own computer software to draw a phase portrait for the equation. Plot enough solution curves to be able to see what is happening in the entire plane.

7. $x'' + 5x' + 6x = 0$

8. $x'' + 4x = 0$

9. $x'' + x' + 5x = 0$

10. $x'' + 2x' + x = 0$

# 4

# Linear Systems of First-order Differential Equations

Real-world problems rarely come in the form of a single differential equation involving a single unknown variable. Biologists modeling population growth need to account for the interaction of many species, chemists can encounter mixing problems with several compartments, and engineers solve problems with large numbers of interacting parts.

The first thing we will do in this chapter is show that almost any model involving ordinary differential equations, no matter how complicated, can be expressed as a system of first-order differential equations. It may be necessary to define new variables to represent derivatives of other variables, but in general it can be done.

There are at least two good reasons for writing the model as a system of first order equations. One is a practical reason: many of the computer programs for solving differential equations require the equations to be in this form. We have already seen this in the case of a second-order equation, in Section 3.7.

A second, even more important reason, is that if it can be shown that the first-order system is linear, with constant coefficients, then it can be solved analytically, using powerful results from linear algebra.

In this chapter we will look carefully at linear systems of first-order equations. Section 4.2 on matrix algebra will give you all of the linear algebra needed for this chapter and the next, and in Section 4.4 you will be given exact formulas for the solution of a constant coefficient linear system.

It turns out there is another important reason for understanding linear systems, which will be made clear in Chapter 5. Knowing what solutions of a linear system look like in a phase plane can greatly add to our understanding of the geometrical behavior of solutions of a system of nonlinear equations.

## 4.1 Introduction to Systems

As in earlier chapters, we will not treat the most general case, but will assume that a **system of first-order differential equations** in $n$ unknown functions $x_1(t)$, $x_2(t)$, ..., $x_n(t)$ can be written explicitly in terms of the derivatives of the $n$ dependent variables; that is, it can be put into the form

$$\begin{aligned} dx_1/dt &= f_1(t, x_1, x_2, \ldots, x_n) \\ dx_2/dt &= f_2(t, x_1, x_2, \ldots, x_n) \\ &\vdots \\ dx_n/dt &= f_n(t, x_1, x_2, \ldots, x_n), \end{aligned} \qquad (4.1)$$

where $f_1, f_2, \ldots, f_n$ are arbitrary functions of $t, x_1, x_2, \ldots, x_n$.

**4.1.1 Writing Differential Equations as a First-order System.** Any single $n$th-order differential equation $x^{(n)} = f(t, x, x', \ldots, x^{(n-1)})$ can be expressed in the form of a system (4.1) by defining new variables for $x(t)$ and its derivatives up to order $n-1$. The technique for doing this is demonstrated in Example 4.1.1.

**Example 4.1.1.** *Write the third-order equation $x''' + 4x' + 3x = t^2 + 1$ as a system of three first-order equations.*

*Solution.* Define new dependent variables $x_1 \equiv x$, $x_2 \equiv x'$ and $x_3 \equiv x''$. Then the system

$$\begin{aligned} x_1' &= x_2 \\ x_2' &= x_3 \\ x_3' &= -3x_1 - 4x_2 + t^2 + 1 \end{aligned}$$

is a first-order system that has exactly the same set of solutions as the single third-order equation; that is, $x_1$ will be the solution $x$, and $x_2$ and $x_3$ will be its first and second derivatives. ■

We needed to define three variables $x_1$, $x_2$ and $x_3$, since the equation was of order 3. By definition, the derivatives of $x_1$ and $x_2$ are, respectively, $x_2$ and $x_3$, and the derivative of $x_3$ is $\frac{d}{dt}(x'') \equiv x'''$. A formula for $x'''$ in terms of $t, x, x'$, and $x''$ can always be obtained from the original third-order equation. It should be clear how this can be extended to an equation of arbitrary order $n$.

Systems of simultaneous higher-order differential equations can also be written as a first-order system. The method for doing this is demonstrated in the next example.

**Example 4.1.2.** *Write a system of first-order differential equations that is equivalent to the simultaneous equations $x'' = y \sin(2t)$, $y'' = x' + y^2$.*

*Solution.* In this case, it is necessary to define four new variables: $x_1 \equiv x$, $x_2 \equiv x'$, $x_3 \equiv y$, and $x_4 \equiv y'$, since the equations in $x$ and $y$ are both of second order. Then the

## 4.1 Introduction to Systems

equivalent system can be written in the form:

$$x_1' = x_2 \equiv f_1(t, x_1, x_2, x_3, x_4)$$
$$x_2' = x_3 \sin(2t) \equiv f_2(t, x_1, x_2, x_3, x_4)$$
$$x_3' = x_4 \equiv f_3(t, x_1, x_2, x_3, x_4)$$
$$x_4' = x_2 + (x_3)^2 \equiv f_4(t, x_1, x_2, x_3, x_4)$$

Make sure that you see exactly how this system was constructed! ∎

We can sometimes perform preliminary operations on a set of equations to put them into a form that can be handled by our methods. The next example shows how this might be done.

**Example 4.1.3.** *Write a system of the form (4.1) that is equivalent to the pair of simultaneous differential equations*

$$x' + 2y' = 2x + t$$
$$3y' + x' = y + \sin(t).$$

*Solution.* These equations are both of first order in $x$ and $y$, but we need to rewrite them in the form $x' = f(x, y)$, $y' = g(x, y)$. In this case, we could use any method for solving the two equations for the two unknowns $x'$ and $y'$. For example, subtracting the first equation from the second gives $y' = -2x + y + \sin(t) - t$, and subtracting 2 times the second equation from 3 times the first gives $x' = 6x + 3t - 2y - 2\sin(t)$. Now we can define $x_1 = x$ and $x_2 = y$, and rewrite the system as

$$x_1' = 6x_1 - 2x_2 + 3t - 2\sin(t)$$
$$x_2' = -2x_1 + x_2 + \sin(t) - t.$$

This is clearly a system of the form (4.1). ∎

### 4.1.2 Linear Systems.

**Definition 4.1.** *If each of the functions $f_i(t, x_1, x_2, \ldots, x_n)$, $i = 1, \ldots, n$, in (4.1) is linear in the dependent variables $x_i$, that is,*

$$f_i(t, x_1, \ldots, x_n) = a_{i1}x_1 + a_{i2}x_2 + \cdots + a_{in}x_n + b_i,$$

*where the $a_{ij}$ and $b_i$ are arbitrary functions of t, then (4.1) is called a **linear system**; that is, an n-**dimensional linear system of first-order differential equations** has the form*

$$x_1' = a_{11}x_1 + a_{12}x_2 + \cdots + a_{1n}x_n + b_1$$
$$x_2' = a_{21}x_1 + a_{22}x_2 + \cdots + a_{2n}x_n + b_2 \qquad (4.2)$$
$$\vdots$$
$$x_n' = a_{n1}x_1 + a_{n2}x_2 + \cdots + a_{nn}x_n + b_n$$

*If all $b_i(t) \equiv 0$, the system is called a **homogeneous linear system**. The number n of equations is referred to as the **dimension** of the system.*

The system in Example 4.1.1 is linear, but not homogeneous. In Example 4.1.2 the system is nonlinear because of the term $(x_3)^2$ in the fourth equation. In Example 4.1.3 the system is linear, but not homogeneous.

The following example illustrates a standard problem from mechanics.

**Example 4.1.4. Double Spring-mass System.** *Two springs A and B, both with spring constant k, are connected to two objects C and D, each of mass M. The objects slide along a frictionless platform. The spring A is attached to the left-hand end of the platform. The diagram below shows the equilibrium position of all the elements in the system. The displacement of C from its equilibrium position at time t is $x_1(t)$ and the displacement of D is $x_2(t)$. Initially the object D is held fixed and C is moved to the left a positive distance ρ, and both objects are then let go. You are going to write an equation of the form "mass × acceleration = sum of forces" for each mass C and D.*

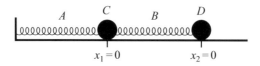

*Solution.* If the positive direction of $x_1$ and $x_2$ is to the right, then for the mass C, the sum of the forces on C is $f_A + f_B$, where $f_A$ and $f_B$ are the forces due to the springs A and B respectively. The elongation of spring A at time t is the value of $x_1(t)$, so the force of A on the mass C is $-kx_1(t)$ (if $x_1$ is positive, the spring is pulling mass C to the left). The elongation of the spring B at time t is $x_2(t) - x_1(t)$, and if it is positive the force on C is positive.

Putting this all together, we see that one equation will be

$$Mx_1''(t) = -kx_1(t) + k(x_2(t) - x_1(t)) = k(x_2(t) - 2x_1(t)).$$

Using a similar analysis for the force on mass D, check that we end up with the 2-dimensional system

$$\begin{aligned} Mx_1'' &= k(x_2 - 2x_1) \\ Mx_2'' &= -k(x_2 - x_1). \end{aligned} \quad (4.3)$$

Rewrite these equations as a system of first-order differential equations. What is the size of the system? Is it linear? Is it homogeneous? What would the initial condition be for each variable in the system?

Since both equations are second-order, it is necessary to use four variables $x_1, x_2, x_3 = x_1'$, and $x_4 = x_2'$. The equations can then be written in the form

$$\begin{aligned} x_1' &= x_3 \\ x_2' &= x_4 \\ x_3' &= \frac{k}{M}(x_2 - 2x_1) \\ x_4' &= -\frac{k}{M}(x_2 - x_1). \end{aligned} \quad (4.4)$$

It is easily seen that system (4.4) is a linear system of dimension 4, and it is homogeneous.

To obtain initial conditions for the four variables, the statement "C is moved ρ units to the left and let go" implies that $x_1(0) = -\rho$, $x_3(0) \equiv x_1'(0) = 0$. Since D is initially held fixed, $x_2(0) = x_4(0) = 0$. ∎

By the end of this chapter, we will have the necessary techniques for finding an analytic solution of the problem in Example 4.1.4.

## 4.1 Introduction to Systems

**Exercises 4.1.** *For equations 1–10, write an equivalent first-order system in the form given by (4.1). Determine the dimension of the system and state whether or not it is linear, and if linear, whether or not it is homogeneous.*

1. $x'' + 5x' + 2x = \sin(t)$
2. $x'' + 2tx' + x = 0$
3. $x''' + 4x'' + 3x = 2 + t$
4. $x^{(4)} = x(x')^2 - tx''$
5. $x' = y, y' = -x$
6. $x'' = x + y, y' = 5x$
7. $x' = xy, y''' = x + y'' + y^2$
8. $x'' = x + z, y' = xy, z' = y + x + 2z$
9. $2y' - x' = x + 3y, x' - y' = 2x - 5y$
   Hint: First solve the two equations for $x'$ and $y'$.
10. $x'' + y' + 2x - 3y = 0, y' + 2y = x''$

11. Two predators X and Y, living on a single prey Z, are interacting in a closed ecosystem. Letting $x(t), y(t)$, and $z(t)$ be the size (in hundreds of individuals) of the three populations at time $t$, the equations of growth are given as

$$x'(t) = -rx(t) + ax(t)z(t)$$
$$y'(t) = -sy(t) - by(t)x(t) + cy(t)z(t) \qquad (4.5)$$
$$z'(t) = qz(t)(1 - z(t)) - z(t)(kx(t) + my(t)),$$

where the parameters $r, s, q, a, b, c, k$, and $m$ are all assumed to be positive.

(a) Is this system linear? Explain.
(b) How does the predator X affect the growth of the prey population?
(c) If the prey dies out, what will happen to the predator X over the long run? Explain.
(d) If the predators both disappear, what will happen to the prey population over the long run? Explain.

12. You are given a mechanical system with three springs A, B, and C and two objects F and G each of mass M. Springs A and C have spring constant $k$, and spring B has spring constant $2k$. The equilibrium positions of all the elements in the system are shown in the diagram below. The spring C is attached to the right-hand end of the platform.

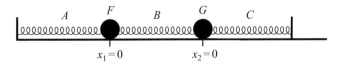

(a) *Write a system of first-order differential equations for the displacements $x_1(t)$ and $x_2(t)$ of the two masses F and G respectively.*
(b) *What is the dimension of the system?*
(c) *Is the system linear?*

## 4.2 Matrix Algebra

To proceed any further in solving linear systems, it is necessary to be able to write them in matrix form. If you have taken a course in linear algebra, this section will be a review. Otherwise, you should read this section very carefully. It contains all of the material needed for the rest of this chapter and also for Chapter 5.

**Definition 4.2.** *A **matrix A** is a rectangular array containing elements arranged in m rows and n columns. When working with differential equations it can be assumed that these elements (also called **scalars**) are either numbers or functions. The notation we will use for a matrix is*

$$\mathbf{A} = \begin{pmatrix} a_{11} & a_{12} & \cdots & a_{1n} \\ a_{21} & a_{22} & \cdots & a_{2n} \\ & & \ddots & \\ a_{m1} & a_{m2} & \cdots & a_{mn} \end{pmatrix}.$$

A matrix with $m$ rows and $n$ columns, for some positive integers $m$ and $n$, is said to be of **size** $m$ by $n$ (written as $m \times n$). In the above notation, $a_{ij}$ represents the element in the $i$th row and $j$th column of **A**. It will sometimes be convenient to use the notation $\mathbf{A} = (a_{ij})$ which shows the form of the general element in **A**.

An $m \times n$ matrix with $n = 1$ has a single column and is called a **column vector**; similarly, an $m \times n$ matrix with $m = 1$ is called a **row vector**. An $m \times n$ matrix with $m = n$ is called a **square matrix** of size $n$. The **zero matrix**, of any size, denoted by **0**, is a matrix with the elements in every row and every column equal to 0.

**Definition 4.3.** *Two matrices $\mathbf{A} = (a_{ij})$ and $\mathbf{B} = (b_{ij})$ are said to be **equal** if they are of the same size $m \times n$, and $a_{ij} = b_{ij}$ for $1 \leq i \leq m, 1 \leq j \leq n$.*

The following three basic algebraic operations are defined for matrices: addition, scalar multiplication, and matrix multiplication.

**Addition of Matrices:** If $\mathbf{A} = (a_{ij})$ and $\mathbf{B} = (b_{ij})$ are both the same size, say $m \times n$, then their sum $\mathbf{C} = \mathbf{A} + \mathbf{B}$ is defined, and $\mathbf{C} = (c_{ij})$ with $c_{ij} = a_{ij} + b_{ij}$ for $1 \leq i \leq m, 1 \leq j \leq n$.

**Example.**

$$\begin{pmatrix} 4 & -6 \\ 3 & 1 \end{pmatrix} + \begin{pmatrix} 2 & 2 \\ 0 & -3 \end{pmatrix} = \begin{pmatrix} 4+2 & -6+2 \\ 3+0 & 1-3 \end{pmatrix} = \begin{pmatrix} 6 & -4 \\ 3 & -2 \end{pmatrix}.$$

Addition of matrices is commutative and associative; that is, if **A**, **B**, and **C** are all of the same size, then $\mathbf{A} + \mathbf{B} = \mathbf{B} + \mathbf{A}$ and $\mathbf{A} + (\mathbf{B} + \mathbf{C}) = (\mathbf{A} + \mathbf{B}) + \mathbf{C}$. The zero matrix acts as an identity for matrix addition; that is, if **A** is any $m \times n$ matrix, then $\mathbf{A} + \mathbf{0} = \mathbf{0} + \mathbf{A} = \mathbf{A}$, where **0** is the zero matrix of size $m \times n$.

## 4.2 Matrix Algebra

**Scalar Multiplication:** For any matrix $\mathbf{A} = (a_{ij})$, the scalar multiple of $\mathbf{A}$ by a scalar $k$ (which can be either a number or a function) is the matrix $k\mathbf{A} = (ka_{ij})$.

**Examples.**

$$2\begin{pmatrix} 1 & 2 & 3 \\ 4 & 5 & 6 \end{pmatrix} = \begin{pmatrix} 2 & 4 & 6 \\ 8 & 10 & 12 \end{pmatrix}, \qquad e^{rt}\begin{pmatrix} 1 & 2t \\ t^2 & 0 \end{pmatrix} = \begin{pmatrix} e^{rt} & 2te^{rt} \\ t^2 e^{rt} & 0 \end{pmatrix}.$$

**Multiplication of Matrices:** If $\mathbf{A} = (a_{ij})$ is an $m \times n$ matrix and $\mathbf{B} = (b_{ij})$ is an $n \times p$ matrix, then the product $\mathbf{C} = \mathbf{AB}$ is defined. The $(i, j)$ element of $\mathbf{C}$ is

$$c_{ij} = \sum_{k=1}^{n} a_{ik} b_{kj},$$

and $\mathbf{C}$ has size $m \times p$. Note that the number of columns of $\mathbf{A}$ has to be the same as the number of rows of $\mathbf{B}$.

**Example.**

$$\begin{pmatrix} 4 & -6 \\ 3 & 1 \end{pmatrix}\begin{pmatrix} 2 & 2 & -5 \\ 0 & -3 & 1 \end{pmatrix} = \begin{pmatrix} 8+0 & 8+18 & -20-6 \\ 6+0 & 6+(-3) & -15+1 \end{pmatrix} = \begin{pmatrix} 8 & 26 & -26 \\ 6 & 3 & -14 \end{pmatrix}.$$

It is easier to remember how to multiply two matrices if you first recall (from physics or multivariable calculus) the definition of the dot product of two vectors. Let $\vec{a} = (a_1, a_2, \ldots, a_n)$ and $\vec{b} = (b_1, b_2, \ldots, b_n)$ be arbitrary vectors, both of length $n$. Then their dot product is defined by

$$\vec{a} \cdot \vec{b} = a_1 b_1 + a_2 b_2 + \cdots + a_n b_n.$$

The $(i, j)$ element in the matrix product $\mathbf{AB}$ is the dot product of the $i$th row of $\mathbf{A}$ and the $j$th column of $\mathbf{B}$.

**Example.** *Let*

$$\mathbf{A} = \begin{pmatrix} 1 & 2 & 3 \\ 0 & 1 & -2 \\ -2 & 4 & 1 \end{pmatrix} \equiv \begin{pmatrix} \vec{R}_1 \\ \vec{R}_2 \\ \vec{R}_3 \end{pmatrix}$$

*and*

$$\mathbf{B} = \begin{pmatrix} 0 & -1 \\ 2 & 1 \\ 3 & 4 \end{pmatrix} \equiv \begin{pmatrix} \vec{C}_1 & \vec{C}_2 \end{pmatrix},$$

*where $\vec{R}_1$, $\vec{R}_2$, and $\vec{R}_3$ are the three rows of $\mathbf{A}$ and where $\vec{C}_1$ and $\vec{C}_2$ are the two columns of $\mathbf{B}$. The product $\mathbf{AB}$ of the $3 \times 3$ matrix $\mathbf{A}$ times the $3 \times 2$ matrix $\mathbf{B}$ is defined and will* be of size 3 × 2. Computing the product consists of computing six dot products, as shown below.

$$\mathbf{AB} = \begin{pmatrix} \vec{R}_1 \cdot \vec{C}_1 & \vec{R}_1 \cdot \vec{C}_2 \\ \vec{R}_2 \cdot \vec{C}_1 & \vec{R}_2 \cdot \vec{C}_2 \\ \vec{R}_3 \cdot \vec{C}_1 & \vec{R}_3 \cdot \vec{C}_2 \end{pmatrix}$$

$$= \begin{pmatrix} 0 + 2 \cdot 2 + 3 \cdot 3 & 1 \cdot -1 + 2 \cdot 1 + 3 \cdot 4 \\ 0 + 1 \cdot 2 + (-2) \cdot 3 & 0 + 1 \cdot 1 + (-2) \cdot 4 \\ 0 + 4 \cdot 2 + 1 \cdot 3 & -2 \cdot -1 + 4 \cdot 1 + 1 \cdot 4 \end{pmatrix}$$

$$= \begin{pmatrix} 13 & 13 \\ -4 & -7 \\ 11 & 10 \end{pmatrix}.$$

*The required condition for multiplication, that the number of columns in* **A** *is the same as the number of rows in* **B**, *guarantees that these dot products are defined.*

**Properties of addition, and scalar and matrix multiplication.** Whenever the dimensions are such that the following operations are defined, then

$$\mathbf{A}(\mathbf{B} + \mathbf{C}) = \mathbf{AB} + \mathbf{AC}$$

$$(\mathbf{A} + \mathbf{B})\mathbf{C} = \mathbf{AC} + \mathbf{BC}$$

$$k(\mathbf{AB}) = (k\mathbf{A})\mathbf{B} = \mathbf{A}(k\mathbf{B})$$

$$\mathbf{ABC} = (\mathbf{AB})\mathbf{C} = \mathbf{A}(\mathbf{BC}).$$

However,

$$\mathbf{AB} \neq \mathbf{BA} \text{ in general.}$$

There is a simple relation between the three algebraic operations, which will be used in later sections. If $\mathbf{A} = (a_{ij})$ is any $m \times n$ matrix and $\mathbf{x} = (x_i)$ is an $n \times 1$ matrix (i.e., a column vector), then

$$\mathbf{Ax} = \begin{pmatrix} a_{11} & a_{12} & \cdots & a_{1n} \\ a_{21} & a_{22} & \cdots & a_{2n} \\ & & \ddots & \\ a_{m1} & a_{m2} & \cdots & a_{mn} \end{pmatrix} \begin{pmatrix} x_1 \\ x_2 \\ \vdots \\ x_n \end{pmatrix}$$

$$= \begin{pmatrix} a_{11}x_1 + a_{12}x_2 + \cdots + a_{1n}x_n \\ a_{21}x_1 + a_{22}x_2 + \cdots + a_{2n}x_n \\ \vdots \\ a_{m1}x_1 + a_{m2}x_2 + \cdots + a_{mn}x_n \end{pmatrix} \qquad (4.6)$$

$$\equiv x_1 \begin{pmatrix} a_{11} \\ a_{21} \\ \vdots \\ a_{m1} \end{pmatrix} + \cdots + x_n \begin{pmatrix} a_{1n} \\ a_{2n} \\ \vdots \\ a_{mn} \end{pmatrix};$$

that is, the matrix product $\mathbf{Ax}$ can be written as a linear combination of the columns of $\mathbf{A}$, scalar multiplied by the corresponding entries in the vector $\mathbf{x}$. Use the properties of matrix multiplication, addition, and equality to check this carefully!

The following definition will be needed when we apply matrix methods to solve differential equations.

## 4.2 Matrix Algebra

**Definition 4.4.** *If a matrix* **A** *has elements that are functions of t, say* $\mathbf{A} = (a_{ij}(t))$, *then the* **derivative** *of* **A** *is defined to be the matrix*

$$\frac{d}{dt}\mathbf{A}(t) \equiv \mathbf{A}'(t) = (a'_{ij}(t)).$$

Example: if $\mathbf{A}(t) = \begin{pmatrix} t^3 & e^{2t} \\ 1/t & \sin(t) \end{pmatrix}$, then $\mathbf{A}'(t) = \begin{pmatrix} 3t^2 & 2e^{2t} \\ -1/t^2 & \cos(t) \end{pmatrix}$.

**Linearly independent vectors.** Given a set of nonzero $n$-dimensional vectors $\mathbf{x}_1, \mathbf{x}_2, \ldots, \mathbf{x}_k$, we want to define what is meant by this being a linearly independent set. Basically, what this means is that no vector in the set can be written as a linear combination of the others.

**Definition 4.5.** *Given a set of k row or column vectors* $\mathbf{x}_1, \mathbf{x}_2, \ldots, \mathbf{x}_k$, *all of the same dimension, the set is said to be* **linearly independent** *if the only linear combination* $c_1\mathbf{x}_1 + c_2\mathbf{x}_2 + \cdots + c_k\mathbf{x}_k$ *which equals the* **0** *vector has coefficients* $c_1 = c_2 = \cdots = c_k = 0$. *If the set is not linearly independent it is called a* **linearly dependent** *set of vectors.*

**Example 4.2.1.** *Show that the set of vectors* $\{(1,0,0),(0,1,0),(0,0,1)\}$ *is linearly independent and that the set* $\{\mathbf{v}_1, \mathbf{v}_2, \mathbf{v}_3\} = \{(1,0,-2),(3,4,1),(5,4,-3)\}$ *is linearly dependent.*

*Solution.* If we set $c_1(1,0,0) + c_2(0,1,0) + c_3(0,0,1) = (0,0,0)$, the left side simplifies to a single $1 \times 3$ row vector $(c_1, c_2, c_3)$, and using the definition for the equality of matrices, it will be equal to the right side if, and only if, $c_1 = 0$, $c_2 = 0$, and $c_3 = 0$.

The second set is linearly dependent. The linear combination $2\mathbf{v}_1 + \mathbf{v}_2 - \mathbf{v}_3$ can be seen to be equal to the zero vector (check it). There are actually infinitely many linear combinations that add up to zero, since the equation $2\mathbf{v}_1 + \mathbf{v}_2 - \mathbf{v}_3 = \mathbf{0}$ can be multiplied by an arbitrary scalar. Also, any one of the three vectors $\mathbf{v}_1$, $\mathbf{v}_2$, or $\mathbf{v}_3$ can be written as a linear combination of the other two; for example $\mathbf{v}_3 = 2\mathbf{v}_1 + \mathbf{v}_2$. ∎

In the case of two nonzero $n$-dimensional vectors $\mathbf{u}$ and $\mathbf{v}$, $\mathbf{u}$ and $\mathbf{v}$ are linearly dependent if and only if they are scalar multiples of each other. A test for linear independence of $n$ vectors of dimension $n$ will be given after the determinant is defined.

**Additional definitions for square matrices.** In our work with systems of differential equations we will usually be dealing with square matrices of some dimension $n$. For $n \times n$ matrices some additional definitions are needed.

**Definition 4.6.** *The* **main diagonal** *of a square matrix*

$$\mathbf{A} = \begin{pmatrix} a_{11} & a_{12} & \cdots & a_{1n} \\ a_{21} & a_{22} & & a_{2n} \\ & & \ddots & \\ a_{n1} & a_{n2} & \cdots & a_{nn} \end{pmatrix}$$

*is the diagonal going from top left to lower right, containing the elements* $a_{11}, a_{22}, \ldots, a_{nn}$.

For $n \times n$ matrices there is an identity for matrix multiplication.

**Definition 4.7.** *The $n \times n$ matrix $\mathbf{I} = (\delta_{ij})$, with $\delta_{ij} = 1$ if $i = j$ and $\delta_{ij} = 0$ if $i \neq j$, is called the n-dimensional **identity matrix**. It satisfies the equations $\mathbf{IA} = \mathbf{AI} = \mathbf{A}$ for any $n \times n$ matrix $\mathbf{A}$.*

An identity matrix has ones along the main diagonal and zeros everywhere else. For example, the $3 \times 3$ identity matrix is

$$\mathbf{I} = \begin{pmatrix} 1 & 0 & 0 \\ 0 & 1 & 0 \\ 0 & 0 & 1 \end{pmatrix}.$$

**Example 4.2.2.** *With $n = 2$, show that $\mathbf{IA} = \mathbf{A}$ for an arbitrary $2 \times 2$ matrix $\mathbf{A}$.*

*Solution.*

$$\mathbf{IA} = \begin{pmatrix} 1 & 0 \\ 0 & 1 \end{pmatrix} \begin{pmatrix} a & b \\ c & d \end{pmatrix} = \begin{pmatrix} 1a + 0c & 1b + 0d \\ 0a + 1c & 0b + 1d \end{pmatrix} = \begin{pmatrix} a & b \\ c & d \end{pmatrix} = \mathbf{A}.$$

Check that multiplying $\mathbf{A}$ and $\mathbf{I}$ in the opposite order also produces the matrix $\mathbf{A}$. ∎

There are two important functions, called the trace and the determinant, that are defined only for square matrices.

**Definition 4.8.** *If $\mathbf{A} = (a_{ij})$ is any $n \times n$ matrix, the **trace of** $\mathbf{A}$, denoted by $\mathbf{tr}(\mathbf{A})$, is defined by*

$$\mathbf{tr}(\mathbf{A}) = a_{11} + a_{22} + \cdots + a_{nn};$$

*that is, the trace of a square matrix is the sum of the elements on its main diagonal.*

**Definition 4.9.** *The **determinant** of a square matrix $\mathbf{A}$ is denoted by $\mathbf{det}(\mathbf{A})$ and is defined for a $2 \times 2$ matrix by*

$$\mathbf{det}\begin{pmatrix} a & b \\ c & d \end{pmatrix} = ad - bc.$$

This simple way of calculating the determinant does not work for matrices of dimension greater than 2. For an $n \times n$ matrix with $n > 2$ the determinant can be found using the **method of cofactors**, which is contained in Appendix C. The definition of the determinant of a square matrix may seem unduly complicated; however, the determinant of $\mathbf{A}$ is simply the quantity which appears in the denominator when solving $n$ linear equations in $n$ unknowns, with coefficient matrix $\mathbf{A}$. Whether the determinant is nonzero or zero, respectively, determines whether or not the system of linear equations has a unique solution for any choice of the right-hand side.

Using the determinant, there is a simple test to see if a set of exactly $n$ vectors of dimension $n$ is linearly independent. If the $n \times n$ matrix containing the vectors as its columns (or rows) has determinant zero, the vectors are linearly dependent; if the determinant is not zero the set is linearly independent.

The following definition will be needed in the next section of this chapter.

**Definition 4.10.** *Given an $n \times n$ matrix $\mathbf{A}$, if there exists a nonzero vector $\vec{\mathbf{u}}$ that satisfies $\mathbf{A}\vec{\mathbf{u}} = \lambda \vec{\mathbf{u}}$ for some scalar $\lambda$, then $\lambda$ is called an **eigenvalue** of $\mathbf{A}$. The vector $\vec{\mathbf{u}}$ is called an*

## 4.2 Matrix Algebra

***eigenvector*** *of* **A** *corresponding to the eigenvalue* $\lambda$. *The pair* $(\lambda, \vec{u})$ *is called an* ***eigenpair*** *of* **A**.

*Note: the use of the Greek letter lambda for an eigenvalue of a matrix is standard in most mathematical texts.*

It turns out that eigenvectors and eigenvalues provide solutions to systems of linear differential equations, and in the next section you will be given methods for computing them. In that section, we will also be slightly nonstandard and use the letter $r$ to represent an eigenvalue. This makes $e^{\lambda t}$ look more like the exponential solution $e^{rt}$ that we used in Chapter 3.

**Linear systems of differential equations in matrix form.** With the above definitions, it is now possible to write any linear system of first-order differential equations (4.2) in matrix form. Consider the matrix equation

$$\begin{pmatrix} x_1(t) \\ x_2(t) \\ \vdots \\ x_n(t) \end{pmatrix}' = \begin{pmatrix} a_{11}(t) & a_{12}(t) & \cdots & a_{1n}(t) \\ a_{21}(t) & a_{22}(t) & \cdots & a_{2n}(t) \\ & & \ddots & \\ a_{n1}(t) & a_{n2}(t) & \cdots & a_{nn}(t) \end{pmatrix} \begin{pmatrix} x_1(t) \\ x_2(t) \\ \vdots \\ x_n(t) \end{pmatrix} + \begin{pmatrix} b_1(t) \\ b_2(t) \\ \vdots \\ b_n(t) \end{pmatrix}. \quad (4.7)$$

Using Definition 4.4, the left-hand side of this equation is the $n \times 1$ matrix (think column vector) $\begin{pmatrix} x_1'(t) \\ x_2'(t) \\ \vdots \\ x_n'(t) \end{pmatrix}$. If the indicated matrix multiplication and addition are carried out on the right-hand side it becomes the $n \times 1$ matrix (column vector)

$$\begin{pmatrix} a_{11}x_1(t) + a_{12}x_2(t) + \cdots + a_{1n}x_n(t) + b_1(t) \\ a_{21}x_1(t) + a_{22}x_2(t) + \cdots + a_{2n}x_n(t) + b_2(t) \\ \ddots \\ a_{n1}x_1(t) + a_{n2}x_2(t) + \cdots + a_{nn}x_n(t) + b_n(t) \end{pmatrix};$$

therefore, using the definition of equality of matrices, setting the right-hand side of (4.7) equal to the left-hand side is exactly equivalent to stating the $n$ equations in the system (4.2). If we define the three matrices

$$\vec{x} = \begin{pmatrix} x_1(t) \\ x_2(t) \\ \vdots \\ x_n(t) \end{pmatrix}, \vec{b} = \begin{pmatrix} b_1(t) \\ b_2(t) \\ \vdots \\ b_n(t) \end{pmatrix}, \text{ and } \mathbf{A} = \begin{pmatrix} a_{11}(t) & a_{12}(t) & \cdots & a_{1n}(t) \\ a_{21}(t) & a_{22}(t) & \cdots & a_{2n}(t) \\ & & \ddots & \\ a_{n1}(t) & a_{n2}(t) & \cdots & a_{nn}(t) \end{pmatrix},$$

then the system (4.7) can be written compactly in matrix form as

$$\vec{x}' = \mathbf{A}\vec{x} + \vec{b}.$$

In the remainder of the book, we will be using matrix methods to solve systems of linear differential equations. For clarity we will use the following notation:

- Boldface capital letters, such as **A**, **B**, ..., will be used to denote square matrices.

- Boldface lowercase letters, with an arrow above them, such as $\vec{x}$ and $\vec{b}$, will be used to denote column vectors (that is, $n \times 1$ matrices).

**Example 4.2.3.** *Write the third-order equation $x''' + 4x' + 3x = t^2 + 1$ in matrix form.*

*Solution.* It was shown in Example 4.1.1 that this equation is equivalent to the system

$$\begin{aligned} x_1' &= x_2 \\ x_2' &= x_3 \\ x_3' &= -3x_1 - 4x_2 + t^2 + 1. \end{aligned}$$

Check carefully that this system is equivalent to the matrix equation

$$\begin{pmatrix} x_1 \\ x_2 \\ x_3 \end{pmatrix}' = \begin{pmatrix} 0 & 1 & 0 \\ 0 & 0 & 1 \\ -3 & -4 & 0 \end{pmatrix} \begin{pmatrix} x_1 \\ x_2 \\ x_3 \end{pmatrix} + \begin{pmatrix} 0 \\ 0 \\ t^2 + 1 \end{pmatrix}, \quad (4.8)$$

which can be shortened to $\vec{x}' = \mathbf{A}\vec{x} + \vec{b}$ if we define the vector of unknowns $\vec{x} = \begin{pmatrix} x_1 \\ x_2 \\ x_3 \end{pmatrix}$, the coefficient matrix $\mathbf{A} = \begin{pmatrix} 0 & 1 & 0 \\ 0 & 0 & 1 \\ -3 & -4 & 0 \end{pmatrix}$, and the right-hand side vector $\vec{b} = \begin{pmatrix} 0 \\ 0 \\ t^2 + 1 \end{pmatrix}$. ∎

In the next example, the equations for the double spring-mass system from Example 4.1.4 in Section 4.1 are put into matrix form.

**Example 4.2.4. Double spring-mass system (continued).**

*Write the system of equations*

$$\begin{aligned} x_1' &= x_3 \\ x_2' &= x_4 \\ x_3' &= \frac{k}{M}(x_2 - 2x_1) \\ x_4' &= -\frac{k}{M}(x_2 - x_1). \end{aligned} \quad (4.9)$$

*in matrix form.*

*Solution.* We will simplify the system by defining a new parameter $\omega^2 = \frac{k}{M}$. In an undamped spring-mass system we saw that the parameter $\omega_0 = \sqrt{k/M}$ determined the natural frequency of oscillation of the system. It will also be true for the double spring-mass system. With this substitution, the matrix equation for the system becomes

$$\vec{x}' \equiv \begin{pmatrix} x_1 \\ x_2 \\ x_3 \\ x_4 \end{pmatrix}' = \begin{pmatrix} 0 & 0 & 1 & 0 \\ 0 & 0 & 0 & 1 \\ -2\omega^2 & \omega^2 & 0 & 0 \\ \omega^2 & -\omega^2 & 0 & 0 \end{pmatrix} \begin{pmatrix} x_1 \\ x_2 \\ x_3 \\ x_4 \end{pmatrix} = \mathbf{A}\vec{x}. \quad (4.10)$$

This is a homogeneous linear system of dimension 4. ∎

**Exercises 4.2.** *Given the matrices*

$$\mathbf{A} = \begin{pmatrix} 2 & 1 \\ -3 & 4 \end{pmatrix}, \quad \mathbf{B} = \begin{pmatrix} -1 & 2 \\ 1 & 3 \end{pmatrix}, \quad \mathbf{C} = \begin{pmatrix} 0 & 1 & 2 \\ -3 & -4 & 1 \end{pmatrix}, \quad \mathbf{D} = \begin{pmatrix} 0 & 1 \\ -3 & -4 \end{pmatrix},$$

*compute each expression 1–6 below, if it is defined; otherwise state why it is not defined.*

## 4.2 Matrix Algebra

1. $\mathbf{A} + \mathbf{B}$

2. $2\mathbf{A} + (-3)\mathbf{B}$

3. $\mathbf{C}(\mathbf{A} + \mathbf{B})$

4. $(\mathbf{A} + \mathbf{B})\mathbf{C}$

5. $\mathbf{ABC}$

6. $\mathbf{A} - 2\mathbf{I}$, where $\mathbf{I}$ is the $2 \times 2$ identity matrix.

7. Show that $\mathbf{AB} \neq \mathbf{BA}$.

8. Find the trace and determinant of the matrix $\mathbf{A}$.

9. Find the trace and determinant of the matrix $\mathbf{B}$.

10. If $\mathbf{A}$ is an $n \times n$ matrix and $\mathbf{AQ} = \mathbf{QA} = \mathbf{I}$ for some nonzero $n \times n$ matrix $\mathbf{Q}$, then $\mathbf{Q}$ is called the multiplicative **inverse** of $\mathbf{A}$, and is denoted by $\mathbf{A}^{-1}$. Show that if $\mathbf{A} = \begin{pmatrix} a & b \\ c & d \end{pmatrix}$ is any $2\times 2$ matrix with nonzero determinant, then $\mathbf{A}^{-1} = \frac{1}{\det \mathbf{A}} \begin{pmatrix} d & -b \\ -c & a \end{pmatrix}$; that is, show that both $\mathbf{AA}^{-1}$ and $\mathbf{A}^{-1}\mathbf{A}$ are equal to the $2 \times 2$ identity matrix. If $\det \mathbf{A} = 0$, the inverse of $\mathbf{A}$ is not defined.

11. Use the formula from Exercise 10 to find inverses of the matrices $\mathbf{A}, \mathbf{B},$ and $\mathbf{D}$. In each case, check that the product of the matrix and its inverse is the identity matrix.

*The inverse matrix can be used to solve a system of linear equations $\mathbf{Ax} = \mathbf{b}$. If $\det \mathbf{A} \neq 0$, then $\mathbf{x} = \mathbf{Ix} = (\mathbf{A}^{-1}\mathbf{A})\mathbf{x} = \mathbf{A}^{-1}(\mathbf{Ax}) = \mathbf{A}^{-1}\mathbf{b}$; that is, the solution of the system is the vector $\mathbf{x} = \mathbf{A}^{-1}\mathbf{b}$. Use this procedure to do problems 12 and 13. You will have to write the system in matrix form first.*

12. Solve the system of linear equations
$$2x + y = 6$$
$$-3x + 4y = 9$$

13. Solve the system of linear equations
$$-x + 2y = 0$$
$$x + 3y = 5$$

14. Show that $\mathbf{x} = \begin{pmatrix} 3 \\ 0 \end{pmatrix}$ is an eigenvector of the matrix $\mathbf{A} = \begin{pmatrix} 2 & 3 \\ 0 & 1 \end{pmatrix}$ by showing that $\mathbf{Ax} = \lambda \mathbf{x}$ for some scalar $\lambda$. What is the eigenvalue $\lambda$?

*Convert each of the linear differential equations below to a first-order system and write the system in matrix form (as in Example 4.2.3).*

15. $2x'' + 4x' + 5x = e^{2t}$

16. $x'' + 12x' + 27x = 0$

17. $x^{(iv)} + 2x''' - 3x'' + x' - x = 0$

18. $x''' + 4x = 0$

## 4.3 Eigenvalues and Eigenvectors

In this section we will develop, step by step, an analytic method for finding solutions of a homogeneous linear system of first-order differential equations

$$\vec{x}' = A\vec{x} \tag{4.11}$$

with constant coefficient matrix $A$.

When we looked for solutions of a single linear differential equation of order $n$ in Chapter 3, we first guessed that the solution would be of the form $x(t) = e^{rt}$ and then "plugged it into the equation." In the case of a linear system with constant matrix $A$, we need a test solution that we can again "plug into" the equation. It is clear that the test solution $\vec{x}$ must now be a vector.

To see what that test solution might look like, consider the simple 2-dimensional system

$$\vec{x}' = \begin{pmatrix} 2 & 0 \\ 1 & 3 \end{pmatrix} \vec{x}, \tag{4.12}$$

where $\vec{x} \equiv \begin{pmatrix} x_1 \\ x_2 \end{pmatrix}$.

We know that solving this system is equivalent to solving the two simultaneous first-order equations

$$\begin{array}{rcl} x_1' & = & 2x_1 \\ x_2' & = & x_1 + 3x_2. \end{array}$$

We have cheated a bit by picking a system in which the first equation doesn't even depend on the variable $x_2$, but the result we obtain from this system, for our test vector solution, is going to be correct.

Since the first equation is independent of $x_2$, our method for separable first-order equations can be used to show that the general solution is $x_1(t) = Ce^{2t}$ (check it). If this value for $x_1$ is substituted into the second equation, it becomes

$$x_2' = 3x_2 + Ce^{2t},$$

which is a linear first-order equation with general solution $x_2(t) = -Ce^{2t} + De^{3t}$. Check it!

Using the formulas we have just found for $x_1$ and $x_2$, the general solution of the system (4.12) can then be written in vector form as

$$\vec{x}(t) \equiv \begin{pmatrix} x_1(t) \\ x_2(t) \end{pmatrix} = \begin{pmatrix} Ce^{2t} \\ -Ce^{2t} + De^{3t} \end{pmatrix} \equiv Ce^{2t} \begin{pmatrix} 1 \\ -1 \end{pmatrix} + De^{3t} \begin{pmatrix} 0 \\ 1 \end{pmatrix},$$

which suggests that (vector) solutions of $\vec{x}' = A\vec{x}$ will be linear combinations of terms of the form $e^{rt}\vec{u}$, where $\vec{u}$ is a constant $n$-dimensional column vector and $r$ is a scalar. We will now show that this assumption is correct.

Let $\vec{x}(t) = e^{rt}\vec{u}$ and substitute $\vec{x}$ and its derivative into (4.11). Using the properties of scalar multiplication and differentiation of matrices defined in the previous section, the left-hand side of (4.11), with $\vec{x} = e^{rt}\vec{u}$, is

$$\vec{x}'(t) \equiv \frac{d}{dt}(e^{rt}\vec{u}) = \frac{d}{dt}\begin{pmatrix} e^{rt}u_1 \\ e^{rt}u_2 \\ \vdots \\ e^{rt}u_n \end{pmatrix} = \begin{pmatrix} re^{rt}u_1 \\ re^{rt}u_2 \\ \vdots \\ re^{rt}u_n \end{pmatrix} = re^{rt}\vec{u}.$$

## 4.3 Eigenvalues and Eigenvectors

Using the properties for matrix and scalar multiplication, the right-hand side of (4.11) can be written as

$$A\vec{x} = A(e^{rt}\vec{u}) = e^{rt}A\vec{u}.$$

To make the two sides equal we need

$$re^{rt}\vec{u} = e^{rt}A\vec{u}.$$

Dividing by the nonzero scalar $e^{rt}$, it can be seen that the two sides of (4.11) will be identical if, and only if, $r$ and $\vec{u}$ can be chosen to satisfy

$$r\vec{u} = A\vec{u}. \tag{4.13}$$

According to Definition 4.10 in the previous section, this says that the vector $\vec{u}$ must be an eigenvector of the matrix $A$, corresponding to an eigenvalue $r$; that is, $(r, \vec{u})$ must be an eigenpair of $A$.

> If, for a given square matrix $A$, we can find a vector $\vec{u}$ and a corresponding scalar $r$ such that $A\vec{u} = r\vec{u}$, then we will have found a solution $\vec{x}(t) = e^{rt}\vec{u}$ of our system $\vec{x}' = A\vec{x}$ of differential equations.

The remainder of this section will show what is involved in finding the eigenvalues and eigenvectors of a real square matrix $A$. Once we know how to do this, we will be able to construct solutions $\vec{x} = e^{rt}\vec{u}$ for the system of differential equations $\vec{x}' = A\vec{x}$. Our examples will concentrate on $2 \times 2$ matrices, since we are going to be most interested in systems of two equations in two unknowns. It will be clear, however, that the method extends to systems of dimension $n > 2$, with constant coefficient matrix $A$.

Solving the matrix equation (4.13) for a matrix $A$ still involves finding both the scalar $r$ and the vector $\vec{u}$; however, an important theorem in linear algebra makes this possible.

**Theorem 4.1.** *Let $M$ be an $n \times n$ constant matrix. The system of linear equations $M\vec{x} = \vec{b}$ has a unique $n \times 1$ solution vector $\vec{x}$ for every choice of the $n \times 1$ column vector $\vec{b}$ if, and only if, the determinant of the matrix $M$ is unequal to $0$.*

This was implied in the previous section when the determinant was defined. We will use this theorem in a negative way to find the eigenvalues of a matrix.

If the eigenvalue problem (4.13) is rewritten in the form

$$A\vec{u} - r\vec{u} = \vec{0},$$

where $\vec{0}$ is the zero vector of length $n$, then using the fact that the identity matrix $I$ satisfies $I\vec{u} = \vec{u}$ for any vector $\vec{u}$, the $n$-dimensional identity matrix $I$ can be inserted on the left, and our rule for distributivity of matrix multiplication over matrix addition can be used to write

$$A\vec{u} - r\vec{u} \equiv A\vec{u} - rI\vec{u} = (A - rI)\vec{u} = \vec{0}.$$

Note that $A - rI$ is just an $n \times n$ matrix of the form:

$$A - rI = \begin{pmatrix} a_{11} & a_{12} & \cdots & a_{1n} \\ a_{21} & a_{22} & \cdots & a_{2n} \\ & & \ddots & \\ a_{n1} & a_{n2} & \cdots & a_{nn} \end{pmatrix} - r \begin{pmatrix} 1 & 0 & \cdots & 0 \\ 0 & 1 & \cdots & 0 \\ & & \ddots & \\ 0 & 0 & \cdots & 1 \end{pmatrix} = \begin{pmatrix} a_{11}-r & a_{12} & \cdots & a_{1n} \\ a_{21} & a_{22}-r & \cdots & a_{2n} \\ & & \ddots & \\ a_{n1} & a_{n2} & \cdots & a_{nn}-r \end{pmatrix}$$

having elements $a_{ii} - r$ on the main diagonal.

The vector $\vec{u} = \vec{0}$ is an obvious solution of the system of equations $(\mathbf{A} - r\mathbf{I})\vec{u} = \vec{0}$, and by Theorem 4.1 it will be the only solution unless the determinant of the matrix $\mathbf{A} - r\mathbf{I}$ is zero. Since an eigenvector $\vec{u}$ is required to be nonzero, this implies that $r$ can be an eigenvalue of $\mathbf{A}$ only if $\det(\mathbf{A} - r\mathbf{I}) = 0$. For an $n \times n$ matrix $\mathbf{A}$, $\det(\mathbf{A} - r\mathbf{I})$ is a polynomial of degree $n$ in $r$. It is called the **characteristic polynomial** of the matrix $\mathbf{A}$ and its zeros will be the eigenvalues of $\mathbf{A}$.

**Example 4.3.1.** *Find the eigenvalues of the matrix* $\mathbf{A} = \begin{pmatrix} 1 & 3 \\ 4 & 2 \end{pmatrix}$.

*Solution.* The eigenvalues are the solutions $r$ of the equation $\det(\mathbf{A} - r\mathbf{I}) = 0$. For our matrix $\mathbf{A}$,

$$\det(\mathbf{A} - r\mathbf{I}) = \det\left(\begin{pmatrix} 1 & 3 \\ 4 & 2 \end{pmatrix} - r\begin{pmatrix} 1 & 0 \\ 0 & 1 \end{pmatrix}\right) = \det\begin{pmatrix} 1-r & 3 \\ 4 & 2-r \end{pmatrix}$$

$$= (1-r)(2-r) - 12 = r^2 - 3r - 10 = (r+2)(r-5).$$

Setting this equal to zero, we see that the two roots $r_1 = -2$ and $r_2 = 5$ are the eigenvalues of $\mathbf{A}$. ∎

To find an associated eigenvector $\vec{u}$, for each eigenvalue $r$ we solve the equation

$$(\mathbf{A} - r\mathbf{I})\vec{u} = \vec{0}. \tag{4.14}$$

First, note that an eigenvalue $r$ makes the determinant of the matrix $\mathbf{A} - r\mathbf{I}$ zero, so the solution vector $\vec{u}$ is not unique. In fact, if $\vec{u}$ is any nonzero vector satisfying (4.14) then so is any scalar multiple $k\vec{u}$. Check this!

**Example 4.3.2.** *Find a family of eigenvectors of* $\mathbf{A} = \begin{pmatrix} 1 & 3 \\ 4 & 2 \end{pmatrix}$ *corresponding to the eigenvalue* $r = -2$.

*Solution.* If $r = -2$, then $\mathbf{A} - r\mathbf{I} = \begin{pmatrix} 1 & 3 \\ 4 & 2 \end{pmatrix} - (-2)\begin{pmatrix} 1 & 0 \\ 0 & 1 \end{pmatrix} = \begin{pmatrix} 3 & 3 \\ 4 & 4 \end{pmatrix}$. Check that this matrix does have determinant equal to zero. Now set

$$(\mathbf{A} - (-2)\mathbf{I})\vec{u} = \begin{pmatrix} 3 & 3 \\ 4 & 4 \end{pmatrix}\begin{pmatrix} u_1 \\ u_2 \end{pmatrix} = \begin{pmatrix} 3u_1 + 3u_2 \\ 4u_1 + 4u_2 \end{pmatrix}$$

equal to the zero vector. This implies that $u_1$ and $u_2$ must satisfy the following two linear equations:

$$3u_1 + 3u_2 = 0$$
$$4u_1 + 4u_2 = 0,$$

but these equations are multiples of each other and they are both satisfied if, and only if, $u_1 = -u_2$. Any nonzero vector $\vec{u} = \begin{pmatrix} u_1 \\ u_2 \end{pmatrix}$ with $u_1 = -u_2$ is an eigenvector of $\mathbf{A}$, corresponding to the eigenvalue $r = -2$. We will arbitrarily choose $\vec{u} = \begin{pmatrix} 1 \\ -1 \end{pmatrix}$, but note that for any real number $k$, $k\begin{pmatrix} 1 \\ -1 \end{pmatrix} \equiv \begin{pmatrix} k \\ -k \end{pmatrix}$ will be an eigenvector corresponding to the eigenvalue $r = -2$. Once an eigenvalue and corresponding eigenvector

## 4.3 Eigenvalues and Eigenvectors

are found, it is easy to check that they are correct. Compute

$$\mathbf{A}\vec{u} = \begin{pmatrix} 1 & 3 \\ 4 & 2 \end{pmatrix} \begin{pmatrix} 1 \\ -1 \end{pmatrix} = \begin{pmatrix} -2 \\ 2 \end{pmatrix} = -2 \begin{pmatrix} 1 \\ -1 \end{pmatrix},$$

and note that $\mathbf{A}\vec{u}$ is equal to $-2\vec{u}$ as required. For practice, show that this procedure can be used to obtain $\vec{v} = k \begin{pmatrix} 3 \\ 4 \end{pmatrix}$ as a family of eigenvectors for $r = 5$. ■

An eigenvalue, together with a corresponding eigenvector, is called an **eigenpair**; therefore, $\mathbf{A}$ has the two eigenpairs $\left(-2, \begin{pmatrix} 1 \\ -1 \end{pmatrix}\right)$ and $\left(5, \begin{pmatrix} 3 \\ 4 \end{pmatrix}\right)$. In this process we have also determined two vector solutions

$$\vec{x}_1(t) = \begin{pmatrix} x(t) \\ y(t) \end{pmatrix} = e^{-2t} \begin{pmatrix} 1 \\ -1 \end{pmatrix} \text{ and } \vec{x}_2(t) = \begin{pmatrix} x(t) \\ y(t) \end{pmatrix} = e^{5t} \begin{pmatrix} 3 \\ 4 \end{pmatrix}$$

of the system of differential equations $\vec{x}' = \mathbf{A}\vec{x}$ with coefficient matrix $\mathbf{A} = \begin{pmatrix} 1 & 3 \\ 4 & 2 \end{pmatrix}$.
At this point you should therefore be able to show that the pair of functions $x(t) = e^{-2t}, y(t) = -e^{-2t}$ and also the pair $x(t) = 3e^{5t}, y(t) = 4e^{5t}$ form solution sets of the system of differential equations

$$x' = x + 3y, \quad y' = 4x + 2y.$$

Since the characteristic polynomial $\det(\mathbf{A} - r\mathbf{I})$ of a $2 \times 2$ matrix $\mathbf{A}$ is a quadratic polynomial, it can have distinct real roots, two equal real roots, or complex roots. Suppose the characteristic polynomial has complex roots. If $\mathbf{A}$ is a real matrix, these complex roots will occur in complex conjugate pairs $\alpha \pm \iota\beta$. It is still possible to find eigenvectors $\vec{z}$ satisfying $(\mathbf{A} - (\alpha \pm \iota\beta)\mathbf{I})\vec{z} = \vec{0}$, but the vector $\vec{z}$ will also be complex; that is, $\vec{z}$ can be written as $\vec{z} = \vec{u} + \iota\vec{v}$ for some real vectors $\vec{u}$ and $\vec{v}$. Furthermore, it can be easily shown, using complex arithmetic, that if $\vec{u} + \iota\vec{v}$ is an eigenvector corresponding to $\alpha + \iota\beta$, then $\vec{u} - \iota\vec{v}$ will be an eigenvector for $\alpha - \iota\beta$.

**Example 4.3.3.** *Find the eigenvalues and eigenvectors of* $\mathbf{A} = \begin{pmatrix} 1 & -4 \\ 1 & 1 \end{pmatrix}$.

*Solution.* Setting the characteristic polynomial equal to zero gives

$$\det(\mathbf{A} - r\mathbf{I}) = \det \begin{pmatrix} 1-r & -4 \\ 1 & 1-r \end{pmatrix} = (1-r)^2 + 4 = r^2 - 2r + 5 = 0,$$

and the quadratic formula can be used to show that the solutions of this quadratic equation are $r = 1 \pm 2\iota$. To find an eigenvector $\vec{z}$ corresponding to the eigenvalue $r = 1 + 2\iota$, set $(\mathbf{A} - r\mathbf{I})\vec{z} = \vec{0}$; that is,

$$(\mathbf{A} - (1+2\iota)\mathbf{I})\vec{z} = \begin{pmatrix} 1-(1+2\iota) & -4 \\ 1 & 1-(1+2\iota) \end{pmatrix} \begin{pmatrix} z_1 \\ z_2 \end{pmatrix}$$

$$= \begin{pmatrix} -2\iota & -4 \\ 1 & -2\iota \end{pmatrix} \begin{pmatrix} z_1 \\ z_2 \end{pmatrix} = \begin{pmatrix} -2\iota z_1 - 4z_2 \\ z_1 - 2\iota z_2 \end{pmatrix} = \begin{pmatrix} 0 \\ 0 \end{pmatrix}.$$

Again, the two rows of the matrix $\mathbf{A} - r\mathbf{I} = \begin{pmatrix} -2\iota & -4 \\ 1 & -2\iota \end{pmatrix}$ are multiples of each other (this is always true for a $2 \times 2$ matrix with zero determinant). Check that each element in the first row is $-\iota$ times the corresponding element in the second row. This means that both final equations will be satisfied if, and only if, $z_1 = 2\iota z_2$. We will arbitrarily let $z_2 = 1$, and pick $\vec{z} = \begin{pmatrix} 2\iota \\ 1 \end{pmatrix}$ as our eigenvector. This can be written as

$$\vec{z} = \begin{pmatrix} 0 + 2\iota \\ 1 + 0\iota \end{pmatrix} = \begin{pmatrix} 0 \\ 1 \end{pmatrix} + \iota \begin{pmatrix} 2 \\ 0 \end{pmatrix} = \vec{u} + \iota \vec{v}.$$

It can be checked that $(r, \vec{z})$ is an eigenpair by showing that if $\vec{z}$ is multiplied by the matrix $\mathbf{A}$, it multiplies $\vec{z}$ by the complex scalar $1 + 2\iota$. You can also check that an eigenvector corresponding to the eigenvalue $1 - 2\iota$ is the complex conjugate of $\vec{z}$; that is, $\vec{u} - \iota \vec{v}$. ∎

In the next section you will be shown how to extract two real solutions of $\vec{x}' = \mathbf{A}\vec{x}$ from a complex pair of eigenvalues and their associated eigenvectors.

The next example shows that our method for finding eigenvalues and eigenvectors works for a $3 \times 3$ matrix; for larger matrices there are sophisticated numerical methods for finding both eigenvalues and eigenvectors which you can learn all about in a course on numerical analysis. These methods are available on computers and also on many scientific calculators. You need to learn how to find eigenvalues and eigenvectors with whatever technology you have available.

**Example 4.3.4.** *Find the eigenvalues of the matrix*

$$\mathbf{A} = \begin{pmatrix} 1 & 1 & 0 \\ -2 & -1 & 3 \\ 0 & 0 & 2 \end{pmatrix},$$

*and find an eigenvector corresponding to each eigenvalue.*

*Solution.* To compute the characteristic polynomial we need to evaluate

$$\det(\mathbf{A} - r\mathbf{I}) = \det \begin{pmatrix} 1-r & 1 & 0 \\ -2 & -1-r & 3 \\ 0 & 0 & 2-r \end{pmatrix}.$$

This can be done by the method of cofactors given in Appendix C:

$$\det(\mathbf{A} - r\mathbf{I}) = (2-r) \det \begin{pmatrix} 1-r & 1 \\ -2 & -1-r \end{pmatrix} = (2-r)(r^2 + 1).$$

This is a polynomial of degree 3, and its three roots $r = 2, \iota,$ and $-\iota$ are the eigenvalues of $\mathbf{A}$. The only real eigenvalue is 2, so we will first look for a vector $\vec{u} = \begin{pmatrix} u \\ v \\ w \end{pmatrix}$ such that

$$(\mathbf{A} - 2\mathbf{I})\vec{u} = \begin{pmatrix} -1 & 1 & 0 \\ -2 & -3 & 3 \\ 0 & 0 & 0 \end{pmatrix} \begin{pmatrix} u \\ v \\ w \end{pmatrix} = \begin{pmatrix} -u + v \\ -2u - 3v + 3w \\ 0 \end{pmatrix}$$

## 4.3 Eigenvalues and Eigenvectors

is equal to the zero vector. This requires that $v = u$ and $3w = 2u + 3v = 5u$; therefore, one eigenvector for $r = 2$ is $\vec{u} = \begin{pmatrix} 3 \\ 3 \\ 5 \end{pmatrix}$ and any scalar multiple of $\vec{u}$ will also be an eigenvector. Check that $\vec{u}$ satisfies $\mathbf{A}\vec{u} = 2\vec{u}$.

To find an eigenvector $\vec{z}$ corresponding to the complex eigenvalue $r = \iota$, set $(\mathbf{A} - \iota\mathbf{I})\vec{z}$ equal to the zero vector; that is, write

$$(\mathbf{A} - \iota\mathbf{I})\vec{z} = \begin{pmatrix} 1 - \iota & 1 & 0 \\ -2 & -1 - \iota & 3 \\ 0 & 0 & 2 - \iota \end{pmatrix} \begin{pmatrix} u \\ v \\ w \end{pmatrix} = \begin{pmatrix} (1 - \iota)u + v \\ -2u - (1 + \iota)v + 3w \\ (2 - \iota)w \end{pmatrix} = \begin{pmatrix} 0 \\ 0 \\ 0 \end{pmatrix},$$

which has the solution $w = 0, v = (-1 + \iota)u$. Therefore the vector

$$\vec{z} = \begin{pmatrix} 1 \\ -1 + \iota \\ 0 \end{pmatrix} = \begin{pmatrix} 1 \\ -1 \\ 0 \end{pmatrix} + \iota \begin{pmatrix} 0 \\ 1 \\ 0 \end{pmatrix} = \vec{u} + \iota\vec{v}$$

is an eigenvector for $r = \iota$, and the complex conjugate $\vec{u} - \iota\vec{v}$ is an eigenvector corresponding to the other eigenvalue $r = -\iota$. ∎

**Exercises 4.3.** *For the matrices 1–6 below, find the eigenvalues. Find an eigenvector corresponding to each eigenvalue. In each case, do this first by hand and then use whatever technology you have available to check your results. Remember that any constant multiple of the eigenvector you find will also be an eigenvector.*

1. $\mathbf{A} = \begin{pmatrix} -2 & 2 \\ 1 & -3 \end{pmatrix}$

2. $\mathbf{B} = \begin{pmatrix} 2 & 1 \\ 0 & -3 \end{pmatrix}$

3. $\mathbf{C} = \begin{pmatrix} -4 & 2 \\ -2 & 0 \end{pmatrix}$

4. $\mathbf{D} = \begin{pmatrix} 1 & -4 \\ 4 & -7 \end{pmatrix}$

5. $\mathbf{E} = \begin{pmatrix} 1 & 3 \\ -3 & 1 \end{pmatrix}$

6. $\mathbf{F} = \begin{pmatrix} 2 & 8 \\ -1 & -2 \end{pmatrix}$

**COMPUTER PROBLEMS.** *To find eigenvalues and eigenvectors in Maple, use the instructions*
with(LinearAlgebra); *to bring in the necessary routines*
A:=<<2,0>|<-3,5>> *to store the matrix* $A = \begin{pmatrix} 2 & -3 \\ 0 & 5 \end{pmatrix}$
v,e:=Eigenvectors(A); *to produce* $v, e \longrightarrow \begin{pmatrix} 2 \\ 5 \end{pmatrix} \begin{pmatrix} 1 & -1 \\ 0 & 1 \end{pmatrix}$, *where $v$ contains the eigenvalues of $A$ and the matrix $e$ has the eigenvectors as its columns.*

*The corresponding instructions in Mathematica are:*

A={{2,0},{-3,5}}
{vals,vecs}=Eigensystem[A]

*The output of this instruction is* {{2,5},{{1,0},{-1,1}}}

*Use technology to find the eigenvalues and eigenvectors of the matrices* **G** *and* **H**. *In each case, try to obtain eigenvectors with integer coefficients.*

7. $\mathbf{G} = \begin{pmatrix} 1 & 0 & 0 \\ -1 & 2 & 0 \\ 4 & 6 & -4 \end{pmatrix}$

8. $\mathbf{H} = \begin{pmatrix} 2 & 0 & 0 \\ 1 & 2 & -1 \\ 1 & 0 & 1 \end{pmatrix}$ *One of the eigenvalues of* **H** *has algebraic multiplicity 2; that is, it appears as a double root of the characteristic polynomial. It may have two linearly independent eigenvectors, or only one. Which seems to be the case here? Explain.*

9. *Most computational algebra systems return eigenvectors that have been "normalized to unit length". This means that if* $\vec{u} = (u_1, u_2, \ldots, u_n)$ *is the eigenvector returned, it will satisfy* $|u_1|^2 + |u_2|^2 + \cdots + |u_n|^2 = 1$. *The table below contains eigenpairs for the matrix* **A** *in Example 4.3.4. Those on the left were calculated by hand and those on the right were numerically generated by a TI-89 calculator, using the functions* eigVl *and* eigVc.

| computed by hand | computed by TI-89 |
|---|---|
| $\left(2, \begin{pmatrix} 3 \\ 3 \\ 5 \end{pmatrix}\right)$ | $\left(2, \begin{pmatrix} 0.45750 \\ 0.45750 \\ 0.76249 \end{pmatrix}\right)$ |
| $\left(\iota, \begin{pmatrix} 1 \\ -1+\iota \\ 0 \end{pmatrix}\right)$ | $\left(\iota, \begin{pmatrix} -0.40825 - 0.40825\iota \\ 0.81650 \\ 0 \end{pmatrix}\right)$ |
| $\left(-\iota, \begin{pmatrix} 1 \\ -1-\iota \\ 0 \end{pmatrix}\right)$ | $\left(-\iota, \begin{pmatrix} -0.40825 + 0.40825\iota \\ 0.81650 \\ 0 \end{pmatrix}\right)$ |

(a) *For each eigenvector on the right, show that it is (approximately) a scalar multiple of the corresponding vector on the left. The scalar may be either a real or a complex number.*

(b) *Show that each eigenvector on the right has been normalized to unit length, as explained above. For a complex number,* $|\alpha + \iota\beta|^2 = \alpha^2 + \beta^2$.

(c) *Use your own technology to compute eigenpairs of the matrix* **A** *in Example 4.3.4, and compare them to those in the table.*

10. *Show that if the linear homogeneous system* $\vec{x}' = \mathbf{A}\vec{x}$ *has a solution of the form* $\vec{x} = e^{rt}\vec{u}$, *then* $\vec{x} = e^{rt}(k\vec{u})$ *is also a solution for any constant k; therefore, it does not matter which eigenvector you choose for a given eigenvalue, since they are all scalar multiples of each other.*

## 4.4 Analytic Solutions of the Linear System $\vec{x}' = A\vec{x}$

In this section we will derive exact formulas for the general solution of the 2-dimensional linear homogeneous constant coefficient system

$$\vec{x}' = A\vec{x} = \begin{pmatrix} a & b \\ c & d \end{pmatrix} \vec{x}. \tag{4.15}$$

For a linear constant coefficient system $\vec{x}' = A\vec{x}$, of any dimension $n$, we already know that if $(r, \vec{u})$ is an eigenpair for $A$, then the vector $\vec{x} = e^{rt}\vec{u}$ is a solution of the system. Furthermore, since the system is linear it is easy to show, using the properties of matrix algebra, that if $\vec{x}_1, \vec{x}_2, ..., \vec{x}_k$ are solutions, then so is any linear combination $c_1\vec{x}_1 + c_2\vec{x}_2 + \cdots + c_k\vec{x}_k$ for any constants $c_1, ..., c_k$.

If the system is 2-dimensional, and $\vec{x}_1$ and $\vec{x}_2$ are two linearly independent vector solutions, then $\vec{x}(t) = c_1\vec{x}_1(t) + c_2\vec{x}_2(t)$ will be called a **general solution**, and it will be shown that it is always possible to find constants $c_1$ and $c_2$ to make $\vec{x}(t)$ satisfy arbitrary initial conditions of the form $\vec{x}(t_0) = \begin{pmatrix} x(t_0) \\ y(t_0) \end{pmatrix} = \begin{pmatrix} x_0 \\ y_0 \end{pmatrix}$.

When $A$ is a $2 \times 2$ matrix, the eigenvalues of $A$ are solutions of the characteristic equation

$$\det(A - rI) = \det \begin{pmatrix} a-r & b \\ c & d-r \end{pmatrix} = (a-r)(d-r) - bc = 0.$$

This quadratic equation can be expressed in terms of the trace and determinant of $A$. Denoting the trace of $A$ by $\text{tr}(A)$ and the determinant by $\det(A)$, we have

$$\det(A - rI) = r^2 - (a+d)r + (ad - bc) \equiv r^2 - \text{tr}(A)r + \det(A) = 0. \tag{4.16}$$

It is also useful to note that if $r_1$ and $r_2$ are the roots of this quadratic (that is, the eigenvalues), then it can be factored into a product of linear factors as

$$r^2 - \text{tr}(A)r + \det(A) = (r - r_1)(r - r_2) = r^2 - (r_1 + r_2)r + r_1 r_2.$$

Comparing coefficients of the quadratic polynomials on the left and right, it can be seen that for any $2 \times 2$ matrix $A$ the sum of the eigenvalues is equal to $\text{tr}(A)$ and the product of the eigenvalues is equal to $\det(A)$. This information will be needed in the next chapter, when we describe the geometric behavior of 2-dimensional systems. It is also a useful way to check your computation of the eigenvalues.

Using the quadratic formula to find the solutions of (4.16), we see that the eigenvalues of $A$ are given by

$$r = \frac{\text{tr}(A) \pm \sqrt{(\text{tr}(A))^2 - 4\det(A)}}{2}. \tag{4.17}$$

There are three cases to consider, depending on the sign of the discriminant $K = (\text{tr}(A))^2 - 4\det(A)$. If $K > 0$ there will be two distinct real eigenvalues, if $K = 0$ there is a single real eigenvalue, and if $K < 0$ the eigenvalues are complex. For each case we will find a formula for the general solution of the system (4.15).

**Case 1** ($K > 0$): Let $r_1$ and $r_2$ be the two distinct real roots defined by (4.17), and let $(r_1, \vec{u}_1)$ and $(r_2, \vec{u}_2)$ be eigenpairs of **A**. It is shown in linear algebra that eigenvectors corresponding to different eigenvalues are linearly independent; therefore, the general solution of (4.15) can be written in the form

$$\vec{x}(t) = c_1 e^{r_1 t} \vec{u}_1 + c_2 e^{r_2 t} \vec{u}_2. \tag{4.18}$$

The constants $c_1$ and $c_2$ make it possible to satisfy arbitrary initial conditions.

**Case 2** ($K < 0$): Let $\alpha \pm \iota\beta$ be the complex conjugate eigenvalues, and let $\vec{u} + \iota\vec{v}$ be an eigenvector corresponding to $\alpha + \iota\beta$. Two complex solutions of (4.15) are given by

$$\vec{z}_1(t) = e^{(\alpha + \iota\beta)t}(\vec{u} + \iota\vec{v}), \quad \vec{z}_2(t) = e^{(\alpha - \iota\beta)t}(\vec{u} - \iota\vec{v}).$$

Using Euler's identity $e^{\iota z} = \cos(z) + \iota \sin(z)$,

$$\vec{z}_1(t) = e^{\alpha t}(\cos(\beta t) + \iota \sin(\beta t))(\vec{u} + \iota\vec{v})$$
$$= e^{\alpha t}(\cos(\beta t)\vec{u} - \sin(\beta t)\vec{v}) + \iota e^{\alpha t}(\cos(\beta t)\vec{v} + \sin(\beta t)\vec{u}).$$

Similarly, we can write

$$\vec{z}_2(t) = e^{\alpha t}(\cos(\beta t)\vec{u} - \sin(\beta t)\vec{v}) - \iota e^{\alpha t}(\cos(\beta t)\vec{v} + \sin(\beta t)\vec{u}).$$

The two real vectors $\frac{1}{2}(\vec{z}_1(t) + \vec{z}_2(t)) = e^{\alpha t}(\cos(\beta t)\vec{u} - \sin(\beta t)\vec{v})$ and $\frac{1}{2\iota}(\vec{z}_1(t) - \vec{z}_2(t)) = e^{\alpha t}(\cos(\beta t)\vec{v} + \sin(\beta t)\vec{u})$ are also solutions of (4.15), and their linear combination gives a formula for the general solution in terms of real vectors:

$$\vec{x}(t) = c_1 e^{\alpha t}[\cos(\beta t)\vec{u} - \sin(\beta t)\vec{v}] + c_2 e^{\alpha t}[\cos(\beta t)\vec{v} + \sin(\beta t)\vec{u}]. \tag{4.19}$$

**Case 3** ($K = 0$): Let $(r, \vec{u}_1)$ be one eigenpair for **A**, corresponding to the single eigenvalue $r$. Then there are two possibilities. It may happen that there is another nonzero vector $\vec{u}_2$, which is not a scalar multiple of $\vec{u}_1$, for which $\mathbf{A}\vec{u}_2 = r\vec{u}_2$. In this case, the general solution of (4.15) can be written as

$$\vec{x}(t) = c_1 e^{rt} \vec{u}_1 + c_2 e^{rt} \vec{u}_2. \tag{4.20}$$

In general this will not be the case, and we will have to find a second linearly independent solution in some other way. If we assume a second solution can be written in the form $e^{rt}(t\vec{u}_1 + \vec{u}^*)$, where $\vec{u}^*$ is to be determined, then the condition for this to be a solution of (4.15) is

$$\frac{d}{dt}\left(e^{rt}(t\vec{u}_1 + \vec{u}^*)\right) \equiv \mathbf{A}\left(e^{rt}(t\vec{u}_1 + \vec{u}^*)\right). \tag{4.21}$$

The product rule for differentiation works for matrices, and we have

$$\frac{d}{dt}\left(e^{rt}(t\vec{u}_1 + \vec{u}^*)\right) = (re^{rt})(t\vec{u}_1 + \vec{u}^*) + (e^{rt})(\vec{u}_1)$$
$$= rte^{rt}\vec{u}_1 + re^{rt}\vec{u}^* + e^{rt}\vec{u}_1.$$

## 4.4 Analytic Solutions of the Linear System $\vec{x}' = A\vec{x}$

This must be equal to

$$A(e^{rt}(t\vec{u}_1 + \vec{u}^*)) = te^{rt}A\vec{u}_1 + e^{rt}A\vec{u}^* = te^{rt}r\vec{u}_1 + e^{rt}A\vec{u}^*.$$

We have used the fact that the eigenvector $\vec{u}_1$ satisfies $A\vec{u}_1 = r\vec{u}_1$. Now, if we set the two sides of (4.21) equal and divide by $e^{rt}$,

$$rt\vec{u}_1 + r\vec{u}^* + \vec{u}_1 = rt\vec{u}_1 + A\vec{u}^*,$$

and it can be seen that the vector $\vec{u}^*$ must satisfy the equation $(A - rI)\vec{u}^* = \vec{u}_1$. The general solution of (4.15) can be written in the form

$$\vec{x}(t) = c_1 e^{rt}\vec{u}_1 + c_2 e^{rt}(t\vec{u}_1 + \vec{u}^*), \quad \text{where } (A - rI)\vec{u}^* = \vec{u}_1. \tag{4.22}$$

The following three examples demonstrate how to solve a 2-dimensional system in each of the three cases.

**Example 4.4.1.** *Solve the initial-value problem*

$$\begin{array}{rcl} x' & = & x + 3y, \quad x(0) = -1, \\ y' & = & 4x + 2y, \quad y(0) = 2. \end{array} \tag{4.23}$$

*Solution.* This system is equivalent to the matrix equation $\vec{x}' = A\vec{x} = \begin{pmatrix} 1 & 3 \\ 4 & 2 \end{pmatrix}\vec{x}$, and in the previous section we found that this matrix $A$ has two real eigenpairs

$$\left(-2, \begin{pmatrix} 1 \\ -1 \end{pmatrix}\right) \quad \text{and} \quad \left(5, \begin{pmatrix} 3 \\ 4 \end{pmatrix}\right).$$

Since the eigenvalues are real and unequal, we use the general solution for Case 1 to write

$$\vec{x}(t) \equiv \begin{pmatrix} x(t) \\ y(t) \end{pmatrix} = c_1 e^{-2t} \begin{pmatrix} 1 \\ -1 \end{pmatrix} + c_2 e^{5t} \begin{pmatrix} 3 \\ 4 \end{pmatrix} = \begin{pmatrix} c_1 e^{-2t} + 3c_2 e^{5t} \\ -c_1 e^{-2t} + 4c_2 e^{5t} \end{pmatrix}.$$

Check that if $x(t) = c_1 e^{-2t} + 3c_2 e^{5t}$ and $y(t) = -c_1 e^{-2t} + 4c_2 e^{5t}$ are substituted into the first-order system (4.23), both equations are identically satisfied for all values of $t$.

To satisfy the initial conditions, we let $t = 0$ in the general solution and write

$$\vec{x}(0) \equiv \begin{pmatrix} x(0) \\ y(0) \end{pmatrix} = c_1 e^0 \begin{pmatrix} 1 \\ -1 \end{pmatrix} + c_2 e^0 \begin{pmatrix} 3 \\ 4 \end{pmatrix} \equiv \begin{pmatrix} 1 & 3 \\ -1 & 4 \end{pmatrix} \begin{pmatrix} c_1 \\ c_2 \end{pmatrix}.$$

The final equality above uses (4.6) in the section on Matrix Algebra.

To obtain $c_1$ and $c_2$, it is necessary to solve the linear system

$$\begin{pmatrix} 1 & 3 \\ -1 & 4 \end{pmatrix} \begin{pmatrix} c_1 \\ c_2 \end{pmatrix} = \begin{pmatrix} x(0) \\ y(0) \end{pmatrix} = \begin{pmatrix} -1 \\ 2 \end{pmatrix}.$$

Remember that such a system has a unique solution if, and only if, the determinant of the $2 \times 2$ coefficient matrix is nonzero. In this case, the coefficient matrix contains eigenvectors of $A$ as its two columns. We know that eigenvectors corresponding to distinct eigenvalues are linearly independent, and we also know that any $2 \times 2$ matrix with linearly independent columns will have a nonzero determinant; thus, the initial-value problem can be completed by using a linear equations solver to find $c_1 = -10/7$, $c_2 = 1/7$. The unique solution of the initial-value problem is given by

$$\vec{x}(t) \equiv \begin{pmatrix} x(t) \\ y(t) \end{pmatrix} = -\frac{10}{7}e^{-2t}\begin{pmatrix} 1 \\ -1 \end{pmatrix} + \frac{1}{7}e^{5t}\begin{pmatrix} 3 \\ 4 \end{pmatrix} = \begin{pmatrix} -\frac{10}{7}e^{-2t} + \frac{3}{7}e^{5t} \\ \frac{10}{7}e^{-2t} + \frac{4}{7}e^{5t} \end{pmatrix}. \quad \blacksquare$$

In the next example the eigenvalues and eigenvectors are complex.

**Example 4.4.2.** *Find the general solution of the system*

$$\vec{x}' = \mathbf{A}\vec{x} = \begin{pmatrix} 1 & -4 \\ 1 & 1 \end{pmatrix}\vec{x}.$$

*Solution.* In Example 4.3.3 it was shown that $\mathbf{A}$ has complex eigenvalues $1 \pm 2\iota$. An eigenvector for $r = 1 + 2\iota$ was found to be $\vec{u} + \iota\vec{v} = \begin{pmatrix} 0 \\ 1 \end{pmatrix} + \iota\begin{pmatrix} 2 \\ 0 \end{pmatrix}$. Using the general solution for Case 2, with $\alpha = 1, \beta = 2, \vec{u} = \begin{pmatrix} 0 \\ 1 \end{pmatrix}$, and $\vec{v} = \begin{pmatrix} 2 \\ 0 \end{pmatrix}$, gives

$$\vec{x}(t) = c_1 e^t \left[\cos(2t)\begin{pmatrix} 0 \\ 1 \end{pmatrix} - \sin(2t)\begin{pmatrix} 2 \\ 0 \end{pmatrix}\right] + c_2 e^t \left[\cos(2t)\begin{pmatrix} 2 \\ 0 \end{pmatrix} + \sin(2t)\begin{pmatrix} 0 \\ 1 \end{pmatrix}\right].$$

If the indicated additions and scalar multiplications are performed, this simplifies to

$$\vec{x}(t) \equiv \begin{pmatrix} x(t) \\ y(t) \end{pmatrix} = c_1 e^t \begin{pmatrix} -2\sin(2t) \\ \cos(2t) \end{pmatrix} + c_2 e^t \begin{pmatrix} 2\cos(2t) \\ \sin(2t) \end{pmatrix}.$$

For practice, you can check that $x(t) = e^t(-2c_1 \sin(2t) + 2c_2 \cos(2t))$ and $y(t) = e^t(c_1 \cos(2t) + c_2 \sin(2t))$ satisfy the simultaneous first-order equations $x' = x - 4y$, $y' = x + y$.

As in the previous case, the constants $c_1$ and $c_2$ can be chosen to make $x$ and $y$ satisfy any initial conditions. $\blacksquare$

The next example demonstrates what can happen when the characteristic polynomial has a double real root.

**Example 4.4.3.** *Find general solutions for the two systems*

$$\text{(a) } \vec{x}' = \begin{pmatrix} 2 & 0 \\ 0 & 2 \end{pmatrix}\vec{x} \qquad \text{(b) } \vec{x}' = \begin{pmatrix} 2 & 1 \\ 0 & 2 \end{pmatrix}\vec{x}$$

*Solution.* Both systems have characteristic equation $(r-2)^2 = 0$, so they have only one eigenvalue $r = 2$. An eigenvector $\vec{u} = \begin{pmatrix} u \\ v \end{pmatrix}$ for the system (a) must satisfy

$$(\mathbf{A} - r\mathbf{I})\vec{u} = \left(\begin{pmatrix} 2 & 0 \\ 0 & 2 \end{pmatrix} - 2\begin{pmatrix} 1 & 0 \\ 0 & 1 \end{pmatrix}\right)\begin{pmatrix} u \\ v \end{pmatrix} = \begin{pmatrix} 0 \\ 0 \end{pmatrix}.$$

In this case, $\mathbf{A} - r\mathbf{I}$ is the zero matrix and any $2 \times 1$ vector will satisfy this equation. This means that we can arbitrarily choose for eigenvectors two vectors that are not scalar multiples of each other, say $\begin{pmatrix} 1 \\ 0 \end{pmatrix}$ and $\begin{pmatrix} 0 \\ 1 \end{pmatrix}$, and the general solution can then be written as

$$\vec{x}(t) = c_1 e^{2t} \begin{pmatrix} 1 \\ 0 \end{pmatrix} + c_2 e^{2t} \begin{pmatrix} 0 \\ 1 \end{pmatrix} = e^{2t} \begin{pmatrix} c_1 \\ c_2 \end{pmatrix}.$$

## 4.4 Analytic Solutions of the Linear System $\vec{x}' = A\vec{x}$

The matrix equation (a) is equivalent to the system $x' = 2x$, $y' = 2y$, and these two differential equations can be solved independently. From this, it is clear that $x(t) = c_1 e^{2t}$, $y(t) = c_2 e^{2t}$ is the general solution.

In equation (b), any eigenvector must satisfy

$$\left(\begin{pmatrix} 2 & 1 \\ 0 & 2 \end{pmatrix} - 2\begin{pmatrix} 1 & 0 \\ 0 & 1 \end{pmatrix}\right)\begin{pmatrix} u \\ v \end{pmatrix} = \begin{pmatrix} 0 & 1 \\ 0 & 0 \end{pmatrix}\begin{pmatrix} u \\ v \end{pmatrix} = \begin{pmatrix} v \\ 0 \end{pmatrix} = \begin{pmatrix} 0 \\ 0 \end{pmatrix}.$$

This says that the only condition on the eigenvector is that its second component is $v = 0$. If we take the eigenvector to be $\vec{u}_1 = \begin{pmatrix} 1 \\ 0 \end{pmatrix}$, then any other eigenvector $k\vec{u}_1$ is a scalar multiple of $\vec{u}_1$. This means that we need to find a second vector $\vec{u}^* = \begin{pmatrix} w \\ z \end{pmatrix}$ that satisfies $(A - 2I)\vec{u}^* = \vec{u}_1$; that is,

$$(A - 2I)\vec{u}^* = \begin{pmatrix} 0 & 1 \\ 0 & 0 \end{pmatrix}\begin{pmatrix} w \\ z \end{pmatrix} = \begin{pmatrix} z \\ 0 \end{pmatrix} = \begin{pmatrix} 1 \\ 0 \end{pmatrix}.$$

The only condition on $\vec{u}^*$ is that its second component $z$ must equal 1. We will arbitrarily take $\vec{u}^* = \begin{pmatrix} 0 \\ 1 \end{pmatrix}$. Now the second formula in Case 3 gives

$$\vec{x}(t) = c_1 e^{2t}\begin{pmatrix} 1 \\ 0 \end{pmatrix} + c_2 e^{2t}\left(t\begin{pmatrix} 1 \\ 0 \end{pmatrix} + \begin{pmatrix} 0 \\ 1 \end{pmatrix}\right) = \begin{pmatrix} c_1 e^{2t} + c_2 t e^{2t} \\ c_2 e^{2t} \end{pmatrix}$$

as a general solution of the system (b). ■

### 4.4.1 Application 1: Mixing Problem with Two Compartments.

In this application you will be shown how to extend the method for solving the single-compartment mixing problem, described in Section 2.3, to problems involving more than one compartment. The case of two compartments leads to two first-order equations in two unknowns, and it can be solved using one of the formulas derived above.

A two-compartment mixing problem is shown in Figure 4.1. Tank A is connected to Tank B by two separate pipes, as shown. Pure water is flowing into Tank A at a rate of 2 gallons/minute. Tank A is initially filled with 50 gallons of water with 5 pounds of salt dissolved in it. Tank B contains 40 gallons of water with 1 pound of salt dissolved in it. The solution flows from Tank A to Tank B at a rate of 3 gallons/minute and from Tank B to Tank A (in a separate pipe) at a rate of 1 gallon/minute. A thoroughly mixed solution is also being drained from Tank B at a rate of 2 gallons/minute. Let $x(t)$ be the amount of salt in Tank A, and $y(t)$ the amount of salt in Tank B, at time $t$. Using "rate in − rate out" for the rate of change of salt in each tank (see Section 2.3), we want to write a system of simultaneous differential equations for $x$ and $y$.

Using the fact that $\frac{dx}{dt}$ is in pounds per minute, the rate equation for Tank A is

$$dx/dt = \text{rate in} - \text{rate out} = \left[2 \cdot 0 + 1 \cdot \frac{y(t)}{40}\right] - \left[3 \cdot \frac{x(t)}{50}\right].$$

Make sure you see where this comes from!

Similarly, the equation for Tank B is

$$dy/dt = \left[3 \cdot \frac{x(t)}{50}\right] - \left[1 \cdot \frac{y(t)}{40} + 2 \cdot \frac{y(t)}{40}\right].$$

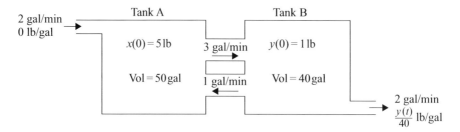

Figure 4.1. Two-tank mixing problem

This leads to the linear system of equations

$$x'(t) = -\frac{3}{50}x(t) + \frac{1}{40}y(t)$$
$$y'(t) = \frac{3}{50}x(t) - \frac{3}{40}y(t).$$
(4.24)

**Example 4.4.4.** *Write the equations (4.24) in matrix form and solve the system. Use the initial conditions $x(0) = 5$, $y(0) = 1$. Keep four decimal accuracy in your calculations of the eigenvalues and eigenvectors. (At this point, you should be using technology to find the eigenpairs.) Also plot $x(t)$ and $y(t)$ for $0 \leq t \leq 100$.*

*Solution.* The system of equations in matrix form is

$$\begin{pmatrix} x \\ y \end{pmatrix}' = \begin{pmatrix} -0.06 & 0.025 \\ 0.06 & -0.075 \end{pmatrix} \begin{pmatrix} x \\ y \end{pmatrix},$$

and approximate eigenpairs of the 2×2 coefficient matrix are $\left(-0.028051, \begin{pmatrix} 0.61625 \\ 0.78755 \end{pmatrix}\right)$ and $\left(-0.10695, \begin{pmatrix} -0.47001 \\ 0.88266 \end{pmatrix}\right)$.

Using (4.18) for the real and unequal eigenvalue case, the solution can be written as

$$\vec{x} = c_1 e^{\lambda_1 t}\vec{u_1} + c_2 e^{\lambda_2 t}\vec{u_2} = c_1 e^{-0.028051t}\begin{pmatrix} 0.61625 \\ 0.78755 \end{pmatrix} + c_2 e^{-0.10695t}\begin{pmatrix} -0.47001 \\ 0.88266 \end{pmatrix}.$$

The constants $c_1$ and $c_2$ can be found by setting $t = 0$ and solving the linear system

$$\vec{x}(0) = \begin{pmatrix} 0.61625 & -0.47001 \\ 0.78755 & 0.88266 \end{pmatrix}\begin{pmatrix} c_1 \\ c_2 \end{pmatrix} = \begin{pmatrix} x(0) \\ y(0) \end{pmatrix} = \begin{pmatrix} 5 \\ 1 \end{pmatrix}.$$

This gives $c_1 = 5.34224$ and $c_2 = -3.63365$. With these values, the solutions are

$$x(t) = 3.29215e^{-0.028051t} + 1.70785e^{-0.10695t},$$
$$y(t) = 4.20728e^{-0.028051t} - 3.20728e^{-0.10695t}.$$

These two functions are plotted in Figure 4.2. As $t \to \infty$ the concentration of salt in both tanks tends to zero since there is no salt in the water coming in. ∎

The mixing problem in Example 4.4.4 resulted in a homogeneous linear system of dimension 2. It would not be homogeneous if there was any salt in the water being pumped into Tank A. Do you see why?

## 4.4 Analytic Solutions of the Linear System $\vec{x}' = A\vec{x}$

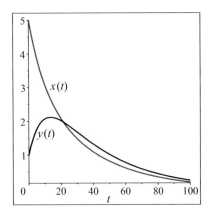

Figure 4.2. Amount of salt in the two tanks

### 4.4.2 Application 2: Double Spring-mass System.
We are now able to completely solve the problem that was first described in Example 4.1.4 in Section 4.1. It was shown that the "double spring-mass problem" could be written as a linear system of first-order equations of dimension 4, and in Example 4.2.4 the system was written in matrix form as

$$\vec{x}' = \begin{pmatrix} x_1' \\ x_2' \\ x_3' \\ x_4' \end{pmatrix} = \begin{pmatrix} 0 & 0 & 1 & 0 \\ 0 & 0 & 0 & 1 \\ -2\omega^2 & \omega^2 & 0 & 0 \\ \omega^2 & -\omega^2 & 0 & 0 \end{pmatrix} \begin{pmatrix} x_1 \\ x_2 \\ x_3 \\ x_4 \end{pmatrix} = A\vec{x}.$$

To find the eigenvalues of $A$, we need to set $\det(A - \lambda I) = 0$ and solve for $\lambda$. Using the method of cofactors (see Appendix C),

$$\det(A - \lambda I) = \det\begin{pmatrix} -\lambda & 0 & 1 & 0 \\ 0 & -\lambda & 0 & 1 \\ -2\omega^2 & \omega^2 & -\lambda & 0 \\ \omega^2 & -\omega^2 & 0 & -\lambda \end{pmatrix}$$

$$= -\lambda \det\begin{pmatrix} -\lambda & 0 & 1 \\ \omega^2 & -\lambda & 0 \\ -\omega^2 & 0 & -\lambda \end{pmatrix} + 1 \cdot \det\begin{pmatrix} 0 & -\lambda & 1 \\ -2\omega^2 & \omega^2 & 0 \\ \omega^2 & -\omega^2 & -\lambda \end{pmatrix},$$

and applying cofactors a second time gives

$$\det(A - \lambda I) = \lambda^4 + 3\omega^2 \lambda^2 + \omega^4 \equiv (\lambda^2)^2 + (3\omega^2)(\lambda^2) + \omega^4 = 0.$$

This is a quadratic equation in $\lambda^2$, and the quadratic formula can be used to write

$$\lambda^2 = \frac{-3\omega^2 \pm \sqrt{9\omega^4 - 4\omega^4}}{2} = \left(\frac{-3 \pm \sqrt{5}}{2}\right)\omega^2.$$

Since the numbers $\frac{-3 \pm \sqrt{5}}{2}$ are negative, there are four pure imaginary eigenvalues. Letting $\alpha^2$ and $\beta^2$ be the positive constants

$$\alpha^2 = \left|\frac{-3 + \sqrt{5}}{2}\right|, \quad \beta^2 = \left|\frac{-3 - \sqrt{5}}{2}\right|,$$

the four eigenvalues can be written as

$$\lambda = \pm \alpha \omega \iota, \ \pm \beta \omega \iota.$$

The eigenvector $\vec{z} \equiv \begin{pmatrix} a \\ b \\ c \\ d \end{pmatrix}$ for $\lambda = \alpha \omega \iota$ is found by solving the equation $(\mathbf{A} - \lambda \mathbf{I})\vec{z} = \vec{0}$.

This is the equation

$$\begin{pmatrix} -\alpha \omega \iota & 0 & 1 & 0 \\ 0 & -\alpha \omega \iota & 0 & 1 \\ -2\omega^2 & \omega^2 & -\alpha \omega \iota & 0 \\ \omega^2 & -\omega^2 & 0 & -\alpha \omega \iota \end{pmatrix} \begin{pmatrix} a \\ b \\ c \\ d \end{pmatrix} = \begin{pmatrix} -\alpha \omega \iota a + c \\ -\alpha \omega \iota b + d \\ -2\omega^2 a + \omega^2 b - \alpha \omega \iota c \\ \omega^2 a - \omega^2 b - \alpha \omega \iota d \end{pmatrix} = \begin{pmatrix} 0 \\ 0 \\ 0 \\ 0 \end{pmatrix},$$

and we see that it requires that $c = (\alpha \omega \iota)a$ and $d = (\alpha \omega \iota)b$. Adding the last two rows of the matrix gives

$$-\omega^2 a - (\alpha \omega \iota)c - (\alpha \omega \iota)d = 0.$$

If the values for $c$ and $d$ are substituted, we have

$$-\omega^2 a + \alpha^2 \omega^2 a + \alpha^2 \omega^2 b = 0,$$

and this can be solved to give

$$b = \left( \frac{1 - \alpha^2}{\alpha^2} \right) a.$$

With the arbitrary choice $a = 1$, the eigenvector is

$$\vec{z} = \vec{u} + \iota \vec{v} = \begin{pmatrix} 1 \\ A \\ 0 \\ 0 \end{pmatrix} + \iota \begin{pmatrix} 0 \\ 0 \\ \alpha \omega \\ \alpha \omega A \end{pmatrix}, \qquad A \equiv \frac{1 - \alpha^2}{\alpha^2}.$$

A similar computation results in an eigenvector

$$\vec{w} = \vec{r} + \iota \vec{s} = \begin{pmatrix} 1 \\ B \\ 0 \\ 0 \end{pmatrix} + \iota \begin{pmatrix} 0 \\ 0 \\ \beta \omega \\ \beta \omega B \end{pmatrix}, \qquad B \equiv \frac{1 - \beta^2}{\beta^2}$$

for $\lambda = \beta \omega \iota$.

Using the fact that eigenvectors for complex conjugate eigenvalues are also complex conjugates, the four eigenpairs for the matrix $\mathbf{A}$ are

$$\left( 0 \pm \alpha \omega \iota, \vec{u} \pm \iota \vec{v} \right), \left( 0 \pm \beta \omega \iota, \vec{r} \pm \iota \vec{s} \right). \tag{4.25}$$

In this problem the complex eigenvalues are distinct so it is possible to use the complex eigenvalue formula (4.19) twice to write the solution in the form

$$\vec{x}(t) = C_1 e^{0t} \left[ \cos(\alpha \omega t)\vec{u} - \sin(\alpha \omega t)\vec{v} \right] + C_2 e^{0t} \left[ \cos(\alpha \omega t)\vec{v} + \sin(\alpha \omega t)\vec{u} \right]$$
$$+ C_3 e^{0t} \left[ \cos(\beta \omega t)\vec{r} - \sin(\beta \omega t)\vec{s} \right] + C_4 e^{0t} \left[ \cos(\beta \omega t)\vec{s} + \sin(\beta \omega t)\vec{r} \right].$$

## 4.4 Analytic Solutions of the Linear System $\vec{x}' = A\vec{x}$

Simplifying the linear combinations of the vectors $\vec{u}$ and $\vec{v}$, and also those of $\vec{r}$ and $\vec{s}$, we can write

$$\vec{x}(t) = C_1 \begin{pmatrix} \cos(\alpha\omega t) \\ A\cos(\alpha\omega t) \\ -\alpha\omega\sin(\alpha\omega t) \\ -A\alpha\omega\sin(\alpha\omega t) \end{pmatrix} + C_2 \begin{pmatrix} \sin(\alpha\omega t) \\ A\sin(\alpha\omega t) \\ \alpha\omega\cos(\alpha\omega t) \\ A\alpha\omega\cos(\alpha\omega t) \end{pmatrix}$$

$$+ C_3 \begin{pmatrix} \cos(\beta\omega t) \\ B\cos(\beta\omega t) \\ -\beta\omega\sin(\beta\omega t) \\ -B\beta\omega\sin(\beta\omega t) \end{pmatrix} + C_4 \begin{pmatrix} \sin(\beta\omega t) \\ B\sin(\beta\omega t) \\ \beta\omega\cos(\beta\omega t) \\ B\beta\omega\cos(\beta\omega t) \end{pmatrix}.$$

Now formula (4.6) for the matrix product can be used to write the solution very compactly in matrix form as

$$\vec{x}(t) = \begin{pmatrix} \cos(\alpha\omega t) & \sin(\alpha\omega t) & \cos(\beta\omega t) & \sin(\beta\omega t) \\ A\cos(\alpha\omega t) & A\sin(\alpha\omega t) & B\cos(\beta\omega t) & B\sin(\beta\omega t) \\ -\alpha\omega\sin(\alpha\omega t) & \alpha\omega\cos(\alpha\omega t) & -\beta\omega\sin(\beta\omega t) & \beta\omega\cos(\beta\omega t) \\ -A\alpha\omega\sin(\alpha\omega t) & A\alpha\omega\cos(\alpha\omega t) & -B\beta\omega\sin(\beta\omega t) & B\beta\omega\cos(\beta\omega t) \end{pmatrix} \begin{pmatrix} C_1 \\ C_2 \\ C_3 \\ C_4 \end{pmatrix}.$$

In this form it is easy to check that lines 3 and 4 give the derivatives of the functions defined by lines 1 and 2.

To find the constants $C_i$, the initial conditions at $t = 0$ are used to write

$$\vec{x}(0) = \begin{pmatrix} 1 & 0 & 1 & 0 \\ A & 0 & B & 0 \\ 0 & \alpha\omega & 0 & \beta\omega \\ 0 & A\alpha\omega & 0 & B\beta\omega \end{pmatrix} \begin{pmatrix} C_1 \\ C_2 \\ C_3 \\ C_4 \end{pmatrix} = \begin{pmatrix} -\rho \\ 0 \\ 0 \\ 0 \end{pmatrix}.$$

This results in two $2 \times 2$ linear systems:

$$\begin{pmatrix} 1 & 1 \\ A & B \end{pmatrix} \begin{pmatrix} C_1 \\ C_3 \end{pmatrix} = \begin{pmatrix} -\rho \\ 0 \end{pmatrix} \quad \text{and} \quad \begin{pmatrix} \alpha\omega & \beta\omega \\ A\alpha\omega & B\beta\omega \end{pmatrix} \begin{pmatrix} C_2 \\ C_4 \end{pmatrix} = \begin{pmatrix} 0 \\ 0 \end{pmatrix},$$

with solution $C_2 = C_4 = 0$ and $C_1 = \frac{-\rho B}{B-A}$, $C_3 = \frac{\rho A}{B-A}$. Using these, the two displacement functions $x_1(t)$ and $x_2(t)$ can be written as

$$x_1(t) = \frac{\rho}{B-A}[A\cos(\beta\omega t) - B\cos(\alpha\omega t)],$$

$$x_2(t) = \frac{\rho AB}{B-A}[\cos(\beta\omega t) - \cos(\alpha\omega t)].$$

These are shown plotted in Figure 4.3. The values used for the parameters are $\omega = \sqrt{\frac{k}{m}} = 1$ and $\rho = 1$. The initial displacement $\rho$ simply multiplies the values of $x_1$ and $x_2$, while $\omega$ determines the frequency of the oscillation.

**Exercises 4.4.** Let $A, B, \ldots, F$ be the six matrices $A = \begin{pmatrix} -2 & 2 \\ 1 & -3 \end{pmatrix}$, $B = \begin{pmatrix} 2 & 1 \\ 0 & -3 \end{pmatrix}$, $C = \begin{pmatrix} -4 & 2 \\ -2 & 0 \end{pmatrix}$, $D = \begin{pmatrix} 1 & -4 \\ 4 & -7 \end{pmatrix}$, $E = \begin{pmatrix} 1 & 3 \\ -3 & 1 \end{pmatrix}$, $F = \begin{pmatrix} 2 & 8 \\ -1 & -2 \end{pmatrix}$. These are the matrices for which eigenpairs were found in Exercises 4.3.
Find general solutions in vector form for the systems 1–3:

1. $\vec{x}' = B\vec{x}$

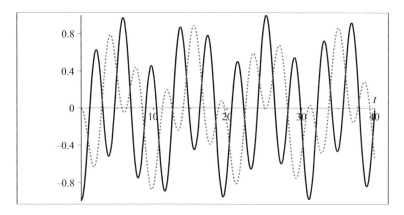

**Figure 4.3.** Oscillatory behavior of the two masses: $x_1(t)$ (solid) and $x_2(t)$ (dotted)

2. $\vec{x}' = \mathbf{D}\vec{x}$

3. $\vec{x}' = \mathbf{F}\vec{x}$

*Solve the following initial-value problems; give the solutions in vector form.*

4. $\vec{x}' = \mathbf{A}\vec{x}, \ \vec{x}(0) = \begin{pmatrix} 3 \\ 0 \end{pmatrix}$

5. $\vec{x}' = \mathbf{C}\vec{x}, \ \vec{x}(0) = \begin{pmatrix} 2 \\ 3 \end{pmatrix}$

6. $\vec{x}' = \mathbf{E}\vec{x}, \ \vec{x}(0) = \begin{pmatrix} 0 \\ 1 \end{pmatrix}$

*Find solutions to the systems in problems 7–11, given in component form. If initial conditions are not given, find a general solution. Give the solution in component form (that is, give expressions for $x(t)$ and $y(t)$).*

7. $x' = y, y' = x$

8. $x' = -y, y' = x$

9. $x' = 2x - y, y' = y$

10. $x' = -x, y' = x, x(0) = 1, y(0) = -1$

11. $x' = x + y, y' = -x + y, x(0) = 0, y(0) = 1$.

12. Solve problem 9 again, without using matrices.

13. Solve problem 10 again, without using matrices.

14. Let $\mathbf{A}$ be an $n \times n$ matrix with constant elements $a_{i,j}$. If $\vec{x}_i' = \mathbf{A}\vec{x}_i$ for each $n$-dimensional vector $\vec{x}_1, \vec{x}_2, \ldots, \vec{x}_k$, show that the vector $\vec{x} = c_1\vec{x}_1 + c_2\vec{x}_2 + \cdots + c_k\vec{x}_k$ also satisfies $\vec{x}' = \mathbf{A}\vec{x}$.

## 4.4 Analytic Solutions of the Linear System $\vec{x}' = A\vec{x}$

*Hint: First use the definitions of vector addition and differentiation to show that*
$$\frac{d}{dt}(a\vec{x}(t) + b\vec{y}(t)) = a\vec{x}'(t) + b\vec{y}'(t).$$

*Exercises 15–19 refer to the two-tank problem in Section 4.4.1.*

15. Does the amount of salt in the two tanks ever become equal? If so, at what value of $t$?

16. What limits are $x$ and $y$ approaching as $t \to \infty$? Justify this in terms of the physical model. *Hint: Think about the concentration of salt in the input to Tank A.*

17. Write equations for the mixing problem if the flow rate from Tank A into Tank B is changed from 3 gal/min to 2 gal/min. The volumes in the two tanks will both be functions of time.

18. Let $\vec{x}' = A\vec{x} + \vec{b}$ be a **nonhomogeneous linear system**, and let $\vec{x}_H$ be a solution of the associated homogeneous system $\vec{x}' = A\vec{x}$. Show that if $\vec{x}_P$ is any vector satisfying the nonhomogeneous system, then the vector $\vec{x} = \vec{x}_H + \vec{x}_P$ satisfies $\vec{x}' = A\vec{x} + \vec{b}$.

The result of Exercise 18. can be used to prove that a **general solution of a nonhomogeneous linear system** can be written in the form

$$\vec{x} = \vec{x}_H + \vec{x}_P,$$

where $\vec{x}_H$ is a general solution of the associated homogeneous system and $\vec{x}_P$ is any particular solution of the nonhomogeneous system.

19. Redo the two-tank problem in Section 4.4.1, assuming the solution entering Tank A from the outside has a concentration of $p$ pounds of salt per gallon.

    (a) Show that the system of equations for $x(t)$ and $y(t)$ is $\vec{x}' = A\vec{x} + \vec{b}$, where $A$ is the coefficient matrix found in Example 4.4.4 and $\vec{b} = \begin{pmatrix} 2p \\ 0 \end{pmatrix}$.

    (b) Assume a particular solution $\vec{x}_P = \begin{pmatrix} a \\ b \end{pmatrix}$ with undetermined constant coefficients $a$ and $b$. Substitute $\vec{x}_P$ into the nonhomogeneous equation and solve for $a$ and $b$. (Hint: The vector $\vec{x}_P{}'$ is the 0-vector).

    (c) Add $\vec{x}_P$ to the general solution found in Example 4.4.4. This will be the general solution of the nonhomogeneous system. Let $p = 0.1$ pounds/gallon, and use the same initial conditions as before, $x(0) = 5, y(0) = 1$, to determine the constants in the general solution.

    (d) Plot $x(t)$ and $y(t)$ for $0 \le t \le 100$.

    (e) How did the results change from those in the problem in Section 4.4.1? Can you still justify the results physically?

*Exercise 20 refers to the double spring-mass problem.*

20. Exercise 12 in Section 4.1 asked you to set up equations for a mechanical system with three springs A, B, and C and two objects F and G, each of mass M, where $x_1(t)$ and $x_2(t)$ are the displacements of the two masses F and G respectively. Springs A and C have spring constant $k$, and spring B has spring constant $2k$. The equilibrium positions of all the elements in the system is shown in the diagram below. The spring C is attached to the right-hand end of the platform.

(a) *If the object G is initially pulled right 1 meter and let go (assuming F is held at its equilibrium position and let go simultaneously), write initial conditions for $x_1$, $x_1'$, $x_2$, and $x_2'$.*

(b) *Solve the initial-value problem, either analytically or using your differential equations software. Draw graphs of $x_1(t)$ and $x_2(t)$.*

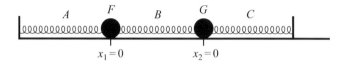

## 4.5 Large Linear Systems; the Matrix Exponential

It was mentioned in the introduction to Chapter 4 that very large linear systems of differential equations appear in many fields of applied mathematics. Finding the eigenvalues and eigenvectors for large systems is not a simple problem. One method for finding the solutions of a large system of linear differential equations with constant coefficients is by means of a matrix called the matrix exponential. The information provided in this section is aimed at making it possible for you to use a matrix exponential routine in Maple or *Mathematica* to solve a large linear system with constant coefficients.

### 4.5.1 Definition and Properties of the Matrix Exponential.

**Definition 4.11.** *The* **matrix exponential** *of an $n \times n$ constant matrix $\mathbf{A}$ is defined by*

$$e^{\mathbf{A}t} \equiv \mathbf{I} + \mathbf{A}t + \mathbf{A}^2 \frac{t^2}{2!} + \mathbf{A}^3 \frac{t^3}{3!} + \cdots. \tag{4.26}$$

This should remind you of the Taylor series for the exponential function $e^{at}$. For any square matrix $\mathbf{A}$ and any value of $t$ this infinite series of matrices can be shown to converge.

While finding the value of $e^{\mathbf{A}t}$ for large matrices $\mathbf{A}$ is a difficult numerical problem, the following example shows that finding the matrix exponential of a diagonal matrix is surprisingly easy.

**Example 4.5.1.** *Find the matrix exponential $e^{\mathbf{A}t}$ for the diagonal matrix*

$$\mathbf{A} = \begin{pmatrix} a & 0 \\ 0 & b \end{pmatrix}.$$

*Solution.* The powers of $\mathbf{A}$ for any positive integer $k$ are of the form

$$\mathbf{A}^k = \overbrace{\begin{pmatrix} a & 0 \\ 0 & b \end{pmatrix}\begin{pmatrix} a & 0 \\ 0 & b \end{pmatrix}\cdots\begin{pmatrix} a & 0 \\ 0 & b \end{pmatrix}}^{k \text{ times}} = \begin{pmatrix} a^k & 0 \\ 0 & b^k \end{pmatrix},$$

## 4.5 Large Linear Systems; the Matrix Exponential

so that the series (4.26) becomes

$$e^{\mathbf{A}t} = \begin{pmatrix} 1 & 0 \\ 0 & 1 \end{pmatrix} + \begin{pmatrix} a & 0 \\ 0 & b \end{pmatrix} t + \begin{pmatrix} a^2 & 0 \\ 0 & b^2 \end{pmatrix} \frac{t^2}{2!} + \begin{pmatrix} a^3 & 0 \\ 0 & b^3 \end{pmatrix} \frac{t^3}{3!} + \cdots,$$

and, using the definitions of scalar multiplication and matrix addition,

$$e^{\mathbf{A}t} = \begin{pmatrix} 1 + at + (at)^2/2! + \cdots & 0 \\ 0 & 1 + bt + (bt)^2/2! + \cdots \end{pmatrix} = \begin{pmatrix} e^{at} & 0 \\ 0 & e^{bt} \end{pmatrix}. \quad (4.27)$$

■

For any constant square matrix $\mathbf{A}$, the matrix exponential has the important property that it satisfies the matrix differential equation

$$\frac{d}{dt}(e^{\mathbf{A}t}) = \mathbf{A}e^{\mathbf{A}t}.$$

To see this, note that since $\mathbf{A}$ is a matrix of constants,

$$\frac{d}{dt}\left(\mathbf{A}^k \frac{t^k}{k!}\right) = \mathbf{A}^k \frac{d}{dt}\left(\frac{t^k}{k!}\right) = \mathbf{A}^k \left(\frac{kt^{k-1}}{k!}\right) = \mathbf{A}^k \frac{t^{k-1}}{(k-1)!}.$$

Using the definition of $e^{\mathbf{A}t}$, we can write

$$\frac{d}{dt}(e^{\mathbf{A}t}) = \frac{d}{dt}\left(\mathbf{I} + \mathbf{A}t + \mathbf{A}^2 \frac{t^2}{2!} + \mathbf{A}^3 \frac{t^3}{3!} + \cdots \right)$$

$$= \mathbf{0} + \mathbf{A} + \mathbf{A}^2 \frac{2t}{2!} + \mathbf{A}^3 \frac{3t^2}{3!} + \cdots = \mathbf{A}(\mathbf{I} + \mathbf{A}t + \mathbf{A}^2 \frac{t^2}{2!} + \cdots) = \mathbf{A}e^{\mathbf{A}t}.$$

The matrix $\mathcal{X} \equiv e^{\mathbf{A}t}$ is called a **fundamental matrix** for the linear system of differential equations $\vec{\mathbf{x}}' = \mathbf{A}\vec{\mathbf{x}}$. If we write $\mathcal{X}$ and $\mathcal{X}'$ in terms of their column vectors,

$$\mathcal{X} = \begin{pmatrix} \vec{\mathbf{x}}_1 & \vec{\mathbf{x}}_2 & \cdots & \vec{\mathbf{x}}_n \\ \downarrow & \downarrow & & \downarrow \end{pmatrix}, \quad \mathcal{X}' = \begin{pmatrix} \vec{\mathbf{x}}'_1 & \vec{\mathbf{x}}'_2 & \cdots & \vec{\mathbf{x}}'_n \\ \downarrow & \downarrow & & \downarrow \end{pmatrix},$$

then the matrix product $\mathbf{A}\mathcal{X}$ can be written as $\mathbf{A}\mathcal{X} = \begin{pmatrix} \mathbf{A}\vec{\mathbf{x}}_1 & \mathbf{A}\vec{\mathbf{x}}_2 & \cdots & \mathbf{A}\vec{\mathbf{x}}_n \\ \downarrow & \downarrow & & \downarrow \end{pmatrix}.$

Make sure you see why this is true (use the definition of matrix multiplication).

If a matrix $\mathcal{X}$ satisfies the differential equation $\mathcal{X}' = \mathbf{A}\mathcal{X}$, we see that each column vector $\vec{\mathbf{x}}_i$ of the matrix $\mathcal{X}$ is a solution of the vector equation $\vec{\mathbf{x}}'_i = \mathbf{A}\vec{\mathbf{x}}_i$.

**Properties of the Matrix Exponential.**

- $e^{\mathbf{0}} = \mathbf{I}$, where $\mathbf{0}$ is the $n \times n$ zero matrix, and $\mathbf{I}$ is the $n \times n$ identity matrix. This is easily seen from the series definition of $e^{\mathbf{A}t}$.

- If $\mathbf{A}$ and $\mathbf{B}$ commute, that is, $\mathbf{AB} = \mathbf{BA}$, then $e^{\mathbf{A}+\mathbf{B}} = e^{\mathbf{A}}e^{\mathbf{B}}$.

- The inverse of $e^{\mathbf{A}t}$ is $e^{-\mathbf{A}t}$.

  This last property holds since $e^{\mathbf{A}t}e^{-\mathbf{A}t} = e^{\mathbf{A}t-\mathbf{A}t} = e^{\mathbf{0}} = \mathbf{I}$.

A theorem in linear algebra tells us that the columns of an invertible matrix are linearly independent; therefore, the column vectors $\vec{\mathbf{x}}_i, i = 1, \ldots, n$, of $\mathcal{X}$ form a linearly independent set of solutions of the system $\vec{\mathbf{x}}' = \mathbf{A}\vec{\mathbf{x}}$. Using (4.6) in the section on matrix algebra, the general solution of this system can be written as

$$\vec{\mathbf{x}} = \mathcal{X}\vec{\mathbf{c}} = \sum_{i=1}^{n} c_i \vec{\mathbf{x}}_i(t) \equiv c_1\vec{\mathbf{x}}_1 + c_2\vec{\mathbf{x}}_2 + \cdots + c_n\vec{\mathbf{x}}_n.$$

The following definition will be needed in the next example.

**Definition 4.12.** *A nonzero square matrix* $\mathbf{N}$ *is called* **nilpotent** *if* $\mathbf{N}^k$ *is the zero matrix for some integer* $k > 1$.

The series for the exponential matrix $e^{\mathbf{N}t}$ of a nilpotent matrix $\mathbf{N}$ has only $k$ terms. Just use the definition: $e^{\mathbf{N}t} = \mathbf{I} + \mathbf{N}t + \mathbf{N}^2 \frac{t^2}{2!} + \cdots + \mathbf{N}^{k-1}\frac{t^{k-1}}{(k-1)!} + \mathbf{0} + \mathbf{0} + \cdots$.

**Example 4.5.2.** *Use the matrix exponential of* $\mathbf{A}$ *to solve the system of differential equations*

$$\vec{x}' = \mathbf{A}\vec{x} = \begin{pmatrix} 2 & 0 \\ 1 & 2 \end{pmatrix} \vec{x}. \tag{4.28}$$

*Solution.* The matrix $\mathbf{A}$ was specially chosen to make finding $e^{\mathbf{A}t}$ relatively easy. In this particular case we can write $\mathbf{A}$ as a sum of two special matrices; that is,

$$\mathbf{A} = \mathbf{D} + \mathbf{N} = \begin{pmatrix} 2 & 0 \\ 0 & 2 \end{pmatrix} + \begin{pmatrix} 0 & 0 \\ 1 & 0 \end{pmatrix},$$

where $\mathbf{D}$ is a diagonal matrix and $\mathbf{N}$ is nilpotent. Check that $\mathbf{N}^2$ is the zero matrix.

We know that $e^{\mathbf{D}t} = \begin{pmatrix} e^{2t} & 0 \\ 0 & e^{2t} \end{pmatrix}$. To find $e^{\mathbf{N}t}$, use the series

$$e^{\mathbf{N}t} = \mathbf{I} + \mathbf{N}t + \mathbf{0} + \mathbf{0} + \cdots = \begin{pmatrix} 1 & 0 \\ 0 & 1 \end{pmatrix} + \begin{pmatrix} 0 & 0 \\ 1 & 0 \end{pmatrix} t = \begin{pmatrix} 1 & 0 \\ t & 1 \end{pmatrix}.$$

Furthermore, $\mathbf{D}$ and $\mathbf{N}$ commute; that is

$$\mathbf{DN} = \begin{pmatrix} 2 & 0 \\ 0 & 2 \end{pmatrix}\begin{pmatrix} 0 & 0 \\ 1 & 0 \end{pmatrix} = \begin{pmatrix} 0 & 0 \\ 2 & 0 \end{pmatrix} = \begin{pmatrix} 0 & 0 \\ 1 & 0 \end{pmatrix}\begin{pmatrix} 2 & 0 \\ 0 & 2 \end{pmatrix} = \mathbf{ND},$$

so

$$e^{\mathbf{A}t} = e^{\mathbf{D}t+\mathbf{N}t} = e^{\mathbf{D}t}e^{\mathbf{N}t} = \begin{pmatrix} e^{2t} & 0 \\ 0 & e^{2t} \end{pmatrix}\begin{pmatrix} 1 & 0 \\ t & 1 \end{pmatrix} = \begin{pmatrix} e^{2t} & 0 \\ te^{2t} & e^{2t} \end{pmatrix} \equiv \mathcal{X}.$$

This means that the general solution of the system (4.28) is

$$\vec{x} = c_1\vec{x}_1 + c_2\vec{x}_2 = c_1\begin{pmatrix} e^{2t} \\ te^{2t} \end{pmatrix} + c_2\begin{pmatrix} 0 \\ e^{2t} \end{pmatrix},$$

where $\vec{x}_1$ and $\vec{x}_2$ are the two columns of the fundamental matrix $\mathcal{X}$. To check, show that the pair of functions $x = c_1 e^{2t}$, $y = c_1 t e^{2t} + c_2 e^{2t}$ satisfies the two simultaneous equations $x' = 2x$, $y' = x + 2y$ for any values of $c_1$ and $c_2$. ∎

### 4.5.2 Using the Matrix Exponential to Solve a Nonhomogeneous System.

The matrix exponential can also be used to solve a nonhomogeneous linear system

$$\vec{x}' = \mathbf{A}\vec{x} + \vec{b}(t),$$

with constant coefficient matrix $\mathbf{A}$.

If the system is written in the form $\vec{x}' - \mathbf{A}\vec{x} = \vec{b}(t)$ and multiplied on the left by $e^{-\mathbf{A}t}$, then

$$e^{-\mathbf{A}t}\vec{x}' - e^{-\mathbf{A}t}\mathbf{A}\vec{x} = e^{-\mathbf{A}t}\vec{b}(t). \tag{4.29}$$

## 4.5 Large Linear Systems; the Matrix Exponential

The multiplication rule for differentiation applies to matrices; that is, if the product of the matrices $\mathbf{Y}$ and $\mathbf{Z}$ is defined, then

$$\frac{d}{dt}(\mathbf{Y}(t)\mathbf{Z}(t)) = \mathbf{Y}(t)\mathbf{Z}'(t) + \mathbf{Y}'(t)\mathbf{Z}(t).$$

Using this and the fact that $\frac{d}{dt}\left(e^{-\mathbf{A}t}\right) = -\mathbf{A}\left(e^{-\mathbf{A}t}\right)$, and that $\mathbf{A}$ and $e^{-\mathbf{A}t}$ commute, (4.29) can be written as

$$\frac{d}{dt}\left(e^{-\mathbf{A}t}\vec{\mathbf{x}}\right) = e^{-\mathbf{A}t}\vec{\mathbf{b}}(t).$$

This last equation can be integrated from 0 to $t$ to give

$$e^{-\mathbf{A}t}\vec{\mathbf{x}} = \int_0^t e^{-\mathbf{A}s}\vec{\mathbf{b}}(s)ds + \vec{\mathbf{c}},$$

and multiplying both sides by $e^{\mathbf{A}t}$, we have

$$\vec{\mathbf{x}}(t) = e^{\mathbf{A}t}\int_0^t e^{-\mathbf{A}s}\vec{\mathbf{b}}(s)ds + e^{\mathbf{A}t}\vec{\mathbf{c}} = \int_0^t e^{\mathbf{A}(t-s)}\vec{\mathbf{b}}(s)ds + e^{\mathbf{A}t}\vec{\mathbf{c}},$$

where the term $e^{\mathbf{A}t}\vec{\mathbf{c}} \equiv \mathcal{X}\vec{\mathbf{c}}$ is the general solution of the corresponding homogeneous system.

Since the algebra in this method can become tedious, we will do a very simple example to illustrate how it works.

**Example 4.5.3.** *Solve the nonhomogeneous system*

$$\vec{\mathbf{x}}' = \mathbf{A}\vec{\mathbf{x}} + \vec{\mathbf{b}} = \begin{pmatrix} 0 & 2 & 0 \\ 0 & 0 & 1 \\ 0 & 0 & 0 \end{pmatrix}\vec{\mathbf{x}} + \begin{pmatrix} t \\ 1 \\ 2 \end{pmatrix}. \tag{4.30}$$

*Solution.* The matrix $\mathbf{A}$ in this example is a nilpotent matrix. In this case $\mathbf{A}^3 = \mathbf{0}$. Because of this, the series expression for $e^{\mathbf{A}t}$ is just the finite sum

$$e^{\mathbf{A}t} = I + \mathbf{A}t + \mathbf{A}^2\frac{t^2}{2!} + \mathbf{0} + \cdots$$

$$= \begin{pmatrix} 1 & 0 & 0 \\ 0 & 1 & 0 \\ 0 & 0 & 1 \end{pmatrix} + \begin{pmatrix} 0 & 2 & 0 \\ 0 & 0 & 1 \\ 0 & 0 & 0 \end{pmatrix}t + \begin{pmatrix} 0 & 0 & 2 \\ 0 & 0 & 0 \\ 0 & 0 & 0 \end{pmatrix}\frac{t^2}{2!} = \begin{pmatrix} 1 & 2t & t^2 \\ 0 & 1 & t \\ 0 & 0 & 1 \end{pmatrix}.$$

Using this, the nonhomogeneous part of the solution is

$$\vec{\mathbf{x}}_p(t) = \int_0^t e^{\mathbf{A}(t-s)}\vec{\mathbf{b}}(s)ds = \int_0^t \begin{pmatrix} 1 & 2(t-s) & (t-s)^2 \\ 0 & 1 & t-s \\ 0 & 0 & 1 \end{pmatrix}\begin{pmatrix} s \\ 1 \\ 2 \end{pmatrix}ds$$

$$= \int_0^t \begin{pmatrix} s + 2(t-s) + 2(t-s)^2 \\ 1 + 2(t-s) \\ 2 \end{pmatrix}ds = \begin{pmatrix} \frac{3t^2}{2} + \frac{2t^3}{3} \\ t + t^2 \\ 2t \end{pmatrix},$$

where we have used the fact that for any matrix $\mathbf{A} = (a_{ij}(t))$, the integral is $\int_0^t \mathbf{A}(s)ds = (\int_0^t a_{ij}(s)ds)$. The general solution of the system can now be written as

$$\vec{x}(t) = e^{\mathbf{A}t}\vec{c} + \vec{x}_p(t) = \begin{pmatrix} 1 & 2t & t^2 \\ 0 & 1 & t \\ 0 & 0 & 1 \end{pmatrix} \begin{pmatrix} c_1 \\ c_2 \\ c_3 \end{pmatrix} + \begin{pmatrix} \frac{3t^2}{2} + \frac{2t^3}{3} \\ t + t^2 \\ 2t \end{pmatrix},$$

where $\vec{c}$ can be taken to be the initial condition vector $\begin{pmatrix} c_1 \\ c_2 \\ c_3 \end{pmatrix} = \begin{pmatrix} x(0) \\ y(0) \\ z(0) \end{pmatrix}$. You can check that $\vec{x}$ has components $x, y$, and $z$ that satisfy the differential equations $x' = 2y + t, y' = z + 1$, and $z' = 2$ for any values of the constants $c_1, c_2$, and $c_3$. ∎

### 4.5.3 Application: Mixing Problem with Three Compartments.
A mixing problem with three tanks is diagrammed below.

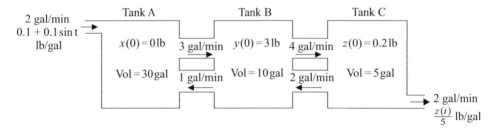

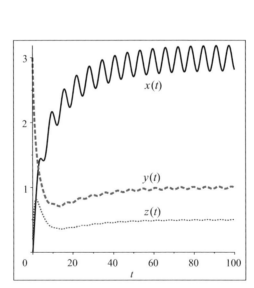

Figure 4.4. Pounds of salt in the three tanks at time $t$

Write out the rate equations for the three tanks, and show that the nonhomogeneous linear system can be written in matrix form as

$$\vec{x}'(t) = \mathbf{A}\vec{x}(t) + \vec{b}(t) = \begin{pmatrix} -0.1 & 0.1 & 0 \\ 0.1 & -0.5 & 0.4 \\ 0 & 0.4 & -0.8 \end{pmatrix} \vec{x}(t) + \begin{pmatrix} 0.2(1 + \sin(t)) \\ 0 \\ 0 \end{pmatrix}.$$

## 4.5 Large Linear Systems; the Matrix Exponential

The matrix exponential $e^{At}$ was found by using a Maple routine, `MatrixExponential`. Once you have $e^{At}$, the solution to the problem is the vector

$$\vec{x}(t) = \begin{pmatrix} x(t) \\ y(t) \\ z(t) \end{pmatrix} = e^{At} \begin{pmatrix} 0 \\ 3 \\ 0.2 \end{pmatrix} + \int_0^t e^{A(t-s)} \begin{pmatrix} 0.2(1 + \sin(s)) \\ 0 \\ 0 \end{pmatrix} ds.$$

Maple was also used to compute $\vec{x}(t)$, and a graph of the resulting functions $x(t)$, $y(t)$, and $z(t)$ is shown in Figure 4.4.

When $t$ gets large, the concentration of pollutant in each of the three tanks should oscillate around the average concentration in the liquid being pumped into the system. Does this happen?

**Exercises 4.5.** *In Exercises 1–6, find the exponential matrix $e^{At}$.*

1. $\mathbf{A} = \begin{pmatrix} 2 & 0 \\ 0 & -4 \end{pmatrix}$

2. $\mathbf{A} = \begin{pmatrix} 0 & 1 \\ 0 & 0 \end{pmatrix}$

3. $\mathbf{A} = \begin{pmatrix} -1 & 2 \\ 0 & -1 \end{pmatrix}$  *Hint: write $\mathbf{A} = \mathbf{D} + \mathbf{N}$, with $\mathbf{D}$ diagonal and $\mathbf{N}$ nilpotent.*

4. $\mathbf{A} = \begin{pmatrix} 0 & 1 \\ 1 & 0 \end{pmatrix}$  *Hint: use the Taylor series for $\sinh(t) = \frac{e^t - e^{-t}}{2}$ and $\cosh(t) = \frac{e^t + e^{-t}}{2}$.*

5. $\mathbf{A} = \begin{pmatrix} 1 & 0 \\ 1 & 0 \end{pmatrix}$

6. $\mathbf{A} = \begin{pmatrix} 1 & 1 \\ 0 & 0 \end{pmatrix}$

*In Exercises 7–12, use the exponential matrix $e^{At}$ (found in Exercises 1–6) to find the solution of $\vec{x}' = \mathbf{A}\vec{x}$, with initial conditions $x(0) = 1$, $y(0) = -1$. Check that the components $x(t)$ and $y(t)$ satisfy the two first-order equations.*

7. $\mathbf{A} = \begin{pmatrix} 2 & 0 \\ 0 & -4 \end{pmatrix}$

8. $\mathbf{A} = \begin{pmatrix} 0 & 1 \\ 0 & 0 \end{pmatrix}$

9. $\mathbf{A} = \begin{pmatrix} -1 & 2 \\ 0 & -1 \end{pmatrix}$  *Hint: write $\mathbf{A} = \mathbf{D} + \mathbf{N}$, with $\mathbf{D}$ diagonal and $\mathbf{N}$ nilpotent.*

10. $\mathbf{A} = \begin{pmatrix} 0 & 1 \\ 1 & 0 \end{pmatrix}$  *See the hint for Exercise 4.*

11. $\mathbf{A} = \begin{pmatrix} 1 & 0 \\ 1 & 0 \end{pmatrix}$

12. $\mathbf{A} = \begin{pmatrix} 1 & 1 \\ 0 & 0 \end{pmatrix}$

13. *Solve the nonhomogeneous IVP, using the exponential matrix found in Exercise 4.*

$$\vec{x}' = \begin{pmatrix} 0 & 1 \\ 1 & 0 \end{pmatrix} \vec{x} + \begin{pmatrix} 2 \\ 0 \end{pmatrix}, \quad \vec{x}(0) = \begin{pmatrix} 1 \\ 1 \end{pmatrix}$$

# 5

# Geometry of Autonomous Systems

In this chapter we will study the geometric method of phase plane analysis for autonomous 2-dimensional systems of first-order differential equations; that is, systems of the form

$$x' = f(x, y)$$
$$y' = g(x, y).$$

We showed in Chapter 3 that it is possible to picture all of the solutions of an autonomous second-order differential equation in a single plane, called the phase plane. It will be shown in Section 5.1 that this can be easily extended to a phase plane for any autonomous system of dimension two.

Section 5.2 examines the phase planes for autonomous 2-dimensional linear systems

$$\vec{x}' = \begin{pmatrix} x' \\ y' \end{pmatrix} = \begin{pmatrix} a & b \\ c & d \end{pmatrix} \begin{pmatrix} x \\ y \end{pmatrix} = \mathbf{A}\vec{x}. \tag{5.1}$$

Since we have already found algebraic formulas for the solutions of these systems in Chapter 4, the behavior of trajectories in the phase plane can be described very precisely. It will be shown that if $\det \mathbf{A} \neq 0$, the behavior of solution curves about the origin $\vec{x} = (0, 0)$ can be classified into six cases, completely determined by the trace and determinant of $\mathbf{A}$.

Using this information, and a very important theorem called the **Hartman-Grobman Theorem**, it is shown in Section 5.3 that much of what happens in the phase plane of an autonomous nonlinear system can be understood by looking at the behavior of solutions around the equilibrium points $(\bar{x}, \bar{y})$, where the slope functions $f(x, y)$ and $g(x, y)$ are both equal to zero.

The importance of the classification of linear systems, using the trace and determinant of the matrix $\mathbf{A}$, becomes even clearer in Section 5.4, where phase plane analysis is used to study bifurcations of autonomous systems of the form

$$x' = f(x, y, \alpha)$$
$$y' = g(x, y, \alpha)$$

containing a single parameter $\alpha$.

Section 5.5 contains two student projects, each using most of the material in the chapter.

## 5.1 The Phase Plane for Autonomous Systems

We are now at the point where we can begin to fully understand the behavior of 2-dimensional autonomous systems. It was shown in Chapter 3 that the entire family of solutions of an autonomous second-order equation $x'' = f(x, x')$ could be displayed in a 2-dimensional plane called a phase plane. You should look back and see that this was accomplished by first writing the equation as a 2-dimensional system $x' = y$, $y' = f(x, y)$, and then picturing solution curves as parametric curves $\{t, x(t), x'(t)\}$, $t_0 \leq t \leq t_{max}$, in the $(x, x')$-plane. We can now simply extend this approach to an arbitrary 2-dimensional autonomous system

$$\begin{aligned} x' &= f(x, y) \\ y' &= g(x, y), \end{aligned} \qquad (5.2)$$

where the solution curves $\mathcal{C}(t) = \{t, x(t), y(t)\}$, $t_0 \leq t \leq t_{max}$, will be plotted as parametric curves in the $(x, y)$-plane.

We know from calculus that the tangent vector to the parametric curve $\mathcal{C}(t)$ through any point $(x, y)$ in the plane is just the vector

$$\vec{T}(x(t), y(t)) = x'(t)\vec{i} + y'(t)\vec{j} \equiv f(x(t), y(t))\vec{i} + g(x(t), y(t))\vec{j}.$$

If the system is autonomous, these vectors can be computed at any point $(x, y)$ in the plane without solving the differential equations (5.2). The tangent vector $\vec{T}$ is just the vector we have been referring to, in Chapter 4, as $\vec{x}'(t)$. Once a grid of tangent vectors is plotted, we will have a picture of how solutions behave in various areas of the plane.

**Definition 5.1.** *A* **phase plane** *for the system* (5.2) *is an* $(x, y)$-*plane in which vectors* $\vec{v} = x'(t)\vec{i} + y'(t)\vec{j} \equiv f(x, y)\vec{i} + g(x, y)\vec{j}$ *have been drawn at a grid of points* $(x_i, y_i)$. *This set of vectors is said to form a* **vector field**, *or* **direction field**, *for the system. A* **phase portrait** *for* (5.2) *is a phase plane with enough solution curves drawn in to show how solutions behave in every part of the plane.*

Figure 5.1 shows three views of a set of solutions of the predator-prey system

$$\begin{aligned} x'(t) &= f(x, y) = x(2 - 3x) - 4xy \\ y'(t) &= g(x, y) = -y + 3xy. \end{aligned} \qquad (5.3)$$

All three graphs can be produced by the same `DEplot` instruction in Maple, by adding the option `scene=[x,y]` for the phase plot, and `scene=[t,x]` or `scene=[t,y]` for the time plots. The default option will produce the phase plot.

Assume that $x(t)$ is the size of the prey population, in thousands, at time $t$, and $y(t)$ is hundreds of predators. In this model it can be seen, from the equation for $x'$,

## 5.1 The Phase Plane for Autonomous Systems

that the prey population is growing logistically with a carrying capacity of $\frac{2}{3} \times 1000$ if no predators are around, and the predator population will die out exponentially in the absence of prey. As one might expect, interaction between predators and prey affect the prey negatively and the predator positively.

It is clear from the middle and right-hand graphs in Figure 5.1 that solutions plotted in the $(t, x)$-plane or the $(t, y)$-plane can intersect, but this is not the case in the $(x, y)$-plane since the system (5.3) satisfies the hypotheses of the following existence and uniqueness theorem.

**Theorem 5.1** (Existence and Uniqueness Theorem for 2-dimensional Systems). *If, in some region R of the $(x, y)$-plane, the system (5.2) satisfies the properties*
  *(1) $f(x, y)$ and $g(x, y)$ are continuous everywhere in R and*
  *(2) the first partial derivatives $\frac{\partial f}{\partial x}, \frac{\partial f}{\partial y}, \frac{\partial g}{\partial x},$ and $\frac{\partial g}{\partial y}$ are continuous for all $(x, y)$ in R,
then given any initial conditions $x(t_0) = x_0$, $y(t_0) = y_0$ at a point $(x_0, y_0)$ in R, there is a unique solution curve through that point and it exists for all values of t for which the solution remains in R.*

We need one more definition before beginning a serious study of the phase plane.

**Definition 5.2.** *An **equilibrium solution** of a 2-dimensional autonomous system (5.2) is a constant solution $x(t) = \bar{x}$, $y(t) = \bar{y}$ for which $f(\bar{x}, \bar{y})$ and $g(x, \bar{y})$ are simultaneously equal to 0.*

In the phase plane, the equilibrium solutions of (5.2) are points $(\bar{x}, \bar{y})$ (variously called **equilibrium points, fixed points, stationary points,** and **critical points**) where $f(\bar{x}, \bar{y}) = g(\bar{x}, \bar{y}) = 0$, which implies that the tangent vector $\vec{x}'$ at $(\bar{x}, \bar{y})$ is the zero vector. If a solution starts at an equilibrium point, it will stay there forever, since the rate of change of both $x$ and $y$ is zero; therefore, these points in the phase plane represent parametrically an entire solution curve $x(t) \equiv \bar{x}$, $y(t) \equiv \bar{y}$, $-\infty < t < \infty$.

For a linear system of the form $\vec{x}' = A\vec{x}$, an equilibrium solution must be a constant vector $\vec{x} = (\bar{x}, \bar{y})$ that satisfies the matrix equation $\vec{x}' = A\vec{x} = \vec{0}$; therefore, if the determinant of the matrix $A$ is not zero, there will be exactly one equilibrium solution $\vec{x} = \vec{0}$, corresponding to the point $(\bar{x}, \bar{y}) = (0, 0)$ in the phase plane.

On the other hand, for a nonlinear autonomous system, there can be any number of equilibrium solutions, and finding them may present a significant problem.

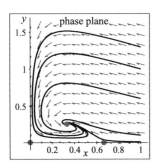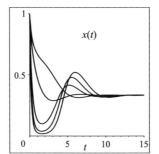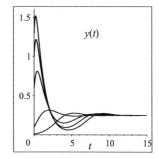

Figure 5.1. Three views of the predator-prey model (5.3)

**Example 5.1.1.** *For the predator-prey system* (5.3), *set $f(x,y)$ and $g(x,y)$ equal to zero and find all of the equilibrium solutions.*

*Solution.* We need to solve the simultaneous equations

$$x' = f(x,y) = x(2-3x) - 4xy = x(2-3x-4y) = 0$$
$$y' = g(x,y) = -y + 3xy = y(-1+3x) = 0.$$

If $y = 0$, the first equation requires $x(2-3x) = 0$. This gives two values for $x$, namely 0 and $\frac{2}{3}$, and produces two equilibrium points $(x,y) = (0,0)$ and $(x,y) = (2/3, 0)$. If $y \neq 0$, then $(-1+3x) = 0$ implies that $x = 1/3$, and from the first equation we must have $4y = 2 - 3x = 1$. This gives a third solution $(x,y) = (1/3, 1/4)$. These points are denoted by circles in the left-hand graph in Figure 5.1. ∎

In Section 5.2 a slightly more organized way of finding the equilibrium points will be given, but it always involves solving a system of nonlinear equations.

Only the positive quadrant of the phase plane for the system (5.3) has been pictured in Figure 5.1, since negative populations are not meaningful biologically. For practice, try to see if you can start at an arbitrary point in the phase plane in Figure 5.1 and describe what happens to the populations $x$ and $y$ as $t$ increases. You can then use the other two graphs to check your answer. In the rest of this chapter it will be important to know how to get as much information as you can from the phase portrait, since it is the graph that allows you to understand the behavior of the entire family of solutions.

**Example 5.1.2.** *In the phase plane for the predator-prey system* (5.3), *redrawn in Figure 5.2, assume initial populations $x(0) = 1$, $y(0) = 1$ and describe what happens as t increases.*

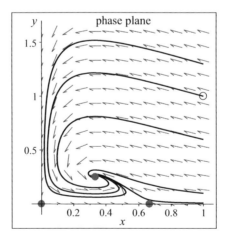

Figure 5.2. Phase plane for the system (5.3)

*Solution.* Following this trajectory in the phase plane, we can see that $x$ first decreases almost to zero and then returns to about 0.5 before heading back to the equilibrium point $(1/3, 1/4)$ as $t \to \infty$. The predator population $y$ starts to increase and then drops almost to zero before increasing again to its equilibrium value $y = 1/4$. What you do

## 5.2 Geometric Behavior of Linear Autonomous Systems

not have in the phase plane is information about the time at which various things like maxima and minima occur, but remember that the information you do have can all be obtained even when the system cannot be solved analytically. ∎

**Exercises 5.1.** *For the systems in Exercises 1–5, use your computer algebra system to draw a phase portrait. Find all the equilibrium solutions and describe how solutions behave in each part of the phase plane. In particular, where do solutions end up as $t \to \infty$?*

1. *This system comes from the mixing problem in Example 4.4.4, and the corresponding Exercise 19 in Section 4.4.*

$$x' = -0.06x + 0.025y + 0.2$$
$$y' = 0.06x - 0.075y$$

2. *An overdamped spring-mass equation in system form:*

$$x' = y$$
$$y' = -4x - 6y$$

3. *A predator-prey system:*

$$x' = 2x(1-x) - xy$$
$$y' = -\frac{1}{2}y + xy$$

4. *The "weak spring" equation:*

$$x' = y$$
$$y' = -3y - (x - x^3)$$

5. *The van der Pol equation (this will be seen later in Section 5.3):*

$$x' = y$$
$$y' = -x + y(1 - x^2)$$

## 5.2 Geometric Behavior of Linear Autonomous Systems

In this section we will describe in detail the phase planes for 2-dimensional constant coefficient linear systems $\vec{x}' = \mathbf{A}\vec{x}$ with $x' = ax + by$, $y' = cx + dy$. For a linear, homogeneous system with constant coefficients it is possible to describe the phase plane behavior in a very precise manner. This may seem like a lot of unnecessary work when we already know how to solve these systems analytically. It will, however, turn out to play an important role in understanding phase planes for nonlinear systems, and we have stressed the fact that most real-world problems are nonlinear.

Linear systems can be classified into six types in terms of the geometric behavior of solutions around the equilibrium solution $(0, 0)$ in the phase plane. There are two special cases that will be treated separately at the end of this section. One special case includes any system for which $\det(\mathbf{A}) = 0$. In Section 4.4 we showed that the product of the eigenvalues of a $2 \times 2$ matrix $\mathbf{A}$ is equal to $\det(\mathbf{A})$, so this special case includes all systems that have at least one zero eigenvalue. The second special case consists of systems for which the two eigenvalues of $\mathbf{A}$ are real and equal.

We will consider, first, systems with distinct real eigenvalues, and then those with complex eigenvalues. The special cases will be considered last.

### 5.2.1 Linear Systems with Real (Distinct, Nonzero) Eigenvalues.

We know from Section 4.4 that when $\mathbf{A}$ has two distinct, real eigenvalues $r_1$ and $r_2$, the general solution of the matrix system $\vec{\mathbf{x}}' = \mathbf{A}\vec{\mathbf{x}}$ can be written in the form

$$\vec{\mathbf{x}}(t) = c_1 e^{r_1 t} \vec{\mathbf{u}}_1 + c_2 e^{r_2 t} \vec{\mathbf{u}}_2,$$

where $(r_1, \vec{\mathbf{u}}_1)$ and $(r_2, \vec{\mathbf{u}}_2)$ are two eigenpairs for the matrix $\mathbf{A}$. We will again need to use the fact, from linear algebra, that eigenvectors corresponding to distinct eigenvalues are linearly independent (that is, not constant multiples of each other).

Let $(r_i, \vec{\mathbf{u}}_i)$ be one of the eigenpairs, with $\vec{\mathbf{u}}_i = \begin{pmatrix} u_{i1} \\ u_{i2} \end{pmatrix}$. Define $l_i$ to be the infinite line (in the phase plane) determined by the points $(0,0)$ and $(u_{i1}, u_{i2})$ (see Figure 5.3). Every nonzero vector that is a scalar multiple of $\vec{\mathbf{u}}_i$ (that is, every eigenvector for $r_i$) lies along the line $l_i$.

**Theorem 5.2.** *Assume the eigenvalues of $\mathbf{A}$ are real. If a line $l_i$ lies along an eigenvector of $\mathbf{A}$, then in the phase plane any solution of $\vec{\mathbf{x}}' = \mathbf{A}\vec{\mathbf{x}}$ that starts at a point $(x, y)$ on the line $l_i$ remains on $l_i$ for all $t$; as $t \to \infty$ it approaches the origin if the eigenvalue $r_i < 0$, or moves away from the origin if $r_i > 0$.*

*Proof.* If $(\bar{x}, \bar{y}) \neq (0,0)$ is any point on the line $l_i$, the position vector $\vec{\mathbf{x}} = \begin{pmatrix} \bar{x} \\ \bar{y} \end{pmatrix}$ from $(0,0)$ to $(\bar{x}, \bar{y})$ is some scalar multiple of the eigenvector $\vec{\mathbf{u}}_i$; that is, $\vec{\mathbf{x}} = c\vec{\mathbf{u}}_i$ for some constant $c$.

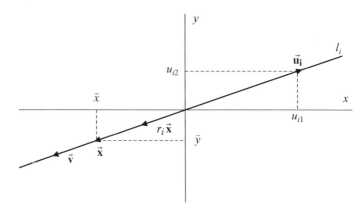

Figure 5.3. Tangent vector $\vec{\mathbf{v}}$ at $(\bar{x}, \bar{y})$, $r_i > 0$

The tangent vector $\vec{\mathbf{v}}$ to the solution curve through $(\bar{x}, \bar{y})$ satisfies

$$\vec{\mathbf{v}} \equiv \vec{\mathbf{x}}' = \mathbf{A}\vec{\mathbf{x}} = \mathbf{A}(c\vec{\mathbf{u}}_i) = c(\mathbf{A}\vec{\mathbf{u}}_i) = c(r_i \vec{\mathbf{u}}_i) = r_i(c\vec{\mathbf{u}}_i) = r_i \vec{\mathbf{x}};$$

therefore, for all values of $t$, the tangent vector to the solution curve through $(\bar{x}, \bar{y})$ is a vector in the direction of $\vec{\mathbf{x}}$ if $r_i > 0$, or in the opposite direction if $r_i < 0$. Since the tangent vector points along the line $l_i$ for all $t$, this means that the solution must move along $l_i$. It will move away from $(0,0)$ if $r_i > 0$ or toward $(0,0)$ if $r_i < 0$.

## 5.2 Geometric Behavior of Linear Autonomous Systems

Once the direction of the two eigenvectors is determined (they cannot have the same direction since they are linearly independent and cannot be constant multiples of each other), the geometry of the phase plane depends only on the signs of the two eigenvalues $r_1$ and $r_2$. Remember that it is being assumed that neither eigenvalue is equal to zero. There are three cases to consider.

**Case 1:** $r_1 < r_2 < 0$. If both eigenvalues are negative, the solutions along both $l_1$ and $l_2$ tend toward $(0,0)$ as $t \to \infty$. A solution $\vec{x}(t) = c_1 e^{r_1 t} \vec{u}_1 + c_2 e^{r_2 t} \vec{u}_2$ that does not lie along either line $l_1$ or $l_2$ will still tend to $(0,0)$ as $t \to \infty$ since $e^{r_1 t}$ and $e^{r_2 t}$ both approach 0; as $t$ increases $e^{r_1 t}$ tends to zero more quickly than $e^{r_2 t}$, and the solution will ultimately approach the origin in a direction tangent to the line $l_2$. A phase plane for the system $\vec{x}' = \begin{pmatrix} -7 & 2 \\ 4 & -5 \end{pmatrix} \vec{x}$, with eigenpairs $(r_1, \vec{u}_1) = \left(-9, \begin{pmatrix} 1 \\ -1 \end{pmatrix}\right)$ and $(r_2, \vec{u}_2) = \left(-3, \begin{pmatrix} 1 \\ 2 \end{pmatrix}\right)$, is shown in Figure 5.4. Six solutions, not along the eigenvector directions, have also been drawn by the Maple routine. The direction arrows in the phase plane indicate that all solutions are moving toward the origin. In this case the origin is a stable equilibrium and is called a **sink**.

**Case 2:** $r_1 > r_2 > 0$. With both eigenvalues positive, the phase plane will look exactly like that in **Case 1**, except all solutions will move away from the origin as $t \to \infty$. In this case the origin is unstable and is called a **source**.

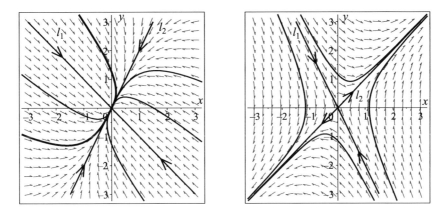

Figure 5.4. Real eigenvalues $r_1 < r_2 < 0$  Figure 5.5. Real eigenvalues $r_1 < 0 < r_2$

**Case 3:** $r_1 < 0 < r_2$. As $t \to \infty$, solutions along the line $l_1$, determined by $\vec{u}_1$, will tend toward the origin and solutions along the line $l_2$, determined by $\vec{u}_2$, will move away from the origin. Solutions of the form $\vec{x}(t) = c_1 e^{r_1 t} \vec{u}_1 + c_2 e^{r_2 t} \vec{u}_2$, with $c_1$ and $c_2$ both unequal to zero, will tend toward the line $l_1$ as $t \to -\infty$ and toward the line $l_2$ as $t \to \infty$. Figure 5.5 shows the phase plane for the system $\vec{x}' = \begin{pmatrix} 0 & 1 \\ 2 & -1 \end{pmatrix} \vec{x}$, with eigenpairs $(r_1, \vec{u}_1) = \left(-2, \begin{pmatrix} -1 \\ 2 \end{pmatrix}\right)$ and $(r_2, \vec{u}_2) = \left(1, \begin{pmatrix} 1 \\ 1 \end{pmatrix}\right)$. Note that no matter how close a solution gets to $(0,0)$, it will move off to infinity as $t \to \infty$, unless it lies exactly on the line $l_1$. In this case the origin is unstable, since almost all solutions tend away from it as $t \to \infty$, and it is called a **saddle point**. The line $l_1$ is called a **separatrix** for the saddle point. As can be seen in Figure 5.5, it separates solutions that ultimately go in different directions. □

**5.2.2 Linear Systems with Complex Eigenvalues.** If a complex eigenpair for $\mathbf{A}$ is $(\alpha + \iota\beta, \vec{\mathbf{u}} + \iota\vec{\mathbf{v}})$, we know that the general solution can be written in the form

$$\vec{x}(t) = c_1 e^{\alpha t}\big(\cos(\beta t)\vec{\mathbf{u}} - \sin(\beta t)\vec{\mathbf{v}}\big) + c_2 e^{\alpha t}\big(\cos(\beta t)\vec{\mathbf{v}} + \sin(\beta t)\vec{\mathbf{u}}\big)$$
$$\equiv e^{\alpha t}\big([c_1\cos(\beta t) + c_2\sin(\beta t)]\vec{\mathbf{u}} + [c_2\cos(\beta t) - c_1\sin(\beta t)]\vec{\mathbf{v}}\big)$$
$$\equiv e^{\alpha t} R\big(\cos(\beta t + \phi)\vec{\mathbf{u}} - \sin(\beta t + \phi)\vec{\mathbf{v}}\big),$$

where $R = \sqrt{c_1^2 + c_2^2}$, $R\cos(\phi) = c_1$, and $R\sin(\phi) = -c_2$. We have used the same trigonometric substitution that was used in Chapter 3 to write a sum of a sine function and a cosine function as a single sine or cosine function.

If $\alpha = 0$, it can be shown that the equation is a parametric equation for an ellipse with center at the origin in the $(x, y)$-plane. Proving this fact is not a trivial task, since the axes of the ellipse may be rotated about the origin.

For the complex eigenvalue case, there are again three different types of solutions in the phase plane, depending on whether the real part $\alpha$ of the eigenvalues is positive, negative, or zero.

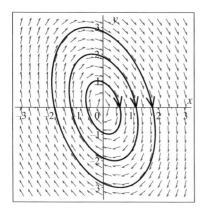

Figure 5.6. Pure imaginary eigenvalues

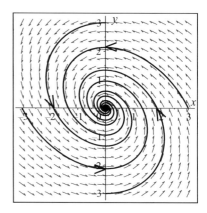

Figure 5.7. Complex eigenvalues with real part $\alpha < 0$

**Case 4:** $r = \alpha \pm \iota\beta$, $\alpha = 0$. Solutions $\vec{x}(t)$ in the phase plane will be concentric ellipses about the origin and will all be periodic with period $\frac{2\pi}{\beta}$. A phase plane for the system $\vec{x}' = \begin{pmatrix} 1 & 2 \\ -5 & -1 \end{pmatrix} \vec{x}$, with eigenvalues $\pm 3\iota$, is shown in Figure 5.6. The equilibrium solution $(0,0)$ in this case is called a **center**.

**Case 5:** $r = \alpha \pm \iota\beta$, $\alpha < 0$. In this case solutions still cycle periodically about the origin but the amplitude decreases exponentially as $t \to \infty$, due to the factor $e^{\alpha t}$. The solution curves are concentric spirals that approach the origin as $t \to \infty$. An example is

## 5.2 Geometric Behavior of Linear Autonomous Systems

shown in Figure 5.7 for the system $\vec{x}' = \begin{pmatrix} -1 & -2 \\ 2 & 0 \end{pmatrix} \vec{x}$, which has complex eigenvalues $-\frac{1}{2} \pm \frac{\sqrt{15}}{2}\iota$. The origin in this case is stable and is called a **spiral sink**.

**Case 6:** $r = \alpha \pm \iota\beta$, $\alpha > 0$. If the complex eigenvalues have positive real part $\alpha$ the solutions will spiral outward as $t \to \infty$. The phase plane looks exactly the same as in Case 5 except all of the arrows are reversed. In this case the origin is unstable and is called a **spiral source**.

In **Cases 4, 5,** or **6** the solutions can rotate either clockwise or counterclockwise about the origin. The direction of rotation is easily determined from the matrix **A**. At $(1, 0)$ in the phase plane, the tangent vector to the solution through $(1, 0)$ is given by

$$\vec{x}' = \begin{pmatrix} x' \\ y' \end{pmatrix} = \begin{pmatrix} a & b \\ c & d \end{pmatrix} \begin{pmatrix} 1 \\ 0 \end{pmatrix} = \begin{pmatrix} a \\ c \end{pmatrix}.$$

If the coefficient $c$ in the matrix is positive, then at the point $(1, 0)$ $y' = c$ is positive and hence $y$ is increasing. This implies that the solution curve through the point $(1, 0)$ will cut across the $x$-axis in an upward direction; therefore, the solution curves will be rotating in a counterclockwise direction. If $c < 0$ the rotation will be clockwise. In Figure 5.6, for the system with $\mathbf{A} = \begin{pmatrix} 1 & 2 \\ -5 & -1 \end{pmatrix}$, the coefficient $c = -5$ and the solutions can be seen to be rotating clockwise. In Figure 5.7, for the system with $\mathbf{A} = \begin{pmatrix} -1 & -2 \\ 2 & 0 \end{pmatrix}$, $c = 2$ and the solutions are rotating counterclockwise.

### 5.2.3 The Trace-determinant Plane.
For any 2-dimensional constant coefficient linear system $\vec{x}' = \mathbf{A}\vec{x}$, the type of equilibrium at the origin in the phase plane can be completely determined from the values of the trace and determinant of **A**, without having to solve for the eigenvalues and eigenvectors. It was previously shown that the sign of $K$, where $K = \text{tr}(\mathbf{A})^2 - 4\det(\mathbf{A})$, determines whether the eigenvalues are real or complex. If $K > 0$ the eigenvalues $r_1$ and $r_2$ are real, and if $\det(\mathbf{A}) \equiv r_1 r_2$ is positive $r_1$ and $r_2$ are of the same sign and if it is negative they are of opposite sign. If they are of the same sign, the sign of $\text{tr}(\mathbf{A}) \equiv r_1 + r_2$ determines whether they are both positive or both negative. If $K < 0$, so that the eigenvalues are complex, then we know that the real part $\alpha$ is equal to $\text{tr}(\mathbf{A})/2$, and therefore the sign of the trace will distinguish between systems with spiral source, spiral sink, or center at the origin. For complex

Table 5.1. Data needed to determine equilibrium type at $(0, 0)$

| Case | equilibrium type | eigenvalues | det **A** | tr(**A**) | tr(**A**)$^2$ − 4 det **A** |
|---|---|---|---|---|---|
| 1 | sink | real, $r_1 < r_2 < 0$ | + | − | + |
| 2 | source | real, $0 < r_1 < r_2$ | + | + | + |
| 3 | saddle | real, $r_1 < 0 < r_2$ | − | arbitrary | + |
| 4 | center | complex, $\alpha = 0$ | + | 0 | − |
| 5 | spiral sink | complex, $\alpha < 0$ | + | − | − |
| 6 | spiral source | complex, $\alpha > 0$ | + | + | − |

conjugate eigenvalues, $\det(\mathbf{A}) = (\alpha + \iota\beta)(\alpha - \iota\beta) = \alpha^2 + \beta^2$ is always positive. This information is summarized in Table 5.1.

This information can be encoded even more succinctly by using a diagram referred to as a **trace-determinant plane**. This is a plane in which the trace of a matrix $\mathbf{A}$ is

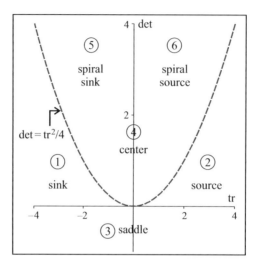

Figure 5.8. The trace-determinant plane

plotted along the horizontal axis and its determinant along the vertical axis (see Figure 5.8). The parabola $\det = \text{tr}^2/4$ (i.e., the set of points where $K = 0$) plotted in this plane separates systems $\vec{x}' = \mathbf{A}\vec{x}$ with real eigenvalues from those having complex eigenvalues. A system for which the point $(\text{tr}\,\mathbf{A}, \det \mathbf{A})$ lies exactly on the parabola (i.e., $\det = \text{tr}^2/4$) will have equal real eigenvalues. The phase plane for this special case will be described in Section 5.2.4. Using the information in Table 5.1, the six cases can be seen to fall into the six regions of the trace-determinant plane similarly labeled in Figure 5.8. Any system with $\det(\mathbf{A}) < 0$ can be seen to have a saddle point at $(0,0)$, since the two (real) eigenvalues will be of opposite sign if, and only if, their product $r_1 r_2 \equiv \det \mathbf{A}$ is negative. Case 4 (center) occurs only if the point $(\text{tr}\,\mathbf{A}, \det \mathbf{A})$ lies exactly on the positive vertical axis. The other special cases, to be considered below, occur when the point $(\text{tr}\,\mathbf{A}, \det \mathbf{A})$ lies on the horizontal axis (i.e., when $\det(\mathbf{A}) = 0$) or when it lies on the parabola ($\det = \text{tr}^2/4$).

The following example shows how useful the trace-determinant plane can be.

**Example 5.2.1.** *You are given the system* $\vec{x}' = \mathbf{A}\vec{x}$ *with* $\mathbf{A} = \begin{pmatrix} 2 & -3-\alpha \\ 1 & -\alpha \end{pmatrix}$ *and asked to ascertain the type of the equilibrium point* $(0,0)$ *for values of the parameter* $\alpha = 1, 2,$ *and 4. You are also asked to find the value of $\alpha$ for which the equilibrium changes from a spiral sink to a sink.*

*Solution.* We can easily compute $\text{tr}(\mathbf{A}) = 2 - \alpha$ and $\det(\mathbf{A}) = 2(-\alpha) - (1)(-3-\alpha) = 3 - \alpha$. Then the points $(\text{tr}(\mathbf{A}), \det(\mathbf{A})) = (2 - \alpha, 3 - \alpha)$ can be plotted in a trace-determinant plane to find the type of the equilibrium at $(0,0)$. This is done in the figure below.

## 5.2 Geometric Behavior of Linear Autonomous Systems

| $\alpha$ | $(2-\alpha, 3-\alpha)$ | type |
|---|---|---|
| 1 | (1, 2) | spiral source |
| 2 | (0, 1) | center |
| 4 | (−2, −1) | saddle |

The parametric curve $(2-\alpha, 3-\alpha)$ in the trace-determinant plane is a straight line with equation $\det = \operatorname{tr} + 1$. To find the value of $\alpha$ where the equilibrium changes from a spiral sink to a sink, it is only necessary to find the intersection of this line with the curve $\det = \operatorname{tr}^2/4$. Make sure you see why. This involves solving a quadratic equation for the trace:

$$\operatorname{tr}^2/4 = \operatorname{tr} + 1, \quad \text{so} \quad \operatorname{tr}^2/4 - \operatorname{tr} - 1 = 0, \quad \text{hence} \quad \operatorname{tr} = 2 - 2\sqrt{2}.$$

Since $\operatorname{tr} = 2 - \alpha$, this means the intersection occurs when $\alpha = 2 - \operatorname{tr} = 2\sqrt{2} \approx 2.83$. The two phase portraits below show the system when $\alpha = 2.5$ and $\alpha = 2.9$.

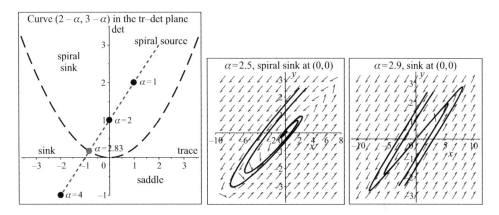

It is clear from the phase portraits that the equilibrium has changed from a spiral sink to a sink. For practice, find the value of $\alpha$ where the equilibrium at $(0,0)$ changes to a saddle point, and draw a phase plane for a value of $\alpha$ that puts the system into that region. ∎

### 5.2.4 The Special Cases.
The goal here is to understand what happens when the point $(\operatorname{tr}(\mathbf{A}), \det(\mathbf{A}))$ lands exactly on the horizontal axis or exactly on the parabola $\det = \operatorname{tr}^2/4$ in the trace-determinant plane.

**Case A: One zero eigenvalue** $r_1 = 0, r_2 \neq 0$. This will occur if the point $(\operatorname{tr}(\mathbf{A}), \det(\mathbf{A}))$ lies on the horizontal axis ($\det = 0$) in the trace-determinant plane. If $r_2 > 0$, the trace of $\mathbf{A}$ will be positive, and the point $(\operatorname{tr}(\mathbf{A}), \det(\mathbf{A}))$ will be in the right half of the trace-determinant plane, and if $r_2 < 0$ the point will be in the left half-plane. The general solution in this case is

$$\vec{x}(t) = c_1 e^{0t} \vec{u}_1 + c_2 e^{r_2 t} \vec{u}_2 = c_1 \vec{u}_1 + c_2 e^{r_2 t} \vec{u}_2.$$

Letting $l_1$ and $l_2$ be the lines containing the eigenvectors corresponding to the eigenvalues $r_1$ and $r_2$, assume first that $r_2 < 0$. Any point $\vec{x} = (\bar{x}, \bar{y})$ on the line $l_1$, that is, any scalar multiple of $\vec{u}_1$, is an equilibrium point, since $\vec{x}' = \mathbf{A}\vec{x} = \mathbf{A} c \vec{u}_1 = c \mathbf{A} \vec{u}_1 = c r_1 \vec{u}_1 = \mathbf{0}$.

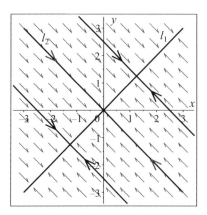

Figure 5.9. One zero eigenvalue

A solution starting on $l_2$ will move toward $(0,0)$ along $l_2$, and a solution starting at any other point will move along a line parallel to $l_2$ toward the equilibrium point at $c_1\vec{u}_1$. Figure 5.9 shows a phase plane for the system $\vec{x}' = \begin{pmatrix} -1 & 1 \\ 1 & -1 \end{pmatrix} \vec{x}$ with eigenpairs $(r_1, \vec{u}_1) = \left(0, \begin{pmatrix} 1 \\ 1 \end{pmatrix}\right)$ and $(r_2, \vec{u}_2) = \left(-2, \begin{pmatrix} 1 \\ -1 \end{pmatrix}\right)$. If $r_2 > 0$, all of the arrows along the lines parallel to $l_2$ are reversed.

The case where both eigenvalues of $\mathbf{A}$ are zero is very interesting and is left for the student (see Exercise 21. at the end of the section).

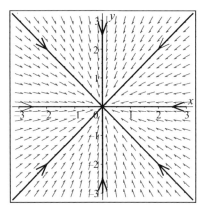

Figure 5.10. Equal real eigenvalues, linearly independent eigenvectors

**Case B: Two equal real (nonzero) eigenvalues** $r_1 = r_2 = r$. This occurs when the point $(\mathrm{tr}(\mathbf{A}), \det(\mathbf{A}))$ lies on the parabola $\det = \mathrm{tr}^2/4$ in the trace-determinant plane. If there are two linearly independent eigenvectors $\vec{u}_1$ and $\vec{u}_2$ corresponding to the eigenvalue $r$, then every vector in the plane can be written as a linear combination $c_1\vec{u}_1 + c_2\vec{u}_2$. This means that every vector in the plane is an eigenvector of $\mathbf{A}$ (see Exercise 20.).

## 5.2 Geometric Behavior of Linear Autonomous Systems

The general solution is $\vec{x}(t) = c_1 e^{rt} \vec{u}_1 + c_2 e^{rt} \vec{u}_2 = e^{rt}(c_1 \vec{u}_1 + c_2 \vec{u}_2)$, and therefore every solution moves along a straight line, either toward the origin if $r < 0$ or away if $r > 0$. Figure 5.10 shows a phase plane for the system $\vec{x}' = \begin{pmatrix} -2 & 0 \\ 0 & -2 \end{pmatrix} \vec{x}$ which has the single eigenvalue $r = -2$, and for which every vector in the plane is an eigenvector.

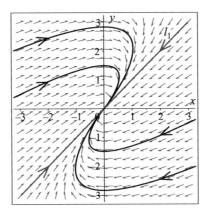

Figure 5.11. Equal real eigenvalues, single eigenvector

If only one linearly independent eigenvector exists, then according to (4.22) the solution is

$$\vec{x}(t) = c_1 e^{rt} \vec{u}_1 + c_2 e^{rt}(t \vec{u}_1 + \vec{u}^*),$$

where $\vec{u}^*$ is any vector satisfying $(\mathbf{A} - r\mathbf{I})\vec{u}^* = \vec{u}_1$. If $r < 0$, solutions along $l_1$ will tend to $(0, 0)$ as $t \to \infty$. Any other solution will be seen to tend to $(0, 0)$ and approach the line $l_1$ as $t \to \infty$. If $r > 0$, the phase plane has exactly the same form, but with the arrows all reversed. Figure 5.11 shows a phase plane for the system $\vec{x}' = \begin{pmatrix} -4 & 2 \\ -2 & 0 \end{pmatrix} \vec{x}$ which has a multiple eigenvalue $r = -2$ and only one linearly independent eigenvector direction. An eigenpair for this matrix is $(r, \vec{u}_1) = \left(-2, \begin{pmatrix} 1 \\ 1 \end{pmatrix}\right)$. The origin, in this case, is often referred to as an **improper node**, and it is stable if $r < 0$ and unstable if $r > 0$.

**Exercises 5.2.** *For each matrix* $\mathbf{A}$ *below, compute the trace and determinant. Use the trace-determinant plane to decide what type of equilibrium the system* $\vec{x}' = \mathbf{A}\vec{x}$ *has at* $(0, 0)$.

1. $\mathbf{A} = \begin{pmatrix} 1 & 2 \\ -2 & -2 \end{pmatrix}$

2. $\mathbf{A} = \begin{pmatrix} 2 & 3 \\ -1 & 0 \end{pmatrix}$

3. $\mathbf{A} = \begin{pmatrix} -4 & 2 \\ -1 & 1 \end{pmatrix}$

4. $\mathbf{A} = \begin{pmatrix} -1 & 1 \\ 1 & -2 \end{pmatrix}$

5. $\mathbf{A} = \begin{pmatrix} 2 & 3 \\ -2 & -2 \end{pmatrix}$

6. $\mathbf{A} = \begin{pmatrix} 1 & 2 \\ 1 & 3 \end{pmatrix}$

For each system in 7–12, find the eigenpairs (you will probably want to use technology for this) and sketch by hand a phase portrait for the system. If the eigenvalues are real, the eigenvectors should be included in the sketch. Put arrows on each solution trajectory to denote the direction of motion.

7. $x' = x + 2y,\ y' = -2x - 2y$

8. $x' = 2x + 3y,\ y' = -x$

9. $x' = -4x + 2y,\ y' = -x + y$

10. $x' = -x + y,\ y' = x - 2y$

11. $x' = 2x + 3y,\ y' = -2x - 2y$

12. $x' = x + 2y,\ y' = x + 3y$

Write each spring-mass equation 13–18 as a system and use the trace-determinant plane or eigenvalues to determine the type of equilibrium at $(0,0)$. How does the type of the equilibrium compare with the type of damping (undamped, underdamped, critically damped, or overdamped)? Can you formulate a general statement about this?

13. $x'' + 4x' + 2x = 0$

14. $x'' + 9x = 0$

15. $3x'' + 2x' + x = 0$

16. $x'' + x' + 2x = 0$

17. $x'' + 5x' + 4x = 0$

18. $x'' + 2x' + x = 0$

19. Write the equation $mx'' + bx' + kx = 0$ in matrix form (i.e., with $y = x'$). Show that the characteristic polynomial $\det(\mathbf{A} - r\mathbf{I})$ of the coefficient matrix $\mathbf{A}$ has the same roots as the characteristic polynomial $mr^2 + br + k$ defined in Chapter 3.

20. If $\vec{u}_1$ and $\vec{u}_2$ are two eigenvectors of a matrix $\mathbf{A}$ for the same eigenvalue $r$, show that for any constants $c_1$ and $c_2$ the vector $c_1\vec{u}_1 + c_2\vec{u}_2$ is also an eigenvector corresponding to the eigenvalue $r$.

21. **Computer Project.** The behavior of solutions of $\vec{x}' = \mathbf{A}\vec{x}$ around the origin in the phase plane has been described for every case except one; that is, when the eigenvalues $r_1$ and $r_2$ of $\mathbf{A}$ are both zero. In this case, both the trace and determinant of $\mathbf{A}$ are zero, and the system is represented in the trace-determinant plane by the point $(0, 0)$; therefore, you might expect the phase portrait to look rather strange. A slight change, no matter how small, in the matrix $\mathbf{A}$ could put the system into any one of the six regions in the trace-determinant plane, causing a distinct change in the phase portrait.

## 5.2 Geometric Behavior of Linear Autonomous Systems

(a) *Show that if both eigenvalues of $\mathbf{A}$ are zero, then either $\mathbf{A} \equiv \mathbf{0}$ (the zero matrix) or else it has the form*

$$\begin{pmatrix} a & -b \\ \dfrac{a^2}{b} & -a \end{pmatrix}, \quad a \text{ arbitrary}, \ b \neq 0. \tag{5.4}$$

*(Hint: Just try to construct a matrix with trace and determinant both zero.)*

(b) *If $\mathbf{A} \equiv \mathbf{0}$, show that every point in the phase plane of $\vec{\mathbf{x}}' = \mathbf{A}\vec{\mathbf{x}}$ is an equilibrium point (remember that the tangent vector at each point is $\vec{\mathbf{x}}'$).*

In the remainder of the problem, assume $\mathbf{A}$ has the form given in (5.4).

(c) *Show that the vector $\vec{\mathbf{u}}_1 = \begin{pmatrix} b \\ a \end{pmatrix}$ is an eigenvector of $\mathbf{A}$ corresponding to the double eigenvalue $r = 0$ and that every other eigenvector must be a scalar multiple of $\vec{\mathbf{u}}_1$.*

(d) *Show that the vector $\vec{\mathbf{u}}^* = \begin{pmatrix} 0 \\ -1 \end{pmatrix}$ satisfies $(\mathbf{A} - 0\mathbf{I})\vec{\mathbf{u}}^* = \vec{\mathbf{u}}_1$, and write out a general solution for the system, using the Case 3 formula*

$$\vec{\mathbf{x}}(t) = c_1 e^{rt} \vec{\mathbf{u}}_1 + c_2 e^{rt} (t\vec{\mathbf{u}}_1 + \vec{\mathbf{u}}^*) = \begin{pmatrix} x(t) \\ y(t) \end{pmatrix}.$$

(e) *Using your solution from part (d), show that at any time $t$, $x(t)$ and $y(t)$ are related by $y(t) = \dfrac{a}{b} x(t) - c_2$; thus, all the trajectories in the phase plane lie along parallel lines of slope $\dfrac{a}{b}$. The constant $c_2$ depends on the initial conditions.*

(f) *Show that every point on the line $y = \dfrac{a}{b} x - c_2$ with $c_2 = 0$ is an equilibrium point. Hint: if $(x, y)$ is any point in the phase plane satisfying $y = \dfrac{a}{b} x$, show that the tangent vector $\vec{\mathbf{x}}'$ at $(x, y)$ must be the zero vector.*

(g) *If $b > 0$, show that if a point $(x, y)$ in the phase plane lies above the line $y = \dfrac{a}{b} x$ (that is, satisfies $y > \dfrac{a}{b} x$), then $x'(t)$ is negative. This means the solution at that point is moving to the left. If $y < \dfrac{a}{b} x$, show that $x'(t)$ is positive, so the solution through that point is moving to the right. On the other hand, if $b < 0$, trajectories above the line $y = \dfrac{a}{b} x$ move right, and those below the line move left.*

(h) *Use everything you have learned above to draw a complete phase portrait for the system*

$$\vec{\mathbf{x}}' = \mathbf{A}\vec{\mathbf{x}} = \begin{pmatrix} 1 & -2 \\ \dfrac{1}{2} & -1 \end{pmatrix} \vec{\mathbf{x}}.$$

*Draw enough trajectories to determine the behavior of every solution in the phase plane. Describe their behavior in your own words.*

**Hard Problem:** *The matrix $\mathbf{A}$ in (h) can be changed slightly to put it into different regions in the trace-determinant plane. For each of the four regions 1, 3, 4, and 5 in Figure 5.8 construct a matrix $\mathbf{A}^*$, very close to $\mathbf{A}$, for which $(\mathrm{tr}(\mathbf{A}^*), \det(\mathbf{A}^*))$ is in the given region. Sketch a phase portrait for the system $\vec{\mathbf{x}}' = \mathbf{A}^* \vec{\mathbf{x}}$. You should try to visualize the phase portrait morphing (gradually changing) from its form when the matrix is $\mathbf{A}$ into the form it takes when the matrix is $\mathbf{A}^*$.*

# 5.3 Geometric Behavior of Nonlinear Autonomous Systems

Using the results from the previous section, we are now in a position to describe geometrically the phase plane behavior of solutions of arbitrary 2-dimensional autonomous systems of the form

$$\begin{aligned} x' &= f(x,y) \\ y' &= g(x,y). \end{aligned} \quad (5.5)$$

In the process of doing this we will refer to three important theorems that have made it possible to begin to understand the behavior of nonlinear systems, and this is where your introduction to dynamical systems really begins.

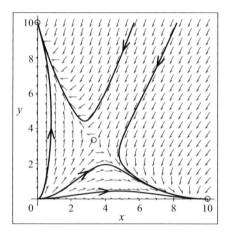

Figure 5.12. Phase portrait for system (5.6)

Figure 5.12 shows a numerically generated phase portrait for the system

$$\begin{aligned} x' &= f(x,y) = x(2-0.2x) - 0.4xy \\ y' &= g(x,y) = y(4-0.4y) - 0.8xy. \end{aligned} \quad (5.6)$$

This system, which is clearly nonlinear, is an example of a **Lotka-Volterra competing species model**, in which $x(t)$ and $y(t)$ represent the populations at time $t$ of two biological species that are competing with each other for some resource in their joint ecosystem. Each population is growing logistically, as defined in Section 2.1, but is negatively influenced by its interaction with the other population (the two terms $0.4xy$ and $0.8xy$ represent this interaction). Nothing new is involved in drawing a phase plane for this system. The direction vectors at each point $(x,y)$ are $\vec{v} = f(x,y)\vec{i} + g(x,y)\vec{j}$, and they can be computed without solving the system. To sketch the trajectories, numerical solutions can be easily computed using the methods described in Section 3.7.

Suppose a biologist looked at Figure 5.12, and then asked what would happen over the long term if initially the populations were around the values $x(0) = 4$, $y(0) = 4$. Could you give him an answer? It appears as though solutions will end up either at $(0,10)$ or $(10,0)$. In the first case the population $x$ has been driven to extinction and only $y$ remains, and in the second case the opposite occurs. Knowing which of these two possibilities will occur could be important to a population biologist. In this section we will learn how to find out precisely what is going on in this phase plane.

## 5.3 Geometric Behavior of Nonlinear Autonomous Systems

Our first problem will be to find all the equilibrium solutions for the system; that is, the points $(x, y)$ in the phase plane where $f(x, y)$ and $g(x, y)$ are both equal to zero. As we stated earlier, this is no longer as simple as it was for linear systems, and one possible technique for finding them will be described below.

In this problem, we will find that one of the equilibrium points is at $(\frac{10}{3}, \frac{10}{3})$. (Check this!) In a small region around this point in the phase plane (indicated by a circle in Figure 5.12), solutions appear to behave very much like the solutions of a linear system around a saddle point. This is no coincidence, and we will show how to make this statement precise.

A nonlinear system such as (5.6) can have any number of equilibrium solutions, but we will be able to describe the behavior of the system in a small neighborhood around each equilibrium by using a very important theorem, called the **Hartman-Grobman Theorem**, that was proved less than 100 years ago (this is a very short time in mathematical history).

Essentially, this theorem says that if an equilibrium point $(\bar{x}, \bar{y})$ is moved to the origin in the phase plane by making a change of dependent variables

$$u = x - \bar{x}, \quad v = y - \bar{y}$$

and if all of the nonlinear terms in $u$ and $v$ are discarded, then close to $(\bar{x}, \bar{y})$ the nonlinear system behaves very much like the linearized system in $u$ and $v$ at $(u, v) = (0, 0)$. Basically, this works because when $u$ and $v$ are close to zero, higher-order terms like $u^2$, $v^2$, $uv$, etc., are much smaller and can essentially be ignored. There is one restriction; the equilibrium of the linearized system must be a **hyperbolic equilibrium**. This means that it cannot be of a type that lies on one of the special-case curves in the trace-determinant plane; that is, the eigenvalues cannot be zero or have zero real part. In these special cases, a small perturbation can change the system from one type to another. A center is considered to be one of the special cases in this sense, in that it can easily become either a spiral sink or a spiral source by a change in the equations, no matter how small.

### 5.3.1 Finding the Equilibrium Points. 

In order to find the equilibrium solutions of the system (5.5), we will require the following definition.

**Definition 5.3.** *The **nullclines** for $x'$, in a system 5.5, are all the curves in the phase plane along which $x' = 0$. These can be found by setting $f(x, y) = 0$ and, if possible, solving explicitly for $y$ as a function of $x$ or for $x$ as a function of $y$. Similarly, the curves where $g(x, y) = 0$ are called **nullclines** for $y'$.*

At any point along a nullcline for $x'$, since $f(x, y) = 0$, the tangent vector $f(x, y)\vec{\mathbf{i}} + g(x, y)\vec{\mathbf{j}}$ in the phase plane will be vertical, and along a nullcline for $y'$ the tangent vector will be horizontal. Make sure you see why this is true!

One way to find all the equilibrium points is to sketch the nullclines for $x'$ and $y'$. Any point of intersection of an $x'$-nullcline with a $y'$-nullcline will be a point where both $f(x, y)$ and $g(x, y)$ are zero; therefore, it will be an equilibrium point.

In our competing species example, to find the nullclines for $x'$, we set $f(x, y) = x(2 - 0.2x - 0.4y) = 0$; therefore, the nullclines for $x'$ consist of the two curves $x \equiv 0$ (the $y$-axis) and the straight line $y = \frac{2-0.2x}{0.4} = 5 - \frac{1}{2}x$. Similarly, the two nullclines for

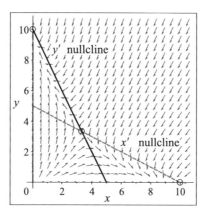

Figure 5.13. Nullclines for the system (5.6)

$y'$ are $y \equiv 0$ and $y = \frac{4 - 0.8x}{0.4} = 10 - 2x$. These curves are shown in the phase plane for the system (5.6) in Figure 5.13. In general the nullclines are not solution curves. In this figure, the tangent vectors in the direction field have been drawn with their midpoints at the grid points, and most of them are not exactly at points on the nullclines. This is why most of the arrows do not appear to be exactly horizontal or vertical when they cross the nullclines.

For the competing species model (5.6) there are four equilibrium points, namely, $(\bar{x}, \bar{y}) = (0,0), (0,10), (10,0)$, and $(\frac{10}{3}, \frac{10}{3})$. The fourth point was found by solving $2 - 0.2x - 0.4y = 0$, $4 - 0.4y - 0.8x = 0$ simultaneously for $x$ and $y$. Make sure that you see that these are all the equilibrium points. Trace along each nullcline in Figure 5.13 and see where it intersects a nullcline of the opposite type. Remember that the $y$-axis is a nullcline for $x'$ and the $x$-axis is a nullcline for $y'$.

### 5.3.2 Determining the Type of an Equilibrium.
The **type** of the equilibrium at $(\bar{x}, \bar{y}) = (0,0)$, for the nonlinear system (5.6), is the easiest to determine. If we substitute $u = x - 0$ and $v = y - 0$ into the equations, and drop the nonlinear terms in $u$ and $v$, the equation $u' = 2u - 0.2u^2 - 0.4uv$ becomes $u' \approx 2u$. Similarly $v' = 4v - 0.4v^2 - 0.8uv$ reduces to $v' \approx 4v$. These linearized equations $u' \approx 2u, v' \approx 4v$ can be written in matrix form as

$$\vec{u}' \equiv \begin{pmatrix} u' \\ v' \end{pmatrix} = A\vec{u} = \begin{pmatrix} 2 & 0 \\ 0 & 4 \end{pmatrix} \vec{u},$$

and plotting the point $(\operatorname{tr}(A), \det(A)) = (6, 8)$ in the trace-determinant plane, we see that this linear system has a source at $(0, 0)$. Check it! The Hartman-Grobman Theorem then tells us that the system (5.6) also has a source at $(x, y) = (0, 0)$.

In Figure 5.14 it can be seen that close to the origin all of the solutions are moving away, and in a small neighborhood of $(0, 0)$ solutions will behave like those around a linear source. This is exactly what the Hartman-Grobman Theorem says. In fact, the eigenvectors for $r = 2$, the smaller eigenvalue of $A$, lie along the $x$-axis (check it!); therefore, as $t$ goes backwards to $-\infty$ solutions tend to $(0, 0)$ in a direction tangent to the $x$-axis. This can be seen clearly in Figure 5.14.

## 5.3 Geometric Behavior of Nonlinear Autonomous Systems

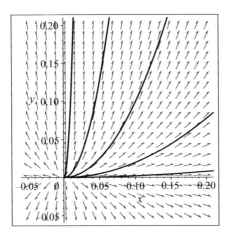

**Figure 5.14.** Trajectories of the system (5.6) close to $(0, 0)$

To determine the behavior of the system (5.6) at any of the other equilibrium points $(\bar{x}, \bar{y})$, it is necessary to substitute $u = x - \bar{x}, v = y - \bar{y}$ into (5.6) and throw away any nonlinear terms in $u$ and $v$. This is a messy calculation, but there is a standard method for doing it that works at any equilibrium point of a system (5.5), if the functions $f$ and $g$ are sufficiently differentiable. Because $\bar{x}$ and $\bar{y}$ are constants, and $u$ and $v$ are functions of $t$, $d(u(t))/dt = d(x(t) - \bar{x})/dt = dx/dt$ and similarly $d(v(t))/dt = d(y(t) - \bar{y})/dt = dy/dt$. If $f$ and $g$ are sufficiently differentiable, they can be expanded in Taylor series about $(\bar{x}, \bar{y})$. For example,

$$f(x(t), y(t)) = f(\bar{x} + u(t), \bar{y} + v(t))$$
$$= \underbrace{f(\bar{x}, \bar{y}) + \frac{\partial f}{\partial x} u(t) + \frac{\partial f}{\partial y} v(t)}_{\text{linear part}} + \text{ higher order terms in } u \text{ and } v,$$

where the partial derivatives are all evaluated at the equilibrium point $(x, y) = (\bar{x}, \bar{y})$, and $g(x, y)$ has a similar Taylor series expansion.

If $(\bar{x}, \bar{y})$ is an equilibrium point, then $f(\bar{x}, \bar{y})$ and $g(\bar{x}, \bar{y})$ are both equal to zero; therefore, when the equation $x' = f(x, y)$ is **linearized** at the equilibrium point, and the nonlinear terms are discarded, it becomes $x' \equiv u' \approx \frac{\partial f}{\partial x} u(t) + \frac{\partial f}{\partial y} v(t)$. Similarly, the equation for $y'$, under the substitution and linearization, becomes $y' \equiv v' \approx \frac{\partial g}{\partial x} u(t) + \frac{\partial g}{\partial y} v(t)$. This gives us an approximating linear system in the new dependent variables $u$ and $v$ of the form

$$\begin{pmatrix} u \\ v \end{pmatrix}' = \mathbf{J}(\bar{x}, \bar{y}) \begin{pmatrix} u \\ v \end{pmatrix} = \begin{pmatrix} \frac{\partial f}{\partial x} & \frac{\partial f}{\partial y} \\ \frac{\partial g}{\partial x} & \frac{\partial g}{\partial y} \end{pmatrix} \Bigg|_{x=\bar{x}, y=\bar{y}} \begin{pmatrix} u \\ v \end{pmatrix}.$$

The matrix $\mathbf{J}$, containing the partial derivatives, is called the **Jacobian matrix** of the system (5.5). You may remember having seen the Jacobian before if you took a course in vector calculus. When the partial derivatives are evaluated at an equilibrium point

$(x, y) = (\bar{x}, \bar{y})$, the trace and determinant of $\mathbf{J}(\bar{x}, \bar{y})$ can be used to determine the type of the equilibrium.

We are now able to give a formal definition for a hyperbolic equilibrium of a nonlinear system.

**Definition 5.4.** *An equilibrium point for an autonomous system of equations is called a* **hyperbolic equilibrium** *if the Jacobian at the equilibrium point has no eigenvalues with zero real part.*

The Hartman-Grobman Theorem tells us that if the point $(\text{tr}(\mathbf{J}(\bar{x}, \bar{y})), \det(\mathbf{J}(\bar{x}, \bar{y})))$ lies inside one of the regions 1, 2, 3, 5, or 6 in the trace-determinant plane, then in a small region about the equilibrium point $(\bar{x}, \bar{y})$ our nonlinear system has the same type of behavior as the corresponding linear system has around $(0, 0)$. Once the Jacobian matrix is computed, all the equilibrium points can be easily tested. The example below shows how this is done.

**Example 5.3.1.** *Compute the Jacobian matrix for the system (5.6) and use it to determine the type of the equilibrium points $(0, 10)$, $(10, 0)$, and $\left(\frac{10}{3}, \frac{10}{3}\right)$.*

*Solution.* If the equations are written in the form
$$x' = f(x, y) = 2x - 0.2x^2 - 0.4xy$$
$$y' = g(x, y) = 4y - 0.4y^2 - 0.8xy,$$

the partial derivatives are $\frac{\partial f}{\partial x} = 2 - 0.4x - 0.4y$, $\frac{\partial f}{\partial y} = -0.4x$, $\frac{\partial g}{\partial x} = -0.8y$, $\frac{\partial g}{\partial y} = 4 - 0.8y - 0.8x$, and the Jacobian matrix at any point $(x, y)$ is

$$\mathbf{J}(x, y) = \begin{pmatrix} 2 - 0.4x - 0.4y & -0.4x \\ -0.8y & 4 - 0.8y - 0.8x \end{pmatrix}.$$

At the equilibrium point $(x, y) = (0, 10)$, the Jacobian is $\mathbf{J}(0, 10) = \begin{pmatrix} -2 & 0 \\ -8 & -4 \end{pmatrix}$, with $(\text{tr}(\mathbf{J}), \det(\mathbf{J})) = (-6, 8)$. Since $\det = 8 < \text{tr}^2/4 = 9$, $(0, 10)$ is a sink.

At $(x, y) = (10, 0)$, $\mathbf{J}(10, 0) = \begin{pmatrix} -2 & -4 \\ 0 & -4 \end{pmatrix}$, with $(\text{tr}(\mathbf{J}), \det(\mathbf{J})) = (-6, 8)$; therefore, $(10, 0)$ is also a sink.

At $(\bar{x}, \bar{y}) = (\frac{10}{3}, \frac{10}{3})$,

$$\mathbf{J}(\bar{x}, \bar{y}) = \begin{pmatrix} -\frac{2}{3} & -\frac{4}{3} \\ -\frac{8}{3} & -\frac{4}{3} \end{pmatrix},$$

with $(\text{tr}(\mathbf{J}), \det(\mathbf{J})) = (-2, -\frac{24}{9})$. Since $\det(\mathbf{J})$ is negative, the point $(\bar{x}, \bar{y}) = (\frac{10}{3}, \frac{10}{3})$ is a saddle point for (5.6).

We could also test the equilibrium point at $(0, 0)$ by computing

$$\mathbf{J}(0, 0) = \begin{pmatrix} 2 & 0 \\ 0 & 4 \end{pmatrix}.$$

Since $(\text{tr}(\mathbf{J}), \det(\mathbf{J})) = (6, 8)$ lies in region 2 of the trace-determinant plane, it confirms our previous statement that $(0, 0)$ is a source for this system.

The sinks $(10, 0)$ and $(0, 10)$ are stable equilibria, and in a small enough region around each one all the solutions will tend toward the equilibrium point. ∎

## 5.3 Geometric Behavior of Nonlinear Autonomous Systems

If a nonlinear system has a saddle point, even more can be learned about the geometry of the phase plane, using a second theorem called the **Stable Manifold Theorem**. Remember that a linear system with a saddle point, having eigenpairs $(r_1, \vec{u}_1)$ and $(r_2, \vec{u}_2)$ where $r_1 < 0 < r_2$, has exactly two trajectories which tend to the saddle point as $t \to \infty$; and all other trajectories eventually tend away from the saddle. The Stable Manifold Theorem says that if $(r_1, \vec{u}_1)$ and $(r_2, \vec{u}_2)$ are eigenpairs for the linearized system about $(u, v) = (0, 0)$, and if $r_1 < 0 < r_2$, then there are exactly two trajectories in the phase plane of the nonlinear system which tend toward $(\bar{x}, \bar{y})$ as $t \to \infty$. They are not necessarily straight-line solutions, since nonlinear terms in $u$ and $v$ have been thrown away, and as we move away from $(u, v) = (0, 0)$ these nonlinear terms become significant. The theorem, however, says that as $t \to \infty$, these two trajectories approach $(\bar{x}, \bar{y})$ in a direction tangent to the eigenvector line $l_1$ corresponding to $\vec{u}_1$. Just as the separatrix for the linear system separates solutions going in different directions, these corresponding curves (called the **stable manifold** of the saddle point) do the same thing for the nonlinear system.

**Example 5.3.2.** *Find eigenpairs for the matrix* $\mathbf{J}(\frac{10}{3}, \frac{10}{3})$ *in Example 5.3.1 and use them to draw an approximation to the stable manifold for the system* (5.6) *about the saddle point.*

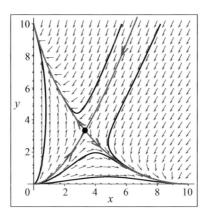

Figure 5.15. Stable and unstable manifolds of the saddle point at $(\frac{10}{3}, \frac{10}{3})$

*Solution.* Approximate eigenpairs for $\mathbf{J} = \begin{pmatrix} -\frac{2}{3} & -\frac{4}{3} \\ -\frac{8}{3} & -\frac{4}{3} \end{pmatrix}$ are $\left(-2.915, \begin{pmatrix} 0.5101 \\ 0.8601 \end{pmatrix}\right)$ and $\left(0.915, \begin{pmatrix} -0.6446 \\ 0.7645 \end{pmatrix}\right)$. If $\varepsilon$ is very small, good approximations to initial points for the two pieces of the stable manifold are given by $(\frac{10}{3} + 0.5101\varepsilon, \frac{10}{3} + 0.8601\varepsilon)$ and $(\frac{10}{3} - 0.5101\varepsilon, \frac{10}{3} - 0.8601\varepsilon)$. Do you see why? The trajectories through these points must be computed numerically for negative values of $t$, since they are heading toward the fixed point and our initial data is at the end of the trajectory. Figure 5.15 shows numerically computed trajectories with $\varepsilon = 0.01$ and $-8 \leq t \leq 0$. Solutions have also been drawn through the initial points $(\frac{10}{3} - 0.6446\varepsilon, \frac{10}{3} + 0.7645\varepsilon)$ and $(\frac{10}{3} + 0.6446\varepsilon, \frac{10}{3} - 0.7645\varepsilon)$.

These two trajectories correspond to the solutions of the linear system along the other eigenvector line $l_2$. They form what is called the unstable manifold of the saddle point, and they are moving away from the equilibrium point. ∎

We see that trajectories in the phase plane have now been clearly separated into two different types: those on the left of the stable manifold end up at the sink $(0, 10)$ and those on the right end up at $(10, 0)$. Proof of this last statement depends on a third theorem, called the **Poincaré-Bendixson Theorem**, which describes the types of limiting behavior (in the phase plane) of solutions of 2-dimensional autonomous systems. Basically, it says that as $t \to \pm\infty$ every solution that remains bounded must approach either an equilibrium point or a closed orbit in the plane. In the case of the system (5.6), it can be proved that solutions that enter the square $0 < x < 10$, $0 < y < 10$ can never leave. The Poincaré-Bendixson Theorem then implies that as $t \to \infty$ every solution inside the square must ultimately approach one of the two stable equilibria at $(0, 10)$ or $(10, 0)$.

In the next example we will see a system that has a single closed orbit called a **limit cycle**. This phenomenon occurs only for nonlinear systems, and it is entirely different from the case of concentric ellipses around a center for a linear system.

### 5.3.3 A Limit Cycle—the Van der Pol Equation.
The nonlinear second-order differential equation

$$x'' + \mu(x^2 - 1)x' + x = 0 \tag{5.7}$$

is named for an electrical engineer, Balthasar Van der Pol (1889–1959), and came out of work he was doing with oscillator circuits for radios, in 1924. In the early twentieth century radios were made using vacuum tubes; transistors had not yet been invented.

In a simple linear RLC-circuit the current $I(t)$ satisfies a differential equation of the form $LI'' + RI' + \frac{1}{C}I = E'(t)$. The term $RI'$ is the derivative of $\Delta V = RI$, the voltage drop across the resistor. In the circuit Van der Pol was studying, this voltage drop was modelled by a nonlinear function of the form $\Delta V = f(I) = bI^3 - aI$. The time derivative of this term, by the chain rule, is $\frac{d(f(I))}{dt} = \frac{df}{dI}\frac{dI}{dt} = (3bI^2 - a)I'(t)$. A change of variables can be used to reduce the resulting homogeneous equation, $LI'' + (3bI^2 - a)I' + \frac{1}{C}I = 0$, to the standard form given in (5.7).

Equation (5.7) is a second-order, nonlinear, autonomous equation, and we know that it can be written as a system by introducing a new variable $y = x'$. In this case, the equivalent system is

$$\begin{aligned} x' &= y & &= f(x, y) \\ y' &= -x - \mu(x^2 - 1)y & &= g(x, y). \end{aligned} \tag{5.8}$$

This system contains a parameter $\mu$, and it will be of interest to see how the behavior of the system depends on the value of $\mu$.

To find the equilibrium solutions of (5.8), we set both $f(x, y)$ and $g(x, y)$ equal to zero. The only point $(x, y)$ in the phase plane for which both $f$ and $g$ are simultaneously zero is the origin $(x, y) = (0, 0)$. (Check it!) To determine the behavior of the system

## 5.3 Geometric Behavior of Nonlinear Autonomous Systems

close to this equilibrium point, we compute the Jacobian matrix

$$\mathbf{J}(x,y) = \begin{pmatrix} \frac{\partial f}{\partial x} & \frac{\partial f}{\partial y} \\ \frac{\partial g}{\partial x} & \frac{\partial g}{\partial y} \end{pmatrix} = \begin{pmatrix} 0 & 1 \\ -1 - 2\mu xy & -\mu(x^2 - 1) \end{pmatrix}.$$

At the equilibrium point, the Jacobian is $\mathbf{J}(0,0) = \begin{pmatrix} 0 & 1 \\ -1 & \mu \end{pmatrix}$, with $(\text{tr}(\mathbf{J}), \det(\mathbf{J})) = (\mu, 1)$. In the physical problem studied by van der Pol the value of $\mu$ is positive, and as a first example we will look at the behavior of solutions of (5.7) with $\mu = 1$. With $(\text{tr}(\mathbf{J}), \det(\mathbf{J})) = (1, 1)$, the origin can be seen to be a spiral source.

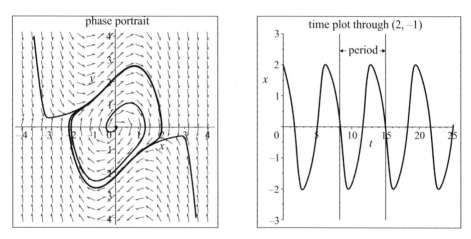

Figure 5.16. Van der Pol's equation with $\mu = 1$

A phase portrait for this equation is shown in Figure 5.16. As expected, solutions can be seen to be spiralling away from the origin. However, as they move away from the origin they all become attracted to a periodic solution called a **limit cycle**. There is actually a way to prove the existence of a limit cycle in this case, but it is too involved to give here. Solutions that start outside this cycle also approach it from outside as $t \to \infty$. We might have expected something like this to occur if we think of the Van der Pol equation as a spring-mass equation with nonconstant damping coefficient $\mu(x^2 - 1)$. Since the damping is positive if $x > 1$, solutions that start with $x$ large are slowed down, but solutions for which $x < 1$ are pushed away from the origin (by the negative damping).

The right-hand graph in Figure 5.16 shows a single solution plotted in the $(t, x)$-plane. As $x(t)$ gets close to the limit cycle, it oscillates in an almost periodic fashion. In the exercises you will be asked to determine how the "period" depends on the parameter $\mu$.

If you are interested in learning more about what can happen in the phase planes of 2-dimensional systems, and about the three theorems mentioned in this section, there is a nice discussion in Part III of a book by Hale and Koçak [J. Hale and H. Koçak, *Dynamics and Bifurcations*, Springer-Verlag, 1991]. For those who would like to learn more about dynamical systems in general, there is an excellent book by Steven Strogatz [Steven H. Strogatz, *Nonlinear Dynamics and Chaos*, Addison Wesley, 1994] which should be very readable for students who have made it to this point in Chapter 5.

**Exercises 5.3.** *Four nonlinear models are described in Exercises 1–4. For each one*

- Use the $x'$ and $y'$ nullclines to find all the equilibrium points in the given region of the phase plane.

- Compute the Jacobian matrix of the system.

- Determine the type of each equilibrium point (if it is a hyperbolic equilibrium).

- Use the above information to sketch a phase portrait by hand. Show as much detail as you can. Then use whatever technology you have available to generate the phase portrait, and compare it to your sketch.

1. **Competing species model.** *The functions $x(t)$ and $y(t)$ represent the population size, at time t, of two competing species in the same ecosystem. Their growth equations are given by*

$$x' = x(1-x) - xy$$
$$y' = y\left(\frac{3}{4} - y\right) - \frac{1}{2}xy.$$

Let $x$ vary between 0 and 1.5, and $y$ between 0 and 1.5.

2. **Damped pendulum model.** *This is similar to the equation that was discussed in Subsection 3.7.2 in Example 3.7.1. If we let $x \equiv \theta$, the second-order pendulum equation $\theta'' + \theta' + 4\sin(\theta) = 0$ is equivalent to the system*

$$x' = y$$
$$y' = -4\sin(x) - y.$$

Let $x$ vary between $-8$ and 8, and $y$ between $-10$ and 10.

3. **Western grasslands model**[1]. *This is a model of the competition between "good" grass and weeds, on a fixed area $\Omega$ of rangeland where cattle are allowed to graze. The two dependent variables $g(t)$ and $w(t)$ represent, respectively, the fraction of the area $\Omega$ colonized by the good (perennial) grass and the weeds at any time t. Letting the region $\Omega$ be the unit square, $\Omega = \{0 \leq g \leq 1, 0 \leq w \leq 1\}$, the model is given by*

$$\frac{dg}{dt} = R_g g\left(1 - g - 0.6w\frac{E+g}{0.31E+g}\right)$$
$$\frac{dw}{dt} = R_w w\left(1 - w - 1.07g\frac{0.31E+g}{E+g}\right).$$

The parameters $R_g$ and $R_w$ represent intrinsic growth rates of the grass and weeds, respectively, and the cattle stocking rate is introduced through the parameter $E$. For this problem, assume $R_g = 0.27, R_w = 0.4$, and $E = 0.3$. With these parameter values, there are 3 equilibrium points, $(0,0), (0,1), (1,0)$, on the boundary of the region $\Omega$; that is, points at which one or both of the species have died out. There are two more equilibrium points in the interior of the region. One of these latter two is a sink, and represents a stable situation in which the grass and the weeds both survive in abundance.

---

[1] For a nice write-up of this problem by Thomas LoFaro, Kevin Cooper, and Ray Huffaker, you can go to their website, **www.sci.wsu.edu/idea/Range**.

4. **Predator-prey model.** In this model, $x(t)$ represents the number of predators at time $t$, and $y(t)$ is the number of prey. The predators are affected positively by their interactions with the prey, while the effect of the interaction on the prey is negative. If no prey are available, the predators will die out exponentially. Notice that in the absence of predators the prey will increase exponentially without limit.

$$x' = -x + xy$$
$$y' = 4y - 2xy.$$

Let $x$ vary between 0 and 5, and $y$ between 0 and 5.

*The following problem involves an interesting 2-predator, 1-prey system. It is a 3-dimensional problem, but we can still find equilibrium points (in $(x, y, z)$-space) by setting each of the derivatives equal to zero.*

5. Two predators $X$ and $Y$, living on a single prey $Z$, are interacting in a closed ecosystem. Letting $x(t)$, $y(t)$, and $z(t)$ be the size of the respective populations at time $t$, the equations of growth can be written as

$$\begin{aligned} x'(t) &= -0.2x(t) + 0.5z(t)x(t) \\ y'(t) &= -0.1y(t) - 0.05x(t)y(t) + 0.4y(t)z(t) \\ z'(t) &= 0.4z(t)\big[1 - z(t)\big] - z(t)\big[0.2x(t) + 0.2y(t)\big]. \end{aligned} \quad (5.9)$$

Find all the equilibrium points $(\bar{x}, \bar{y}, \bar{z})$. Is there any equilibrium point where all three species coexist? Explain.

*The three problems below refer to the Van der Pol equation in Subsection 5.3.3.*

6. Think about what the phase portrait should look like if $\mu = 3$. Use technology to construct it, and describe, as precisely as you can, how it differs from the phase portrait shown in Figure 5.16.

7. For $\mu = 0.5, 1.0, 2.0, 4.0,$ and $7.0$ draw time plots, through the initial point $(x(0), y(0)) = (2, -1)$, showing the solution of (5.7) on the interval $0 \le t \le 25$. For each value of $\mu$, find the approximate period of oscillation (refer to Figure 5.16). State as precisely as you can how it varies with $\mu$.

8. Decide what you think should happen when $\mu$ is negative. Construct a phase portrait for $\mu = -1$ and describe it in your own words.

## 5.4 Bifurcations for Systems

In Definition 2.7 in Section 2.7 we defined what is meant by a bifurcation of an autonomous first-order differential equation with respect to a parameter. Definition 2.7 can also be applied to autonomous 2-dimensional systems. Basically, a system bifurcates at a value $\alpha = \alpha^*$ of a parameter if there is a sudden qualitative change in the behavior of solutions as $\alpha$ passes through the value $\alpha^*$. This kind of change is most easily seen in the phase plane.

The theory of bifurcations in dynamical systems is still evolving, and we will limit our discussion to second-order autonomous systems containing a single parameter $\alpha$;

that is, systems that can be written in the form

$$x' = f(x, y, \alpha) \tag{5.10}$$
$$y' = g(x, y, \alpha).$$

Even with these limitations the subject can get extremely involved; therefore, we are going to simply present examples of three familiar models in which bifurcations can occur. In each case the trace-determinant plane will be used to study how the solutions of the system change as the parameter $\alpha$ passes through its bifurcation values.

### 5.4.1 Bifurcation in a Spring-mass Model. 
We have already seen one simple example of a bifurcation in an autonomous 2-dimensional system. Given an unforced spring-mass equation of the form $mx'' + bx' + kx = 0$, we know that when the damping coefficient $b$ passes through the critical damping value $b^2 = 4mk$, the solutions of the system change character. To see how this shows up in our new way of looking at things, consider the equation

$$x'' + bx' + 4x = 0, \tag{5.11}$$

where $m = 1$, $k = 4$, and the damping coefficient $b$ is allowed to vary.

If we write this as a system

$$\begin{aligned} x' &= y &&= f(x, y) \\ y' &= -4x - by &&= g(x, y, b), \end{aligned} \tag{5.12}$$

then because the system is linear, it can be written in matrix form as

$$\vec{x}' = \begin{pmatrix} 0 & 1 \\ -4 & -b \end{pmatrix} \vec{x} = \mathbf{A}\vec{x},$$

where $\vec{x}$ is the vector $\begin{pmatrix} x \\ y \end{pmatrix}$.

We know that since $\det \mathbf{A} = 4 \neq 0$, the only equilibrium point is the origin $\vec{x} = \vec{0}$. The Jacobian matrix $\mathbf{J}(x, y) = \begin{pmatrix} \frac{\partial f}{\partial x} & \frac{\partial f}{\partial y} \\ \frac{\partial g}{\partial x} & \frac{\partial g}{\partial y} \end{pmatrix} = \begin{pmatrix} 0 & 1 \\ -4 & -b \end{pmatrix}$. For a linear system with constant coefficients, the Jacobian is just the matrix $\mathbf{A}$, and it is independent of $x$ and $y$. In Figure 5.17 a parametric curve is plotted, showing the point $(\operatorname{tr}(\mathbf{J}(0,0)), \det(\mathbf{J}(0,0))) = (-b, 4)$ in the trace-determinant plane as the value of the parameter $b$ varies. In this case the curve is just a horizontal line since the determinant of $\mathbf{A}$ is constant at 4.

As $b$ goes from $-\infty$ to $+\infty$ the critical point at $(0, 0)$ bifurcates three times, at $b = -4, 0$, and $4$. At $b = -4$ it changes from a source to a spiral source. At $b = 0$ it is a center (it would not necessarily be a center if the system (5.12) were nonlinear, since $(0, 0)$ is not a hyperbolic equilibrium when $b = 0$). For $b > 0$ the point becomes a spiral sink, and then changes into a sink as $b$ passes through the value 4. For a spring-mass equation, where the damping is assumed to be positive, we see that there is a single bifurcation at $b = 4$, where the damping changes from underdamped to overdamped. Remember that in the underdamped case, solutions in the phase plane are spiralling in to zero (that is, $(0, 0)$ is a spiral sink) and for the over damped case ($b > 4$), the solutions all tend to zero exponentially (that is, $(0, 0)$ is a sink).

## 5.4 Bifurcations for Systems

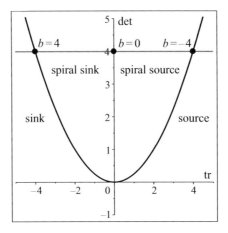

Figure 5.17. Trace-determinant plane for $x'' + bx' + 4x = 0$

### 5.4.2 Bifurcation of a Predator-prey Model.
We will now look at a slightly more complicated model that is nonlinear and can have more than one equilibrium solution.

The following system of equations is an example of a classical **Lotka-Volterra Predator-prey model**. It is used to describe the growth of two interacting populations, one of which is a predator, and the other its prey. We will pick a model containing a single variable parameter $N$:

$$\begin{aligned} x' &= -x + xy &&= f(x,y) \\ y' &= 4y\left(1 - \frac{y}{N}\right) - 2xy &&= g(x,y,N). \end{aligned} \quad (5.13)$$

The predator population $x(t)$ is affected positively by its interaction with the prey (denoted by the term $xy$), but will die out if no prey are available. The prey population $y(t)$ is affected negatively by the interaction (denoted by the term $-2xy$), and prey are limited in growth by their carrying capacity $N$. The equilibrium points, that is, points $(x,y)$ where both $f(x,y)$ and $g(x,y,N)$ are zero, can be seen to be

$$(0,0), (0,N), \left(2 - \frac{2}{N}, 1\right).$$

To find them, it helps to sketch the $x'$ and $y'$ nullclines.

The Jacobian matrix is

$$\mathbf{J}(x,y) = \begin{pmatrix} \frac{\partial f}{\partial x} & \frac{\partial f}{\partial y} \\ \frac{\partial g}{\partial x} & \frac{\partial g}{\partial y} \end{pmatrix} = \begin{pmatrix} y - 1 & x \\ -2y & 4 - \frac{8}{N}y - 2x \end{pmatrix}.$$

At the critical point $(0,0)$, $\mathbf{J}(0,0) = \begin{pmatrix} -1 & 0 \\ 0 & 4 \end{pmatrix}$, and it is easily seen that since $\det(\mathbf{J}(0,0)) = -4$, this point is a saddle for any value of $N$.

At the critical point $(0,N)$, $\mathbf{J}(0,N) = \begin{pmatrix} N - 1 & 0 \\ -2N & -4 \end{pmatrix}$, with $(\operatorname{tr}(\mathbf{J}), \det(\mathbf{J})) = (N - 5, 4(1 - N))$. This point will also be a saddle point when $N > 1$. When $N < 1$, the trace is negative and $(0,N)$ will either be a sink or a spiral sink.

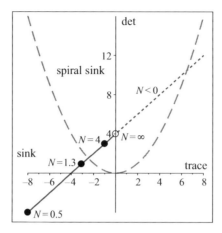

Figure 5.18. Trace-determinant plane for the predator-prey model, showing the path of the critical point $(2 - 2/N, 1)$ as $N$ varies. The path crosses the trace axis when $N = 1$.

The point we are most interested in is the third equilibrium point $(\bar{x}, \bar{y}) = (2 - \frac{2}{N}, 1)$ where the two species can coexist. This point will lie in the positive quadrant only if $N > 1$. The Jacobian at $(\bar{x}, \bar{y})$ is $\mathbf{J}(2 - \frac{2}{N}, 1) = \begin{pmatrix} 0 & 2 - \frac{2}{N} \\ -2 & -\frac{4}{N} \end{pmatrix}$, with $(\mathrm{tr}(\mathbf{J}), \det(\mathbf{J})) = (-\frac{4}{N}, -\frac{4}{N} + 4)$.

Figure 5.18 shows the path of this point in the trace-determinant plane, as $N$ goes from 0.5 to $\infty$. The point is a saddle for $N < 1$ and changes to a sink at the bifurcation value $N = 1$. To find the value of $N$ where the point bifurcates to become a spiral sink, we need to solve

$$\det \mathbf{J} = \frac{[\mathrm{tr}(\mathbf{J})]^2}{4} \implies 4 - \frac{4}{N} = \frac{(-4/N)^2}{4}.$$

This gives two nonzero values $N = \frac{1 \pm \sqrt{5}}{2}$. Therefore, the bifurcation from sink to spiral sink occurs at $N = \frac{1+\sqrt{5}}{2} \approx 1.618$. Since $N$ is the carrying capacity of the prey, $N < 0$

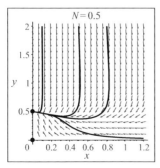 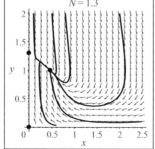 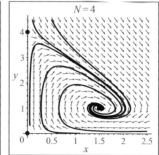

Figure 5.19. Predator-prey model with $N = 0.5, 1.3,$ and $4.0$

## 5.4 Bifurcations for Systems

is not a biologically realistic value. (For $N < 0$ this point is either a source or spiral source.)

The bifurcations that are biologically meaningful are at $N = 1$ where the point switches from a saddle to a sink, and at $N = \frac{1+\sqrt{5}}{2}$ where it changes from a sink to a spiral sink. In Figure 5.19 three phase portraits are shown, with values of $N = 0.5, 1.3$, and $4$. In the left-hand graph, with $N = 0.5$, the critical point $(\bar{x}, \bar{y})$ has moved out of the positive quadrant, and all solutions can be seen to be converging to the sink at $(0, N)$ where the predator becomes extinct. The middle and right-hand graphs show the change in solution behavior when $(\bar{x}, \bar{y})$ changes from a sink to a spiral sink. For a biologist this might be interpreted to mean that if the carrying capacity of the prey population is large enough, one might observe cycles in the size of the prey and predator populations before they stabilize at their equilibrium values.

### 5.4.3 Bifurcation Analysis Applied to a Competing Species Model.
The simplest form of a **competing species model** is

$$x' = f(x, y) = x(a - bx) - cxy \equiv x(a - bx - cy)$$
$$y' = g(x, y) = y(d - ey) - pxy \equiv y(d - ey - px), \quad (5.14)$$

containing six positive parameters $a, b, c, d, e$, and $p$. The populations $x$ and $y$ are assumed to be growing logistically, with intrinsic growth rates $a$ and $d$ and carrying capacities $a/b$ and $d/e$. The constants $c$ and $p$ determine the negative effect of the interaction between the two species on their respective growth rates.

In the competing species example studied in Section 5.3, it was found that there was no stable equilibrium where both species could coexist. In the long term either one or both species went extinct.

We want to pose the question: "is there a region of parameter space where coexistence is possible?" This would require a stable equilibrium point $(\bar{x}, \bar{y})$ where both populations are strictly greater than zero. If both populations are strictly positive, then $(\bar{x}, \bar{y})$ must satisfy

$$a - b\bar{x} - c\bar{y} = 0$$
$$d - e\bar{y} - p\bar{x} = 0. \quad (5.15)$$

Using Cramer's Rule to solve the equations in (5.15),

$$\bar{x} = \frac{ae - dc}{be - cp} \quad (5.16)$$
$$\bar{y} = \frac{bd - ap}{be - cp}.$$

From (5.16), it is clear that there are several ways to vary the parameters so that both $\bar{x}$ and $\bar{y}$ are positive; therefore, we will look first at the Jacobian at $(\bar{x}, \bar{y})$ to see if it tells us what needs to be true for stability at the point.

The formulas for $\frac{\partial f}{\partial x}$ and $\frac{\partial g}{\partial y}$ at $(\bar{x}, \bar{y})$ can be simplified, using (5.15):

$$\frac{\partial f}{\partial x} = a - 2b\bar{x} - c\bar{y} = \underbrace{(a - b\bar{x} - c\bar{y})}_{0} - b\bar{x} = -b\bar{x}$$

and
$$\frac{\partial g}{\partial y} = d - 2e\bar{y} - p\bar{x} = \underbrace{(d - e\bar{y} - p\bar{x})}_{0} - e\bar{y} = -e\bar{y},$$

and this means that at $(\bar{x}, \bar{y})$ the Jacobian has the simplified form

$$\mathbf{J}(\bar{x}, \bar{y}) = \begin{pmatrix} \frac{\partial f}{\partial x} & \frac{\partial f}{\partial y} \\ \frac{\partial g}{\partial x} & \frac{\partial g}{\partial y} \end{pmatrix} = \begin{pmatrix} -b\bar{x} & -c\bar{x} \\ -p\bar{y} & -e\bar{y} \end{pmatrix}, \text{ with } \det \mathbf{J}(\bar{x}, \bar{y}) = (be - cp)\bar{x}\bar{y}.$$

Since stability at $(\bar{x}, \bar{y})$ requires that $\det \mathbf{J}(\bar{x}, \bar{y})$ be positive (remember that a saddle point is not a stable equilibrium), we must have $(be - cp) > 0$.

With $(be - cp) > 0$, it can be seen from (5.16) that the point $(\bar{x}, \bar{y})$ will be in the positive quadrant if and only if

$$ae > dc \text{ and } bd > ap. \tag{5.17}$$

Assuming that these conditions are satisfied, the trace of $\mathbf{J}$ is $-(b\bar{x} + e\bar{y})$ and it is negative whenever both $\bar{x}$ and $\bar{y}$ are positive; therefore, the only question is whether $(\bar{x}, \bar{y})$ is a sink or a spiral sink.

To answer this question, we need a simple lemma.

**Lemma 5.1.** *For any real numbers A and B,* $\frac{(A+B)^2}{4} \geq AB$.

*Proof.* $(A - B)^2 = A^2 + B^2 - 2AB \geq 0$, since the square of a real number is always nonnegative. This implies that $A^2 + B^2 \geq 2AB$, and therefore

$$\frac{(A+B)^2}{4} = \frac{(A^2 + B^2) + 2AB}{4} \geq \frac{2AB + 2AB}{4} = AB. \qquad \square$$

For $A$ and $B$ positive numbers, this is equivalent to stating the fact that the arithmetic mean $\frac{A+B}{2}$ is at least as large as the geometric mean $\sqrt{AB}$.

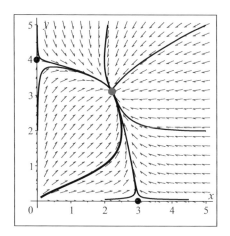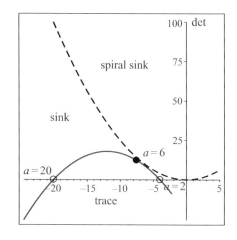

Figure 5.20. The competing species model with coexistence

This lemma is just what is needed to show that at $(\bar{x}, \bar{y})$, with both $\bar{x}$ and $\bar{y}$ positive,

$$\frac{(\text{tr}(\mathbf{J})^2)}{4} = \frac{\left(-(b\bar{x} + e\bar{y})\right)^2}{4} = \frac{(b\bar{x} + e\bar{y})^2}{4} \geq (b\bar{x})(e\bar{y}) > (b\bar{x})(e\bar{y}) - (c\bar{x})(p\bar{y}) = \det \mathbf{J}.$$

## 5.4 Bifurcations for Systems

Since $\det \mathbf{J} < \frac{(\text{tr}(\mathbf{J}))^2}{4}$, the point $(\bar{x}, \bar{y})$ will be a sink.

We are going to let $a$ be our variable parameter, and arbitrarily choose values for the other parameters that will satisfy all the above conditions. The values $b = 2$, $e = 1$, $c = 0.5$, and $p = 0.4$ make $be - cp = 1.8 > 0$. The parameter $d$ can take any positive value, since the other two conditions (5.17), requiring that $\frac{dc}{e} < a < \frac{bd}{p}$, will give a positive range for $a$ if, and only if, $\frac{c}{e} < \frac{b}{p}$; but this is equivalent to the condition $be > cp$ which has already been satisfied. Letting $d = 4$, we will then need $a > dc/e = 2$ and $a < bd/p = 20$.

The Jacobian matrix can be written as a function of $a$, using the formulas (5.16) for $\bar{x}$ and $\bar{y}$:

$$\mathbf{J}(\bar{x}, \bar{y}) = \begin{pmatrix} -b\bar{x} & -c\bar{x} \\ -p\bar{y} & -e\bar{y} \end{pmatrix} = \begin{pmatrix} \frac{\partial f}{\partial x} & \frac{\partial f}{\partial y} \\ \frac{\partial g}{\partial x} & \frac{\partial g}{\partial y} \end{pmatrix} = \frac{1}{1.8}\begin{pmatrix} 4 - 2a & 1 - 0.5a \\ 0.16a - 3.2 & 0.4a - 8 \end{pmatrix}.$$

The right-hand graph in Figure 5.20 shows the parametric curve $(\text{tr}(\mathbf{J}(\bar{x}, \bar{y}, a)), \det(\mathbf{J}(\bar{x}, \bar{y}, a)))$ drawn in a trace-determinant plane. The dashed curve is a graph of $\det = \frac{(\text{tr})^2}{4}$. It can be seen that there is no bifurcation at the point $(\bar{x}, \bar{y})$ for values of $a$ between 2 and 20. At either end of this interval, the determinant becomes negative, and the point $(\bar{x}, \bar{y})$ bifurcates from a sink to a saddle at the two bifurcation values $a = 2$ and $a = 20$. The left-hand graph shows a phase portrait for the system with $a = 6$; that is, the system

$$x' = x(6 - 2x) - 0.5xy, \quad y' = y(4 - y) - 0.4xy.$$

It is clear that all solutions in the positive quadrant converge to the stable sink at $(\bar{x}, \bar{y}) \approx (2.222, 3.111)$ as $t \to \infty$. This represents an equilibrium where the two species coexist, and for any initial conditions $(x(0), y(0))$ with both $x(0)$ and $y(0)$ strictly greater than zero, the solution will tend to $(\bar{x}, \bar{y})$ in the limit as $t \to \infty$. With $a = 6$ there are two other equilibrium points at $(3, 0)$ and $(0, 4)$ which can both be shown to be saddles. Think carefully about how solutions behave in a small neighborhood of each of these points.

**Exercises 5.4.** *In the first three exercises you should be using your computer algebra system to draw phase portraits.*

1. *In many predator-prey models, like the one in Section 5.4.2, the prey population is assumed to grow logistically, so that in the absence of predators the prey tends to its carrying capacity. In this problem it will be assumed that the prey has a level below which it tends to die out, and only above that level tends to its carrying capacity. (In Exercise 8 at the end of Section 2.7 you were asked to draw a phase line for an equation of this type.) Consider the predator-prey model*

$$x' = x(ax - 1)(1 - x) - \varepsilon xy = f(x, y)$$
$$y' = -0.5y + xy = g(x, y),$$

   *where $a$ is a positive number greater than 2 and $\varepsilon > 0$ measures the negative effect of the predator population Y on the prey population X.*

   (a) *Find the four equilibrium solutions $(\bar{x}, \bar{y})$ for this system, in terms of the two parameters $a$ and $\varepsilon$.*

(b) *Use the Jacobian to find the type of the equilibrium points. Show that the only one that can change type for $2 < a < \infty$ is the point $(\bar{x}, \bar{y}) = \left(\frac{1}{2}, \frac{a-2}{4\varepsilon}\right)$. How does the parameter $\varepsilon$ affect the equilibrium points?*

(c) *Determine the bifurcation value $a^*$ for the parameter $a$.*

(d) *Assuming $\varepsilon = 0.5$, sketch a phase portrait for a value of $a$ between 2 and $a^*$, and one for a value of $a > a^*$.*

(e) *Explain the differences in these two phase planes, in terms meaningful to a biologist.*

2. *Consider the competing species model*

$$x' = x(2 - 2x) - 0.5xy$$
$$y' = y\left(1 - \frac{1}{2}y\right) - Pxy$$

*with parameter P.*

(a) *Compute the Jacobian matrix $\mathbf{J}$ for this system.*

(b) *Find all equilibrium solutions. Let $(\bar{x}, \bar{y})$ denote the equilibrium where the two species coexist.*

(c) *For the equilibrium $(\bar{x}, \bar{y})$ found in part (b), plot the curve $(\text{tr}(\mathbf{J}(\bar{x}, \bar{y})), \det(\mathbf{J}(\bar{x}, \bar{y})))$ in a trace-determinant plane for values of the parameter $P$ from 0 to 2. Describe any bifurcations that occur.*

3. *In a competing species model the growth rates of both populations are negatively affected by their interaction. Suppose the two populations are cooperating, instead of competing. The following equations represent such a model:*

$$x' = 0.2x\left(1 - \frac{x}{N}\right) + 0.1xy$$
$$y' = 0.6y\left(1 - \frac{y}{3}\right) + 0.05xy.$$

*Both interaction terms are now positive.*

(a) *What do you expect might happen to the populations X and Y over the long term?*

(b) *Let the carrying capacity N of the population X be equal to 4, and find all equilibrium points for the system. Is there an equilibrium where the populations coexist?*

(c) *Evaluate the Jacobian matrix at each equilibrium and determine its type. If the initial conditions $x(0)$ and $y(0)$ are both positive, what must happen to a solution as $t \to \infty$? Explain.*

(d) *Let $N = 10$, and again find an equilibrium where the two populations coexist. What happens now to a solution starting in the positive quadrant? Explain.*

## 5.5 Student Projects

Both of the projects in this section are drawn directly from the scientific literature; one of the papers was written in 1972 and the other in 2013. At this point it would be profitable to look at the papers. You should be able to understand what the problems are about by reading the abstracts. Both projects will require you to use all the things you have learned in Chapter 5. To do these projects, it will be necessary to have a good computer algebra system and know how to use it.

### 5.5.1 The Wilson-Cowan Equations.
A student project at the end of Chapter 2 looked at the single neuron equation, which models the activity in a collection of nerve cells receiving inputs from neighboring cells in the brain. If you did not do that project, you should go back and look at the information given in the introduction. In this application we are going to look at an extension of the model, which was first introduced by two University of Chicago mathematicians, and published in an oft-cited paper[2] in 1972. The Wilson-Cowan system of equations models the behavior of two interconnected populations of neurons; a population $E$ of excitatory cells (these have a positive effect on the cells to which they are connected) and a population $I$ of inhibitory cells (which have a negative effect on cells to which they are connected).

We will look at a simplified form of this system which can be written as

$$x'(t) = -x(t) + S(ax(t) - by(t) - \theta_x) = f(x,y)$$
$$y'(t) = -y(t) + S(cx(t) - dy(t) - \theta_y) = g(x,y), \quad (5.18)$$

where $x(t)$ and $y(t)$ are the percent of cells active at time $t$ in the populations $E$ and $I$, respectively. The function $S(z)$ determines the average response of neurons to a given amount $z$ of synaptic input, and we will use the response function

$$S(z) = \frac{1}{1 + e^{-z}},$$

which is the same function $S_a(z)$ used in Chapter 2, with $a = 1$. The function $S$ increases monotonically from 0 to 1 as $t$ goes from $-\infty$ to $+\infty$. The positive constants $a, b, c,$ and $d$ in (5.18) can be used to vary the magnitude of the effect of neurons from one of the populations on those in the other population or on those in their own population. For example, $b$ is a measure of the inhibitory (i.e., negative) effect of neurons in population $I$ on those in population $E$. The constants $\theta_x$ and $\theta_y$ represent threshold values for the excitatory and inhibitory cells, respectively. Total input to the cell population must be close to, or above, the threshold value in order to produce a significant change in activity level.

As an example, consider the following system:

$$x'(t) = -x(t) + S(11x(t) - 5.5y(t) - 3.1) = f(x,y)$$
$$y'(t) = -y(t) + S(8x(t) - 3.3y(t) - 3.3) = g(x,y). \quad (5.19)$$

The nullcline for $x'$ is the curve defined by $x = S(11x - 5.5y - 3.1)$. In Chapter 2 we showed that the response function $S(z)$ has the inverse $z = \ln(\frac{S}{1-S})$; therefore, the nullcline for $x'$ can be written in the form

$$11x - 5.5y - 3.1 = S^{-1}(x) = \ln\left(\frac{x}{1-x}\right).$$

---

[2] H. R. Wilson and J. D. Cowan, Excitatory and inhibitory interactions in localized populations of model neurons, *Biophysical Journal*, **12**(1972), pp. 1–24.

Solving this for $y$,

$$y = \frac{1}{5.5}\left(11x - 3.1 - \ln\left(\frac{x}{1-x}\right)\right). \tag{5.20}$$

Similarly, by setting $g(x, y) = 0$, the nullcline for $y'$ can be written in the form

$$x = \frac{1}{8}\left(\ln\left(\frac{y}{1-y}\right) + 3.3y + 3.3\right). \tag{5.21}$$

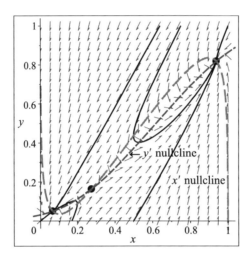

Figure 5.21. Phase plane for the Wilson-Cowan system (5.19)

The two nullclines are shown (as dashed curves) in the phase plane for (5.19) in Figure 5.21. For the values of the parameters in (5.19), there can be seen to be three equilibrium solutions; that is, points of intersection of the $x'$ and $y'$ nullclines. By replacing $x$ in (5.20) with its value in (5.21), a numerical solver can be used to find the three equilibrium points:

$$P_1 \approx (0.06588, 0.05027)$$
$$P_2 \approx (0.2756, 0.1633)$$
$$P_3 \approx (0.9381, 0.8183).$$

To determine the type of each equilibrium, we need the Jacobian matrix of the system (5.18). In Chapter 2 it was shown that $\frac{d}{dz}S(z) = S(z)(1-S(z))$, so using the chain rule,

$$\frac{\partial}{\partial x}\left(S(ax - by - \theta_x)\right) = S'(ax - by - \theta_x)\frac{\partial}{\partial x}(ax - by - \theta_x) = aS'(ax - by - \theta_x)$$
$$= aS(ax - by - \theta_x)(1 - S(ax - by - \theta_x)).$$

Using this information, and the fact that at any equilibrium point $(\bar{x}, \bar{y})$ we have $\bar{x} = S(a\bar{x} - b\bar{y} - \theta_x)$ and $\bar{y} = S(c\bar{x} - d\bar{y} - \theta_y)$, show that the Jacobian matrix can be written as

$$\mathbf{J}(\bar{x}, \bar{y}) = \begin{pmatrix} -1 + a\bar{x}(1-\bar{x}) & -b\bar{x}(1-\bar{x}) \\ c\bar{y}(1-\bar{y}) & -1 - d\bar{y}(1-\bar{y}) \end{pmatrix}. \tag{5.22}$$

## 5.5 Student Projects

Write out complete answers to problems 1–6. You will need a computer algebra system to do the numerical work and construct the phase portrait.

(1) Use (5.22), with the values of the parameters in system (5.19), to compute the Jacobian matrix at the three equilibrium points $P_1$, $P_2$, and $P_3$. Determine the type of each of these equilibria.

(2) If the equilibrium $(\bar{x}, \bar{y})$ is a saddle point, find eigenpairs for the matrix $\mathbf{J}(\bar{x}, \bar{y})$. Use them to sketch the stable manifold of the saddle point.

(3) Using your computer algebra system and the information obtained in problems 1 and 2, draw a complete phase portrait for the system (5.19). Let $x$ and $y$ both range from 0 to 1. Describe, as precisely as you can, how solutions behave in every part of the unit square. Especially note where solutions end up as $t \to \infty$.

(4) Give an argument to say whether or not solutions that start in the unit square can ever leave it.

(5) Suppose the excitatory cells are receiving additional negative input from outside the two populations of cells. This could be modeled by adding a quantity $-e(t)$ to the argument of the response function in the equation for $x'(t)$; that is, by writing $x'(t) = -x(t) + S(ax(t) - by(t) - \theta_x - e(t))$. Assuming $e(t) = e$ for some positive constant $e$, how would this change the position of the $x'$ nullcline, given by (5.20)?

(6) Estimate how big $e$ must be to cause a bifurcation of the system (remember that the value of $e$ is being subtracted). If $e$ is increased beyond the bifurcation value, how many equilibrium solutions will there be? Will the resulting level of activity in the two populations of neurons converge to a limit closer to 0 or to 100%? Explain, and justify your explanation in terms of the biological model.

One question that might be of interest to a neuroscientist is whether a system of this type has a stable limit cycle. A limit cycle is stable if trajectories are approaching it from both inside and outside. If it does have a stable cycle, then for certain initial values the activity in the two populations will oscillate in a nearly periodic fashion.

In the above example no limit cycle exists, so we will look at a second system:

$$\begin{aligned} x'(t) &= -x(t) + S(12x(t) - 7y(t) - 2.5) = f(x, y) \\ y'(t) &= -y(t) + S(20x(t) - 2y(t) - 7 + P) = g(x, y), \end{aligned} \quad (5.23)$$

where the parameter $P$ in the second equation represents an external source of constant input (either positive or negative) to the inhibitory population. For any value of $P$ between $-3.0$ and $9.0$ it can be shown that the system (5.23) has a single equilibrium point $(\bar{x}, \bar{y})$. For any fixed value of $P$ this point can be found by solving simultaneous equations similar to (5.20) and (5.21).

(7) For each $P = -3, -1, 1, 2, 5, 7,$ and $9$

  (a) Find the single equilibrium point $(\bar{x}, \bar{y})_P$ of the system (5.23). Be sure to use the correct values of the parameters.
  (b) Evaluate the Jacobian at $(\bar{x}, \bar{y})_P$, and let $T_P = (\text{tr}(\mathbf{J}), \det(\mathbf{J}))$ at $(\bar{x}, \bar{y})_P$.
  (c) Plot the seven points $T_P$ in a labeled trace-determinant plane, and estimate the value of $P$ when the critical point changes from a spiral sink to a spiral source, and the value where it changes from a spiral source back to a spiral sink.

This is characteristic of the way in which limit cycles appear. The bifurcation that occurs when an equilibrium with complex eigenvalues has its real part go through zero is called an **Andronov-Hopf bifurcation**. It produces limit cycles on one side of the bifurcation value, and they can be either stable or unstable limit cycles.

(8) Draw completely labeled phase portraits for the system (5.23) for the three values $P = -1, 2$, and 5. Determine, in each case, if a limit cycle exists. One way to do this is to plot some time curves $x(t)$ with initial points close to the equilibrium point, and see if $x$ moves toward the equilibrium or away from it. To see if a limit cycle is stable from both sides, you should also plot $x(t)$ starting at a point far from the equilibrium point.

**5.5.2 A New Predator-prey Equation—Putting It All Together.** This problem is an excellent illustration of how scientists from different disciplines can work together to obtain interesting results. In a recently published paper[3], written jointly by a mathematician and a biologist, the following predator-prey model is studied:

$$x' = x(a - \lambda x) - yP(x)$$
$$y' = -\delta y - \mu y^2 + cyP(x). \tag{5.24}$$

The parameters $a, \lambda, \delta, \mu$, and $c$ are all assumed to be positive.

This is very similar to previous predator-prey models we have looked at, except for the interaction terms $-yP(x)$ and $cyP(x)$. In the paper, the biologist first gives a game-theoretic argument to show that if the prey are able to use group defense behavior, the predators' attack rate $P(x)$ may actually decrease when the prey population gets large enough. The function $P(x)$ is a nonnegative function of $x$ formally referred to by population biologists as the predator's functional response, and in the paper it is shown that if the prey's group behavior is taken into account, $P$ will satisfy the conditions

(1) $P(0) = 0$,

(2) $P(x) \geq 0$, for $x > 0$,

(3) there exists a positive value $x = x_c$ such that
$$P'(x) > 0 \text{ if } 0 \leq x < x_c, \text{ and } P'(x) \leq 0 \text{ for } x \geq x_c.$$

The function chosen by the authors to represent $P$ is a rational function of the form
$$P(x) = \frac{mx}{\alpha x^2 + \beta x + 1}.$$

The general idea behind the paper is to show that with only these minimum assumptions on the predator's response function, the dynamics of the resulting system can be shown to be limited to a small number of possibilities.

It is first shown that if $K > \frac{c(\delta+a)^2}{4\lambda\delta}$, then all trajectories starting in the positive quadrant will ultimately enter the triangular region $T$ of the $(x, y)$-plane, bounded by the lines $x = 0$, $y = 0$, and $y = K - cx$. Once inside, they cannot exit from the closed region $T$ (that is, the region $T$ with its boundary).

The term $-\mu y^2$ in the second equation in (5.24) represents intraspecific competition among the predators. For example, this might be due to fighting over the food

---
[3]V. Naudot and E. Noonburg, Predator-prey systems with a general nonmonotonic functional response, *Physica D: Nonlinear Phenomena*, **253**(2013), pp. 1–11.

## 5.5 Student Projects

supply. If $\mu$ is small enough, it is shown (using a fair amount of algebra) that there can be at most two equilibrium points in the interior of $T$. Using dynamical systems theory, the mathematician is able to show that in the case where all the equilibrium points are hyperbolic (remember this means their Jacobian has no eigenvalues with zero real part), and no limit cycles exist, there are exactly four types of phase planes possible. A diagram showing them is included in the paper, and looking at this part of the paper should be very profitable. More mathematics is used to show that under certain further conditions there can be a limit cycle.

Write out complete answers to problems 1–8. Let $\mu = 0.005$ and let $P(x)$ be the rational function
$$P(x) = \frac{x}{0.2x^2 + 0.5x + 1}.$$

(1) Check that $(0, 0)$ and $(\frac{a}{\lambda}, 0)$ are the only equilibrium points of (5.24) where at least one of the populations has died out.

(2) Sketch a graph of $P(x)$ and show that it satisfies the required properties. Differentiate $P$, and find $x \equiv x_c$, where $P(x)$ takes its maximum value.

(3) Show that the Jacobian matrix for (5.24) is
$$\mathbf{J}(x, y) = \begin{pmatrix} a - 2\lambda x - yP'(x) & -P(x) \\ cyP'(x) & -\delta - 2\mu y + cP(x) \end{pmatrix},$$
and use it to show that $(0, 0)$ is always a saddle point of the system and $(\frac{a}{\lambda}, 0)$ can be either a sink or a saddle point.

(4) Show that if $(\bar{x}, \bar{y})$ is any equilibrium point of (5.24) in the interior of the triangle $T$, it must satisfy the two equations
$$\bar{y} = \frac{cP(\bar{x}) - \delta}{\mu}$$
$$\bar{y} = \frac{\bar{x}(a - \lambda\bar{x})}{P(\bar{x})} = (a - \lambda\bar{x})(0.2\bar{x}^2 + 0.5\bar{x} + 1).$$
(5.25)

(5) With $\mu = 0.005$, find positive values of $a$, $\lambda$, $\delta$, and $c$ such that the two curves defined in (5.25) have no intersections; that is, no interior equilibrium points exist. Using these values of the parameters, draw a phase portrait and describe what happens to the predator and prey populations as $t \to \infty$. Use several initial conditions with $x(0)$ and $y(0)$ both positive.

(6) Find positive values of $a$, $\lambda$, $\delta$, and $c$ such that the two curves defined in (5.25) have exactly one intersection. Use the Jacobian to determine the type of the interior equilibrium solution. Is the equilibrium $(\frac{a}{\lambda}, 0)$ a sink or a saddle point? Draw a phase portrait and describe what happens to the two populations as $t \to \infty$.

(7) Find positive values of $a$, $\lambda$, $\delta$, and $c$ such that the two curves defined in (5.25) have exactly two intersections. Use the Jacobian to determine the type of each of the equilibrium points (in this case you should have either a source and a saddle, or a sink and a saddle). Draw a phase portrait and describe what happens to trajectories in each region of the positive quadrant.

(8) Let $\lambda = 0.6$, $\delta = 0.2$, $c = 0.5$, and $\mu = 0.005$. For values of $a$ between 0.32 and 2.0, there is a single interior equilibrium point $C$. Using the Jacobian, show that $C$ changes from a spiral sink to a spiral source at $a = \bar{a} \approx 1.472$, so that a Hopf bifurcation occurs at this value of $a$ and a limit cycle exists when $a > \bar{a}$. Draw phase portraits for $a = 0.5$ and $a = 2$, and for both cases describe what happens to the two populations in terms that a biologist might use.

# 6

# Laplace Transforms

The method of Laplace transforms gives us another way to solve linear differential equations. It works for linear, constant coefficient equations of any order and can also be applied to systems of linear differential equations. While at first it may not seem to provide anything new, it does have one very nice feature. The driving function of the linear equation does not need to be continuous. With the characteristic polynomial method a differential equation with a discontinuous driving function would have to be solved in steps. At each point of discontinuity a new equation would have to be solved, using the final values of the previous equation as its initial values. With Laplace transforms it will be shown that this type of problem can be done in a single step.

The Laplace transform is just one of several different kinds of integral transforms used in applied mathematics. Its most important property, from our point of view, is that it converts a problem in differential equations into a problem in algebra.

## 6.1 Definition and Some Simple Laplace Transforms

The technique of solving differential equations by Laplace transforms, as described in this chapter, will be seen to apply to initial-value problems where the initial values are given at $t = 0$. For this reason the functions involved only need to be defined on the interval $[0, \infty)$. A change of independent variable of the form $\tau = t - t_0$ can always be used to put the initial point at zero.

With the exception of the impulse function, defined in Section 6.5, the functions that we will be able to transform have to satisfy two properties. They must be at least piecewise continuous and they must be bounded in a certain way as $t \to \infty$. The following two definitions make these properties precise.

**Definition 6.1.** *A function $f(t)$ is called* **piecewise continuous** *on $[0, \infty)$ if it is continuous except possibly at a finite number of points $t_k$ on any finite subinterval of $[0, \infty)$,*

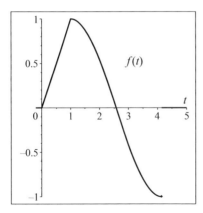 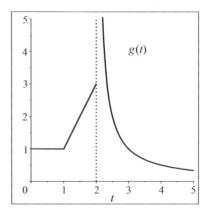

Figure 6.1. Piecewise defined functions $f$ and $g$

and at each $t_k$ both of the limits

$$\lim_{t \to t_k^+} f(t) \quad \text{and} \quad \lim_{t \to t_k^-} f(t)$$

exist. This means that the function cannot go to $\pm\infty$ on either side of a point of discontinuity.

**Example 6.1.1.** *The functions $f$ and $g$ in Figure 6.1 are examples of piecewise defined functions.*

The formula for $f$ is

$$f(t) = \begin{cases} t & \text{if } 0 \leq t \leq 1 \\ \cos(t-1) & \text{if } 1 < t \leq \pi+1 \\ 0 & \text{if } t > \pi+1. \end{cases}$$

It can be seen that $f$ satisfies all the properties in Definition 6.1; therefore, $f(t)$ is piecewise continuous on $[0, \infty)$. At $t = 1$, $f$ is actually continuous, but its derivative $f'$ is not.

The function $g$ has the formula

$$g(t) = \begin{cases} 1 & \text{if } 0 \leq t \leq 1 \\ -1 + 2t & \text{if } 1 < t < 2 \\ \dfrac{1}{t-2} & \text{if } t > 2. \end{cases}$$

Because $\lim_{t \to 2^+} g(t) = \infty$, $g$ is not piecewise continuous. ∎

**Definition 6.2.** *A function $f(t)$ is said to be of **exponential order** if there exist constants $M$, $a$, and $T$ such that $|f(t)| < Me^{at}$ for all $t > T$.*

Most functions used in engineering and scientific applications are of exponential order. An example of a function that grows so fast that it does not satisfy Definition 6.2 is $f(t) = e^{t^2}$. To see this, we assume there exist finite constants $M, a,$ and $T$ such that $e^{t^2} < Me^{at}$ for all $t > T$. Dividing both sides by $e^{t^2}$ would imply that $1 < Me^{at-t^2}$ for all $t > T$. This is clearly false, since $\lim_{t \to \infty} e^{t(a-t)} = 0$ for any finite constant $a$.

## 6.1 Definition and Some Simple Laplace Transforms

In the definition below, it will be seen that the Laplace transform of a function $f(t)$ involves an improper integral, and requiring $f(t)$ to be piecewise continuous and of exponential order is sufficient to show convergence of this integral where necessary.

**Definition 6.3.** *If $f(t)$ is a piecewise continuous function of exponential order, the* **Laplace transform** *of $f(t)$ is defined as*

$$\mathcal{L}(f(t)) = \int_0^\infty e^{-st} f(t) dt \equiv F(s).$$

It will be a useful convention in this chapter to label functions of $t$ by lowercase letters, and their Laplace transform with the corresponding capital letter. It can be seen from the definition that the transform will be a function of $s$, where here $s$ is assumed to be a real variable. It is only necessary for the transform of $f$ to exist for all $s$ greater than some finite value $a$, and requiring the function $f$ to be of exponential order guarantees that this will be true.

**6.1.1 Four Simple Laplace Transforms.** To get started, we will compute four simple transforms. The first will be done very carefully, in the way you were taught to evaluate improper integrals in your calculus course. Because the function $f$ is of exponential order, we will never have to worry about the convergence of the improper integral, and after this calculation we will simply evaluate the functions at infinity without using the limit notation.

To find the transform of the constant function $f(t) \equiv 1$, we write

$$\mathcal{L}(1) = \int_0^\infty e^{-st}(1) dt = \lim_{B \to \infty} \int_0^B e^{-st} dt$$

$$= \lim_{B \to \infty} \left[ \frac{e^{-st}}{-s} \Big|_0^B \right] = \lim_{B \to \infty} \left[ \frac{e^{-sB}}{-s} - \frac{e^0}{-s} \right] = 0 + \frac{1}{s} = \frac{1}{s}, \text{ for any } s > 0.$$

The transform of $f(t) = e^{at}$ is also simple to compute:

$$\mathcal{L}(e^{at}) = \int_0^\infty e^{-st}(e^{at}) dt = \int_0^\infty e^{-(s-a)t} dt$$

$$= \left[ \frac{e^{-(s-a)t}}{-(s-a)} \Big|_0^\infty \right] = 0 - \frac{e^0}{-(s-a)} = \frac{1}{s-a}, \text{ if } s > a.$$

The transforms of the sine and cosine functions could be found in the same way by evaluating the integrals $\int_0^\infty e^{-st} \sin(bt) dt$ and $\int_0^\infty e^{-st} \cos(bt) dt$. Since the integrals are rather messy, we are going to use an easier method that involves Euler's formula and integration of the complex function $f(t) = e^{bti}$. Writing

$$\mathcal{L}(e^{bti}) = \mathcal{L}(\cos(bt) + \iota \sin(bt))$$

$$= \int_0^\infty e^{-st} (\cos(bt) + \iota \sin(bt)) dt$$

$$= \int_0^\infty e^{-st} \cos(bt) dt + \iota \int_0^\infty e^{-st} \sin(bt) dt,$$

we see that the real part of the Laplace transform of $e^{bıt}$ will be the transform of $\cos(bt)$ and the imaginary part will be the transform of $\sin(bt)$. Using the definition of the transform,

$$\mathcal{L}(e^{bıt}) = \int_0^\infty e^{-st}(e^{bıt})dt$$

$$= \int_0^\infty e^{-(s-bı)t}dt = \left.\frac{e^{-(s-bı)t}}{-(s-bı)}\right|_0^\infty = 0 - \frac{e^0}{-(s-bı)} = \frac{1}{(s-bı)}.$$

To find the real and imaginary parts of $\frac{1}{(s-bı)}$, we use the complex conjugate to write

$$\frac{1}{(s-bı)} \cdot \frac{(s+bı)}{(s+bı)} = \frac{s+bı}{s^2+b^2} \equiv \left(\frac{s}{s^2+b^2}\right) + \left(\frac{b}{s^2+b^2}\right)ı.$$

This gives us the formulas

$$\mathcal{L}(\cos(bt)) = \Re\left(\frac{s+bı}{s^2+b^2}\right) = \frac{s}{s^2+b^2},$$

$$\mathcal{L}(\sin(bt)) = \Im\left(\frac{s+bı}{s^2+b^2}\right) = \frac{b}{s^2+b^2}.$$

All you really need from this are the final two formulas (which are real).

**6.1.2 Linearity of the Laplace Transform.** The Laplace transform is a linear operator; that is,

$$\mathcal{L}(af(t) + bg(t)) = a\mathcal{L}(f(t)) + b\mathcal{L}(g(t)).$$

To prove that this is true, we use the definition of the transform and the linearity property of the integral to write

$$\mathcal{L}(af(t) + bg(t)) = \int_0^\infty e^{-st}(af(t) + bg(t))dt$$

$$= a\int_0^\infty e^{-st}f(t)dt + b\int_0^\infty e^{-st}g(t)dt = a\mathcal{L}(f(t)) + b\mathcal{L}(g(t)).$$

**Example 6.1.2.** Use linearity and the four formulas derived in the previous subsection to find the Laplace transform of each of

(1) $7e^{2t}$

(2) $5\sin(2t) + 3\cos(2t)$

(3) $2 + 6e^{-4t}$

(4) $\sin(2t + 5)$

Write the answer in the form of a rational function $\frac{P(s)}{Q(s)}$, where $P$ and $Q$ are polynomials in $s$.

Solution.

(1) $\mathcal{L}(7e^{2t}) = 7\mathcal{L}(e^{2t}) = 7\left(\frac{1}{s-2}\right) = \frac{7}{s-2}.$

(2) $\mathcal{L}(5\sin(2t) + 3\cos(2t)) = 5\mathcal{L}(\sin(2t)) + 3\mathcal{L}(\cos(2t)) = 5\frac{2}{s^2+4} + 3\frac{s}{s^2+4} = \frac{10+3s}{s^2+4}.$

## 6.1 Definition and Some Simple Laplace Transforms

(3) $\mathcal{L}(2 + 6e^{-4t}) = 2\mathcal{L}(1) + 6\mathcal{L}(e^{-4t}) = 2\frac{1}{s} + 6\frac{1}{s+4} = \frac{2}{s} + \frac{6}{s+4} = \frac{2(s+4)+6s}{s(s+4)} = \frac{8s+8}{s(s+4)}.$

(4) Use the formula for $\sin(A + B)$ to write

$$\mathcal{L}(\sin(2t + 5)) = \mathcal{L}(\sin(2t)\cos(5) + \cos(2t)\sin(5))$$

$$= \cos(5)\left(\frac{2}{s^2+4}\right) + \sin(5)\left(\frac{s}{s^2+4}\right) = \frac{\sin(5)s + 2\cos(5)}{s^2+4}. \blacksquare$$

**6.1.3 Transforming the Derivative of $f(t)$.** The formula that makes Laplace transforms so useful for solving differential equations is the formula for the transform of the derivative of a function. It is proved by using integration by parts: $\int u\,dv = uv - \int v\,du$.

For any piecewise continuous function $f$, of exponential order, the Laplace transform of its derivative $f'(t)$ is computed as follows:

$$\mathcal{L}(f'(t)) = \int_0^\infty e^{-st} f'(t)\,dt = e^{-st} f(t)\Big|_0^\infty - \int_0^\infty (-se^{-st}) f(t)\,dt$$

$$= (0 - f(0)) + s\int_0^\infty e^{-st} f(t)\,dt = s\mathcal{L}(f(t)) - f(0).$$

It can be seen that the transform of a derivative of $f$ of arbitrary order can also be obtained by using this formula. For example,

$$\mathcal{L}(f''(t)) = \mathcal{L}((f'(t))') = s\mathcal{L}(f'(t)) - f'(0)$$

$$= s(s\mathcal{L}(f(t)) - f(0)) - f'(0) = s^2\mathcal{L}(f(t)) - sf(0) - f'(0).$$

Everything we have done in this section can now be summarized in our first short table of Laplace transforms. Remember that by convention $F(s)$ denotes the Laplace transform of $f(t)$.

Table 6.1

| function | Laplace transform |
|---|---|
| 1 | $\frac{1}{s}$, if $s > 0$ |
| $e^{at}$ | $\frac{1}{s-a}$, if $s > a$ |
| $\sin(bt)$ | $\frac{b}{s^2+b^2}$ |
| $\cos(bt)$ | $\frac{s}{s^2+b^2}$ |
| $af(t) + bg(t)$ | $aF(s) + bG(s)$ (linearity) |
| $f'(t)$ | $sF(s) - f(0)$ |
| $f''(t)$ | $s^2F(s) - sf(0) - f'(0)$ |

**Exercises 6.1.** *Use linearity and the formulas in Table 6.1 to find the Laplace transform of the functions in Problems 1–8. Write the transform as a rational function $\frac{P(s)}{Q(s)}$.*

1. $e^{3t}$

2. $\sin(5t)$

3. $2 + \cos(t)$

4. $-4 + e^{-t}$

5. $2e^t - 3\cos(t)$

6. $e^{4t} - 10 + \sin(t)$

7. $\sin(3t + 2)$ *Hint: use the formula* $\sin(A \pm B) = \sin(A)\cos(B) \pm \cos(A)\sin(B)$.

8. $\cos(2t - 5)$ *Hint: use the formula* $\cos(A \pm B) = \cos(A)\cos(B) \mp \sin(A)\sin(B)$.

*Use the integral definition of the Laplace transform to find the transforms in Problems 9–12:*

9. $\mathcal{L}(\sin(bt))$ *Hint: use the formula* $\int e^{at}\sin(bt)\,dt = e^{at}\left(\frac{a\sin(bt) - b\cos(bt)}{a^2 + b^2}\right)$.

10. $\mathcal{L}(\cos(bt))$ *Hint: use the formula* $\int e^{at}\cos(bt)\,dt = e^{at}\left(\frac{a\cos(bt) + b\sin(bt)}{a^2 + b^2}\right)$.

11. $\mathcal{L}(t)$

12. $\mathcal{L}(t^2)$

13. *Find a formula for $\mathcal{L}(f'''(t))$ in terms of $F(s) = \mathcal{L}(f(t))$. Assume initial conditions $f(0), f'(0),$ and $f''(0)$ are given.*

*Sketch each function defined below and determine whether or not it is piecewise continuous on $[0, \infty)$.*

14.
$$f(t) = \begin{cases} 0 & \text{if } 0 \leq t < 1 \\ \sin(2\pi t) & \text{if } 1 \leq t < 2 \\ 0 & \text{if } t \geq 2 \end{cases}$$

15.
$$g(t) = \begin{cases} 1 - t & \text{if } 0 \leq t \leq 1 \\ t - 1 & \text{if } 1 < t < 2 \\ 1 & \text{if } t \geq 2 \end{cases}$$

16. $k(t) = t - n$ for $n < t \leq n + 1$, $n = 0, 1, 2, \ldots$
    ($k(t)$ is called a "sawtooth" function).

17.
$$h(t) = \begin{cases} 0 & \text{if } 0 \leq t \leq \frac{1}{2} \\ \frac{2^n - 1}{2^n} & \text{if } \frac{2^n - 1}{2^n} < t \leq \frac{2^{n+1} - 1}{2^{n+1}}, \text{ for } n = 1, 2, \ldots \\ 1 & \text{if } t > 1 \end{cases}$$

## 6.2 Solving Equations, the Inverse Laplace Transform

How do we go about solving differential equations with Laplace transforms? To get the idea, we will consider a simple initial-value problem. Suppose you are asked to find the solution of the IVP

$$x' + 2x = e^{-3t}, \quad x(0) = 1. \tag{6.1}$$

It should be clear that if two functions are identical, then their Laplace transforms are identical, so we can take the transform of both sides of (6.1) and set them equal:

$$\mathcal{L}(x' + 2x) = \mathcal{L}(e^{-3t}).$$

Using the linearity property of the transform, and our formula $\mathcal{L}(e^{at}) = \frac{1}{s-a}$ with $a = -3$, we can write

$$\mathcal{L}(x') + 2\mathcal{L}(x) = \frac{1}{s+3}.$$

If we denote $\mathcal{L}(x)$ by $X(s)$, and use the formula for $\mathcal{L}(x')$, the equation becomes

$$(sX(s) - x(0)) + 2X(s) = \frac{1}{s+3}.$$

Now insert the initial condition $x(0) = 1$ and solve algebraically for the transform $X(s)$:

$$(s+2)X(s) = x(0) + \frac{1}{s+3} = 1 + \frac{1}{s+3} = \frac{s+4}{s+3}.$$

Dividing by $(s+2)$,

$$X(s) = \frac{s+4}{(s+3)(s+2)}. \tag{6.2}$$

All that is needed to finish the initial-value problem is a way to find the function $x(t)$ for which the Laplace transform $\mathcal{L}(x(t)) = X(s) = \frac{s+4}{(s+3)(s+2)}$. The function $x(t)$ will be the solution to our IVP. The initial condition is already incorporated into the solution.

Finding a function $x(t)$ that has a Laplace transform $X(s)$ is called inverting the Laplace transform, and it basically amounts to reading Table 6.1 from right to left.

**Definition 6.4.** *The function $f(t)$ that has Laplace transform $F(s)$ is called the **inverse Laplace transform** of $F(s)$ and is denoted by $\mathcal{L}^{-1}(F(s))$. As long as we are working with piecewise continuous functions of exponential order, the inverse function is unique except possibly at points of discontinuity. For any continuous function $f$, of exponential order, $\mathcal{L}^{-1}(\mathcal{L}(f(t))) \equiv f(t)$.*

**Theorem 6.1.** *The inverse Laplace transform $\mathcal{L}^{-1}$ is a linear operator.*

*Proof.* If $f$ and $g$ are piecewise continuous and of exponential order, and if $\mathcal{L}(f) = F(s)$ and $\mathcal{L}(g) = G(s)$, then

$$\mathcal{L}^{-1}(aF(s) + bG(s)) \equiv \mathcal{L}^{-1}(a\mathcal{L}(f) + b\mathcal{L}(g))$$
$$= \mathcal{L}^{-1}(\mathcal{L}(af + bg)) \text{ by linearity of } \mathcal{L}$$
$$= af + bg \equiv a\mathcal{L}^{-1}(F(s)) + b\mathcal{L}^{-1}(G(s)).$$

Therefore, the operator $\mathcal{L}^{-1}$ is linear. □

Unfortunately, in our small Table 6.1 there is no function on the right that has the form of the right-hand side of (6.2), but remember from calculus that there is a way to write $\frac{s+4}{(s+3)(s+2)}$ in the form $\frac{A}{s+3} + \frac{B}{s+2}$. The process for doing this is called expansion by partial fractions. In this simple case you can check that the function $X(s)$ can be written in the form

$$X(s) = \frac{s+4}{(s+3)(s+2)} \equiv \frac{-1}{s+3} + \frac{2}{s+2}.$$

This means that the solution of the initial value problem (6.1) is

$$x(t) = \mathcal{L}^{-1}\left(\frac{-1}{s+3} + \frac{2}{s+2}\right) = (-1)\mathcal{L}^{-1}\left(\frac{1}{s+3}\right) + (2)\mathcal{L}^{-1}\left(\frac{1}{s+2}\right),$$

and using Table 6.1,

$$x(t) = -e^{-3t} + 2e^{-2t}.$$

You can check this answer by showing that $x(t)$ satisfies both the differential equation and the initial condition in (6.1).

Before going any further, a review will be given of partial fraction expansions.

**6.2.1 Partial Fraction Expansions.** To expand a rational function $R(s) = \frac{P(s)}{Q(s)}$ with partial fractions, where $P$ and $Q$ are polynomials and the degree of $P$ is less than the degree of $Q$,

- Factor the denominator $Q(s)$. A theorem in algebra states that any polynomial with real coefficients can be factored into a product of integral powers of linear factors $(s-a)^m$ and irreducible quadratic factors $(s^2+bs+c)^n$, where irreducible means that $b^2 < 4c$ so that the quadratic has complex roots. The integers $m$ and $n$ are greater than or equal to one, and they represent the multiplicity with which the factor appears in the polynomial $Q(s)$.

- For each linear factor $(s-a)$ of multiplicity 1 in $Q(s)$, include the term $\frac{A}{s-a}$ in the partial fraction expansion.

- For each linear factor $(s-a)^m$ of multiplicity $m > 1$, include in the partial fraction expansion the $m$ terms
$$\frac{A_1}{s-a} + \frac{A_2}{(s-a)^2} + \cdots + \frac{A_m}{(s-a)^m}.$$

- For each irreducible quadratic factor $(s^2+bs+c)^n$ of multiplicity $n \geq 1$, include the $n$ terms
$$\frac{B_1 s + C_1}{s^2+bs+c} + \frac{B_2 s + C_2}{(s^2+bs+c)^2} + \cdots + \frac{B_n s + C_n}{(s^2+bs+c)^n}.$$

If the terms are all put back over a common denominator, which should be the original denominator $Q(s)$, the constants $A_i$, $B_j$, and $C_j$ can be found by making the coefficients of like powers of $s$ in the numerator equal on both sides.

Most computer algebra systems, and also some calculators, have a function called expand that will produce a partial fraction expansion for a given rational function.

We are going to do one simple example by hand, and assume that for a complicated rational function you have either a calculator or computer available with an EXPAND function which can do the partial fraction expansion for you.

## 6.2 Solving Equations, the Inverse Laplace Transform

**Example 6.2.1.** *Find the partial fraction expansion of the rational function*

$$R(s) = \frac{P(s)}{Q(s)} = \frac{2s+3}{(s+1)(s^2+2s+4)}.$$

*Solution.* First check the degree of the polynomials $P$ and $Q$. The degree of $P$ is one and the degree of $Q$ is three. If the degree of $Q$ had been less than or equal to the degree of $P$, you would have had to divide the polynomials and work with the remainder.

The term $s^2 + 2s + 4$ is an irreducible quadratic factor since $b^2 - 4c = 2^2 - 16 = -12 < 0$; therefore, the partial fraction expansion contains the two terms

$$R(s) = \frac{A}{s+1} + \frac{Bs+C}{s^2+2s+4}.$$

Putting them over a common denominator, we need to find $A$, $B$, and $C$ to make

$$\frac{A}{s+1} + \frac{Bs+C}{s^2+2s+4} \equiv \frac{A(s^2+2s+4) + (Bs+C)(s+1)}{(s+1)(s^2+2s+4)}$$

equal to the rational function

$$\frac{2s+3}{(s+1)(s^2+2s+4)}.$$

The denominators in the two terms are equal, so it remains to make the numerators identical; that is, we need

$$A(s^2+2s+4) + (Bs+C)(s+1) \equiv 2s+3.$$

Write the left side as a polynomial in $s$:

$$(A+B)s^2 + (2A+C+B)s + (4A+C) \equiv (0)s^2 + (2)s + 3.$$

This will be an identity if and only if the coefficient of each power of $s$ is the same on the two sides. This leads to three equations in three unknowns,

$$A+B = 0, \quad 2A+C+B = 2, \quad 4A+C = 3,$$

with solution $A = \frac{1}{3}$, $B = -\frac{1}{3}$, $C = \frac{5}{3}$; therefore, the partial fraction expansion of $R(s)$ is

$$R(s) = \frac{1/3}{s+1} + \frac{(-1/3)s + (5/3)}{s^2+2s+4}.$$

This example can be used to check the result you get by expanding $R(s)$ with your calculator or computer. ■

It is now possible to solve a simple linear constant-coefficient equation using Laplace transforms. The general method is described below.

> **To solve a differential equation using Laplace transforms**:
> - Take the Laplace transform of both sides of the equation and set them equal.
> - Enter the initial conditions, if any.
> - Solve the equation algebraically for the transform $X(s)$ of the unknown function $x(t)$.
> - Invert $X(s)$ to find $x(t)$.

The following example will show how this is done.

**Example 6.2.2.** *Solve the initial value problem*

$$x'' + 6x' + 5x = 3e^{-2t}, \quad x(0) = 1, \quad x'(0) = -1. \tag{6.3}$$

*Solution.* It can be seen that, considered as a spring-mass equation, this is an overdamped system ($b^2 - 4mk = 36 - 20 > 0$) with a decaying exponential driving function, so we expect the solution to be a sum of decaying exponential terms.

Transforming both sides of the equation,

$$(s^2 X(s) - sx(0) - x'(0)) + 6(sX(s) - x(0)) + 5X(s) = 3\frac{1}{s+2}.$$

Subtract the terms not involving $X(s)$ from both sides:

$$(s^2 + 6s + 5)X(s) = sx(0) + x'(0) + 6x(0) + \frac{3}{s+2} = 5 + s + \frac{3}{s+2} = \frac{(5+s)(s+2) + 3}{s+2}.$$

Solve for $X(s)$ and expand in partial fractions:

$$X(s) = \frac{s^2 + 7s + 13}{(s+2)(s^2 + 6s + 5)} = \frac{s^2 + 7s + 13}{(s+2)(s+1)(s+5)} = \frac{A}{s+2} + \frac{B}{s+1} + \frac{C}{s+5}.$$

To make the numerators equal, set

$$s^2 + 7s + 13 = A(s+1)(s+5) + B(s+2)(s+5) + C(s+2)(s+1).$$

When all the terms in the denominator are linear terms of multiplicity one, as in this case, it is easy to find the coefficients by letting $s$ take values that make one of the linear factors equal to zero. Letting $s = -2$,

$$(-2)^2 + 7(-2) + 13 = A(-2+1)(-2+5) + B(0) + C(0) \Rightarrow 3 = -3A \Rightarrow A = -1.$$

Letting $s = -1$,

$$(-1)^2 + 7(-1) + 13 = A(0) + B(-1+2)(-1+5) + C(0) \Rightarrow 7 = 4B \Rightarrow B = \frac{7}{4}.$$

Letting $s = -5$,

$$(-5)^2 + 7(-5) + 13 = A(0) + B(0) + C(-5+2)(-5+1) \Rightarrow 3 = 12C \Rightarrow C = \frac{1}{4}.$$

Using these values, we can write

$$X(s) = -\frac{1}{s+2} + \frac{7}{4}\left(\frac{1}{s+1}\right) + \frac{1}{4}\left(\frac{1}{s+5}\right),$$

and, using Table 6.1,

$$x(t) = \mathcal{L}^{-1}(X(s)) = -e^{-2t} + \frac{7}{4}e^{-t} + \frac{1}{4}e^{-5t}.$$

Check that $x(t)$ satisfies (6.3) and the given initial conditions. ■

**Exercises 6.2.** *In each of the problems 1–6, use Table 6.1 and the linearity of the inverse transform to find $f(t) = \mathcal{L}^{-1}(F(s))$:*

1. $F(s) = \frac{2}{s+3}$.

2. $F(s) = \frac{2s}{s^2+9}$.

3. $F(s) = \frac{2}{s} + \frac{3}{s+5}$.

4. $F(s) = \frac{1}{s-1} + \frac{3}{s^2+4}$.

## 6.2 Solving Equations, the Inverse Laplace Transform

5. $F(s) = \frac{2s+3}{s^2+9}$. First write it as a sum of two terms in the table.

6. $F(s) = \frac{5}{s^2+4}$. Hint: multiply and divide by the same number.

In problems 7–10, expand the function F(s), using partial fractions, and find $f(t) = \mathcal{L}^{-1}(F(s))$:

7. $F(s) = \frac{s+3}{s^2+3s+2}$.

8. $F(s) = \frac{3}{s^2+4s}$.

9. $F(s) = \frac{s+5}{(s+1)(s+2)(s+4)}$.

10. $F(s) = \frac{2s-1}{(s+2)^2(s+3)}$.

In problems 11–14, transform both sides of the differential equation and solve for X(s). Write X(s) as a ratio of polynomials in s.

11. $x' + 2x = e^{3t}$, $x(0) = 5$.

12. $2x' + 5x = 1 + \sin(t)$, $x(0) = 0$.

13. $x'' + 4x' + 2x = \cos(3t)$, $x(0) = 1$, $x'(0) = -1$.

14. $x'' + 2x' + x = \sin(t+2)$, $x(0) = x'(0) = 0$.

Solve the following initial-value problems completely, and check your answer by comparing it to the solution using the methods of Chapter 2 or Chapter 3.

15. $2x' + 5x = e^{-t}$, $x(0) = 1$.

16. $x' + 3x = 2$, $x(0) = -2$.

17. $x'' + 3x' + 2x = 4e^{-3t}$, $x(0) = x'(0) = 0$.

18. $x'' + 4x' + 3x = 1 + e^{-2t}$, $x(0) = 0$, $x'(0) = 1$.

**COMPUTER PROBLEMS.** Computer algebra systems have instructions that will find Laplace transforms and inverse Laplace transforms, and this is a good time to learn how to use them. In Maple it is necessary to load the package of integral transforms. This is done by executing the instruction

```
with(inttrans);
```

If you end the instruction with a semicolon, it will print out the names of all the routines it is loading. The instructions that you need are

```
laplace(f(t),t,s);
```

which returns the transform of the function f(t) as a function of s, and

```
invlaplace(F(s),s,t);
```

which returns the inverse transform of the function F(s).
The equivalent instructions in Mathematica are

```
LaplaceTransform[y[t],t,s]
```

and
$$\text{InverseLaplaceTransform[F[s],s,t]}.$$
Try checking your answers to problems 1–10 using these instructions.

6.3 Extending the Table

Before solving any more differential equations, it is necessary to add a few entries to our table of transforms. For example, we saw that a partial fraction expansion results in terms of the form
$$\frac{A}{s-a}, \frac{A}{(s-a)^n}, n > 1, \quad \text{and} \quad \frac{Bs+C}{(s^2+bs+c)^m}, m \geq 1,$$
and at this point the only one that appears in the right-hand column of Table 6.1 is the first one.

There are two important theorems that will make it possible to add the needed formulas.

Theorem 6.2. *For any piecewise continuous function f, of exponential order,*
$$\mathcal{L}\left(e^{at} f(t)\right) = F(s-a), \quad \text{where} \quad F(s) = \mathcal{L}\left(f(t)\right).$$

Proof. Using the definition,
$$\mathcal{L}\left(e^{at} f(t)\right) = \int_0^\infty e^{-st} \left(e^{at} f(t)\right) dt = \int_0^\infty e^{-(s-a)t} f(t) dt.$$
The integral on the right is just $F(s)$ with the argument s replaced by $s-a$; therefore, $\mathcal{L}(e^{at} f(t)) = F(s-a)$. □

Theorem 6.3. *For any piecewise continuous function f, of exponential order,*
$$\mathcal{L}\left(t f(t)\right) = -\frac{d}{ds} F(s), \quad \text{where} \quad F(s) = \mathcal{L}\left(f(t)\right).$$

Proof. Under appropriate conditions, which are satisfied in this case due to the assumptions on $f(t)$, one may interchange the operations of differentiation and integration to write
$$\frac{d}{ds} F(s) = \frac{d}{ds} \left(\int_0^\infty e^{-st} f(t) dt\right) = \int_0^\infty \frac{d}{ds} (e^{-st}) f(t) dt$$
$$= \int_0^\infty -t(e^{-st}) f(t) dt = -\int_0^\infty (e^{-st}) (t f(t)) dt = -\mathcal{L}\left(t f(t)\right). \quad □$$

Theorem 6.3 gives us a way to find $\mathcal{L}(t), \mathcal{L}(t^2), \ldots$ by writing
$$\mathcal{L}(t) = \mathcal{L}(t \cdot 1) = -\frac{d}{ds}\mathcal{L}(1) = -\frac{d}{ds}\left(\frac{1}{s}\right) = -\frac{d}{ds}(s^{-1}) = \frac{1}{s^2}.$$
In the exercises you will be asked to show that
$$\mathcal{L}(t^n) = \frac{n!}{s^{n+1}} \quad \text{and} \quad \mathcal{L}\left(t^n f(t)\right) = (-1)^n \frac{d^n}{ds^n} \left(F(s)\right).$$

Now, using Theorems 6.2 and 6.3, we can add the following entries to our table of Laplace transforms.

6.3 Extending the Table

Table 6.2

function	Laplace transform
t	$\dfrac{1}{s^2}$
$tf(t)$	$-\dfrac{d}{ds}F(s)$
t^n	$\dfrac{n!}{s^{n+1}}$
te^{at}	$\dfrac{1}{(s-a)^2}$
$e^{at}f(t)$	$F(s-a)$
$e^{at}\sin(bt)$	$\dfrac{b}{(s-a)^2+b^2}$
$e^{at}\cos(bt)$	$\dfrac{s-a}{(s-a)^2+b^2}$
$te^{at}\sin(bt)$	$\dfrac{2b(s-a)}{((s-a)^2+b^2)^2}$
$te^{at}\cos(bt)$	$\dfrac{(s-a)^2-b^2}{((s-a)^2+b^2)^2}$

6.3.1 Inverting a Term with an Irreducible Quadratic Denominator.
Partial fraction expansions can contain terms of the form

$$\frac{As+B}{s^2+cs+d} \tag{6.4}$$

where the denominator is an irreducible quadratic. We now have the necessary formulas for finding the inverse of this rational function. Since an irreducible quadratic $s^2 + cs + d$ has complex roots, we know that the discriminant $c^2 - 4d$ is less than zero, and this makes it possible to complete the square by writing

$$s^2 + cs + d = \left(s + \frac{c}{2}\right)^2 + \left(d - \frac{c^2}{4}\right) = \left(s + \frac{c}{2}\right)^2 + \left(\sqrt{d - c^2/4}\right)^2.$$

Check it! By completing the square in the denominator, and using Table 6.2, it appears that it should be possible to write (6.4) as a linear combination of the transforms of $e^{at}\cos(bt)$ and $e^{at}\sin(bt)$ as follows:

$$\frac{As+B}{s^2+cs+d} = C_1 \mathcal{L}\left(e^{at}\cos(bt)\right) + C_2 \mathcal{L}\left(e^{at}\sin(bt)\right) \tag{6.5}$$

$$= C_1 \frac{(s-a)}{(s-a)^2+b^2} + C_2 \frac{b}{(s-a)^2+b^2}.$$

The next example shows how this can be done, and then a formula is derived to handle the general case.

Example 6.3.1. *Find the inverse Laplace transform of*

$$\frac{As+B}{s^2+cs+d} = \frac{3s+2}{s^2+2s+10}. \tag{6.6}$$

Solution. Since $c^2 - 4d = 2^2 - 4(10) < 0$, the first step is to complete the square in the denominator:
$$s^2 + 2s + 10 \equiv (s+1)^2 + 10 - 1 = (s+1)^2 + 3^2 \equiv (s-a)^2 + b^2,$$
where $a = -1$ and $b = 3$. Now the numerator has to be written in the form:
$$3s + 2 \equiv C_1(s-a) + C_2 b = C_1(s+1) + C_2 \cdot 3 = C_1 s + (C_1 + 3C_2).$$
Solving this for C_1 and C_2, we find $C_1 = 3$ and $C_1 + 3C_2 = 2$ implies that $C_2 = -\frac{1}{3}$.

Using (6.5), we can write
$$\frac{3s+2}{s^2+2s+10} = 3\left(\frac{s+1}{(s+1)^2+3^2}\right) - \frac{1}{3}\left(\frac{3}{(s+1)^2+3^2}\right)$$
$$= 3\mathcal{L}\left(e^{-t}\cos(3t)\right) - \frac{1}{3}\mathcal{L}\left(e^{-t}\sin(3t)\right).$$

The linearity of the inverse transform and the fact that for a continuous function f, $\mathcal{L}^{-1}(\mathcal{L}(f(t))) \equiv f(t)$, then gives the result
$$\mathcal{L}^{-1}\left(\frac{3s+2}{s^2+2s+10}\right) = 3\left(e^{-t}\cos(3t)\right) - \frac{1}{3}\left(e^{-t}\sin(3t)\right). \qquad \blacksquare$$

The method used in Example 6.3.1 can be applied to any rational function of the form
$$\frac{As+B}{s^2+cs+d}$$
that has an irreducible quadratic as its denominator. Check that if the algebra is worked out, it results in the formula:

if $c^2 < 4d$, then $\mathcal{L}^{-1}\left(\dfrac{As+B}{s^2+cs+d}\right) = Ae^{at}\cos(bt) + \left(\dfrac{aA+B}{b}\right)e^{at}\sin(bt)$, \qquad (6.7)

where $a = -\frac{c}{2}$, $b = \sqrt{d - c^2/4}$.

This formula can be added to our table, as a special entry for inverting transforms of this type in a single step.

We are now ready to solve an arbitrary linear constant coefficient second-order differential equation of the type we studied in Chapter 3.

Example 6.3.2. *Use the method of Laplace transforms to solve the initial value problem*
$$x'' + x' + x = 3\sin(2t), \quad x(0) = 1, \quad x'(0) = 0.$$

This is the same IVP solved in Example 3.4.3 in Section 3.4, so you can compare not only the answers, but also the amount of work required by the two methods.

Solution. First take the transform of both sides of the equation
$$\mathcal{L}(x'' + x' + x) = \mathcal{L}(3\sin(2t))$$
and use linearity and the formula for the transform of $\sin(2t)$ to write
$$\mathcal{L}(x'') + \mathcal{L}(x') + \mathcal{L}(x) = 3\left(\frac{2}{s^2+4}\right).$$
Use the formulas for the transform of x' and x'' to write
$$(s^2 X(s) - sx(0) - x'(0)) + (sX(s) - x(0)) + X(s) = \frac{6}{s^2+4}.$$

6.3 Extending the Table

Subtract the terms that do not involve $X(s)$ from both sides, and enter the initial values:

$$(s^2 + s + 1)X(s) = sx(0) + x'(0) + x(0) + \frac{6}{s^2 + 4}$$

$$= s + 1 + \frac{6}{s^2 + 4} = \frac{(s+1)(s^2+4) + 6}{s^2 + 4}.$$

Now solve for $X(s)$ by dividing by s^2+s+1, and expand the result using a partial fraction expansion (by all means, use an expand function here):

$$X(s) = \frac{s^3 + s^2 + 4s + 10}{(s^2+4)(s^2+s+1)} \equiv \frac{-\frac{6}{13}s - \frac{18}{13}}{s^2 + 4} + \frac{\frac{19}{13}s + \frac{37}{13}}{s^2 + s + 1}. \tag{6.8}$$

The formulas for the Laplace transform of $\cos(2t)$ and $\sin(2t)$ suggest that we write the first term as

$$\frac{-\frac{6}{13}s - \frac{18}{13}}{s^2 + 4} = -\frac{6}{13}\left(\frac{s}{s^2+4}\right) - \frac{18}{13}\left(\frac{1}{2}\right)\left(\frac{2}{s^2+4}\right)$$

$$= -\frac{6}{13}\mathcal{L}\left(\cos(2t)\right) - \frac{9}{13}\mathcal{L}\left(\sin(2t)\right).$$

Notice that a 2 was needed in the numerator of the sine transform, so it had to be multiplied and divided by 2. Now, from Table 6.1 we see that (6.8) can be written as

$$X(s) = -\frac{6}{13}\mathcal{L}\left(\cos(2t)\right) - \frac{9}{13}\mathcal{L}\left(\sin(2t)\right) + \frac{\frac{19}{13}s + \frac{37}{13}}{s^2 + s + 1}. \tag{6.9}$$

Since the denominator $s^2 + s + 1$ in the final term is an irreducible quadratic, that term can be handled by formula (6.7) for $\mathcal{L}^{-1}\left(\frac{As+B}{s^2+cs+d}\right)$. We first complete the square to write

$$s^2 + s + 1 = \left(s + \frac{1}{2}\right)^2 + \left(1 - \frac{1}{4}\right) \equiv (s-a)^2 + b^2, \text{ with } a = -\frac{1}{2},\ b = \sqrt{\frac{3}{4}} = \frac{\sqrt{3}}{2},$$

and then use (6.7) with $A = \frac{19}{13}$ and $B = \frac{37}{13}$. This shows that the final term in (6.9) can be written as

$$\mathcal{L}\left(\frac{19}{13}e^{-\frac{1}{2}t}\cos\left(\frac{\sqrt{3}}{2}t\right) + \frac{-\frac{19}{26} + \frac{37}{13}}{\frac{\sqrt{3}}{2}}e^{-\frac{1}{2}t}\sin\left(\frac{\sqrt{3}}{2}t\right)\right).$$

Finally,

$$x(t) = \mathcal{L}^{-1}\left(X(s)\right)$$

$$= \mathcal{L}^{-1}\left(-\frac{6}{13}\mathcal{L}\left(\cos(2t)\right) - \frac{9}{13}\mathcal{L}\left(\sin(2t)\right)\right.$$

$$\left. + \mathcal{L}\left(e^{-\frac{1}{2}t}\left[\frac{19}{13}\cos\left(\frac{\sqrt{3}}{2}t\right) + \frac{55}{13\sqrt{3}}\sin\left(\frac{\sqrt{3}}{2}t\right)\right]\right)\right)$$

and therefore, using the properties of the inverse transform,

$$x(t) = -\frac{6}{13}\cos(2t) - \frac{9}{13}\sin(2t) + e^{-\frac{1}{2}t}\left(\frac{19}{13}\cos\left(\frac{\sqrt{3}}{2}t\right) + \frac{55}{13\sqrt{3}}\sin\left(\frac{\sqrt{3}}{2}t\right)\right). \tag{6.10}$$

The answer to this problem, solved in Example 3.4.3 by the characteristic polynomial method, is

$$x(t) = 2.8465 e^{-\frac{1}{2}t} \sin\left(\frac{\sqrt{3}}{2}t + 0.53920\right) + 0.83205 \sin(2t + 3.7296). \quad (6.11)$$

To compare the two answers (6.10) and (6.11) we need to write the functions

$$f_1(t) = -\frac{6}{13}\cos(2t) - \frac{9}{13}\sin(2t)$$

and

$$f_2(t) = \frac{19}{13}\cos\left(\frac{\sqrt{3}}{2}t\right) + \frac{55}{13\sqrt{3}}\sin\left(\frac{\sqrt{3}}{2}t\right)$$

as single sine functions, using the usual trigonometric identity.

If f_1 is written as $R \sin(2t + \phi)$, then

$$R = \sqrt{\left(\frac{6}{13}\right)^2 + \left(\frac{9}{13}\right)^2} \approx 0.83205, \qquad \phi = \tan^{-1}\left(\frac{-6/13}{-9/13}\right) + \pi \approx 3.7296,$$

so the first term can be replaced by

$$f_1(t) = -\frac{6}{13}\cos(2t) - \frac{9}{13}\sin(2t) \approx 0.83205 \sin(2t + 3.7296).$$

Similarly, to write f_2 as $\rho \sin\left(\frac{\sqrt{3}}{2}t + \psi\right)$, we need

$$\rho = \sqrt{\left(\frac{19}{13}\right)^2 + \left(\frac{55}{13\sqrt{3}}\right)^2} \approx 2.8465, \qquad \psi = \tan^{-1}\left(\frac{19}{13} \div \frac{55}{13\sqrt{3}}\right) \approx 0.53920,$$

so the second term in (6.10) becomes

$$e^{-\frac{1}{2}t} f_2(t) \approx 2.8465 e^{-\frac{1}{2}t} \sin\left(\frac{\sqrt{3}}{2}t + 0.53920\right).$$

By comparing $x(t) = e^{-\frac{1}{2}t} f_2(t) + f_1(t)$ with (6.11), it can be seen that the two answers to the initial-value problem are the same. ■

6.3.2 Solving Linear Systems with Laplace Transforms.
The method of Laplace transforms can also be applied to solve a system of linear equations with constant coefficients, and this will be a good place to practice your skill with matrix operations.

We are going to illustrate this method on 2-dimensional nonhomogeneous linear constant coefficient systems of the form

$$\begin{aligned} x'(t) &= ax(t) + by(t) + f(t) \\ y'(t) &= cx(t) + dy(t) + g(t), \end{aligned} \quad (6.12)$$

with initial conditions $x(0) = x_0$, $y(0) = y_0$.

Take Laplace transforms of both equations in (6.12):

$$\begin{aligned} sX(s) - x_0 &= aX(s) + bY(s) + F(s) \\ sY(s) - y_0 &= cX(s) + dY(s) + G(s), \end{aligned} \quad (6.13)$$

6.3 Extending the Table

and define the vector $\vec{X} = \begin{pmatrix} X(s) \\ Y(s) \end{pmatrix}$ and the matrix $\mathbf{A} = \begin{pmatrix} a & b \\ c & d \end{pmatrix}$. Now the system (6.13) can be written in matrix form as

$$s\vec{X} - \begin{pmatrix} x_0 \\ y_0 \end{pmatrix} = \mathbf{A}\vec{X} + \begin{pmatrix} F(s) \\ G(s) \end{pmatrix}.$$

Subtracting $\mathbf{A}\vec{X}$ and adding $\begin{pmatrix} x_0 \\ y_0 \end{pmatrix}$ on both sides, and inserting the 2×2 identity matrix $\mathbf{I}$, the equation becomes

$$(s\mathbf{I} - \mathbf{A})\vec{X} = \begin{pmatrix} F(s) + x_0 \\ G(s) + y_0 \end{pmatrix}. \tag{6.14}$$

In Exercises 10. and 11. at the end of Section 4.2 we gave a formula for the multiplicative inverse of a 2×2 matrix and showed that the solution of a system of linear equations of the form $\mathbf{A}\vec{X} = \vec{b}$ can be written as $\vec{X} = \mathbf{A}^{-1}\vec{b}$. For the system (6.14), we can write

$$\vec{X} = (s\mathbf{I} - \mathbf{A})^{-1} \begin{pmatrix} F(s) + x_0 \\ G(s) + y_0 \end{pmatrix}.$$

Using the formula in Exercise 10., the inverse matrix is

$$(s\mathbf{I} - \mathbf{A})^{-1} = \begin{pmatrix} s - a & -b \\ -c & s - d \end{pmatrix}^{-1} = \frac{1}{\det(s\mathbf{I} - \mathbf{A})} \begin{pmatrix} s - d & b \\ c & s - a \end{pmatrix}$$

where $\det(s\mathbf{I} - \mathbf{A}) = (s - a)(s - d) - bc = s^2 - (a + d)s + (ad - bc)$.

We can now write the vector of transforms as

$$\vec{X} \equiv \begin{pmatrix} X(s) \\ Y(s) \end{pmatrix} = \frac{1}{\det(s\mathbf{I} - \mathbf{A})} \begin{pmatrix} s - d & b \\ c & s - a \end{pmatrix} \begin{pmatrix} F(s) + x_0 \\ G(s) + y_0 \end{pmatrix}. \tag{6.15}$$

Performing the indicated matrix multiplication will result in formulas for $X(s) = \mathcal{L}(x(t))$ and $Y(s) = \mathcal{L}(y(t))$. This is definitely a problem where a computer algebra system should be used to invert these transforms to find the solutions $x(t)$ and $y(t)$. The example below will show how all of this works in practice.

Example 6.3.3. Use Laplace transforms to solve the system

$$\begin{aligned} x' &= x + 4y + t \\ y' &= -x - 3y + e^{-2t} \end{aligned} \tag{6.16}$$

with the initial conditions $x(0) = 0$, $y(0) = 1$.

Solution. The matrix $\mathbf{A} = \begin{pmatrix} 1 & 4 \\ -1 & -3 \end{pmatrix}$ and $(s\mathbf{I} - \mathbf{A}) = \begin{pmatrix} s - 1 & -4 \\ 1 & s + 3 \end{pmatrix}$, with

$$\det((s\mathbf{I} - \mathbf{A})) = (s - 1)(s + 3) + 4 \equiv (s + 1)^2.$$

The functions F and G are

$$F(s) = \mathcal{L}(f(t)) = \mathcal{L}(t) = 1/s^2$$

and

$$G(s) = \mathcal{L}(g(t)) = \mathcal{L}(e^{-2t}) = 1/(s + 2).$$

Using (6.15), we can write

$$\begin{pmatrix} X(s) \\ Y(s) \end{pmatrix} = \frac{1}{(s + 1)^2} \begin{pmatrix} s + 3 & 4 \\ -1 & s - 1 \end{pmatrix} \begin{pmatrix} 1/s^2 \\ 1/(s + 2) + 1 \end{pmatrix}.$$

Performing the indicated matrix multiplication and simplifying gives
$$X(s) = \frac{4s^3 + 13s^2 + 5s + 6}{(s+1)^2 s^2 (s+2)} \quad \text{and} \quad Y(s) = \frac{s^4 + 2s^3 - 3s^2 - s - 2}{(s+1)^2 s^2 (s+2)}.$$
Inversion of these two transforms, using `invlaplace` in Maple produces the result
$$x(t) = \mathcal{L}^{-1}(X(s)) = -5 + 3t + 4e^{-2t} + (1 + 10t)e^{-t}$$
$$y(t) = \mathcal{L}^{-1}(Y(s)) = 2 - t - 3e^{-2t} + (2 - 5t)e^{-t},$$
and these can be checked by substituting them into the original pair of equations. It should also be checked that they satisfy the initial conditions. ∎

Exercises 6.3. *Find the Laplace transform of the functions in 1–8.*

1. $e^{-t}\cos(2t)$

2. $e^{-2t}\sin(3t)$

3. $t\sin(5t)$

4. $t\cos(3t)$

5. $t^2 - 6t + 4$

6. $2t^2 - t + 1$

7. $te^{-t}\sin(2t)$

8. $te^{-2t}\cos(t)$

9. Assuming that $\mathcal{L}(t^n) = \frac{n!}{s^{n+1}}$ for some integer $n \geq 1$, use the formula for $\mathcal{L}(tf(t))$ in Table 6.2 to show that $\mathcal{L}(t^{n+1}) = \frac{(n+1)!}{s^{n+2}}$. Since we know the formula is true for $n = 1$, this actually proves the formula true for any positive integer n (this kind of proof is called "proof by induction").

10. Write a simple formula for $\mathcal{L}^{-1}\left(\frac{1}{s^n}\right)$. Hint: start with $\mathcal{L}^{-1}\left(\frac{(n-1)!}{s^n}\right) = t^{n-1}$, and use linearity.

Find $\mathcal{L}^{-1}(F(s))$ for the functions $F(s)$ in 11–16.

11. $F(s) = \dfrac{1}{(s-2)^2} + \dfrac{3}{s^2}$

12. $F(s) = \dfrac{3}{(s+1)^2} + \dfrac{1}{s^3}$

13. $F(s) = \dfrac{s+3}{(s-2)^2}$ Hint: Use partial fractions to write this as $\dfrac{A}{s-2} + \dfrac{B}{(s-2)^2}$.

14. $F(s) = \dfrac{2s+5}{s^2}$

15. $F(s) = \dfrac{s+5}{(s+3)(s^2+1)}$

16. $F(s) = \dfrac{6}{s} + \dfrac{7}{s^2} + \dfrac{3}{s^3}$

Use (6.7) to invert the functions in 17 and 18.

17. $F(s) = \dfrac{s+3}{s^2+4s+8}$

18. $G(s) = \dfrac{2s+3}{s^2+2s+26}$

Use Laplace transforms to solve the initial value problems 19 and 20. Check your answer by solving the equation using the characteristic polynomial method.

19. Solve $x'' + 4x' + 8x = 0$, $x(0) = 1$, $x'(0) = -1$. *(You can use the inverse transform in problem 17..)*

20. Solve $x'' + 2x' + 26x = 0$, $x(0) = 2$, $x'(0) = -1$. *(You can use the inverse transform in problem 18..)*

21. Use Laplace transforms to solve the IVP
$$x'' + 2x' + 10x = 5, \quad x(0) = 0, \quad x'(0) = 1.$$

22. Use Laplace transforms to solve the IVP
$$x'' + 3x' + 2.5x = t, \quad x(0) = x'(0) = 0.$$

23. Use Laplace transforms to solve the system of equations
$$x' = -2x + 2y + 1$$
$$y' = x - 3y + \sin(t)$$
with initial conditions $x(0) = 1$, $y(0) = 0$.

6.4 The Unit Step Function

In the beginning of this chapter it was emphasized that Laplace transform methods can be applied to equations with discontinuous driving functions, and so far everything we have done has been in terms of continuous functions. To apply the methods to piecewise continuous functions we need a way to write such functions as a single expression, and this requires a simple function with a single jump discontinuity.

Definition 6.5. *The* **unit step function** *at $t = 0$, also called the* **Heaviside function**, *is defined as*
$$u(t) = \begin{cases} 0 & \text{if } t < 0 \\ 1 & \text{if } t > 0. \end{cases}$$

It is not necessary to define the value of any of our piecewise continuous functions at their points of discontinuity, since solving differential equations involves integration, and the value of a function at a single point does not affect its integral.

The translated unit step function $u(t - c)$ can be used to provide a jump at an arbitrary point $t = c$, for any real number c.

Definition 6.6. *The* **unit step function** *at $t = c$, for an arbitrary real number c, is defined as*
$$u(t - c) = \begin{cases} 0 & \text{if } t < c \\ 1 & \text{if } t > c. \end{cases}$$

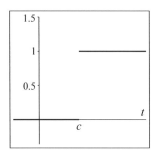

Figure 6.2. The unit step function $u(t-c)$

A graph of $u(t-c)$ for a value of $c > 0$ is shown in Figure 6.2.

When solving initial-value problems with Laplace transforms, it is generally assumed that initial values are given at $t = 0$. If an initial-value problem is given with initial conditions at some other value t_0, it can be made to start at zero by the change of independent variable $\tau = t - t_0$.

It is now possible to write any piecewise continuous function $f(t)$ as a sum of continuous functions, each multiplied by an appropriate unit step function. The following example shows how this is done for a simple step function.

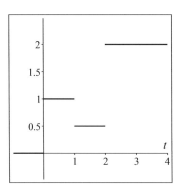

Figure 6.3. The function $q(t)$

Example 6.4.1. *Given the piecewise continuous function*

$$q(t) = \begin{cases} 1 & \text{if } 0 < t < 1 \\ \frac{1}{2} & \text{if } 1 < t < 2 \\ 2 & \text{if } t > 2 \end{cases}$$

shown in Figure 6.3, write a single formula for q, using unit step functions.

Solution. We start with the value of q at $t = 0$, and then add terms at each jump point, multiplied by the appropriate unit step function:

$$q(t) = 1 + u(t-1)\left(\frac{1}{2} - 1\right) + u(t-2)\left(2 - \frac{1}{2}\right) = 1 - \frac{1}{2}u(t-1) + \frac{3}{2}u(t-2).$$

6.4 The Unit Step Function

At each jump point the appropriate unit step function is multiplied by the signed vertical distance of the jump. To see that this works, try evaluating $q(t)$ at values of t in each interval where q is constant. For example,

$$q\left(\frac{3}{2}\right) = 1 - \frac{1}{2}u\left(\frac{3}{2}-1\right) + \frac{3}{2}u\left(\frac{3}{2}-2\right) = 1 - \frac{1}{2}u\left(\frac{1}{2}\right) + \frac{3}{2}u\left(-\frac{1}{2}\right).$$

Since $u(\frac{1}{2}) = 1$ and $u(-\frac{1}{2}) = 0$, this gives the correct value $q(\frac{3}{2}) = 1 - \frac{1}{2} \cdot 1 + \frac{3}{2} \cdot 0 = \frac{1}{2}$. ∎

Now consider the general case. An arbitrary piecewise continuous function with n jumps, and continuous between jumps, given by

$$f(t) = \begin{cases} f_1(t) & \text{if } 0 < t < t_1 \\ f_2(t) & \text{if } t_1 < t < t_2 \\ \cdots \\ f_n(t) & \text{if } t_{n-1} < t < \infty \end{cases}$$

can be written as

$$f(t) = f_1(t) + u(t - t_1)(f_2(t) - f_1(t)) + u(t - t_2)(f_3(t) - f_2(t)) + \cdots + u(t - t_{n-1})(f_n(t) - f_{n-1}(t)).$$

This is again easiest to see by working out an example.

Example 6.4.2. *Write the function graphed in Figure 6.4 in terms of unit step functions.*

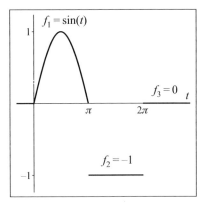

Figure 6.4. The function $f(t)$

Solution. We start by letting $f(t) \equiv f_1(t) = \sin(t)$. At $t = t_1 = \pi$, the function must be changed to the constant value -1. This is done by adding the term

$$u(t - \pi)(f_2(t) - f_1(t)) = u(t - \pi)(-1 - \sin(t)).$$

Since $u(t - \pi)$ is zero up to $t = \pi$, this new term does not change the value of the function on the interval $0 < t < \pi$. At $t = \pi$, the step function $u(t - \pi)$ becomes equal to one, and the two terms add to give the correct value $\sin(t) + (-1 - \sin(t)) = -1$ in the interval $\pi < t < 2\pi$. At $t = 2\pi$ the function must return to zero, and this is done by adding the term

$$u(t - 2\pi)(f_3(t) - f_2(t)) = u(t - 2\pi)(0 - (-1)) \equiv u(t - 2\pi).$$

The result is
$$f(t) = \sin(t) + u(t-\pi)(-1 - \sin(t)) + u(t - 2\pi). \tag{6.17}$$
∎

To convince yourself that $f(t)$ in Example 6.4.2 is the function in Figure 6.4, a calculator can be used to draw the graph. If the Heaviside function is not available on the calculator, it probably has a function called SIGN(t), which is equal to -1 for $t < 0$ and $+1$ for $t > 0$. Using this function, the unit step function can be written as
$$u(t) \equiv \frac{1}{2}(\text{SIGN}(t) + 1).$$
Check that this function has the value 0 if $t < 0$ and 1 if $t > 0$.

To apply the method of Laplace transforms to problems containing piecewise continuous functions, two new formulas are needed.

Theorem 6.4. *The Laplace transform of the unit step function $u(t - c)$, for any $c \geq 0$, is*
$$\mathcal{L}(u(t-c)) = \frac{e^{-cs}}{s}.$$

Proof. For $c \geq 0$,
$$\mathcal{L}(u(t-c)) = \int_0^\infty e^{-st} u(t-c) dt = \int_0^c e^{-st}(0) dt + \int_c^\infty e^{-st}(1) dt$$
$$= 0 + \left. \frac{e^{-st}}{-s} \right|_c^\infty = 0 + 0 - \frac{e^{-sc}}{-s} = \frac{e^{-cs}}{s}. \qquad \square$$

The next theorem gives a formula for the transform of the product of $u(t-c)$ times a function for which we already know the transform.

Theorem 6.5. *The Laplace transform of $u(t-c)f(t-c)$ is*
$$\mathcal{L}(u(t-c)f(t-c)) = e^{-cs} F(s),$$
where $F(s) = \mathcal{L}(f(t))$.

Proof.
$$\mathcal{L}(u(t-c)f(t-c)) = \int_0^\infty e^{-st} u(t-c) f(t-c) dt = \int_c^\infty e^{-st}(1) f(t-c) dt.$$
Making the change of variable $\tau = t - c$,
$$\int_c^\infty e^{-st} f(t-c) dt = \int_0^\infty e^{-s(\tau+c)} f(\tau) d\tau$$
$$= e^{-cs} \int_0^\infty e^{-s\tau} f(\tau) d\tau = e^{-cs} \mathcal{L}(f(t)) = e^{-cs} F(s). \qquad \square$$

We will see that this formula is most useful for inverting terms of the form $e^{-cs} F(s)$. A third formula will be given below to make it easier to transform a term written in the form $u(t-c)f(t)$. To transform a term of the form $u(t-c)f(t)$, we simply define $g(t) \equiv f(t+c)$, which also means that $f(t) \equiv g(t-c)$. Using Theorem 6.5, we can now write
$$\mathcal{L}(u(t-c)f(t)) \equiv \mathcal{L}(u(t-c)g(t-c)) = e^{-cs} G(s), \tag{6.18}$$

6.4 The Unit Step Function

where $G(s) = \mathcal{L}(g(t)) = \mathcal{L}(f(t+c))$.

The table of Laplace transforms can now be extended by adding the three new formulas in Table 6.3.

Table 6.3

function	Laplace transform
$u(t-c)$	$\dfrac{e^{-cs}}{s}$
$u(t-c)f(t-c)$	$e^{-cs}F(s)$, where $F(s) = \mathcal{L}(f(t))$
$u(t-c)f(t)$	$e^{-cs}G(s)$, where $G(s) = \mathcal{L}(f(t+c))$

The first formula in the table is just a special case of the second formula with $f(t) \equiv 1$; that is, with $\mathcal{L}(f(t)) = \dfrac{1}{s}$.

Example 6.4.3. *Find the Laplace transform of $u(t-1)f(t)$, where $f(t) = t^2 + 2$.*

Solution. We will use (6.18):
$$\mathcal{L}\left(u(t-1)(t^2+2)\right) = e^{-1 \cdot s}G(s),$$
where
$$G(s) = \mathcal{L}\left(f(t+1)\right) = \mathcal{L}\left((t+1)^2 + 2\right) = \mathcal{L}(t^2 + 2t + 3) = \frac{2}{s^3} + 2\frac{1}{s^2} + 3\frac{1}{s};$$
therefore,
$$\mathcal{L}\left(u(t-1)(t^2+2)\right) = e^{-s}\left(\frac{2}{s^3} + \frac{2}{s^2} + \frac{3}{s}\right). \blacksquare$$

The next two examples show how problems involving discontinuous driving functions are handled.

Example 6.4.4. *A harmonic oscillator with natural frequency $\omega_0 = 2$, initially at rest, is forced by the ramp function*
$$f(t) = \begin{cases} t & \text{if } 0 < t < 1 \\ 0 & \text{if } t > 1. \end{cases}$$

Solve the IVP
$$x'' + 4x = f(t), \quad x(0) = x'(0) = 0.$$

Solution. Using the unit step function, we see that the forcing function f can be written in the form $f(t) = t - u(t-1)t$; therefore, the Laplace transform of the differential equation can be written as
$$s^2 X(s) + 4X(s) = \mathcal{L}(t - u(t-1)t) = \mathcal{L}(t) - \mathcal{L}(u(t-1)t) = \frac{1}{s^2} - e^{-s}G(s), \quad (6.19)$$

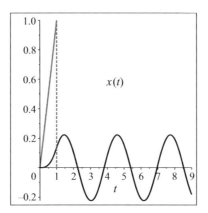

Figure 6.5. The forcing function $f(t)$ and the output $x(t)$

where $G(s) = \mathcal{L}(t+1) = \frac{1}{s^2} + \frac{1}{s}$. Dividing (6.19) by $s^2 + 4$ gives

$$X(s) = \frac{1}{s^2(s^2+4)} - e^{-s}\left(\frac{1}{s^2(s^2+4)} + \frac{1}{s(s^2+4)}\right) \quad (6.20)$$
$$\equiv Y(s) - e^{-s}(Y(s) + Z(s)).$$

We need to invert the two functions $Y(s) = \frac{1}{s^2(s^2+4)}$ and $Z(s) = \frac{1}{s(s^2+4)}$. Using partial fractions,

$$y(t) = \mathcal{L}^{-1}(Y(s)) = \mathcal{L}^{-1}\left(\frac{1}{s^2(s^2+4)}\right) = \mathcal{L}^{-1}\left(\frac{1/4}{s^2} - \frac{1/4}{s^2+4}\right) = \frac{1}{4}t - \frac{1}{8}\sin(2t).$$

Similarly,

$$z(t) = \mathcal{L}^{-1}(Z(s)) = \mathcal{L}^{-1}\left(\frac{1}{s(s^2+4)}\right) = \mathcal{L}^{-1}\left(\frac{1/4}{s} - \frac{(1/4)s}{s^2+4}\right) = \frac{1}{4} - \frac{1}{4}\cos(2t).$$

Using our formula for $\mathcal{L}^{-1}(e^{-as}F(s))$, the inverse of (6.20) can be written as

$$x(t) = \mathcal{L}^{-1}(Y(s) - e^{-s}Y(s) - e^{-s}Z(s))$$
$$= y(t) - u(t-1)y(t-1) - u(t-1)z(t-1)$$
$$= \left(\frac{1}{4}t - \frac{1}{8}\sin(2t)\right)$$
$$- u(t-1)\left(\frac{1}{4}(t-1) - \frac{1}{8}\sin(2(t-1)) + \frac{1}{4} - \frac{1}{4}\cos(2(t-1))\right).$$

Writing this back in the form of a piecewise defined function,

$$x(t) = \begin{cases} \frac{1}{4}t - \frac{1}{8}\sin(2t) & \text{if } 0 < t < 1 \\ -\frac{1}{8}\sin(2t) + \frac{1}{8}\sin(2(t-1)) + \frac{1}{4}\cos(2(t-1)) & \text{if } t > 1. \end{cases}$$

Check carefully the second part of this function, where the two pieces are added together. A graph of $x(t)$, together with the input function, is shown in Figure 6.5. ∎

To see how problems of this type might arise in a real-world situation, consider the following one-compartment mixing problem where the concentration of pollutant in the input solution is changed suddenly.

6.4 The Unit Step Function

Example 6.4.5. *A 25-gallon tank is initially filled with water containing one pound of salt dissolved in it. It is desired to increase the salt concentration from $\frac{1}{25}$ pounds per gallon to 0.2 pounds per gallon. With x(t) equal to the pounds of salt in the tank at time t, this would require that x(t) be 5 pounds. For 5 minutes, a solution containing 0.5 pound of salt per gallon is allowed to run into the tank at a rate of one gallon per minute, and the solution in the tank is allowed to drain out at the same rate. This seems to be taking too long, so the operators decide to dump 5 pounds of salt into the input water, and this well-stirred solution is all fed in over the next minute. At the end of 6 minutes, the concentration is too high, so pure water is run in to lower it. Find the time at which the concentration is back down to 0.2 lbs/gal.*

Solution. As before, we first draw a diagram of the tank showing the input and output.

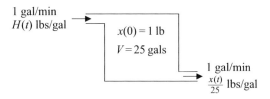

Letting $x(t)$ represent the amount of salt in the tank at time t, we can write

$$x'(t) = \text{rate in} - \text{rate out} = (1.0\text{gal/min})(H(t)\text{lbs/gal}) - (1.0\text{gal/min})\left(\frac{x(t)}{25}\text{lbs/gal}\right),$$

where the function $H(t)$, representing pounds of salt per gallon in the input solution, is shown in the left-hand graph in Figure 6.6.

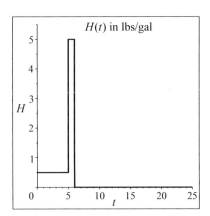 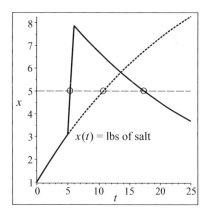

Figure 6.6. Input $H(t)$, in pounds per gallon, and output $x(t)$ in pounds

Writing $H(t)$ in terms of unit step functions, check that

$$H(t) = \frac{1}{2} + u(t-5)(5-0.5) + u(t-6)(0-5) = \frac{1}{2} + \frac{9}{2}u(t-5) - 5u(t-6).$$

The initial-value problem that must be solved is

$$x'(t) + 0.04x(t) = H(t) = \frac{1}{2} + \frac{9}{2}u(t-5) - 5u(t-6), \quad x(0) = 1.$$

Using Laplace transforms,

$$\mathcal{L}(x' + 0.04x) = \mathcal{L}\left(\frac{1}{2} + \frac{9}{2}u(t-5) - 5u(t-6)\right) \Longrightarrow$$

$$(sX(s) - x(0)) + 0.04X(s) = \left(\frac{1}{2}\right)\frac{1}{s} + e^{-5s}\frac{9}{2s} - e^{-6s}\frac{5}{s} \Longrightarrow$$

$$(s + 0.04)X(s) = \left(1 + \frac{1}{2s}\right) + e^{-5s}\frac{9}{2s} - e^{-6s}\frac{5}{s} \Longrightarrow$$

$$X(s) = \frac{1}{s + 0.04} + \frac{1/2}{s(s + 0.04)} \qquad (6.21)$$
$$+ e^{-5s}\frac{9/2}{s(s + 0.04)} - e^{-6s}\frac{5}{s(s + 0.04)}.$$

Define the function $F(s) = \frac{1}{s(s+0.04)}$, and using partial fractions write

$$F(s) = \frac{A}{s} + \frac{B}{s + 0.04} = \frac{25}{s} - \frac{25}{s + 0.04}.$$

This function has inverse Laplace transform

$$f(t) = \mathcal{L}^{-1}(F(s)) = 25 - 25e^{-0.04t}.$$

Now the solution $x(t)$ can be written as the inverse of $X(s)$ in (6.21):

$$x(t) = e^{-0.04t} + 0.5f(t) + \frac{9}{2}u(t-5)f(t-5) - 5u(t-6)f(t-6)$$

$$= e^{-0.04t} + 0.5\left(25 - 25e^{-0.04t}\right) + \frac{9}{2}u(t-5)\left(25 - 25e^{-0.04(t-5)}\right)$$

$$- 5u(t-6)\left(25 - 25e^{-0.04(t-6)}\right).$$

This function is the solid curve in the right-hand graph in Figure 6.6.

To have the desired concentration of 0.2 pounds of salt per gallon, there needs to be 5 pounds of salt in the tank. To find the time when $x(t) = 5$, check that $x(t)$ can be written as a piecewise defined function:

$$x(t) = \begin{cases} 12.5 - 11.5e^{-0.04t} & \text{if } 0 < t < 5 \\ 125.0 - 11.5e^{-0.04t} - 112.5e^{-0.04(t-5)} & \text{if } 5 < t < 6 \\ -11.5e^{-0.04t} - 112.5e^{-0.04(t-5)} + 125e^{-0.04(t-6)} \equiv 9.9983e^{-0.04t} & \text{if } 6 < t. \end{cases}$$

Make sure you see how each piece of this function was computed by adding the appropriate terms in each time interval.

Setting $x(t) = 5$, in each of the time intervals, we see that the process needed to be stopped at either $t \approx 5.40$ min, or if that time was missed, at $t \approx 17.32$ min. If the correction had not been made, and the solution with concentration 1 pound per gallon kept running in, the formula $12.5 - 11.5e^{-0.04t}$ for $x(t)$ is correct for all $t > 0$, and $x(t) = 5$ would have occurred at $t = 10.7$ min. ∎

In the final example of this section we will show that Laplace transforms can also exist for functions with infinitely many jumps. To be piecewise continuous, a function must have only finitely many jumps on any finite interval, but the piecewise linear function $h(t)$ graphed in Figure 6.7 satisfies that condition, even though it is assumed to have a jump at each positive integer value of t.

6.4 The Unit Step Function

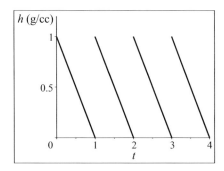

Figure 6.7. Function $h(t)$ with infinitely many jumps

Check carefully that $h(t)$ can be written as
$$h(t) = (1-t)u(t) + u(t-1) + u(t-2) + \cdots + u(t-k) + \cdots$$
$$= (1-t)u(t) + \sum_{k=1}^{\infty} u(t-k).$$

Since h is piecewise continuous and bounded, it has a Laplace transform, which we can write as
$$\mathcal{L}(h(t)) = H(s) = \frac{1}{s} - \frac{1}{s^2} + \sum_{k=1}^{\infty} \frac{e^{-ks}}{s}.$$

We are going to use h, commonly referred to as a sawtooth function, to model a situation in which a patient is receiving a daily dose of a slow-release drug. The amount of the drug in the patient's blood (in g/cc) at any time t (in days) is given by $h(t)$. We then consider the amount $x(t)$ of the drug in an organ in the body using a one-compartment mixing model.

Example 6.4.6. *Assume the volume of blood in an organ of the body is 100 cc, and blood containing $h(t)$ g/cc of a drug flows in and out of the organ at a rate of 100 cc per day. The diagram below shows the flow of blood.*

```
100 cc/day          x(0) = 0           100 cc/day
h(t) g/cc   →    V = 100 cc    →      x(t)/100 g/cc
```

Using the equation
$$x'(t) = \text{rate in} - \text{rate out},$$
we need to solve the initial-value problem
$$x'(t) = 100h(t) - 100\left(\frac{x(t)}{100}\right), \quad x(0) = 0;$$
that is, we must solve the differential equation
$$x' + x = 100h(t) = 100\left((1-t)u(t) + \sum_{k=1}^{\infty} u(t-k)\right), \tag{6.22}$$
and this equation can be solved using Laplace transforms.

Assuming initially there is no drug in the blood, the Laplace transform of (6.22) *is*

$$\mathcal{L}(x' + x) = sX(s) - x(0) + X(s) = (s+1)X(s) = 100\left(\frac{1}{s} - \frac{1}{s^2} + \sum_{k=1}^{\infty} \frac{e^{-ks}}{s}\right).$$

If each term of this equation is divided by $s + 1$, $X(s)$ has the form

$$X(s) = 100\left(\frac{1}{s(s+1)} - \frac{1}{s^2(s+1)} + \sum_{k=1}^{\infty} \frac{e^{-ks}}{s(s+1)}\right).$$

Define the functions $F(s) = \frac{1}{s(s+1)}$ and $G(s) = \frac{1}{s^2(s+1)}$. These are easily inverted, using partial fractions, to show that

$$f(t) = \mathcal{L}^{-1}(F(s)) = 1 - e^{-t}, \quad g(t) = \mathcal{L}^{-1}(G(s)) = e^{-t} + t - 1.$$

Using linearity, and the formula $\mathcal{L}^{-1}(e^{-ks}F(s)) = u(t-k)f(t-k)$, the inverse transform of $X(s)$ is

$$x(t) = \mathcal{L}^{-1}(X(s)) = 100\left(f(t) - g(t) + \sum_{k=1}^{\infty} u(t-k)f(t-k)\right)$$

$$= 100\left(2 - 2e^{-t} - t + \sum_{k=1}^{\infty} u(t-k)\left(1 - e^{-(t-k)}\right)\right).$$

Although $x(t)$ is a continuous function everywhere, its derivative is discontinuous at each integer value $t = 1, 2, \ldots$.

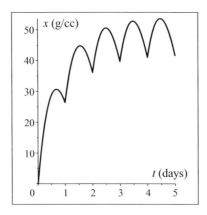

Figure 6.8. Amount of drug in the organ at time t

For any integer $N > 0$, a closed formula for $x(t)$ on the interval $N < t < N + 1$ can be obtained by using the formula for the partial sum of a geometric series with ratio r between 1 and -1:

$$1 + r + r^2 + \cdots + r^N = \frac{1 - r^{N+1}}{1 - r}. \tag{6.23}$$

6.4 The Unit Step Function

Assume $N < t < N+1$; then $u(t-k) = 1$ for $k = 1, 2, \ldots, N$ and $u(t-k) = 0$ for $k \geq N+1$. The formula for $x(t)$ in this interval is

$$\begin{aligned} x(t) &= 100\left(2 - 2e^{-t} - t + (1 - e^{-t+1}) + (1 - e^{-t+2}) + \cdots + (1 - e^{-t+N})\right) \\ &= 100\left(2 - 2e^{-t} - t + N - e^{-t+N}\left[1 + e^{-1} + e^{-2} + \cdots + e^{-(N-1)}\right]\right) \quad (6.24) \\ &= 100\left(2 - 2e^{-t} - t + N - e^{-t+N}\left[\frac{1 - e^{-N}}{1 - e^{-1}}\right]\right); \end{aligned}$$

where (6.23) applies to the sum since the ratio $r \equiv e^{-1} < 1$.

Equation (6.24) is used to plot the graph of $x(t)$ shown in Figure 6.8. It can also be used to show that as $N \to \infty$ the amount of drug in the organ is tending to a fixed daily pattern, given by

$$x(t) = 100\left(2 + (N - t) - e^{N-t}\left(\frac{1}{1 - e^{-1}}\right)\right), \quad N < t < N + 1.$$

Using this formula, a doctor could determine the average amount of the drug in the organ over a 24-hour period, and even determine the time of day when the amount is a maximum. One of the exercises will ask you to compute these values. ■

Exercises 6.4. For Exercises 1–6, sketch a graph of $f(t)$, and using unit step functions $u(t - c)$, write $f(t)$ as a single expression.

1.
$$f(t) = \begin{cases} 2 & \text{if } 0 < t < 1 \\ -1 & \text{if } 1 < t < 2 \\ 1 & \text{if } 2 < t < 5 \\ 0 & \text{if } t > 5 \end{cases}$$

2.
$$f(t) = \begin{cases} 0 & \text{if } 0 < t < 1 \\ 1 & \text{if } 1 < t < 2 \\ 2 & \text{if } 2 < t < 3 \\ 3 & \text{if } t > 3 \end{cases}$$

3.
$$f(t) = \begin{cases} \cos(t) & \text{if } 0 < t < \pi \\ -1 & \text{if } \pi < t < 2\pi \\ \sin(t) & \text{if } 2\pi < t < 3\pi \\ 0 & \text{if } t > 3\pi \end{cases}$$

4.
$$f(t) = \begin{cases} 1 - (t-1)^2 & \text{if } 0 < t < 2 \\ 0 & \text{if } t > 2 \end{cases}$$

5. (sawtooth function)
$$f(t) = \begin{cases} t & \text{if } 0 < t < 1 \\ t - 1 & \text{if } 1 < t < 2 \\ t - 2 & \text{if } 2 < t < 3 \\ 0 & \text{if } t > 3 \end{cases}$$

6.
$$f(t) = \begin{cases} \sin(t) & \text{if } 0 < t < \pi \\ -\sin(t) & \text{if } \pi < t < 2\pi \\ \sin(t) & \text{if } 2\pi < t < 3\pi \\ 0 & \text{if } t > 3\pi \end{cases}$$

For Exercises 7–10, find the Laplace transform of $f(t)$.

7. $f(t) = u(t-2)(1+t^2)$

8. $f(t) = u(t-1)\sin(t)$

9. The function $f(t)$ in Exercise 3

10. The function $f(t)$ in Exercise 4

For Exercises 11–14, find the inverse Laplace transform of $F(s)$.

11. $F(s) = e^{-3s}\left(\dfrac{s}{s^2+1}\right)$

12. $F(s) = e^{-2s}\left(\dfrac{1}{s+1}\right)$

13. $F(s) = \dfrac{1}{s^2+4} + e^{-s}\left(\dfrac{s}{s^2+4}\right)$

14. $F(s) = e^{-s}\left(\dfrac{1}{s}\right) + e^{-3s}\left(\dfrac{1}{s^2}\right)$

Use Laplace transforms to solve the following initial value problems. Sketch a graph of the solution, together with a graph of the driving function.

15. $x' + 2x = 2u(t-3)$, $x(0) = 1$

16. $x' + x = 2 + u(t-1)e^{-t}$, $x(0) = 0$

17. $x'' + 4x' + 4x = u(t-2)$, $x(0) = 0$, $x'(0) = 1$

18. $x'' + 2x' + 5x = 2 + u(t-2)$, $x(0) = 0$, $x'(0) = 0$

19. In this problem we consider an RC-circuit with input voltage given by the function $h(t) = |\sin(t)|$ shown below. This is the kind of input one might have when alternating current is rectified to direct current.

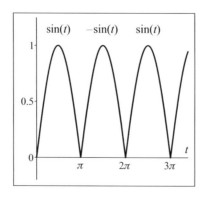

(a) *Write the piecewise function h(t) in terms of unit step functions.*
(b) *Show that the Laplace transform of h(t) can be written as*

$$H(s) = \mathcal{L}(h(t)) = \frac{1}{s^2+1} + 2\sum_{k=1}^{\infty} \frac{e^{-k\pi s}}{s^2+1}.$$

Remember that $\sin(t - k\pi) \equiv \sin(t)$ *if k is even, and is* $-\sin(t)$ *if k is odd.*

(c) *Use the formula for H(s) to solve the RC-circuit problem*

$$i'(t) + i(t) = h(t), \quad i(0) = 0.$$

(d) *Draw a graph of the solution on the interval* $0 \leq t \leq 6\pi$.

20. *In Example 6.4.6, show that the function for x(t), when t is large, is approaching a limiting function that has the form*

$$\chi(\tau) = 100 \left(\frac{e^{1-\tau}}{1-e} + 2 - \tau \right), \quad 0 \leq \tau \leq 1,$$

where $\tau = t - N$. *Assuming that the patient receives the drug daily at 8 AM,*

(a) *Find the time of day when the amount of drug in the organ is a maximum, and determine that maximum.*
(b) *Using the average value formula*

$$\text{ave}(\chi)_{a \leq \tau \leq b} = \frac{1}{b-a} \int_a^b \chi(u) du,$$

find the average amount of drug in the organ during a 24-hour period.

6.5 Convolution and the Impulse Function

In this section two related concepts will be defined: convolution of two functions, and a new "function," called the unit impulse, which essentially exists only in terms of its Laplace transform. Together, these allow us to also define the impulse response of an autonomous dynamical system.

6.5.1 The Convolution Integral. Although we know how to use the linearity of the inverse transform to invert a sum of transforms $F(s) + G(s)$, we do not have a formula for inverting the product of two transforms $F(s) \cdot G(s)$. It turns out that the inverse of the product is not the product of the inverses, it is a more complicated type of product called a convolution.

Convolution of functions appears in several areas of applied mathematics, and there are different definitions. The definition needed here is the following:

Definition 6.7. *Given two integrable functions f(t) and g(t), the* **convolution** *of f and g, denoted by* $f(t) \star g(t)$, *is defined as*

$$f(t) \star g(t) = \int_0^t f(\tau)g(t-\tau)d\tau.$$

By making a change of variable in the integral, it is easy to show that the operation of convolution is commutative; that is,

$$f(t) \star g(t) = g(t) \star f(t).$$

For our purposes the convolution operation is going to provide a way to invert a function of s that we recognize as being the product $F(s) \cdot G(s)$ of two functions in our Laplace transform table.

Theorem 6.6. *Given $f(t)$ and $g(t)$ with $\mathcal{L}(f(t)) = F(s)$ and $\mathcal{L}(g(t)) = G(s)$,*

$$\mathcal{L}(f(t) \star g(t)) = F(s) \cdot G(s).$$

Proof. The proof involves changing the order of integration in a double integral, which is a topic in multivariable calculus. By definition,

$$\mathcal{L}(f(t) \star g(t)) = \int_0^\infty e^{-st}(f(t) \star g(t))\,dt = \int_0^\infty e^{-st}\left(\int_0^t f(\tau)g(t-\tau)d\tau\right)dt.$$

In this double integral, t varies from 0 to ∞, and for each value of t, τ varies from 0 to t. If the order of integration is switched, τ must go from 0 to ∞, and for each τ, t goes from τ to ∞; therefore,

$$\int_0^\infty e^{-st}\left(\int_0^t f(\tau)g(t-\tau)d\tau\right)dt = \int_0^\infty \left(\int_\tau^\infty e^{-st}f(\tau)g(t-\tau)dt\right)d\tau$$

$$= \int_0^\infty f(\tau)\left(\int_\tau^\infty e^{-st}g(t-\tau)dt\right)d\tau,$$

where $f(\tau)$ has been factored out of the inner integral, since $f(\tau)$ is not a function of t.

If the change of variable $u = t - \tau$ is made in the inner integral,

$$\int_0^\infty f(\tau)\left(\int_\tau^\infty e^{-st}g(t-\tau)dt\right)d\tau = \int_0^\infty f(\tau)\left(\int_0^\infty e^{-s(u+\tau)}g(u)du\right)d\tau$$

$$= \int_0^\infty f(\tau)e^{-s\tau}\underbrace{\left(\int_0^\infty e^{-su}g(u)du\right)}_{G(s)}d\tau$$

$$= \left(\int_0^\infty e^{-s\tau}f(\tau)d\tau\right)G(s) = F(s) \cdot G(s). \qquad \square$$

This formula is usually used for inverting transforms; that is, in the form

$$\mathcal{L}^{-1}(F(s) \cdot G(s)) = f(t) \star g(t).$$

Example 6.5.1. *Use convolution to find the inverse Laplace transform of $H(s) = \frac{1}{s^2(s+1)}$.*

Solution. It is easy to see that one possible way to write $H(s)$ is as the product $F(s) \cdot G(s)$ where

$$F(s) = \frac{1}{s^2} = \mathcal{L}(t), \quad G(s) = \frac{1}{s+1} = \mathcal{L}(e^{-t});$$

therefore,

$$h(t) = \mathcal{L}^{-1}(H(s)) = \mathcal{L}^{-1}(F(s) \cdot G(s)) = f(t) \star g(t),$$

6.5 Convolution and the Impulse Function

where $f(t) = t$ and $g(t) = e^{-t}$. Choosing the simplest form of the convolution integral,

$$h(t) = f(t) \star g(t) = \int_0^t f(\tau)g(t-\tau)d\tau$$

$$= \int_0^t \tau\left(e^{-(t-\tau)}\right) d\tau = e^{-t} \int_0^t \tau e^\tau d\tau = t - 1 + e^{-t}.$$

This answer can be easily checked by computing the Laplace transform of $t - 1 + e^{-t}$.

∎

6.5.2 The Impulse Function. In applied problems, it is sometimes useful to be able to represent a "point mass" or "point charge," and this can be done with a pseudo-function referred to as the unit impulse or delta function. Engineers have long used the unit impulse as a way of incorporating a sudden blow to a mechanical system (such as a hammer hitting the system) or in the case of an electrical circuit, using it to represent a sudden surge in the current. Mathematicians were slow to believe that this so-called function could be used to do legitimate mathematics, but in the 1950s a basis for its use was presented in theoretical terms, called the theory of distributions.

In a certain sense the unit impulse at time $t = c$, written as $\delta(t-c)$, can be thought of as the derivative of the unit step function $u(t - c)$. Of course, this does not make mathematical sense, since the derivative of a discontinuous function is not defined at its point of discontinuity. In the following definition you will see how this problem is handled.

Definition 6.8. A **unit impulse** at time $t = c$ is denoted by $\delta(t - c)$ where

$$\delta(t - c) = \begin{cases} 0 & \text{if } t \neq c \\ \infty & \text{at } t = c, \end{cases} \text{ in such a way that } \int_a^b \delta(t - c)dt = 1 \text{ if } a < c < b.$$

The easiest way to understand an impulse function is to think of it as the limit, as $\varepsilon \to 0$, of a sequence of functions $\delta_\varepsilon(t - c)$ defined by

$$\delta_\varepsilon(t - c) = \begin{cases} 0 & \text{if } t < c - \varepsilon \\ \frac{1}{2\varepsilon} & \text{if } c - \varepsilon < t < c + \varepsilon \\ 0 & \text{if } t > c + \varepsilon. \end{cases}$$

The graph in Figure 6.9 shows $\delta_\varepsilon(t - c)$ at $c = 2.5$, for $\varepsilon = 0.5$, 0.2, and 0.1. In each case it can be seen that the area of the rectangle, with base $[c - \varepsilon, c + \varepsilon]$ and height $\frac{1}{2\varepsilon}$, is equal to one. In the limit as $\varepsilon \to 0$ the height of the rectangle obviously tends to ∞, but the integral $\int_{c-\varepsilon}^{c+\varepsilon} \delta_\varepsilon(t - c)dt$ is always equal to one.

Using this definition, it is quite easy to find the Laplace transform of $\delta_\varepsilon(t - c)$ for any ε, and define the transform of the unit impulse at $t = c$ to be the limit of the Laplace transform of $\delta_\varepsilon(t - c)$ as $\varepsilon \to 0$.

Theorem 6.7. *The Laplace transform of $\delta(t - c)$ is e^{-cs}.*

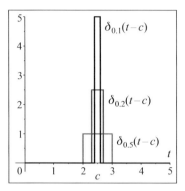

Figure 6.9. The functions $\delta_\varepsilon(t - 2.5)$ for $\varepsilon = 0.5, 0.2, 0.1$

Proof. Writing the transform of $\delta(t - c)$ as the limit of the transform of $\delta_\varepsilon(t - c)$,

$$\mathcal{L}\left(\delta(t-c)\right) = \lim_{\varepsilon \to 0} \left(\mathcal{L}(\delta_\varepsilon(t-c))\right) = \lim_{\varepsilon \to 0} \int_0^\infty e^{-st} \delta_\varepsilon(t-c) dt$$

$$= \lim_{\varepsilon \to 0} \int_{c-\varepsilon}^{c+\varepsilon} e^{-st} \frac{1}{2\varepsilon} dt = \lim_{\varepsilon \to 0} \frac{1}{2\varepsilon} \frac{e^{-st}}{-s} \bigg|_{c-\varepsilon}^{c+\varepsilon}$$

$$= \lim_{\varepsilon \to 0} \frac{1}{2\varepsilon s} \left(-e^{-s(c+\varepsilon)} + e^{-s(c-\varepsilon)}\right) = e^{-cs} \lim_{\varepsilon \to 0} \left(\frac{e^{s\varepsilon} - e^{-s\varepsilon}}{2\varepsilon s}\right).$$

Using the Taylor series for the exponential function, we can write

$$e^{s\varepsilon} - e^{-s\varepsilon} = \left(1 + s\varepsilon + \frac{(s\varepsilon)^2}{2!} + \frac{(s\varepsilon)^3}{3!} + \cdots\right) - \left(1 - s\varepsilon + \frac{(s\varepsilon)^2}{2!} - \frac{(s\varepsilon)^3}{3!} + \cdots\right)$$

$$= 2s\varepsilon + 2\frac{(s\varepsilon)^3}{3!} + 2\frac{(s\varepsilon)^5}{5!} + \cdots = 2s\varepsilon + \mathcal{O}(\varepsilon^3).$$

Therefore,

$$\mathcal{L}\left(\delta(t-c)\right) = e^{-cs} \lim_{\varepsilon \to 0} \left(\frac{2s\varepsilon + \mathcal{O}(\varepsilon^3)}{2\varepsilon s}\right) = e^{-cs} \lim_{\varepsilon \to 0}(1 + \mathcal{O}\left(\varepsilon^2\right)) = e^{-cs}. \quad (6.25)$$

□

This is the first time we have found a transform with no function of s in its denominator. Our formula for the transform of the derivative of a function would imply that the transform of the derivative of $u(t-c)$ should be $s\mathcal{L}(u(t-c)) - u(-c) = s\frac{e^{-cs}}{s} - u(-c)$, and this is just what we have found for the transform of the delta function. You do not need to worry about the mathematics when using this formula, because it has been completely justified in a way that makes mathematicians happy.

Example 6.5.2. *The current in a sinusoidally driven RLC-circuit with resistance = 2 ohms, inductance = 1 henry, and capacitance = $\frac{1}{5}$ farad is hit by a power surge of magnitude Q at time $t = a > 0$. Assume that at $t = 0$ the current $i(0)$ and its derivative $i'(0)$ are both equal to zero. The initial-value problem for the current $i(t)$ is*

$$i'' + 2i' + 5i = 4\sin(3t) + Q\delta(t-a), \quad i(0) = 0, \ i'(0) = 0. \quad (6.26)$$

6.5 Convolution and the Impulse Function

Solve the IVP and plot a graph of the output current i(t) given that an impulse of magnitude Q = 3 occurs at t = 2 sec. Also plot the solution for a surge of magnitude Q = 3 occurring at t = 3 sec.

Solution. Using methods of Chapter 3, the homogeneous solution of (6.26) is

$$i_h(t) = C_1 e^{-t}\cos(2t) + C_2 e^{-t}\sin(2t).$$

Check it!

Lemma 3.3 at the end of Chapter 3.3 can be used to show that the solution of (6.26) can be written as

$$i(t) = i_h(t) + i_f(t) + i_g(t),$$

where i_f is a particular solution of (6.26) with the right-hand side equal to $f(t) = 4\sin(3t)$ and i_g is a particular solution with the right-hand side $g(t) = Q\delta(t-a)$.

Using the methods of Chapter 3, check that $i_f(t) = -\frac{4}{13}\sin(3t) - \frac{6}{13}\cos(3t)$.

To find $i_g(t)$ we will use Laplace transforms to solve

$$i_g'' + 2i_g' + 5i_g = Q\delta(t-a).$$

The initial conditions at $t = 0$ are still both zero.

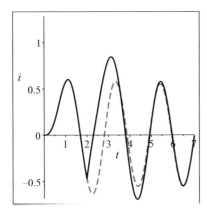 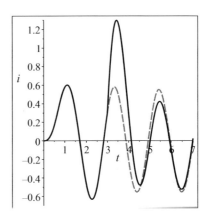

Figure 6.10. Solutions $i(t)$ of (6.26) with $Q = 3$, $a = 2$ (left) and $a = 3$ (right). The dashed curve is the solution if no impulse occurs.

Taking the transform of both sides,

$$(s^2 + 2s + 5)I_g(s) = Qe^{-as}$$

and

$$I_g(s) = Qe^{-as}\frac{1}{s^2 + 2s + 5} = \frac{Q}{2}e^{-as}\frac{2}{(s+1)^2 + 2^2}.$$

Inversion of $I_g(s)$ gives

$$i_g(t) = \mathcal{L}^{-1}(I_g(s)) = \frac{Q}{2}u(t-a)p(t-a),$$

where $p(t) = \mathcal{L}^{-1}\left(\frac{2}{(s+1)^2+2^2}\right) = e^{-t}\sin(2t)$; therefore,

$$i_g(t) = \frac{Q}{2}u(t-a)e^{-(t-a)}\sin(2(t-a))$$

and

$$i(t) = i_h + i_f + i_g$$
$$= C_1 e^{-t} \cos(2t) + C_2 e^{-t} \sin(2t) - \frac{4}{13} \sin(3t) - \frac{6}{13} \cos(3t)$$
$$+ \frac{Q}{2} u(t-a) e^{-(t-a)} \sin(2(t-a)).$$

To find the values of the constants C_1 and C_2, we can ignore the term multiplied by $u(t-a)$, since it is identically equal to zero on the interval $0 < t < a$. Differentiating, at $t = 0$ the value of $i'(t)$ is

$$i'(t) = C_1 \left(-e^{-t} \cos(2t) - 2e^{-t} \sin(2t) \right) + C_2 \left(-e^{-t} \sin(2t) + 2e^{-t} \cos(2t) \right)$$
$$- \frac{12}{13} \cos(3t) + \frac{18}{13} \sin(3t);$$

therefore, $i(0) = C_1 - \frac{6}{13} = 0$ and $i'(0) = -C_1 + 2C_2 - \frac{12}{13} = 0$ imply that $C_1 = \frac{6}{13}$ and $C_2 = \frac{9}{13}$.

The solution of the IVP (6.26) is

$$i(t) = \frac{6}{13} e^{-t} \cos(2t) + \frac{9}{13} e^{-t} \sin(2t) - \frac{4}{13} \sin(3t) - \frac{6}{13} \cos(3t)$$
$$+ \frac{Q}{2} u(t-a) e^{-(t-a)} \sin(2(t-a)).$$

In Figure 6.10 we see that the impulse simply adds the function

$$\frac{Q}{2} u(t-a) e^{-(t-a)} \sin(2(t-a))$$

to the sinusoidal solution (which is the dashed curve shown in both graphs in Figure 6.10). The maximum current therefore depends on the time the impulse occurs. The impulse can cause the derivative of $i(t)$ to be discontinuous at $t = a$. This is most noticeable in the left-hand graph, where the impulse occurred at time $t = 2$. ■

Two more formulas can now be added to our table of Laplace transforms.

Table 6.4

function	Laplace transform
$\delta(t-c)$	e^{-cs}
$f(t) \star g(t)$	$F(s)G(s)$, where $F(s) = \mathcal{L}(f(t)), G(s) = \mathcal{L}(g(t))$

A complete table of transforms, containing all the formulas from Tables 6.1, 6.2, 6.3, and 6.4 is given in Appendix F.

6.5.3 Impulse Response of a Linear, Time-invariant System.
The impulse response of a system modeled by an autonomous linear differential equation is its output when it is forced by a unit impulse at time $t = 0$, with the initial conditions all equal to zero.

6.5 Convolution and the Impulse Function

Definition 6.9. *Given a time-invariant linear system modeled by the differential equation*

$$a_n x^{(n)} + a_{n-1} x^{(n-1)} + \cdots + a_1 x' + a_0 x = \delta(t),$$

*with the forcing function equal to the unit impulse at $t = 0$, and with initial conditions all equal to zero, the output $x(t)$, usually denoted by $h(t)$, is called the **impulse response** of the system.*

If the differential equation is of second-order, we are looking at an initial-value problem of the form

$$ax'' + bx' + cx = \delta(t), \quad x(0) = x'(0) = 0.$$

Using Laplace transforms to solve this IVP,

$$as^2 X(s) + bs X(s) + c X(s) = (as^2 + bs + c) X(s) = \mathcal{L}(\delta(t - 0)) \equiv e^{0t} = 1.$$

Solving for $X(s)$,

$$X(s) = \frac{1}{as^2 + bs + c}.$$

This function $X(s)$ is usually denoted by $H(s)$ and is referred to as the impulse response in the s-domain. Its inverse Laplace transform

$$h(t) = \mathcal{L}^{-1}(H(s)) = \mathcal{L}^{-1}\left(\frac{1}{as^2 + bs + c}\right)$$

is the impulse response function $h(t)$ defined above.

Once $h(t)$ is known, the forced equation

$$ax'' + bx' + cx = f(t), \quad x(0) = x'(0) = 0$$

can be solved for any integrable driving function $f(t)$ by using convolution. This is easily seen by again taking Laplace transforms and writing

$$as^2 X(s) + bs X(s) + c X(s) = \mathcal{L}(f(t)) = F(s).$$

Now solving for $X(s)$ gives

$$X(s) = \left(\frac{1}{as^2 + bs + c}\right) F(s) = H(s) F(s),$$

so that $x(t)$ is the convolution

$$x(t) = \mathcal{L}^{-1}(H(s) F(s)) = h(t) \star f(t). \tag{6.27}$$

In the next example, we will use (6.27) to solve an underdamped spring-mass equation and locate its resonant frequency. In Section 3.5 of Chapter 3 we showed that a spring-mass system with damping constant $b^2 < 2mk$ has a resonant frequency ω^*, such that the output of the system forced by sinusoidal forcing $\sin(\omega t)$ has maximum amplitude when $\omega = \omega^*$.

Example 6.5.3. *Consider a spring-mass system with $m = 1$ kg, $b = 0.5$ N s/m, and $k = 0.625$ N/m. Find the impulse response $h(t)$ for this system and use it to write a formula for the steady-state solution of*

$$x'' + 0.5 x' + 0.625 x = \sin(\omega t). \tag{6.28}$$

Use this formula to find the gain function $G(\omega)$. The gain function was defined in Section 3.5 as the amplitude of the steady-state part of the solution divided by the amplitude of *the forcing function. Sketch a graph of $G(\omega)$ showing the value of ω where the gain is a maximum.*

Solution. The impulse response of (6.28) is the inverse Laplace transform of $H(s) = \frac{1}{s^2+0.5s+0.625}$. Since the quadratic in the denominator of $H(s)$ satisfies $b^2 - 4mk = 0.25 - 2.5 < 0$, we can complete the square to write

$$h(t) = \mathcal{L}^{-1}\left(\frac{1}{s^2 + 0.5s + 0.625}\right)$$

$$= \frac{1}{0.75}\mathcal{L}^{-1}\left(\frac{0.75}{(s+0.25)^2 + (0.75)^2}\right) = \frac{4}{3}e^{-0.25t}\sin(0.75t).$$

To solve (6.28) with the forcing function $f(t) = \sin(\omega t)$, Maple was used to evaluate the convolution integral

$$x(t) = h(t) \star \sin(\omega t) = \int_0^t \frac{4}{3}e^{-0.25\tau}\sin(0.75\tau)\sin(\omega(t-\tau))\,d\tau.$$

In doing this, the initial conditions have been assumed to be zero, but remember that this does not affect the steady-state solution. Since the convolution is the solution of the forced equation, it contains terms that represent the homogeneous solution, as well as terms due to the forcing function. We know that the homogeneous equation is underdamped, so all terms in the homogeneous solution will be multiplied by negative exponentials. The Maple solution, after deleting the terms multiplied by negative exponentials, is

$$\int_0^t \frac{4}{3}e^{-0.25\tau}\sin(0.75\tau)\sin(\omega(t-\tau))\,d\tau$$

$$= x_h + x_p = x_h + \left(\frac{8}{3}\right)\left(\frac{-12\omega\cos(\omega t) - 24\omega^2\sin(\omega t) + 15\sin(\omega t)}{(5 + 8\omega^2 - 12\omega)(5 + 8\omega^2 + 12\omega)}\right).$$

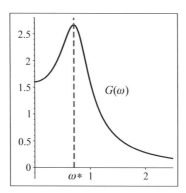

Figure 6.11. Gain function for the system $x'' + 0.5x' + 0.625x = \sin(\omega t)$

To find the gain $G(\omega)$, the steady state solution $x_p(t)$ must be written as a single sine term. If this is done in the usual manner, we find

$$x_p(t) = \left(\frac{8}{3}\right)\left(\frac{\sqrt{(12\omega)^2 + (15 - 24\omega^2)^2}}{(5 + 8\omega^2 - 12\omega)(5 + 8\omega^2 + 12\omega)}\right)\sin(\omega t + \phi).$$

6.5 Convolution and the Impulse Function

The input to the system was $f(t) = \sin(\omega t)$ with amplitude 1, so the gain function is just the amplitude of x_p; that is,

$$G(\omega) = \left(\frac{8}{3}\right)\left(\frac{\sqrt{(12\omega)^2 + (15 - 24\omega^2)^2}}{(5 + 8\omega^2 - 12\omega)(5 + 8\omega^2 + 12\omega)}\right).$$

A plot of this function is shown in Figure 6.11, with the maximum marked at $\omega = \omega_*$. It was shown at the end of Section 3.5 that the maximum should occur at $\omega_* = \sqrt{\frac{k}{m} - \frac{b^2}{2m^2}} = \sqrt{0.625 - 0.25/2} \approx 0.7071$, and this agrees exactly with our result. ∎

Exercises 6.5. *In Exercises 1–6, write $F(s)$ as a product $G(s)H(s)$ and use convolution to find the inverse $f(t) = \mathcal{L}^{-1}(F(s))$. Use your computer algebra system to check the answers.*

1. $F(s) = \dfrac{1}{(s+2)^2}$

2. $F(s) = \dfrac{1}{s(s+3)}$

3. $F(s) = \dfrac{1}{s^2(s+5)}$

4. $F(s) = \dfrac{2}{s(s^2+4)}$

5. $F(s) = \dfrac{1}{(s^2+1)^2}$

6. $F(s) = \dfrac{2s}{(s^2+1)^2}$

Use Laplace transforms to solve the initial-value problems 7 and 8. Sketch the solution on the interval $0 < t < 6$.

7. $x'' + 3x' + 2x = \delta(t - 2), \quad x(0) = 0, \quad x'(0) = 1$

8. $x'' + 4x' + 4x = 3\delta(t - 2), \quad x(0) = 1, \quad x'(0) = -1$

9. (a) *Find the impulse function $h(t)$ for the system modeled by the equation*
$$x'' + 4x' + 4x = 0.$$
   (b) *Use convolution to solve the IVP*
$$x'' + 4x' + 4x = 1 + 2e^{-t}, \quad x(0) = 0, \quad x'(0) = 0.$$
   (c) *What is the steady state solution?*

7

Introduction to Partial Differential Equations

In real world problems it often becomes necessary to solve differential equations in which the unknown function depends on more than one independent variable. Such an equation is called a **partial differential equation** (PDE) and is, in general, more difficult to solve than the ODEs encountered in the first six chapters of this book.

In Chapter 7 we will describe one method for solving PDEs, again called a method of separation of variables, which makes it possible to convert a single partial differential equation into a system of two or more ordinary differential equations. Section 7.1 gives a general outline of the method of separation of variables for finding series solutions for the partial differential equations that appear in most undergraduate physics texts. In order to apply this method it is necessary to have an understanding of orthogonal series of functions, and of boundary-value problems for linear second-order ordinary differential equations. These two topics are covered in Sections 7.2 and 7.3. Appendix G contains a review of partial derivatives and the notation which will be used for them in Chapters 7 and 8.

In 7.3 a very simple connection between function spaces and vector spaces is also introduced by showing how the Sturm-Liouville boundary-value problem for linear second order ODEs is related to the eigenvalue problem for linear operators (matrices) on a vector space. It is hoped that students who have already completed a course in linear algebra will be enticed to look into this connection more deeply and discover how functional analysis can be used to generalize something they have already learned in their study of finite-dimensional vector spaces.

7.1 Solving Partial Differential Equations

When the unknown function in a differential equation depends on more than one independent variable, such as time and one or more space dimensions, the equation becomes a **partial differential equation**. To give a simple example of how such an equation might arise, consider the logistic equation for population growth

$$P'(t) = rP(t)\left(1 - \frac{P(t)}{N}\right). \tag{7.1}$$

This equation is an **ordinary differential equation** modelling the growth of a population $P(t)$ as a function depending only on time t. The parameters r and N are, respectively, the intrinsic growth rate and carrying capacity of the particular population.

Now consider modelling the growth of an algae population along a 1-dimensional strip across a pond. Given the initial population of the algae at each point along the strip, we could apply the above equation to determine the population at that point at any time t. It might be more realistic, however, to assume that there is an interaction between members of the population that lie close to each other along the strip. This might be modelled by adding a diffusion term to the equation:

$$\frac{\partial P}{\partial t} = K\frac{\partial^2 P}{\partial x^2} + rP(x,t)\left(1 - \frac{P(x,t)}{N}\right). \tag{7.2}$$

An excellent discussion of the derivation of this equation, called a *reaction-diffusion* equation, is contained in section 3.5.2 of a book by Gurney & Nisbet.[1] In this equation the parameter K measures the rate at which diffusion occurs, and K is called the diffusion coefficient. In the reference the use of the second derivative term is justified in the case where the population is moving randomly about any point x. Note that the unknown function P now depends on *two* variables, distance x along the strip and the time t.

It is also possible to model the algae growth in a 2-dimensional region, in which case P would become a function of three variables, $P(x, y, t)$, and the diffusion could occur in both the x and y direction.

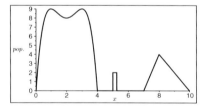

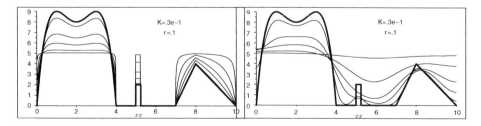

Figure 7.1. Growth of algae on a strip across a lake. Graphs of the population size at times $T = 0, 1, 5, 10, 20$ and 40 using (7.1) are shown on the left, and those using (7.2) on the right.

---

[1] W. S. C. Gurney and R. M. Nisbet, *Ecological Dynamics*, Oxford University Press, 1998.

7.1 Solving Partial Differential Equations

Figure 7.1 shows the growth over time of an algae population along a 10-meter path across a pond, where the initial population is shown in the graph above. If equation (7.1) with parameter values $r = 0.1, N = 5.0$ is assumed to hold at each point along the 10-meter path, the values of the population at times $T = 0, 1, 5, 10, 20,$ and 40 are shown in the lower left-hand figure. The right-hand graph shows the population at the same times, where equation (7.2) is used, with a diffusion coefficient $K = 0.03$. In both cases, as $t \to \infty$ the population tends to the carrying capacity $N = 5$. With diffusion, the interval where the initial population was zero also fills in and slowly tends to the value N.

The first-order ordinary differential equation (7.1) can be solved by separation of variables, and it was shown in Chapter 2 that it has the analytic solution

$$P(t) = \frac{N}{1 + \left(\frac{N}{P_0} - 1\right)e^{-rt}} \tag{7.3}$$

if the initial population $P(0) = P_0$ is strictly greater than zero at the point x, and that it is zero for all t if $P_0 = 0$. How to solve a partial differential equation such as (7.2) is the subject of Chapter 8, and it will be found that since equation (7.2) is nonlinear in P, numerical methods will have to be used to obtain a solution.

Even in the case of *linear* partial differential equations with constant coefficients, closed formulas for solutions are rarely obtainable, and the solution may involve an infinite series of *special functions* called orthogonal functions.

7.1.1 An Overview of the Method of Separation of Variables.
In order to motivate the material in the next two sections, a general description will be given here of a method that we are going to be using. If a partial differential equation is linear in the unknown variable and its derivatives, it is often possible to solve it, using a method again referred to as separation of variables. The following example will give you a very brief idea of how this method works.

A simple partial differential equation which appears in physics texts is the 1-dimensional heat equation (also called a diffusion equation)

$$\frac{\partial u}{\partial t} = a \frac{\partial^2 u}{\partial x^2}, \tag{7.4}$$

where $u(x, t)$ is the temperature at time t along a thin insulated rod of length L, and a is a positive physical parameter. There are many good videos on the web that give a derivation of this partial differential equation, and one of these derivations is presented in detail in Section 8.2.

If it is assumed that the temperature at both ends of the rod is held at $0°C$, the **boundary conditions** for (7.4) are

$$u(0, t) = 0, \ u(L, t) = 0, \ \text{for all } t > 0. \tag{7.5}$$

It is also necessary to specify the temperature everywhere along the rod at some initial time, say $t_0 = 0$. This results in an **initial condition** of the form

$$u(x, 0) = f(x), \ 0 \leq x \leq L, \tag{7.6}$$

where it can be shown that f can be an arbitrary piecewise continuous function defined on the interval $[0, L]$.

The PDE (7.4) is linear in the unknown function u. As a first try at a solution, we assume that there are product solutions of the form

$$u(x,t) = X(x)T(t), \qquad (7.7)$$

and see what happens. Notice that both X and T are now functions of a single variable. In this form, the partial derivatives become ordinary derivatives:

$$\frac{\partial u}{\partial t} = \frac{\partial}{\partial t}(X(x)T(t)) = X(x)T'(t) \qquad (7.8)$$

$$\frac{\partial^2 u}{\partial x^2} = \frac{\partial^2}{\partial x^2}(X(x)T(t)) = X''(x)T(t). \qquad (7.9)$$

To make $u(x,t)$ satisfy (7.4), we must have $X(x)T'(t) \equiv aX''(x)T(t)$, and dividing both sides by $aX(x)T(t)$,

$$\frac{1}{a}\frac{T'(t)}{T(t)} \equiv \frac{X''(x)}{X(x)}. \qquad (7.10)$$

This separates the variables x and t, and to have a function of t identical to a function of x implies that both functions must be equal to the same constant; therefore,

$$\frac{1}{a}\frac{T'(t)}{T(t)} \equiv \frac{X''(x)}{X(x)} \equiv -\lambda. \qquad (7.11)$$

The original partial differential equation has thus been reduced to a *system* of two ordinary differential equations, linked by the parameter λ:

$$T'(t) + a\lambda T(t) = 0 \qquad (7.12)$$
$$X''(x) + \lambda X(x) = 0. \qquad (7.13)$$

The next step is to use the given boundary and initial conditions to obtain conditions that allow us to solve one of these two ordinary differential equations. Notice that the boundary conditions $u(0,t) = u(L,t) = 0$ for all t, together with (7.7), imply that

$$\left. \begin{array}{l} u(0,t) \equiv X(0)T(t) = 0 \\ u(L,t) \equiv X(L)T(t) = 0 \end{array} \right\} \text{ for } t > 0; \qquad (7.14)$$

and since we do not want $T(t)$ to be identically zero (for this would imply $u(x,t) \equiv 0$), this requires $X(0) = X(L) = 0$. This gives two conditions that must be satisfied by the solution of (7.13), but these are not in the form of initial conditions required for a unique solution of a second-order ODE. The problem

$$X''(x) + \lambda X(x) = 0, \quad X(0) = 0, \quad X(L) = 0$$

is an example of a **boundary-value problem**. In Section 7.3 it will be shown that, in general, such a problem has an infinite set of solutions $\{X_n(x)\}_{n=0}^{\infty}$ corresponding to an infinite sequence of parameters λ_n. Once the λ_n are determined, the corresponding first-order equations (7.12) for T_n can be seen to have solution

$$T_n(t) = C_n e^{-a\lambda_n t}.$$

This, in turn, produces an infinite sequence of product solutions

$$u_n(x,t) = X_n(x)T_n(t) = X_n(x)C_n e^{-a\lambda_n t}$$

7.1 Solving Partial Differential Equations

of (7.4). Because (7.4) is linear in u, any linear combination of solutions, of the form $\sum_n X_n(x)T_n(t)$, will also satisfy the equation. If the solution is written as an infinite series

$$u(x,t) = \sum_{n=0}^{\infty} X_n(x)T_n(t) = \sum_{n=0}^{\infty} C_n X_n(x) e^{-a\lambda_n t}, \qquad (7.15)$$

the initial condition $u(x,0) = f(x)$ on $0 \le x \le L$ can then be used to write

$$u(x,0) = \sum_{n=0}^{\infty} C_n X_n(x) e^0 \equiv \sum_{n=0}^{\infty} C_n X_n(x) \equiv f(x).$$

In Section 7.3 it will be shown that the infinite set of solutions $X_n(x)$ of (7.13) possesses a special property called orthogonality, which makes it possible to easily determine the constants C_n which make (7.15) the unique solution of (7.4) satisfying the given initial and boundary conditions.

Using the above technique, we will be able to write the solution of several PDEs in terms of infinite series. In the example above, the differential equation (7.13) for which we need a general solution is the equation $x'' + \lambda x = 0$. With the type of boundary conditions we will be considering, solutions of this boundary-value problem are sine and cosine functions, and the type of orthogonal series that will result is called a trigonometric Fourier series. A complete discussion of the trigonometric Fourier series is contained in the next section.

Exercises 7.1. *For each of the partial differential equations in Exercises 1–6, assume a product solution $u(x,t) = X(x)T(t)$. Try to separate the variables and, if possible, find ordinary differential equations satisfied by $X(x)$ and $T(t)$.*

1. $u_t = au_{xx} + bu_x$

2. $u_t = tu_{xx}$

3. $u_{tt} = bu_{xx} + xu$

4. $u_t = au_{xx} + xu_x$

5. $u_t = au_{xx} + bu + cu^2$ (this is the form of equation (7.2))

6. $u_t = uu_x$

7. Find all values of the parameter λ for which the second-order ODE

$$X''(x) + \lambda X(x) = 0$$

has nonzero solutions $X(x)$ satisfying the two conditions $X(0) = X(1) = 0$. Remember that there are three different forms of the general solution, depending on whether $\lambda < 0, \lambda = 0,$ or $\lambda > 0$. These three forms of the general solution are contained in Table 3.1 in Chapter 3. In each of the three cases, simply write out the solution and see if it can be made to satisfy the two boundary conditions for any appropriate values of λ.

7.2 Orthogonal Functions and Trigonometric Fourier Series

In the early 1800s, Jean-Baptiste Joseph Fourier, a French mathematician and physicist, claimed to show that any piecewise continuous function could be represented on a finite interval in terms of an infinite series of periodic functions. It took several years to convince other mathematicians of the correctness of his work. In this section, the type of series that he worked with, called trigonometric Fourier series, is described.

7.2.1 Orthogonal Families of Functions.

Definition 7.1. *An infinite collection of functions* $S = \{\phi_0(t), \phi_1(t), \phi_2(t), \ldots\} \equiv \{\phi_n(t)\}_{n=0}^{\infty}$ *is said to be an* **orthogonal set** *of functions on the interval* $[\alpha, \beta]$ *if*

$$\int_{\alpha}^{\beta} \phi_m(t)\phi_n(t)dt = 0, \quad \text{whenever} \quad m \neq n. \tag{7.16}$$

Suppose $S = \{\phi_n(t)\}_{n=0}^{\infty}$ is an orthogonal set of functions on $[\alpha, \beta]$, and we want to represent an arbitrary function $f(t)$ defined on $[\alpha, \beta]$ by an infinite series of the form

$$f(t) = \sum_{n=0}^{\infty} a_n \phi_n(t) = a_0 \phi_0(t) + a_1 \phi_1(t) + \cdots. \tag{7.17}$$

If we can assume that this series converges to $f(t)$ on the interval $\alpha \leq t \leq \beta$, it is then easy to find the coefficients a_m. For any $m = 0, 1, 2, \ldots$, using equation (7.16) and omitting the difficult proof of a theorem that allows one to interchange integration and the summation of infinite series, we have

$$\int_{\alpha}^{\beta} f(t)\phi_m(t)dt = \int_{\alpha}^{\beta} \left(\sum_{n=0}^{\infty} a_n \phi_n(t)\right)\phi_m(t)dt = \sum_{n=0}^{\infty} a_n \int_{\alpha}^{\beta} \phi_n(t)\phi_m(t)dt$$

$$= a_m \int_{\alpha}^{\beta} \phi_m^2(t)dt;$$

and therefore, a formula for the mth coefficient in the Fourier series for $f(t)$ is

$$a_m = \frac{\int_{\alpha}^{\beta} f(t)\phi_m(t)dt}{\int_{\alpha}^{\beta} \phi_m^2(t)dt}. \tag{7.18}$$

We want to be able to find Fourier approximations for arbitrary functions $f(t)$ in terms of different sets of orthogonal functions. In computing the coefficients by integration, life is sometimes simplified if we know that the function f satisfies certain properties.

Review of Even and Odd Functions.

Definition 7.2. *A function $f(t)$ is called an* **even function** *if $f(-t) = f(t)$ for all t in the domain of f; it is called an* **odd function** *if $f(-t) = -f(t)$ for all t in the domain of f.*

7.2 Orthogonal Functions and Trigonometric Fourier Series

If f and g are both odd functions, or both even functions, then the product $f(t)g(t)$ is even; and if one is even and the other is odd, then the product is odd.

Example 7.2.1. *Show that if f and g are both odd functions, the product fg is even.*

Solution. The product function fg is defined as $(fg)(t) \equiv f(t)g(t)$. Assuming f and g are both odd, to show that fg is an even function we can write $(fg)(-t) = f(-t)g(-t) = (-f(t))(-g(t)) = f(t)g(t) \equiv (fg)(t)$. ∎

Even functions are symmetric about the y-axis. Some examples of even functions are constants, $\cos(\omega t)$, and any polynomial all of whose terms have even degrees; for example, t^2 or $t^4 + 3t^8$. For any even function $f(t)$, $\int_{-L}^{L} f(t)dt = 2\int_{0}^{L} f(t)dt$.

Odd functions are symmetric about the origin. Examples are $t, 2t^3 - 5t^{11}$, and $\sin(\omega t)$. If $g(t)$ is an odd function, $\int_{-L}^{L} g(t)dt = 0$.

Two functions, which you may not have seen before, should be added to this list. They are the hyperbolic sine, $\sinh(t)$, and hyperbolic cosine, $\cosh(t)$. We will be using these functions when we solve partial differential equations. They are defined as follows:

$$\sinh(t) = \frac{e^t - e^{-t}}{2}, \quad \cosh(t) = \frac{e^t + e^{-t}}{2}.$$

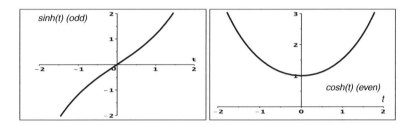

Figure 7.2. The hyperbolic functions $\sinh(t)$ and $\cosh(t)$

Check that $\sinh(t)$ is an odd function, and $\cosh(t)$ is even. They are easily seen to satisfy the differentiation formulas

$$\frac{d}{dt}\sinh(t) = \cosh(t), \quad \frac{d}{dt}\cosh(t) = \sinh(t).$$

One other property of these hyperbolic functions, that will be needed is that $\cosh(t)$ is never equal to 0, and $\sinh(t) = 0$ only if $t = 0$. Graphs of $\cosh(t)$ and $\sinh(t)$ are shown in Figure 7.2.

Trigonometric Fourier Series. We are now ready to study the approximation of functions in terms of the particular orthogonal set

$$\mathcal{S} = \left\{1, \cos\left(\frac{\pi t}{L}\right), \sin\left(\frac{\pi t}{L}\right), \cos\left(\frac{2\pi t}{L}\right), \sin\left(\frac{2\pi t}{L}\right), \ldots, \cos\left(\frac{n\pi t}{L}\right), \sin\left(\frac{n\pi t}{L}\right), \ldots\right\}.$$

If it exists, the series for f in terms of this set is characteristically written in the form

$$f(t) = \frac{a_0}{2} + \sum_{n=1}^{\infty} a_n \cos\left(\frac{n\pi t}{L}\right) + b_n \sin\left(\frac{n\pi t}{L}\right). \tag{7.19}$$

It is called the **trigonometric Fourier series** for $f(t)$ on the interval $[-L, L]$. Note that each function in the set S has period $2L$; that is, $\phi(t + 2L) = \phi(t)$ for all t; therefore, if $f(t)$ can be represented by the series (7.19), it will be a periodic function with period $2L$.

To show that the set $S = \{1, \cos(\frac{\pi t}{L}), \sin(\frac{\pi t}{L}), \ldots\}$ is an orthogonal set on $[-L, L]$, we need to show that $\int_{-L}^{L} \phi_n(t)\phi_m(t)dt = 0$ whenever ϕ_n and ϕ_m are two different functions in S.

Showing that $\int_{-L}^{L} \sin(\frac{n\pi t}{L}) \cdot 1 dt = 0$ and $\int_{-L}^{L} \sin(\frac{n\pi t}{L})\cos(\frac{m\pi t}{L})dt = 0$ for any integers m and n is easy since in both cases the integrand is an *odd* function. Since $\cos\left(\frac{n\pi t}{L}\right)$ is an even function,

$$\int_{-L}^{L} \cos\left(\frac{n\pi t}{L}\right) \cdot 1 dt = 2\int_{0}^{L} \cos\left(\frac{n\pi t}{L}\right) dt = \frac{2L}{n\pi}(\sin(n\pi) - \sin(0)) = 0.$$

Showing that the product of two different sine functions or two different cosine functions integrates to zero is done using trig substitutions and is left to the exercises.

In the next subsection we will have a theorem stating which functions $f(t)$ have convergent trigonometric Fourier series. For these functions, the coefficient of $\phi_n(t)$ in the series is equal to $\frac{\int_{-L}^{L} f(t)\phi_n(t)dt}{\int_{-L}^{L} \phi_n^2(t)dt}$. For the trigonometric Fourier series this implies that for $n \geq 1$,

$$a_n = \frac{\int_{-L}^{L} f(t)\cos(\frac{n\pi t}{L})dt}{\int_{-L}^{L}(\cos(\frac{n\pi t}{L}))^2 dt}, \quad b_n = \frac{\int_{-L}^{L} f(t)\sin(\frac{n\pi t}{L})dt}{\int_{-L}^{L}(\sin(\frac{n\pi t}{L}))^2 dt}.$$

You are going to show in the exercises that $\int_{-L}^{L}(\cos(\frac{n\pi t}{L}))^2 dt = \int_{-L}^{L}(\sin(\frac{n\pi t}{L}))^2 dt = L$ for any $n \geq 1$; therefore, the formulas for the coefficients a_n and b_n are

$$a_n = \frac{1}{L}\int_{-L}^{L} f(t)\cos\left(\frac{n\pi t}{L}\right)dt, \quad b_n = \frac{1}{L}\int_{-L}^{L} f(t)\sin\left(\frac{n\pi t}{L}\right)dt. \quad (7.20)$$

When $n = 0$, the coefficient of the function $\phi_0(t) \equiv 1$ in the Fourier series is $\frac{\int_{-L}^{L} f(t)dt}{\int_{-L}^{L} 1^2 dt}$; and since $\int_{-L}^{L} 1 dt = 2L$, if we use the formula for a_n to compute the constant coefficient, it must be divided by 2. Note that the Fourier series (7.19) starts with the term $a_0/2$. It is also helpful to recognize that if $f(t)$ is equal to its Fourier series on $[-L, L]$, then, since $\frac{1}{2L}\int_{-L}^{L} \sin\left(\frac{n\pi t}{L}\right)dt$ and $\frac{1}{2L}\int_{-L}^{L} \cos\left(\frac{n\pi t}{L}\right)dt$ are both zero for any integer $n > 0$, the constant $\frac{a_0}{2}$ must be equal to the *average value* of the function $f(t)$ on $[-L, L]$; that is, $\frac{1}{2L}\int_{-L}^{L} f(t)dt = \frac{a_0}{2} + 0 + 0 + \cdots = \frac{a_0}{2}$.

In the next subsection we will summarize properties of the trigonometric Fourier series, and state for which functions $f(t)$ these series converge; but first we can do a simple example.

Example 7.2.2. *Find a trigonometric Fourier series for the piecewise continuous function*

$$f(t) = \begin{cases} -1 & \text{if } -L \leq t < 0 \\ 1 & \text{if } 0 \leq t \leq L \end{cases}$$

7.2 Orthogonal Functions and Trigonometric Fourier Series

Solution. First, notice that f is an odd function, and therefore the coefficients $a_n = \frac{1}{L}\int_{-L}^{L} f(t)\cos(\frac{n\pi t}{L})dt$ are all zero. The coefficients b_n satisfy

$$b_n = \frac{1}{L}\int_{-L}^{L} f(t)\sin\left(\frac{n\pi t}{L}\right)dt = \frac{2}{L}\int_{0}^{L} f(t)\sin\left(\frac{n\pi t}{L}\right)dt$$

$$= \frac{2}{L}\int_{0}^{L} 1\cdot\sin\left(\frac{n\pi t}{L}\right)dt = \frac{2}{n\pi}(-\cos(n\pi) + \cos(0)) = \begin{cases} 0 & \text{if } n \text{ is even} \\ \dfrac{4}{n\pi} & \text{if } n \text{ is odd} \end{cases}$$

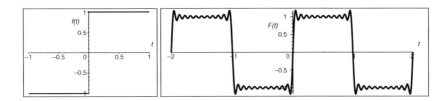

Figure 7.3. The function $f(t)$ and its Fourier approximation $F(t)$

The Fourier series for $f(t)$ can be written in the form

$$f(t) = \sum_{n=0}^{\infty} \frac{4}{(2n+1)\pi}\sin\left(\frac{(2n+1)\pi t}{L}\right)$$

$$= \frac{4}{\pi}\left(\sin\left(\frac{\pi t}{L}\right) + \frac{1}{3}\sin\left(\frac{3\pi t}{L}\right) + \frac{1}{5}\sin\left(\frac{5\pi t}{L}\right) + \cdots\right).$$

For a particular value of L, say $L = 1$, an approximation to $f(t)$ can be obtained by taking terms in the sum out to $n = 10$. A graph of this finite approximation is shown in Figure 7.3. It shows clearly that the Fourier series is a periodic function of period $2L = 2$. ∎

7.2.2 Properties of Fourier Series, Cosine and Sine Series.
Any function $f(t)$ which is at least piecewise continuous on an interval $[-L, L]$ can be expanded in a convergent trigonometric Fourier series

$$\frac{a_0}{2} + \sum_{n=1}^{\infty} a_n \cos\left(\frac{n\pi t}{L}\right) + b_n \sin\left(\frac{n\pi t}{L}\right),$$

$$a_n = \frac{1}{L}\int_{-L}^{L} f(t)\cos\left(\frac{n\pi t}{L}\right)dt, \quad n = 0, 1, \ldots,$$

$$b_n = \frac{1}{L}\int_{-L}^{L} f(t)\sin\left(\frac{n\pi t}{L}\right)dt \quad n = 1, 2, \ldots. \tag{7.21}$$

We know that piecewise continuous means that f has at most a finite number of jump discontinuities on the finite interval $-L \le t \le L$, say at $t = t_1, t_2, \ldots, t_n$, and that the limits $f(t_i^-) = \lim_{t \to t_i^-} f(t)$ and $f(t_i^+) = \lim_{t \to t_i^+} f(t)$ both exist at each point t_i.

The next theorem states an important property of the trigonometric Fourier series.

Theorem 7.1. *If $f(t)$ is a piecewise continuous function on $[-L, L]$, periodic with period $2L$, then the Fourier series (7.21) for $f(t)$ converges to $f(t)$ at every point t where f is continuous, and converges to*

$$\frac{f(t_i^+) + f(t_i^-)}{2}$$

where t_i is a point of discontinuity.

In the next example we will find a Fourier series for a function which is specified on a finite interval $[-L, L]$ and assumed to be periodic of period $2L$.

Example 7.2.3. Let $f(t) = t$ on the interval $-1 \le t \le 1$, and assume f is periodic of period 2. Draw a graph of the periodically extended function f on the interval $[-3, 3]$. Find the Fourier series for f and sketch a graph of the finite sum $\frac{a_0}{2} + \sum_{n=1}^{7} a_n \cos(\frac{n\pi t}{L}) + b_n \sin(\frac{n\pi t}{L})$ which approximates the function f.

Solution. Use the formula for f to sketch a graph of the straight-line function on the interval $[-1, 1]$. To extend it so it is periodic with period 2, the graph on the intervals $[-3, -1]$ and $[1, 3]$ must look exactly like the graph on $[-1, 1]$.

All of the coefficients a_n in the Fourier series are zero, since f is an odd function on its domain; that is, $f(-t) = -t = -f(t)$. The coefficients b_n are found from the formula

$$b_n = \frac{1}{L}\int_{-L}^{L} f(t)\sin\left(\frac{n\pi t}{L}\right)dt = \frac{2}{L}\int_{0}^{L} f(t)\sin\left(\frac{n\pi t}{L}\right)dt$$

$$= 2\int_{0}^{1} t\sin\left(\frac{n\pi t}{1}\right)dt = -\frac{2}{n\pi}(\cos(n\pi)).$$

Figure 7.4 shows the graph of the periodically extended function $f(t)$ with the graph of $\sum_{1}^{7}\left(-\frac{2}{n\pi}\cos(n\pi)\right)\sin(n\pi t) = \frac{2}{\pi}(\frac{\sin(\pi t)}{1} - \frac{\sin(2\pi t)}{2} + \cdots + \frac{\sin(7\pi t)}{7})$ superimposed on it. ∎

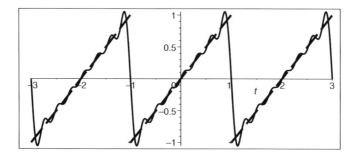

Figure 7.4. The periodically extended function $f(t)$ with its Fourier series approximation overlaid

7.2 Orthogonal Functions and Trigonometric Fourier Series

Cosine and Sine Series. When solving partial differential equations, it is often necessary to approximate a given function f on a **half-interval** $[0, L]$ by a trigonometric series which contains just sine functions, or just cosine functions.

Suppose you are given a function $f(t)$ defined on $[0, L]$ and want to approximate it on that half-interval by a finite sum of the form

$$f(t) \approx \sum_{n=1}^{N} b_n \sin\left(\frac{n\pi t}{L}\right). \tag{7.22}$$

Since it does not matter what the series converges to on $[-L, 0]$, we can assume that the function f is extended as an *odd* function on $[-L, L]$. Then its full Fourier series will contain only sine terms (the coefficients a_n are all zero if f is odd). For $t \in [0, L]$ the series (7.22), with $b_n = \frac{1}{L}\int_{-L}^{L} f(t) \sin\left(\frac{n\pi t}{L}\right) dt = \frac{2}{L}\int_{0}^{L} f(t) \sin\left(\frac{n\pi t}{L}\right) dt$, will converge to $f(t)$ as desired. Note that we never have to define $f(t)$ on $[-L, 0]$, but just assume that f is odd. The infinite series $\sum_{n=1}^{\infty} b_n \sin\left(\frac{n\pi t}{L}\right)$ is called a **Fourier sine series** for $f(t)$ on $[0, L]$, and the coefficients are

$$b_n = \frac{2}{L} \int_0^L f(t) \sin\left(\frac{n\pi t}{L}\right) dt.$$

Similarly, to approximate $f(t)$ on $[0, L]$ by a finite sum of the form

$$f(t) \approx \frac{a_0}{2} + \sum_{n=1}^{N} a_n \cos\left(\frac{n\pi t}{L}\right),$$

we can assume that f is extended to $[-L, L]$ as an *even* function of period $2L$ and use the formula

$$a_n = \frac{2}{L} \int_0^L f(t) \cos\left(\frac{n\pi t}{L}\right) dt, \quad n = 0, 1, 2, \ldots.$$

The infinite series $\frac{a_0}{2} + \sum_{n=1}^{\infty} a_n \cos\left(\frac{n\pi t}{L}\right)$ is called a **Fourier cosine series** for $f(t)$ on $[0, L]$.

In Example 7.2.3 we found a full Fourier series for the function $f(t) = t$ on $-1 \leq t \leq 1$. This, in fact, produced a sine series for $f(t) = t$ on $[0, 1]$, since the function $f(t) = t$ is an odd function on $[-1, 1]$.

Example 7.2.4. *Find a cosine series for $f(t) = t$ on $[0, 1]$.*

Solution. In this case we need to assume that f is periodic of period 2 and is extended as an *even* function on $[-1, 1]$. A graph of the periodically extended function $f(t)$ is shown in Figure 7.5. To find the coefficients in the cosine series

$$\frac{a_0}{2} + \sum_{n=1}^{\infty} a_n \cos\left(\frac{n\pi t}{L}\right),$$

set $L = 1$ and compute

$$a_0 = \frac{2}{L} \int_0^L f(t) dt, \quad a_n = \frac{2}{L} \int_0^L f(t) \cos\left(\frac{n\pi t}{L}\right) dt, \quad n = 1, 2, \ldots.$$

Therefore, $a_0 = 2\int_0^1 t\,dt = 1$, and for $n > 0$

$$a_n = 2\int_0^1 t\cos(n\pi t)dt = \frac{2}{n^2\pi^2}[\cos(n\pi) - 1] = \begin{cases} 0 & \text{if } n \text{ is even} \\ \frac{-4}{n^2\pi^2} & \text{if } n \text{ is odd.} \end{cases}$$

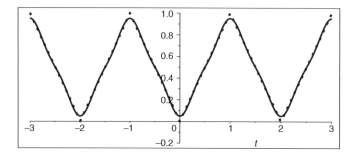

Figure 7.5. The periodically extended even function $f(t)$ with its Fourier series overlaid

The resulting cosine series is

$$f(t) = \frac{1}{2} - \frac{4}{\pi^2}\sum_{n=0}^{\infty}\frac{\cos((2n+1)\pi t)}{(2n+1)^2} = \frac{1}{2} - \frac{4}{\pi^2}\left(\frac{\cos(\pi t)}{1} + \frac{\cos(3\pi t)}{9} + \frac{\cos(5\pi t)}{25} + \cdots\right).$$

A graph of the periodic extension of $f(t)$ on $[-3, 3]$, and the sum of the first three terms in the cosine series, are shown in Figure 7.5. Note that the cosine series converges to $f(t)$ much more quickly than the sine series did. This is because the even extension of the function $f(t) = t$ is *continuous* everywhere. ∎

In both Figure 7.3 and Figure 7.4 it can be seen that, at the points of discontinuity of a piecewise continuous function f, the graph of any finite number of terms in the Fourier series tends to overshoot and undershoot the graph of f on either side of the point of discontinuity. This is referred to as the **Gibbs phenomenon**. You should look this up in a book[2] or on the Web (Wikipedia contains a very nice explanation) and read more about it. The Gibbs phenomenon provides a perfect example of nonuniform convergence. There is always a significant error in the series approximation to the function. The point where it occurs just gets closer and closer to the jump. This has led engineers, and mathematicians working with engineers, to find ways to correct for this error when using Fourier approximations in their work.

Exercises 7.2. 1. *Determine whether each of the following functions is even, odd, or neither.*

(a) $1 + t^2$

(b) $\sin(2t) + 6t$

(c) te^t

(d) $1 + \cos(t)$

(e) $\cosh(t)$

(f) $t + \sinh(t)$

(g) $t\sinh(t)$

(h) $\frac{t}{1+t^2}$

---

[2] A good elementary description of the Gibbs phenomenon is given in Section 7.5.2 of *Partial Differential Equations* by M. Shearer and R. Levy, Princeton University Press, 2015.

7.2 Orthogonal Functions and Trigonometric Fourier Series

2. In Chapter 3 it was shown that the general solution of the ODE $X''(x) - K^2 X(x) = 0$, for any real number $K \neq 0$, can be written in the form $X(x) = c_1 e^{Kx} + c_2 e^{-Kx}$. Show that this can also be written in the form $X(x) = A\cosh(Kx) + B\sinh(Kx)$. Write the constants A and B in terms of c_1 and c_2.

3. Find all real numbers K for which $X''(x) - K^2 X(x) = 0$ has solutions satisfying the boundary conditions $X(0) = 0, X(1) = 1$. Hint: Write $X(x) = A\cosh(Kx) + B\sinh(Kx)$ and try to make it satisfy the two conditions.

4. Redo the previous exercise with the boundary conditions $X(0) = 0, X(1) = 0$.

5. Use the trig identity $\cos(A)\cos(B) = \frac{1}{2}(\cos(A-B) + \cos(A+B))$ to show that $\int_{-L}^{L} \cos(\frac{n\pi t}{L}) \cos(\frac{m\pi t}{L}) dt = 0$ for any integers $m \neq n$.

6. Use the trig identity $\sin(A)\sin(B) = \frac{1}{2}(\cos(A-B) - \cos(A+B))$ to show that $\int_{-L}^{L} \sin(\frac{n\pi t}{L}) \sin(\frac{m\pi t}{L}) dt = 0$ for any integers $m \neq n$.

7. Show that $\int_{-L}^{L}(\cos(\frac{n\pi t}{L}))^2 dt = \int_{-L}^{L}(\sin(\frac{n\pi t}{L}))^2 dt = L$ for any integer $n > 0$.

For each function 8.-12. below, defined on the interval $-1 \leq t \leq 1$ and assumed to be periodic of period 2,

(i) Draw a graph of the function on $-3 \leq t \leq 3$.

(ii) Find a formula for the coefficients of the Fourier series for $f(t)$.

(iii) Using all terms out to $n = 5$, sketch a graph of the Fourier series approximation to $f(t)$ on the interval $[-3, 3]$.

8. $f(t) = t^2$

9. $f(t) = 1 + t$

10. $f(t) = e^t$

11. $f(t) = \sin(t)$

12. $f(t) = \begin{cases} 0 & \text{if } -1 \leq t < -\frac{1}{2} \\ 1 & \text{if } -\frac{1}{2} \leq t \leq \frac{1}{2} \\ 0 & \text{if } \frac{1}{2} \leq t \leq 1 \end{cases}$

13. Approximating a function $f(t)$ by a Maclaurin series $\sum_{0}^{\infty} \frac{f^{(n)}(0)}{n!} t^n$ is equivalent to using the set of functions $\mathcal{S} = \{1, t, t^2, t^3, \ldots\}$. Is this set of functions orthogonal on $[-1, 1]$? Either prove that it is or find two integers $m \neq n$ such that $\int_{-1}^{1} t^m \cdot t^n dt \neq 0$.

14. There is a set $\mathcal{S} = \{P_0(t), P_1(t), P_2(t), \ldots\}$ of polynomials, called the **Legendre polynomials**, which forms an orthogonal set on the interval $[-1, 1]$. For each integer $n = 0, 1, 2, \ldots$, $P_n(t)$ is a polynomial of degree n. Look up Legendre polynomials on the web, or in a textbook, and find formulas for $P_0(t), P_1(t), \ldots, P_5(t)$. Also find a recursion formula that allows you to use $P_{n-1}(t)$ and $P_n(t)$ to determine $P_{n+1}(t)$.

15. Using the Legendre polynomials from the previous exercise, find the first 5 terms in the series approximation $f(t) \approx \sum_{n=0}^{4} c_n P_n(t)$ to the function $f(t) = e^t$ on the interval $-1 \leq t \leq 1$. Remember that the orthogonality condition implies that

$$c_n = \frac{\int_{-1}^{1} f(t) P_n(t) dt}{\int_{-1}^{1} (P_n(t))^2 dt}.$$

Graph the function $y = e^t$ together with the series approximation on $[-1, 1]$. Compare this graph to the one found in Exercise 10..

16. Find the first 4 nonzero terms in the sine series for the function

$$f(t) = \begin{cases} 2t & 0 \leq t < \frac{1}{2} \\ 2 - 2t & \frac{1}{2} \leq t \leq 1 \end{cases}$$

Graph the 4-term approximation to the series on the interval $[-1, 1]$.

17. Find the first 4 nonzero terms in the cosine series for the function $q(t) = \sin(t)$ on $[0, \pi]$. Graph it on the interval $[-3\pi, 3\pi]$. What continuous function does this series converge to on the entire real line?

18. Find the first 4 nonzero terms in the cosine series for the function $p(t) = e^t$ on $[0, 1]$. Graph the function and its approximation on the interval $-1 \leq t \leq 1$.

19. In some of the references to the Gibbs phenomenon, it is stated that if a piecewise continuous function $f(t)$ has a jump of size T at a point t^*, its Fourier series out to $n = N$ terms will overshoot and undershoot the value of the function by approximately 9% of T at $t = t^* \pm \frac{1}{N+\frac{1}{2}}$. Use the Fourier series found for the step function in Example 7.2.2 to test this hypothesis. Let N take the values 3, 11, and 21, and in each case compare the sum of N terms in the series to the value of f at the two given points near the jump at $t^* = 0$.

7.3 Boundary-Value Problems: Sturm-Liouville Equations

It was shown in Section 7.1.1 that at one point in the solution of a PDE by separation of variables it becomes necessary to find nonzero solutions of a second-order ODE, where the initial conditions are given at two different values of the independent variable. None of the methods in Chapters 1–6 apply to problems of this type, and if you worked Exercise 7 at the end of Section 7.1 you may have recognized that this is not an *easy* problem.

In this section you will be shown how to tackle problems of this type, called boundary-value problems. The definition of one important type of boundary-value problem is given below.

Definition 7.3. *A* **regular Sturm-Liouville boundary-value problem** *consists of a second-order linear differential equation which can be written in the form*

$$\frac{d}{dt}[p(t)y'(t)] + (\lambda w(t) - q(t))y(t) = 0, \tag{7.23}$$

7.3 Boundary-Value Problems: Sturm-Liouville Equations

with p, p′, and w continuous functions, q at least piecewise continuous, on the interval $a \leq t \leq b$, and p and w both positive functions on $[a,b]$. The boundary conditions assumed for $y(t)$ are called **homogeneous unmixed boundary conditions** and are given at two points a and b, in the form:

$$\begin{cases} c_1 y(a) + c_2 y'(a) = 0; & c_1, c_2 \text{ not both zero} \\ c_3 y(b) + c_4 y'(b) = 0; & c_3, c_4 \text{ not both zero} \end{cases} \quad (7.24)$$

There are other types of Sturm-Liouville problems, where the equation has one or more singular points, or the boundary conditions are mixed or periodic, but the above definition describes the type of problem that occurs most often in the partial differential equations we will be solving.

The form of the equation in (7.23) is less restrictive than it looks. All of the linear second-order ODEs that we will need to solve can be put into this form by multiplying the equation by an appropriate factor if necessary.

The trivial solution $y(t) \equiv 0$ always satisfies (7.23) and (7.24), but it will be necessary to find nonzero solutions. If the equation (7.23) is rewritten in the form

$$-\frac{1}{w(t)} [p(t) y''(t) + p'(t) y'(t) - q(t) y(t)] = \lambda y(t),$$

it can be seen to be analogous to a matrix equation of the form $\mathbf{A}\mathbf{x} = \lambda \mathbf{x}$, where the linear operator (i.e. the matrix $\mathbf{A}$) is replaced by a linear *differential* operator

$$-\frac{1}{w(t)} \left(p(t) \frac{d^2}{dt^2} + p'(t) \frac{d}{dt} - q(t) \right)$$

which can be applied to any function $y(t)$ contained in a certain set of functions, called a function space. Since the early 1800s, Sturm-Liouville theory has been extensively studied, and it has been shown that for (7.23) with the boundary conditions (7.24), nonzero solutions of a Sturm-Liouville problem only exist if the parameter λ belongs to a certain discrete set of real numbers λ_n, called **eigenvalues** of the particular Sturm-Liouville problem. There is a smallest eigenvalue λ_1, with $\lambda_1 < \lambda_2 < \cdots < \lambda_n < \cdots$ and $\lim_{n \to \infty} \lambda_n = \infty$. When $\lambda = \lambda_n$, the corresponding solution $y(t) = \phi_n(t)$ of the boundary-value problems (7.23), (7.24) is called an **eigenfunction** corresponding to the eigenvalue λ_n; and each ϕ_n is unique up to constant multiples. The sequence of eigenfunctions $S = \{\phi_n(t)\}_{n=1}^{\infty}$ is complete in the sense that any function $f(t)$ in the associated function space can be represented by an infinite series of the form $\sum_1^{\infty} c_n \phi_n(t)$. In addition, the eigenfunctions $\phi_n(t)$ can be shown to be orthogonal with respect to the *weight* function $w(t)$; that is,

$$\int_a^b \phi_n(t) \phi_m(t) w(t) dt = 0, \text{ if } m \neq n.$$

As an example, Theorem 7.2 below gives a simple proof of the orthogonality of the eigenfunctions of the Sturm-Liouville problem given by (7.23) and (7.24).

Theorem 7.2. *The infinite family of eigenfunctions* $S = \{\phi_1, \phi_2, \ldots, \phi_n, \ldots\}$ *of a Sturm-Liouville problem (7.23) with boundary conditions (7.24) is an* **orthogonal family** *of functions on the interval* $[a,b]$, *with respect to the weight function $w(t)$ in equation (7.23);*

that is,
$$\int_a^b w(t)\phi_n(t)\phi_m(t)dt = 0, \quad \text{if } m \neq n.$$

Proof. Using the form of the differential equation (7.23), if λ_i is any eigenvalue with eigenfunction ϕ_i, the fact that ϕ_i satisfies the differential equation allows us to write
$$(p\phi_i')' = (q - \lambda_i w)\phi_i.$$
Letting m and n be any two indices with $\lambda_m \neq \lambda_n$,
$$(p\phi_m')'\phi_n - (p\phi_n')'\phi_m$$
$$= (q - \lambda_m w)\phi_m\phi_n - (q - \lambda_n w)\phi_n\phi_m = (\lambda_n - \lambda_m)w\phi_m\phi_n.$$
Taking the integral of both sides,
$$(\lambda_n - \lambda_m)\int_a^b w\phi_m\phi_n dt = \int_a^b (p\phi_m')'\phi_n dt - \int_a^b (p\phi_n')'\phi_m dt.$$
Integration by parts, on the right, gives
$$(\lambda_n - \lambda_m)\int_a^b w\phi_m\phi_n dt$$
$$= \left(p\phi_m'\phi_n\big|_a^b - \int_a^b p\phi_m'\phi_n' dt\right) - \left(p\phi_n'\phi_m\big|_a^b - \int_a^b p\phi_n'\phi_m' dt\right)$$
$$= p(t)[\phi_m'(t)\phi_n(t) - \phi_n'(t)\phi_m(t)]\big|_a^b.$$
Now check that the form of the boundary conditions (7.24) ensures that, at both $t = a$ and $t = b$,
$$\phi_m'(t)\phi_n(t) - \phi_n'(t)\phi_m(t) = 0.$$
Therefore, since $\lambda_n \neq \lambda_m$, we have shown that
$$\int_a^b w(t)\phi_m(t)\phi_n(t)dt = 0. \qquad \square$$

To solve a particular Sturm-Liouville problem, one must first find a general solution of the differential equation and then check to see for which parameters λ the boundary conditions can be satisfied. The next example shows how this can be done.

Example 7.3.1. *Show that the boundary-value problem*
$$y''(t) + \lambda y(t) = 0, \quad y(0) = 0, \quad y'(1) = 0 \tag{7.25}$$
is a Sturm-Liouville problem, and find all eigenvalues and a corresponding set of orthogonal eigenfunctions.

Solution. We can write $y'' + \lambda y = 0$ in the form $(1 \cdot y')' + \lambda \cdot 1 \cdot y = 0$, so this *is* a Sturm-Liouville equation with $p(t) = w(t) \equiv 1$, and $q(t) \equiv 0$. The given boundary conditions can be written in the form
$$\begin{cases} 1 \cdot y(0) + 0 \cdot y'(0) = 0 \\ 0 \cdot y(1) + 1 \cdot y'(1) = 0. \end{cases} \tag{7.26}$$

7.3 Boundary-Value Problems: Sturm-Liouville Equations

To find all of the nontrivial solutions, we first determine the general solution of the differential equation, and then check to see for which values of λ the given boundary conditions can be satisfied. Since the characteristic equation of the linear constant-coefficient differential equation (7.25) is $r^2 + \lambda = 0$, there will be three different cases depending on whether the two roots r are real and distinct, real and equal, or complex conjugates. We must consider each of these three cases separately.

Case 1: $\lambda < 0$.

Let $\lambda = -K^2$ for some nonzero real number K. Then the roots of the characteristic polynomial $r^2 - K^2 = 0$ are $r = \pm K$. The general solution in this case can be written as $y(t) = c_1 e^{Kt} + c_2 e^{-Kt}$, but we will find it more convenient to write it in the equivalent form $y(t) = A\cosh(Kt) + B\sinh(Kt)$. Now, to satisfy the two boundary conditions, we need to find A and B such that $y(0) = A\cosh(K \cdot 0) + B\sinh(K \cdot 0) = A = 0$ and $y'(1) = AK\sinh(K \cdot 1) + BK\cosh(K \cdot 1) = 0$. Since A has to be zero, and $K \neq 0$, the condition on $y'(1)$ implies that $B = 0$. This means that the only solution is the trivial solution $y(t) \equiv 0$; therefore, there are no negative eigenvalues. Remember that λ is an eigenvalue only if there exists a solution $y(t)$ that is *not* identically zero.

Case 2: $\lambda = 0$.

In this case, the characteristic polynomial is $r^2 = 0$, with a double root $r = 0$. The general solution is $y(t) = c_1 e^{0 \cdot t} + c_2 t e^{0\,t} = c_1 + c_2 t$, with derivative $y'(t) = c_2$. To satisfy the two boundary conditions, $y(0) = c_1 = 0$ and $y'(1) = c_2 = 0$, so again the only solution is the zero solution; therefore, $\lambda = 0$ is *not* an eigenvalue.

Case 3: $\lambda > 0$.

Assume $\lambda = K^2$ for some nonzero K. This is the case where the characteristic polynomial $r^2 + K^2 = 0$ has complex conjugate roots $r = \pm Ki$. The general solution in this case is $y(t) = c_1 e^{0 \cdot t} \cos(Kt) + c_2 e^{0 \cdot t}\sin(Kt) = c_1 \cos(Kt) + c_2 \sin(Kt)$ with derivative $y'(t) = -Kc_1 \sin(Kt) + Kc_2 \cos(Kt)$. To satisfy the boundary conditions, we need $y(0) = c_1 = 0$ and $y'(1) = Kc_2 \cos(K) = 0$. Since $K \neq 0$, in order to have $c_2 \neq 0$, it is necessary that $\cos(K) = 0$. There exists an infinite sequence

$$K = \frac{\pi}{2}, \frac{3\pi}{2}, \ldots, \frac{(2n+1)\pi}{2}, \ldots,$$

for which this is true. Letting $K_n = \frac{(2n+1)\pi}{2}$, the eigenvalue λ_n is

$$\lambda_n = K_n^2 = \left(\frac{(2n+1)\pi}{2}\right)^2, \quad n = 0, 1, 2, \ldots.$$

The corresponding eigenfunctions are $y_n(t) = C_n \sin\left(\frac{(2n+1)\pi t}{2}\right)$, $n = 0, 1, \ldots$. ∎

Note that this gives us a new orthogonal family of functions:

$$\bar{S} = \left\{\sin\left(\frac{\pi t}{2}\right), \sin\left(\frac{3\pi t}{2}\right), \ldots, \sin\left(\frac{(2n+1)\pi t}{2}\right), \ldots\right\},$$

which is *not* the entire family that we used to generate the trigonometric Fourier series. The family $\bar{S}$ is defined on the interval $[0,1]$, and since the weight function $w(t)$ in the Sturm-Liouville equation $x'' + \lambda x = 0$ is identically equal to one, we know from

Theorem 7.2 that the functions in $\bar{S}$ satisfy the orthogonality condition

$$\int_0^1 \sin\left(\frac{(2m+1)\pi t}{2}\right) \sin\left(\frac{(2n+1)\pi t}{2}\right) dt = 0$$

if $m \neq n$; and simple integration can be used to show that

$$\int_0^1 \left(\sin\left(\frac{(2n+1)\pi t}{2}\right)\right)^2 dt = \frac{1}{2}.$$

In the next example we will expand a function $z(t)$, defined on the interval $[0, 1]$, in terms of the orthogonal family $\bar{S}$. This will be done by writing

$$z(t) = \sum_{n=0}^{\infty} a_n \phi_n(t) = \sum_{n=0}^{\infty} a_n \sin\left(\frac{(2n+1)\pi t}{2}\right),$$

with the coefficients a_n given by the formula in (7.18):

$$a_n = \frac{\int_0^1 z(t)\phi_n(t)dt}{\int_0^1 (\phi_n(t))^2 dt} = \frac{\int_0^1 z(t)\sin\left(\frac{(2n+1)\pi t}{2}\right)dt}{\int_0^1 \left(\sin\left(\frac{(2n+1)\pi t}{2}\right)\right)^2 dt} = 2\int_0^1 z(t)\sin\left(\frac{(2n+1)\pi t}{2}\right)dt.$$

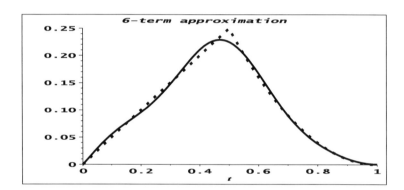

Figure 7.6. Graphs of $z(t)$ and the 6-term series approximation.

Example 7.3.2. *For the piecewise continuous function*

$$z(t) = \begin{cases} 0.5t & \text{if } 0 \leq t \leq 0.5 \\ (t-1)^2 & \text{if } 0.5 < t \leq 1.0 \end{cases}$$

find the first six nonzero terms in the series approximation

$$z(t) \approx \sum_{n=0}^{5} a_n \sin\left(\frac{(2n+1)\pi t}{2}\right),$$

and graph the function $z(t)$ together with the finite series approximation on the interval $0 \leq t \leq 1$.

Solution. Using the formula $a_n = 2\int_0^1 z(t)\sin\left(\frac{(2n+1)\pi t}{2}\right)dt$, the coefficients $a_0, a_1, \ldots, a_5$ can be found numerically to be 0.12997, 0.12256, -0.028551, -0.019673, 0.0096130, 0.0076536. Figure 7.6 shows the function and its six-term series approximation on the interval $0 \leq t \leq 1$. ∎

7.3 Boundary-Value Problems: Sturm-Liouville Equations

A Sturm-Liouville Problem Involving Bessel Functions. The only Sturm-Liouville equation we have worked with so far is $x'' + \lambda x = 0$, and this is partly due to the fact that it is easy to obtain its general solution. A more interesting Sturm-Liouville problem arises when the wave equation is expressed in terms of cylindrical coordinates r, θ, z.

The PDE

$$u_{tt} = u_{rr} + \frac{1}{r}u_r$$

can be used to model the displacement $u(r, t)$, at time t, of a circular drumhead given an initial displacement and/or initial velocity possessing radial symmetry. The symmetry causes the problem to be independent of the variable θ, and since the drumhead is 2-dimensional, there is no dependence on z. If the outer radius of the drum is $r = 1$ and the drumhead is assumed to be fixed around its rim, then the two boundary conditions for $u(r, t)$ are (1) for all $t > 0$, $u(1, t) = 0$ and (2) for all $t > 0$, $u(0, t)$ is finite. Letting $u(r, t) = R(r)T(t)$ and dividing by $R(r)T(t)$ to separate the variables,

$$\frac{RT''}{RT} = \frac{R''T + \frac{1}{r}R'T}{RT} \implies \frac{T''}{T} = \frac{R''}{R} + \frac{1}{r}\frac{R'}{R} = -\lambda$$

leads to the 2-dimensional system of ODEs:

$$T'' + \lambda T = 0 \tag{7.27}$$

$$R'' + \frac{1}{r}R' + \lambda R = 0. \tag{7.28}$$

The given boundary conditions on u require that $R(1) = 0$ and $R(0)$ must be finite. It remains to be shown that the equation (7.28), together with the two boundary conditions on R, *is* a Sturm-Liouville problem. If (7.28) is multiplied by r, it can be seen that

$$0 = r(R'' + \frac{1}{r}R' + \lambda R) \equiv (rR')' + r\lambda R; \tag{7.29}$$

therefore, this is a Sturm-Liouville equation with $p(r) = w(r) = r$ and $q(r) = 0$. Even though the equation is not defined at $r = 0$, and the boundary conditions are not the kind we have been working with, it can still be shown that there exist an infinite sequence of eigenvalues $\lambda_1 < \lambda_2 < \cdots$ and a corresponding set of eigenfunctions $S = \{\phi_n(t)\}_{n=1}^{\infty}$ satisfying the orthogonality condition

$$\int_0^1 w(r)\phi_n(r)\phi_m(r)dr = 0, n \neq m, \quad \text{where } w(r) = r.$$

To find the eigenvalues, we first need the general solution of (7.29). By making the change of variables $x = \sqrt{\lambda}r$ and letting $X(x(r)) \equiv R(r)$, we can write

$$R'(r) = \frac{dR}{dr} = \frac{d}{dx}(X(x))\frac{dx}{dr} = \sqrt{\lambda}X'(x)$$

and

$$R''(r) = \frac{d}{dr}(R'(r)) = \frac{d}{dx}(\sqrt{\lambda}X'(x))\frac{dx}{dr} = \lambda X''(x).$$

Substituting $r = x/\sqrt{\lambda}$ and the formulas for R' and R'' into the equation $rR'' + R' + \lambda rR = 0$ results in the ODE

$$xX''(x) + X'(x) + xX(x) = 0 \tag{7.30}$$

which is Bessel's equation of order 0.

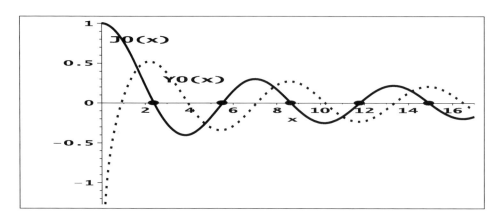

Figure 7.7. The Bessel functions $J_0(x)$ and $Y_0(x)$ with the zeros of J_0 marked on the x-axis

In Chapter 3 we found one series solution for this equation. This solution $X(x) = J_0(x) = \sum_{n=0}^{\infty} \frac{(-1)^n (x^2)^n}{2^{2n}(n!)^2}$ is called the Bessel function of order zero **of the first kind**. There is a second solution $Y_0(x)$, the **Bessel function of order zero of the second kind**, which is infinite at $x = 0$. The general solution of (7.30) is $X(x) = c_1 J_0(x) + c_2 Y_0(x)$; therefore, it can be seen that the general solution of our second-order Sturm-Liouville equation is

$$R(r) = X(\sqrt{\lambda} r) = c_1 J_0(\sqrt{\lambda} r) + c_2 Y_0(\sqrt{\lambda} r).$$

Because $Y_0(x)$ is infinite at $x = 0$, the condition that $R(0)$ be finite implies that the constant c_2 must be zero. Using the other condition $R(1) = 0$ implies that $J_0(\sqrt{\lambda}) = 0$, and this means that $\sqrt{\lambda}$ must be a zero of the Bessel function J_0. It is well known that Bessel functions act very much like decaying sinusoids, and their zeroes have been tabulated and are available in the popular packages like Maple and Mathematica. Figure 7.7 contains a graph of the two functions J_0 and Y_0. Both functions cross the x-axis infinitely often, and the zeros of J_0 form an infinite sequence $z_1 < z_2 < \cdots$ with $\lim_{n \to \infty} z_n = \infty$. It has also been proven that as x increases the function $J_0(x)$ gets closer and closer to a positive multiple of the function $\cos(x - \frac{\pi}{4})$, so its zeros get closer and closer to $(n - \frac{1}{4})\pi$ for n large. Since each $\lambda_n = z_n^2$ is positive, it can be seen that the solution of the equation $T_n'' + \lambda_n T_n = 0$ is $T_n(t) = a_n \cos(\sqrt{\lambda_n} t) + b_n \sin(\sqrt{\lambda_n} t)$ so that the full solution of the PDE is

$$u(r, t) = \sum_{n=1}^{\infty} J_0(z_n r)(a_n \cos(z_n t) + b_n \sin(z_n t)).$$

We now have another set of orthogonal functions $\hat{S} = \{J_0(z_1 r), J_0(z_2 r), \ldots\}$. Given any piecewise continuous function $f(r)$ on $[0, 1]$, we should be able to express it in the form

$$f(r) = \sum_{n=1}^{\infty} a_n J_0(z_n r), \text{ with } a_n = \frac{\int_0^1 r f(r) J_0(z_n r) dr}{\int_0^1 r (J_0(z_n r))^2 dr}.$$

7.3 Boundary-Value Problems: Sturm-Liouville Equations

Example 7.3.3. *A drumhead of radius 1 is initially pressed down in the center, and then released. The function*

$$f(r) = -0.05\cos(\pi r) - 0.05$$

is assumed to be the initial displacement $u(r,0)$ along each radius of the drumhead (see Figure 7.8). Find the first 5 coefficients in the orthogonal series approximation

$$f(r) \approx F_5(r) = \sum_{n=1}^{5} a_n J_0(z_n r),$$

and plot $f(r)$ and $F_5(r)$, $0 \le r \le 1$ on the same set of axes.

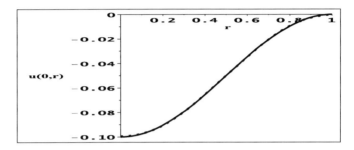

Figure 7.8. Graphs of $f(r)$ and the 5-term series approximation $F_5(r)$ (the dotted curve is F_5)

Solution. In Maple the Bessel function $J_0(x)$ is BesselJ(0,x). To find the zeros, use the instruction

```
for n from 1 to 5 do z[n]:=fsolve(BesselJ(0,x)=0,
                                   x=3.14*(n-0.5)..3.14*n); od:
```

The range specified for the nth zero uses the fact that the zeros of $J_0(x)$ are very close to the zeros of the function $\cos(x - \frac{\pi}{4})$ for large values of x; therefore, the nth zero of J_0 approaches the value $(n - \frac{3}{4})\pi$ as $n \to \infty$.

Due to the orthogonality of the functions $J_0(z_n r)$, the coefficient a_n has the value

$$a_n = \frac{\int_0^1 r f(r) J_0(z_n r) dr}{\int_0^1 r (J_0(z_n r))^2 dr}.$$

Remember that the weight function in the Sturm-Liouville equation (7.29) is $w(r) = r$. A Maple computation gives $a_1 \approx -0.077932$, $a_2 \approx -0.026771$, $a_3 \approx 0.0065626$, $a_4 \approx -0.0028092$, and $a_5 \approx 0.0015091$. The graphs of $f(r)$ and $F_5(r)$ are shown in Figure 7.8. The 5-term Bessel function approximation lies nearly on top of the function $f(r)$. ∎

Exercises 7.3. *For the Sturm-Liouville problems 1–3, find all eigenvalues and corresponding eigenfunctions:*

1. $x'' + \lambda x = 0$, $x(0) = 0$, $x(\pi) = 0$.

2. $x'' + \lambda x = 0$, $x'(0) = 0$, $x(L) = 0$.

3. $x'' + \lambda x = 0$, $x'(0) = 0$, $x'(\pi) = 0$.

4. For the Sturm-Liouville problem $x'' + \lambda x = 0$, $x(0) = 0$, $x(1) + x'(1) = 0$,

   (a) Show that there are no negative eigenvalues.
   (b) Show that $\lambda = 0$ is NOT an eigenvalue.
   (c) Show that the positive eigenvalues are of the form $\lambda_n = K_n^2$, where K_n is the nth solution of the equation $\tan(K_n) = -K_n$, and that the corresponding eigenfunctions are $\phi_n(t) = C_n \sin(K_n t)$.
   (d) Find numerical values for λ_1, λ_2, and λ_3.
   (e) Graph the corresponding three eigenfunctions ϕ_1, ϕ_2 and ϕ_3 (assume the constant multiplier $C_n = 1$). How many zeroes does each of these functions have in the open interval $(0, 1)$?
   (f) Find the numerical value of the integral $\int_0^1 \phi_2(t)\phi_3(t)dt$. What should it equal? Why?

5. Show that the equation $aX''(t) + bX'(t) + \lambda X(t) = 0$ can be turned into a Sturm-Liouvile equation by multiplying it by e^{ct} for a particular value of the constant c; that is, find the constant c such that $e^{ct}(aX'' + bX' + \lambda X) = 0$ has the form of a Sturm-Liouville equation.

6. Using the results of Exercise 5., find all of the eigenvalues and corresponding eigenfunctions of the Sturm-Liouville problem:
$$x'' + 2x' + \lambda x = 0, \quad x(0) = x(1) = 0.$$

7. Find the first 10 terms in the series approximation to $z(t)$ in Example 7.3.2. Plot the approximation together with the curve $z(t)$. Does this give a better fit at the point where the derivative is discontinuous? Explain.

8. In a real n-dimensional vector space V, an $n \times n$ matrix $\mathbf{A}$ is said to be self-adjoint if and only if $\mathbf{A}$ is equal to its transpose. If the inner product in V is defined by $(\vec{x}, \vec{y}) \equiv \sum_1^n x_i y_i$, then $\mathbf{A}$ is self-adjoint if and only if $(\mathbf{A}\vec{x}, \vec{y}) = (\vec{x}, \mathbf{A}\vec{y})$ for all $\vec{x}, \vec{y} \in V$. Self-adjoint matrices are special in the sense that their eigenvalues $\{\lambda_i\}_{i=1}^n$ can be shown to be real and distinct, and the corresponding eigenvectors $\{\vec{v}_i\}_{i=1}^n$ form an orthogonal basis for V; that is, every vector $\vec{x} \in V$ can be written uniquely in the form $\vec{x} = \sum_1^n c_i \vec{v}_i$ for some coefficients c_i.

Consider a simple Sturm-Liouville problem $(px')' + (\lambda - q)x = 0$, with $w(t) \equiv 1$ and homogeneous unmixed boundary conditions (7.24). If we write the equation in the form
$$-(px')' + qx = \lambda x \tag{7.31}$$
and define the linear operator $\mathcal{L}x = -(px')' + qx$, then the equation (7.31) is in the form of an eigenvalue problem
$$\mathcal{L}x = \lambda x,$$
on a space $\mathcal{V}$ of functions $x(t)$ on the interval (a, b) satisfying the given boundary conditions, and such that $\int_a^b (x(t))^2 dt$ exists. Assuming the inner product on $\mathcal{V}$ is defined by $(x, y) = \int_a^b x(t)y(t)dt$, it can be shown that the operator $\mathcal{L}$ is a self-adjoint

7.3 Boundary-Value Problems: Sturm-Liouville Equations

operator by proving that for all $x, y \in \mathcal{V}, (\mathcal{L}x, y) = (x, \mathcal{L}y)$. This is equivalent to showing that

$$\int_a^b (-(px')' + qx)y\,dt = \int_a^b x(-(py')' + qy)\,dt. \tag{7.32}$$

The fact that this linear operator is self-adjoint makes it possible to prove that many of the properties mentioned above for eigenvalues and eigenvectors of self-adjoint matrices hold for eigenvalues and eigenfunctions of a Sturm-Liouville problem.

(a) *Show that with the boundary conditions (7.24), for any functions $x(t), y(t)$ in $\mathcal{V}$, the following term is equal to zero: $[(x(t)y'(t) - x'(t)y(t))]\,|_a^b$. You will need to use the fact that in the specification of the boundary conditions, at least one of c_1, c_2 must be unequal to zero and at least one of c_3, c_4 must be unequal to zero.*

(b) *Using integration by parts (for example, $\int_a^b (px')' y\,dt = (px')y|_a^b - \int_a^b px' y'\,dt$), prove that (7.32) is true.*

COMPUTER PROBLEM. Notice that by the end of Chapter 7 we have actually obtained the full solution of the PDE

$$u_{tt} = u_{rr} + \frac{1}{r}u_r$$

in the form of the series

$$u(r, t) = \sum_{n=1}^{\infty} J_0(z_n r)(a_n \cos(z_n t) + b_n \sin(z_n t)).$$

Figure 7.9. Position of the drumhead at times $t = 0, 5$, and 10

We were able to find the constants a_n by using the initial condition $u(r, 0) = f(r)$. We will see in the next chapter that two initial conditions are required to completely determine the solution of this PDE. Given an initial velocity of the drumhead, $u_t(r, 0) = g(r)$, it is possible to differentiate the series for $u(r, t)$ with respect to t and get a formula for b_n. Show that if the initial velocity $u_t(r, 0)$ is identically zero, the b_n's are all zero. This means that the function

$$u(r, t) = \sum_{n=1}^{\infty} a_n J_0(z_n r) \cos(z_n t)$$

models the behavior of the drumhead if it is displaced at time $t = 0$ and released with zero initial velocity: $u_t(x, 0) \equiv 0, 0 \leq x \leq 1$. To see what the drumhead looks like at any time $t > 0$, given the initial displacement modelled by the function $f(r)$, the following Maple program can be used.

```
> P := evalf(Pi,10); N:=5; dt:=0.1;
> y :=t->BesselJ(0,t); f:=r->-0.05*cos(P*r)-0.05;
> for n from 1 to N do
    z[n]:=fsolve(BesselJ(0,x)=0,x=P*(n-0.5)..P*n);
    a[n]:=
    evalf((int(s*f(s)*y(z[n]*s),s=0..1))/(int(s*y(z[n]*s)^2,s=0..1)));
      od;
> u :=(r,t)-> sum(y(z[m]*r)*a[m]*cos(z[m]*t),m=1..N);
> with(plots); dt:=0.1;
> Ldrum:=[]; for q from 0 to 100 do
  Ldrum:=[op(Ldrum),cylinderplot([r,th,u(r,q*dt)],r=0..1,th=0..2.0*Pi,
       orientation=[-144,21])];od;
> display3d(op(Ldrum),insequence=true);
```

The final instruction will display a graph of the drumhead at time $t = 0$, and clicking on this graph will automatically set up the animation routine which you can use to view the oscillating behavior of the drumhead out to time $t = 100 \cdot dt = 10$. Figure 7.9 shows three views of the drumhead at times $t = 0, 5$, and 10. Note that the cylinderplot function acts to rotate the radial depression function around the drum's center from $\theta = 0$ to $\theta = 2\pi$.

8

Solving Second-order Partial Differential Equations

This chapter focuses on second-order partial differential equations in two variables. These include the classical equations of physics: the heat equation, wave equation, and Laplace's equation. It will be seen that these three equations are canonical examples of what are referred to as parabolic, hyperbolic, and elliptic PDEs. All three of these equations are linear and can be solved by the method of separation of variables. Different initial values and boundary conditions will be shown to require slightly different techniques.

To solve problems encountered in the biological sciences it is often necessary to work with nonlinear PDEs and, for these, numerical methods are often required. A numerical method will be given for each of the three types of linear PDEs, and it will be shown that these also apply to simple nonlinear PDEs.

In the final section of the chapter a student project involving a PDE appearing in the mathematical biology literature is treated in detail.

8.1 Classification of Linear Second-order Partial Differential Equations

Most of the partial differential equations we will consider will be linear equations of second order; where the unknown function $u(x, y)$ is a function of two independent variables x and y, or sometimes t and x, when the variable t is used to represent time. The most general **second-order linear partial differential equation in two independent variables x and y** can be written in the form

$$Au_{xx} + 2Bu_{xy} + Cu_{yy} + Du_x + Eu_y + Fu = G \qquad (8.1)$$

where $A, B, \ldots, G$ can be arbitrary functions of x and y. The coefficient of u_{xy} is written as $2B$ since we are assuming that for the functions u that we will be dealing with, the mixed partial derivatives u_{xy} and u_{yx} will be equal. Any equation that can be put in this form is called **linear**, and if $G(x, y) \equiv 0$ it is called a **homogeneous linear PDE**. Appendix G contains a review of partial derivatives, and this would be a good time to read this appendix in order to be clear about the notation for partial derivatives that will be used in this chapter.

We will mainly be concerned with homogeneous equations in which the coefficients of the second-order derivatives, A, B, and C, are constants. These equations can be classified into three different types as follows:

- if $AC - B^2 > 0$, the equation is called an **elliptic** PDE
- if $AC - B^2 = 0$, the equation is called a **parabolic** PDE
- if $AC - B^2 < 0$, the equation is called a **hyperbolic** PDE

To see why the equations are given these names, we are going to show how the homogeneous constant coefficient equation (8.1) can be turned into one of three simpler forms.

The terms in the left-hand side of (8.1) are in 1-1 correspondence with the terms in a quadratic polynomial of the form

$$P(x, y) = Ax^2 + 2Bxy + Cy^2 + Dx + Ey + F. \tag{8.2}$$

It is well known that a certain linear change of independent variables

$$x = ar + bs \tag{8.3}$$
$$y = cr + ds \tag{8.4}$$

can be used to write

$$P(x(r, s), y(r, s)) = p(r, s) = \lambda_1 r^2 + \lambda_2 s^2 + \alpha r + \beta s + \gamma,$$

where the rs term has been eliminated and λ_1 and λ_2 are the eigenvalues of the symmetric matrix $\mathbf{M} = \begin{pmatrix} A & B \\ B & C \end{pmatrix}$. In the exercises at the end of this section you will be led through a proof of this fact.

In the resulting form it can easily be checked that the graph of $p(r, s)$ is an ellipse if λ_1 and λ_2 are of the same sign, a hyperbola if they have different signs, and a parabola if one of them is zero. Since the product of the eigenvalues of $\mathbf{M}$ is equal to the $\det(\mathbf{M}) = AC - B^2$, the names for the three different types of equations can be seen to be appropriate. The fact that the same transformation works on the PDE to eliminate the mixed partial derivative term can be seen by writing it in "operator" form as

$$\left(AD_xD_x + 2BD_xD_y + CD_yD_y + DD_x + ED_y + F\right)u = 0,$$

where D_x and D_y represent partial differentiation with respect to x and y, respectively.

The three simplest PDEs, one of each type, that we will examine in detail are:

(1) Parabolic PDE: $u_{tt} - a^2 u_x = 0$ (the heat equation);

(2) Hyperbolic PDE: $u_{tt} - b^2 u_{xx} = 0$ (the wave equation);

(3) Elliptic PDE: $u_{xx} + u_{yy} = 0$ (Laplace's equation).

8.1 Classification of Linear Second-order Partial Differential Equations

Much more can be said about how this classification separates PDEs into different types, in terms of physical properties (see, for example, *Partial Differential Equations; an Introduction* by Walter A. Strauss, John Wiley & Sons, 1992), but we will use it mainly to classify equations having different methods of solution.

Exercises 8.1. *In Exercises 1–4, find all first and second partial derivatives of the given function and check that the mixed partial terms are equal:*

1. $f(x, y) = e^{2x+3y}$

2. $g(x, y) = xy^4 + 2x^3 y + 10x - 5y + 20$

3. $h(x, y) = \sin(4x - 3y)$

4. $q(x, y) = 2x \sin(3y) - 4y \cos(5x)$

5. Show that the function $u(x, t) = e^{-Kt} \sin(ax)$ satisfies the heat equation $u_t = u_{xx}$ if $K = a^2$.

6. Show that the function $u(x, y) = \sin(2x) \cosh(2y)$ satisfies Laplace's equation $u_{xx} + u_{yy} = 0$.

In Exercises 7–10, classify the given partial differential equations as elliptic, parabolic, or hyperbolic:

7. $u_{tt} + 2u_t = 4u_{xx}$

8. $u_{xx} + 2u_{yy} + u_x + u_y = x + y$

9. $u_t = u_{xx} + bu_x + u$

10. $u_{tt} = c^2 u_{xx} - bu_t - ku$ (the "telegraph" equation)

For those who are mathematically inquisitive and want to get some practice with matrices, the following exercises will take you step by step through the elimination of the xy term in a second-degree polynomial in two independent variables.
Let $Q(x, y) = Ax^2 + 2Bxy + Cy^2$ be the quadratic terms in the polynomial $P(x, y)$. The polynomial Q, containing only the second-order terms, is referred to as a *quadratic form*.

11. Show, using ordinary matrix multiplication, that

$$Q(x, y) = Ax^2 + 2Bxy + Cy^2 \equiv \begin{pmatrix} x & y \end{pmatrix} \begin{pmatrix} A & B \\ B & C \end{pmatrix} \begin{pmatrix} x \\ y \end{pmatrix}.$$

It does not matter where you start multiplying, since matrix multiplication is associative, i.e. if **L** is an $m \times n$ matrix, **M** is an $n \times p$ matrix, and **N** is a $p \times q$ matrix, then

$$\mathbf{LMN} = (\mathbf{LM})\mathbf{N} = \mathbf{L}(\mathbf{MN}).$$

Remember that the product of two matrices is defined if, and only if, the number of columns in the one on the left is the same as the number of rows in the one on the right.

12. The matrix $\mathbf{M} = \begin{pmatrix} A & B \\ B & C \end{pmatrix}$ is a symmetric matrix, i.e. $\mathbf{M} = \mathbf{M}^T$, and you will learn in linear algebra that its eigenvalues are all real and its eigenvectors are orthogonal. This means that if $\vec{v}_1$ and $\vec{v}_2$ are two linearly independent eigenvectors of $\mathbf{M}$, then the dot product of $\vec{v}_1$ and $\vec{v}_2$ is zero. We can also take $\vec{v}_1$ and $\vec{v}_2$ to be vectors of length 1. This means that $\vec{v}_1 \cdot \vec{v}_1 = 1$ and $\vec{v}_2 \cdot \vec{v}_2 = 1$. Let $\vec{v}_1$ and $\vec{v}_2$ be the orthogonal eigenvectors of a symmetric matrix $\mathbf{M}$, each of length one. If $\mathbf{U} = \begin{pmatrix} \vec{v}_1 & \vec{v}_2 \\ \downarrow & \downarrow \end{pmatrix}$ is the matrix with $\vec{v}_1$ and $\vec{v}_2$ as its columns, show that $\mathbf{U}^T \mathbf{U}$ is the identity matrix $\mathbf{I} = \begin{pmatrix} 1 & 0 \\ 0 & 1 \end{pmatrix}$. The fact that $\mathbf{U}^T \mathbf{U} = \mathbf{I}$ implies that $\mathbf{U}^T$ is the multiplicative inverse of $\mathbf{U}$.

13. Now show that $\mathbf{M U}$ is the same as the matrix $\mathbf{U}$ multiplied on the right by the diagonal matrix $\mathbf{D} = \begin{pmatrix} \lambda_1 & 0 \\ 0 & \lambda_2 \end{pmatrix}$, where λ_1 and λ_2 are the eigenvalues of $\mathbf{M}$; that is,
$$\mathbf{M} \begin{pmatrix} \vec{v}_1 & \vec{v}_2 \\ \downarrow & \downarrow \end{pmatrix} = \begin{pmatrix} \vec{v}_1 & \vec{v}_2 \\ \downarrow & \downarrow \end{pmatrix} \begin{pmatrix} \lambda_1 & 0 \\ 0 & \lambda_2 \end{pmatrix}.$$
Check this very carefully, by letting $\vec{v}_1$ and $\vec{v}_2$ have components and seeing that it works. This is the most important step. Remember that $\mathbf{M}\vec{v}_1 = \lambda_1 \vec{v}_1$ and $\mathbf{M}\vec{v}_2 = \lambda_2 \vec{v}_2$.

14. Multiply both sides of the matrix equation in Exercise 13. on the left by $\mathbf{U}^T = \mathbf{U}^{-1}$ and show that $\mathbf{U}^T \mathbf{M} \mathbf{U} = \begin{pmatrix} \lambda_1 & 0 \\ 0 & \lambda_2 \end{pmatrix}$.

15. If we make the change of variables given by
$$\begin{pmatrix} x \\ y \end{pmatrix} = \mathbf{U} \begin{pmatrix} r \\ s \end{pmatrix}$$
and use the fact that the transpose of a product of two matrices is the product of the transposes in reverse order, i.e. $(\mathbf{PQ})^T = \mathbf{Q}^T \mathbf{P}^T$ whenever the product $\mathbf{PQ}$ is defined, then we end up with the following quadratic form:
$$Q \equiv \begin{pmatrix} x & y \end{pmatrix} \begin{pmatrix} A & B \\ B & C \end{pmatrix} \begin{pmatrix} x \\ y \end{pmatrix} = \left(\mathbf{U} \begin{pmatrix} r \\ s \end{pmatrix} \right)^T \mathbf{M} \left(\mathbf{U} \begin{pmatrix} r \\ s \end{pmatrix} \right) \qquad (8.5)$$
$$= \begin{pmatrix} r \\ s \end{pmatrix}^T \mathbf{U}^T \mathbf{M} \mathbf{U} \begin{pmatrix} r \\ s \end{pmatrix}.$$
Check that putting this together with the result of Exercise 14. gives us the following quadratic form in r and s, with the rs term eliminated:
$$Q(x(r,s), y(r,s)) = \begin{pmatrix} r & s \end{pmatrix} \begin{pmatrix} \lambda_1 & 0 \\ 0 & \lambda_2 \end{pmatrix} \begin{pmatrix} r \\ s \end{pmatrix} = \lambda_1 r^2 + \lambda_2 s^2.$$

For Exercises 16. and 17., write the quadratic Q in matrix form $\mathbf{U}^T \mathbf{M} \mathbf{U}$ and use the eigenvalues of the symmetric square matrix $\mathbf{M}$ to determine whether the form is elliptic, hyperbolic, or parabolic. With the matrix $\mathbf{U}$ of eigenvectors of $\mathbf{M}$, make the linear change of variables $\begin{pmatrix} x \\ y \end{pmatrix} = \mathbf{U} \begin{pmatrix} r \\ s \end{pmatrix}$ to write $Q(x,y)$ in terms of r and s.

16. $Q = 2x^2 + 4xy + 5y^2$

17. $Q = x^2 - 6xy + y^2$

18. If D_x and D_y are considered as operators of partial differentiation by x and y, respectively, then the PDE $2u_{xx} + 4u_{xy} + 5u_{yy} = 0$ can be written in matrix form as

$$\begin{pmatrix} D_x & D_y \end{pmatrix} \begin{pmatrix} 2 & 2 \\ 2 & 5 \end{pmatrix} \begin{pmatrix} D_x \\ D_y \end{pmatrix} u(x,y) = 0.$$

If the linear change of variables $r = ax + by, s = cx + dy$ is made, then

$$D_x u(x,y) = \frac{\partial}{\partial x} u(r,s) = \frac{\partial u}{\partial r}\frac{dr}{dx} + \frac{\partial u}{\partial s}\frac{ds}{dx} = (aD_r + cD_s)u,$$

and in operator notation,

$$\begin{pmatrix} D_x \\ D_y \end{pmatrix} = \begin{pmatrix} a & c \\ b & d \end{pmatrix} \begin{pmatrix} D_r \\ D_s \end{pmatrix} = \mathbf{U} \begin{pmatrix} D_r \\ D_s \end{pmatrix}.$$

We can then write $\begin{pmatrix} D_x & D_y \end{pmatrix} \mathbf{M} \begin{pmatrix} D_x \\ D_y \end{pmatrix} \equiv \begin{pmatrix} D_r & D_s \end{pmatrix} (\mathbf{U}^T \mathbf{M} \mathbf{U}) \begin{pmatrix} D_r \\ D_s \end{pmatrix}$. If the linear change of variables matrix is chosen so $\mathbf{U}$ diagonalizes $\mathbf{M}$, we will have a much simpler PDE of the form $\lambda_1 u_{rr} + \lambda_2 u_{ss} = 0$. Use this process to simplify the PDE $2u_{xx} + 4u_{xy} + 5u_{yy} = 0$ (see Exercise 16.).

8.2 The 1-dimensional Heat Equation

Two of the classic PDEs found in undergraduate physics texts are the heat equation and the wave equation. The 1-dimensional heat equation $u_t = au_{xx}$, with $a > 0$, is the simplest example of a parabolic PDE. If this equation is written in the form $au_{xx} - u_t = 0$ it can be seen to be a linear second-order PDE with constant coefficients $A = a, B = C = 0$. This means that $B^2 - AC = 0$; and, therefore, this satisfies the definition of a parabolic PDE.

We begin by giving a simple derivation of the equation. Let $u(x,t)$ denote the temperature, at time t and position x, along the length of a thin rod with uniform cross-sectional area $\mathcal{A}$, density ρ, and length L. The sides of the rod are perfectly insulated so heat only flows in the x-direction.

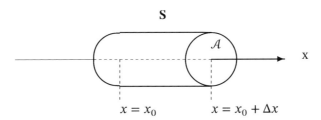

Consider a small segment S of the rod. First note that if the temperature is decreasing at x, then heat flows in the positive x-direction; therefore, the partial derivative of the temperature u with respect to x is negative. If κ is the thermal conductivity of the material in the rod, the heat flow at time t across the face at x_0 is $-\kappa A u_x(x_0, t)$ and across the face at $x_0 + \Delta x$ is $-\kappa A u_x(x_0 + \Delta x, t)$; therefore

$$Q \equiv \text{net heat flow into } S = \kappa A[u_x(x_0 + \Delta x, t) - u_x(x_0, t)],$$

and the total amount of heat entering the volume S during the time interval Δt is

$$Q\Delta t = \kappa A[u_x(x_0 + \Delta x, t) - u_x(x_0, t)]\Delta t.$$

Let Δu be the average change in temperature in S over the time interval Δt. It is proportional to the amount of heat introduced into S and inversely proportional to the mass Δm of S; therefore,

$$\Delta u = \frac{Q\,\Delta t}{s\,\Delta m} = \frac{Q\,\Delta t}{s\,A\,\rho\,\Delta x},$$

where the constant of proportionality s is the specific heat of the material in the rod. This average temperature change occurs at some point inside S, say at $x_0 + \theta \Delta x$, where $0 \leq \theta \leq 1$; therefore,

$$\Delta u = u(x_0 + \theta \Delta x, t + \Delta t) - u(x_0 + \theta \Delta x, t) = \frac{Q\,\Delta t}{s\,A\,\rho\,\Delta x}.$$

Equating the two expressions for $Q\Delta t$:

$$s\,\rho\,A\,\Delta x[u(x_0 + \theta \Delta x, t + \Delta t) - u(x_0 + \theta \Delta x, t)] = \kappa A[u_x(x_0 + \Delta x, t) - u_x(x_0, t)]\Delta t.$$

Now divide both sides of this equation by $A\Delta x \Delta t$ and let Δx and Δt both approach zero; then

$$\lim_{\Delta x, \Delta t \to 0} s\rho \left[\frac{u(x_0 + \theta \Delta x, t + \Delta t) - u(x_0 + \theta \Delta x, t)}{\Delta t}\right]$$

$$= \kappa \lim_{\Delta x \to 0} \left[\frac{u_x(x_0 + \Delta x, t) - u_x(x_0, t)}{\Delta x}\right].$$

Using our formulas for partial derivatives in Appendix G, this simplifies to

$$u_t(x, t) = \left(\frac{\kappa}{s\rho}\right) u_{xx}(x, t);$$

that is, $u_t(x, t) = a u_{xx}(x, t)$ for the positive constant $a = \frac{\kappa}{s\rho}$.

This equation can be generalized to any number of space dimensions; for example, in 3-space the heat equation is

$$\frac{\partial u}{\partial t} = a \nabla^2 u,$$

where $\nabla^2 u$ is the **laplacian** of u, defined by

$$\nabla^2 u = \frac{\partial^2 u}{\partial x^2} + \frac{\partial^2 u}{\partial y^2} + \frac{\partial^2 u}{\partial z^2}.$$

8.2 The 1-dimensional Heat Equation

8.2.1 Solution of the Heat Equation by Separation of Variables. In Section 7.1 a brief overview of the method of separation of variables was given and was applied to the heat equation. It will now be assumed that the function $u(x,t)$ is the temperature at a point x, at time t, in an insulated rod. The rod is L meters long, totally insulated except for the two ends at $x = 0$ and $x = L$. The density ρ of the rod, its thermal conductivity K, and specific heat s are all assumed to be constant along its length. Under these conditions it has been shown that the temperature $u(x, t)$ in the rod will satisfy the heat equation

$$u_t(x,t) = a u_{xx}(x,t),$$

where a is the positive constant $\dfrac{K}{s\rho}$.

In order to obtain a unique solution to such an equation, it can be shown that two types of conditions must be specified.

(1) <u>Boundary Conditions:</u>

Conditions on the temperature function $u(x, t)$ must be specified at both ends of the rod, for all values of $t > 0$. There are different ways to do this. One way is to specify the temperature at each end and assume it remains constant for all $t > 0$. Another condition results if it is assumed that one or both of the ends are insulated. If, for example, the end at $x = L$ is insulated, the condition $u_x(L, t) = 0$ for all $t > 0$ is used to simulate the fact that no heat is flowing across that end. One can also model the case where heat is being dissipated at an end by either convection or radiation.

Names are given to certain types of boundary conditions. Dirichlet boundary conditions imply that the function u is specified on the boundary of the region of interest, while Neumann conditions imply that values of the normal derivative of u are given on the boundary. Robin boundary conditions refer to the case where a linear combination of u and its derivative are given.

We will start by assuming the simplest condition; that is, that the temperature at each end of the rod is held at $0°$ for all $t > 0$.

(2) <u>Initial Conditions:</u>

To guarantee a unique solution of the equation, it is necessary to specify the temperature along the entire length of the rod at some initial time, say $t = 0$. The initial temperature, that is, $u(x, 0)$, must be specified as a function $f(x)$ on the interval $0 \leq x \leq L$, where L is the length of the rod. The function $f(x)$ needs to be at least piecewise continuous on $0 \leq x \leq L$.

The problem we are going to solve can now be summarized as follows:

$$\begin{cases} u_t(x,t) = a u_{xx}(x,t), \\ u(0,t) = u(L,t) = 0 & \text{for } t > 0, \\ u(x,0) = f(x) & \text{for } 0 \leq x \leq L. \end{cases}$$

In Section 7.1 it was shown that by letting $u(x,t) = X(x)T(t)$, the PDE becomes

$$X(x)T'(t) = aX''(x)T(t).$$

Dividing by aXT separates the variables, so that

$$\frac{T'}{aT} = \frac{X''}{X} = -\lambda.$$

where λ must be a constant. Thus the PDE can be written in the form of two simultaneous ordinary differential equations:

$$\begin{cases} T'(t) = -\lambda a T(t), \\ X''(x) + \lambda X(x) = 0. \end{cases}$$

The equation for $X(x)$ turns out to be the Sturm-Liouville equation we considered in Example 7.3.1. To obtain boundary conditions on X, we use the given boundary conditions on $u(x, t)$; that is, since $u(x, t) \equiv X(x)T(t)$,

$$\begin{cases} 0 = u(0, t) \equiv X(0)T(t) \\ 0 = u(L, t) \equiv X(L)T(t) \end{cases}$$

has to be true for all $t > 0$. The only way this can be true, without making $T(t)$ identically zero, is to require that $X(0) = X(L) = 0$. This leads to a Sturm-Liouville problem for X of the following form:

$$X'' + \lambda X = 0, \quad X(0) = 0, \quad X(L) = 0.$$

As was done in Section 7.3, we treat the same three cases, $\lambda < 0, \lambda = 0$, and $\lambda > 0$. Each one must be considered to see for what values of λ there exist nonzero solutions satisfying the boundary conditions $X(0) = X(L) = 0$.

It is easily shown that if $\lambda = -K^2 < 0$ only the zero solution exists, so there are no negative eigenvalues. Similarly, if $\lambda = 0$, the general solution is $X(x) = Cx + D$, so we need $X(0) = D = 0$ and $X(L) = C \cdot L = 0$, which implies that $C = 0$. Therefore, zero is not an eigenvalue.

If $\lambda = K^2 > 0$, the general solution is $X(x) = E\cos(Kx) + F\sin(Kx)$. Setting $X(0) = E = 0$ and $X(L) = F\sin(KL) = 0$ shows that KL must be equal to $n\pi$ for some integer n; that is, $K_n = \frac{n\pi}{L}$. This means that the eigenvalues are $\lambda_n = K_n^2 = \left(\frac{n\pi}{L}\right)^2$ for $n = 1, 2, \ldots$, and with $E = 0$ the general solution gives us the eigenfunctions $X_n(x) = C_n \sin\left(\frac{n\pi x}{L}\right)$.

To find the function $T_n(t)$ corresponding to $X_n(x)$, it is necessary to solve the ODE

$$T_n'(t) = -\lambda_n a T_n(t) = -\frac{n^2 \pi^2 a}{L^2} T_n(t).$$

This is a separable first-order equation with general solution

$$T_n(t) = c_n e^{-\frac{n^2 \pi^2 a}{L^2} t}.$$

We now have an *infinite family of solutions* of our heat equation; namely

$$u_n(x, t) = X_n(x)T_n(t) = b_n \sin\left(\frac{n\pi x}{L}\right) e^{-\frac{n^2 \pi^2 a}{L^2} t},$$

where the arbitrary constant b_n is the product of the two arbitrary constants C_n and c_n. Note that each u_n satisfies the two boundary conditions $u_n(0, t) = u_n(L, t) = 0$ for all $t > 0$. So far we have not used the initial condition.

8.2 The 1-dimensional Heat Equation

Because the heat equation is linear, linear combinations of solutions are also solutions. If we let

$$u(x,t) = \sum_{n=1}^{\infty} b_n \sin\left(\frac{n\pi x}{L}\right) e^{-\frac{n^2\pi^2 a}{L^2}t},$$

then setting $t = 0$ and using the initial condition $u(x,0) = f(x)$ results in the requirement that

$$u(x,0) = \sum_{n=1}^{\infty} b_n \sin\left(\frac{n\pi x}{L}\right) = f(x), \quad 0 \leq x \leq L.$$

Since this is in the form of a Fourier sine series for the function $f(x)$ on $0 \leq x \leq L$, the constants b_n must be the coefficients of that sine series. Therefore, the complete solution of our heat equation is given by

$$u(x,t) = \sum_{n=1}^{\infty} b_n \sin\left(\frac{n\pi x}{L}\right) e^{-\frac{n^2\pi^2 a}{L^2}t}, \quad b_n = \frac{2}{L}\int_0^L f(x)\sin\left(\frac{n\pi x}{L}\right)dx. \tag{8.6}$$

Example 8.2.1. *A rod of length 10 meters is initially heated to a temperature of $100°C$ on the left half $0 \leq x \leq 5$ and to a temperature of $40°C$ on the right half $5 \leq x \leq 10$. If the rod is completely insulated except at the two ends, and the temperature at both ends is held constant at $0°C$ for all $t > 0$, find the temperature $u(x,t)$ in the rod for $0 \leq x \leq 10$, $t > 0$. Assume the constant $a = \dfrac{K}{s\rho} = 4$.*

Solution. We know that the general solution is given by

$$u(x,t) = \sum_{n=1}^{\infty} b_n \sin\left(\frac{n\pi x}{10}\right) e^{-\frac{4n^2\pi^2}{100}t}.$$

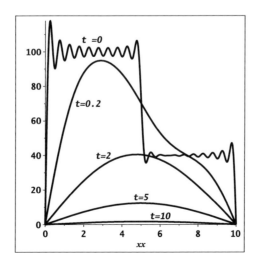

Figure 8.1. Temperature across the rod at times $t = 0, 0.2, 2, 5,$ and 10

Writing the piecewise continuous function $f(x)$ in the form

$$f(x) = \begin{cases} 100 & \text{for } 0 \le x \le 5 \\ 40 & \text{for } 5 < x \le 10 \end{cases}$$

the coefficients are given by

$$b_n = \frac{2}{10} \int_0^{10} f(x) \sin\left(\frac{n\pi x}{10}\right) dx = \frac{1}{5}\int_0^5 100 \sin\left(\frac{n\pi x}{10}\right) dx + \frac{1}{5}\int_5^{10} 40 \sin\left(\frac{n\pi x}{10}\right) dx$$

$$= \frac{200}{n\pi}\left(1 - \cos\left(\frac{n\pi}{2}\right)\right) + \frac{80}{n\pi}\left(\cos\left(\frac{n\pi}{2}\right) - \cos(n\pi)\right).$$

Figure 8.1 shows a Maple plot of the resulting function $u(x,t)$ at five different times $t = 0, 0.2, 2.0, 5.0$, and 10. Even though forty terms were used in the Fourier series it can be seen that the approximation of the piecewise continuous initial temperature function at time $t = 0$ is very rough. Since the series is a Fourier series for an odd function, it must make a jump from -100 to 100 at $x = 0$, so the overshoot due to the Gibbs phenomenon is very large. For $t > 0$ the exponential functions tend to smooth the solution. Note that, as expected, the temperature along the entire rod tends to 0 as $t \to \infty$. ∎

8.2.2 Other Boundary Conditions for the Heat Equation.

We have seen that the 1-dimensional heat equation can be used to model the temperature in a rod which is insulated around its sides. The first case we solved assumed that both ends of the rod were held at $0°$ for all time $t > 0$. Other boundary conditions can be modelled, and we will treat two different cases.

Heat Equation with both ends insulated. If the ends of the rod are *insulated*, the boundary conditions on the partial differential equation $u_t = au_{xx}$ must be changed to $u_x(0,t) = u_x(L,t) = 0$ for all $t > 0$. This implies that no heat can flow across either end of the rod. The method of separation of variables proceeds exactly as before, except that the boundary conditions on the Sturm-Liouville problem $X'' + \lambda X = 0$ must be changed. With $u(x,t) = X(x)T(t)$, the condition $u_x(0,t) \equiv X'(0)T(t) = 0$, for all $t > 0$ implies that $X'(0) = 0$, and similarly the condition $u_x(L,t) = 0$ implies that $X'(L) = 0$.

To solve the Sturm-Liouville problem $X'' + \lambda X = 0$ with $X'(0) = X'(L) = 0$, we again need to treat the three cases $\lambda < 0, \lambda = 0$, and $\lambda > 0$.

(1) If $\lambda < 0$, let $\lambda = -K^2$. Then the general solution can be written as

$$X(x) = A\cosh(Kx) + B\sinh(Kx)$$

$$X'(x) = AK\sinh(Kx) + BK\cosh(Kx)$$

$$X'(0) = BK = 0 \Rightarrow B = 0, \text{ and } X'(L) = AK\sinh(KL) = 0 \Rightarrow A = 0.$$

This means that there are no nontrivial solutions for $\lambda < 0$.

(2) If $\lambda = 0$, then the general solution is $X(x) = C + Dx$, and $X'(x) = D$. Both conditions $X'(0) = X'(L) = 0$ are satisfied if, and only if, the constant $D = 0$. This means that $X(x) \equiv C$, where C is an arbitrary constant, is a solution for eigenvalue $\lambda = 0$.

8.2 The 1-dimensional Heat Equation

(3) If $\lambda > 0$, let $\lambda = K^2$. Then the general solution is

$$X(x) = E\cos(Kx) + F\sin(Kx)$$

$$X'(x) = -KE\sin(Kx) + KF\cos(Kx)$$

and

$$X'(0) = KF = 0 \Rightarrow F = 0.$$

The condition $X'(L) = -KE\sin(KL) = 0$ can be satisfied if $KL = \pi, 2\pi, \ldots, n\pi, \ldots$; therefore, the eigenvalues and corresponding eigenfunctions are $\lambda_0 = 0$, $X_0(x) = $ constant, and

$$\lambda_n = K_n^2 = \frac{n^2\pi^2}{L^2}, \quad X_n(x) = E_n \cos\left(\frac{n\pi x}{L}\right), \quad n = 1, 2, \ldots.$$

For each eigenvalue λ_n, $n = 0, 1, \ldots$, the equation $T_n' = -a\lambda_n T_n$ must be solved. For $\lambda_0 = 0$, the solution $T_0(t)$ is an arbitrary constant; and for $n > 0$, the solution is

$$T_n(t) = C_n e^{-\frac{n^2\pi^2 at}{L^2}}.$$

The general solution for u can now be written as

$$u(x,t) = \frac{a_0}{2} + \sum_{n=1}^{\infty} a_n \cos\left(\frac{n\pi x}{L}\right) e^{-\frac{n^2\pi^2 at}{L^2}}. \tag{8.7}$$

If the initial function is $u(x, 0) = f(x)$, then

$$u(x,0) = \frac{a_0}{2} + \sum_{n=1}^{\infty} a_n \cos\left(\frac{n\pi x}{L}\right) \equiv f(x),$$

and this is a Fourier cosine series for $f(x)$, with coefficients

$$a_n = \frac{2}{L}\int_0^L f(x)\cos\left(\frac{n\pi x}{L}\right)dx, \quad \text{for } n = 0, 1, 2, \ldots. \tag{8.8}$$

Example 8.2.2. *Assume a rod of length 10m, with $a = \frac{K}{s\rho} = 1$, is heated to an initial temperature $f(x) = (50 - 2(x-5)^2)°C$. The rod is totally insulated on its ends, as well as around the sides. Determine the temperature $u(x,t)$ in the rod at time $t > 0$.*

Solution. Using equation (8.7), we can write the temperature as

$$u(x,t) = \frac{a_0}{2} + \sum_{n=1}^{\infty} a_n \cos\left(\frac{n\pi x}{10}\right) e^{-\frac{n^2\pi^2 t}{100}}.$$

A Maple program was written to compute the coefficients given by equation (8.8), for n from 0 to 30. The resulting graphs of $u(x,t)$ at times $t = 0, 0.5, 2, 5,$ and 20 seconds are shown in Figure 8.2.

Note that, as you would expect, $u(x,t)$ tends to a constant temperature along the entire rod, where the constant is the *average value* of the initial temperature function $f(x)$ on the interval $[0, L]$. No heat can escape from the rod. ∎

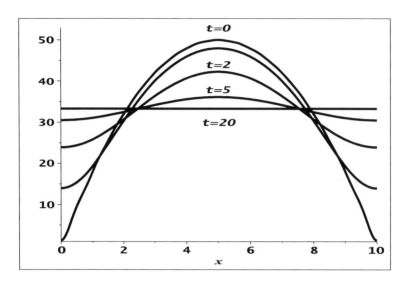

Figure 8.2. Temperature along the rod at times $0, 0.5, 2, 5$, and 20 seconds

Nonhomogeneous Heat Equation. The term nonhomogeneous, in this case, means that the boundary conditions on the heat equation are nonhomogeneous; that is, we will solve the 1-dimensional heat equation assuming that the temperature at each end of the rod can have an arbitrary (but constant) value. To solve this problem, it will be assumed that the temperature $u(x,t)$ is the sum of a function $w(x,t)$ satisfying homogeneous boundary conditions, plus a function $\sigma(x)$ which represents the steady-state temperature to which $u(x,t)$ tends as $t \to \infty$.

The function $\sigma(x)$ must satisfy the heat equation $\frac{\partial \sigma}{\partial t} = a \frac{\partial^2 \sigma}{\partial x^2}$, and since σ is not dependent on t, $\frac{\partial \sigma}{\partial t} \equiv 0$, implying that $\frac{\partial^2 \sigma}{\partial x^2} = \sigma''(x) \equiv 0$. Therefore, σ is a linear function of x which we can write as

$$\sigma(x) = c_1 + c_2 x.$$

If we assume that the two end temperatures are given values $u(0,t) = T_1$ and $u(L,t) = T_2$ for all $t > 0$, then $\sigma(0) = T_1$ and $\sigma(L) = T_2$ completely determine the straight-line function

$$\sigma(x) = T_1 + \frac{T_2 - T_1}{L} x.$$

Now, assuming $u(x,t) = w(x,t) + \sigma(x)$, we can write

$$w(x,t) = u(x,t) - \sigma(x) \longrightarrow$$

$$w_t(x,t) = u_t(x,t) - 0 = a u_{xx}(x,t) = a(w_{xx}(x,t) + \sigma''(x)),$$

and since $\sigma''(x) = 0$, we see that the function w also satisfies the heat equation; that is,

$$w_t(x,t) = a w_{xx}(x,t).$$

In addition, $w(0,t) = u(0,t) - \sigma(0) = T_1 - T_1 = 0$ and $w(L,t) = u(L,t) - \sigma(L) = T_2 - T_2 = 0$. This means that $w(x,t)$ is a solution of the heat equation with *homogeneous*

8.2 The 1-dimensional Heat Equation

boundary conditions $w(0, t) = w(L, t) = 0$ for all $t > 0$. This is the first version of the heat equation that we solved, and the solution was given by

$$w(x, t) = \sum_{n=1}^{\infty} b_n \sin\left(\frac{n\pi x}{L}\right) e^{-\frac{n^2\pi^2 at}{L^2}}.$$

To find the coefficients b_n, we need the initial conditions on $w(x, t)$. But $w(x, 0) = u(x, 0) - \sigma(x) = f(x) - \left(T_1 + \frac{T_2 - T_1}{L}x\right) = f(x) - \sigma(x)$; therefore,

$$b_n = \frac{2}{L} \int_0^L (f(x) - \sigma(x)) \sin\left(\frac{n\pi x}{L}\right) dx, \quad n = 1, 2, \ldots.$$

Now the complete solution of the nonhomogeneous problem can be written in the form

$$u(x, t) = T_1 + \frac{T_2 - T_1}{L} x + \sum_{n=1}^{\infty} b_n \sin\left(\frac{n\pi x}{L}\right) e^{-\frac{n^2\pi^2 at}{L^2}}$$

$$b_n = \frac{2}{L} \int_0^L \left(f(x) - \left[T_1 + \frac{T_2 - T_1}{L} x\right]\right) \sin\left(\frac{n\pi x}{L}\right) dx. \tag{8.9}$$

Example 8.2.3. *Assume that a rod of length 4m is heated so that its temperature at time $t = 0$ is given by*

$$f(x) = \begin{cases} 0 & \text{for } 0 \leq x < 1 \\ 100 & \text{for } 1 \leq x < 3 \\ 0 & \text{for } 3 \leq x \leq 4 \end{cases}$$

The rod is insulated around the sides, and for all $t > 0$ the temperature is kept at constant values $T_1 = 80$ at the left end and $T_2 = 20$ at the right end. Find the temperature in the rod for all $t > 0$. Assume that $a = \frac{K}{s\rho} = 0.1$.

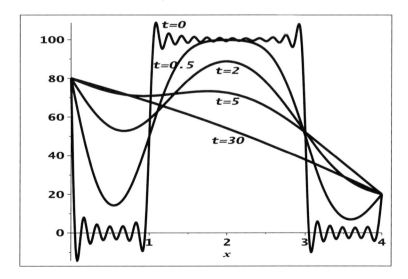

Figure 8.3. Temperature along the rod at times 0, 0.5, 2, 5, and 30 seconds

Solution. The steady state temperature to which the temperature in the rod converges over time is $\sigma(x) = T_1 + \frac{T_2-T_1}{4}x = 80 - 15x$; therefore,

$$u(x,t) = 80 - 15x + \sum_{n=1}^{\infty} b_n \sin\left(\frac{n\pi x}{4}\right) e^{-\frac{n^2\pi^2 0.1t}{16}},$$

where

$$b_n = \frac{2}{4}\int_0^4 (f(x) - \sigma(x))\sin\left(\frac{n\pi x}{4}\right)dx, \; n = 1, 2, \ldots.$$

Figure 8.3 shows the temperature $u(x,t)$ in the rod at times $t = 0, 0.5, 2, 5,$ and 30.

Even with 50 terms in the Fourier series for $f(x)$, the graph of u at time $t = 0$ is wavy; however, as t increases, the graphs become much smoother because of the negative exponentials in each term of the series. As $t \to \infty$, $u(x,t)$ approaches the straight-line steady state solution.

Exercises 8.2. *The starred exercises require a computer.*

1. ★ Solve the heat equation in a rod of length 2 meters, if the ends of the rod at $x = 0$ and $x = 2$ are held at 0, and the initial temperature is given by $f(x) = 20x, \; 0 \le x \le 2$. Assume $a = 1$. Plot solutions at several values of t.

2. Derive the general solution of the heat equation $u_t = au_{xx}$ if the end of the rod at $x = 0$ is held constant at $0°$ and the end at $x = L$ is insulated; that is, $u(0,t) = u_x(L,t) = 0$ for all $t > 0$.

3. ★ Redo Exercise 1., assuming the left end of the rod is insulated and the right end is held at 0.

4. ★ Solve the heat equation for a rod of length 10m, with $a = 1$. Assume $u(0,t) = u_x(10,t) = 0$ for $t > 0$, and the initial temperature is given by $f(x) = 2x(10-x)$, for $0 \le x \le 10$. Plot graphs of the temperature along the rod at times $t = 0, 1, 5, 10,$ and 20.

5. ★ In Exercise 4., describe what is happening to the temperature at the right-hand end of the rod. Use your function $u(x,t)$ to plot a graph of the temperature at $x = 10$ for $0 \le t \le 30$.

6. Derive a series solution for the heat equation $u_t = 0.5u_{xx}$ with initial temperature $u(x,0) \equiv 0$ and nonhomogeneous boundary conditions $u(0,t) = -20°, u(10,t) = 30°$ for all $t > 0$.

7. Show that the equation $\frac{\partial P}{\partial t} = K\frac{\partial^2 P}{\partial x^2} + rP\left(1 - \frac{P}{N}\right)$ is a parabolic PDE. Is it a linear PDE?

8. Can the equation in Exercise 7. be solved by separation of variables? Give a reason for your answer.

8.3 The 1-dimensional Wave Equation

In this section we will derive another linear second-order partial differential equation, by modelling the oscillation of a string stretched taut between its two ends. It is called the 1-dimensional wave equation and is our simplest example of a **hyperbolic** partial differential equation.

Derivation of the Wave Equation $\frac{\partial^2 u}{\partial t^2} = \beta^2 \frac{\partial^2 u}{\partial x^2}$. If one looks up the derivation of the 1-dimensional wave equation on the web, it appears that there is no standard method for doing it. A most interesting derivation (especially for physicists) is given in a video by an MIT physics professor, Walter Lewin. The method below is not nearly as interesting, but is one found in some of the standard mathematical texts.

A thin elastic string is stretched taut between two points. It may be given an initial displacement and/or an initial velocity; for example, by "plucking" the string or hitting it with a piano key. Let $u(x, t)$ denote the vertical displacement of the string at position x at time $t > 0$. The displacement is assumed small enough so that movement only occurs in the vertical direction. The vector $\vec{T}(x, t)$ denotes the tension and it always acts in the tangential direction along the length of the string.

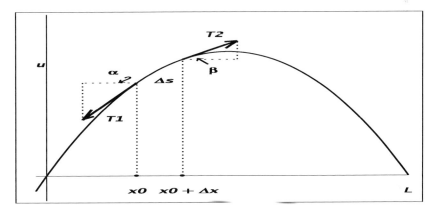

Since we are assuming that there is no movement in the horizontal direction, the horizontal forces are equal; that is,

$$|\vec{T}_1| \cos(\alpha) = |\vec{T}_2| \cos(\beta) = T = \text{ constant}, \qquad (8.10)$$

where $\vec{T}_1$ denotes the tension vector at $x = x_0$ and $\vec{T}_2$ is the tension vector at $x = x_0 + \Delta x$. The notation $|\vec{T}_i|$ denotes the length of the vector $\vec{T}_i$.

For a small segment of the string of length Δs, lying above the interval $[x_0, x_0 + \Delta x]$, Newton's Law implies that the mass times the acceleration of the segment is equal to the sum of the vertical forces:

$$\text{mass} \times \text{acceleration} = (\rho \Delta s) u_{tt}(\bar{x}, t) = |\vec{T}_2| \sin(\beta) - |\vec{T}_1| \sin(\alpha), \qquad (8.11)$$

where ρ is the mass per unit length in the segment and $\bar{x}$ is a value of x between x_0 and $x_0 + \Delta x$. Note that if the vertical displacement of the string is small, $\Delta s \approx \Delta x$. Now divide each term in equation (8.11) by the corresponding term in equation (8.10):

$$\frac{|\vec{T}_2| \sin(\beta)}{|\vec{T}_2| \cos(\beta)} - \frac{|\vec{T}_1| \sin(\alpha)}{|\vec{T}_1| \cos(\alpha)} = \frac{\rho \Delta s}{T} u_{tt}(\bar{x}, t),$$

$$\tan(\beta) - \tan(\alpha) = \frac{\rho \Delta s}{T} u_{tt}(\bar{x}, t).$$

Using the fact that $\tan(\alpha)$ and $\tan(\beta)$ are the slopes of the string (in the x-direction) and that $\Delta s \approx \Delta x$, we can write

$$\frac{T}{\rho} \left[\frac{u_x(x_0 + \Delta x, t) - u_x(x_0, t)}{\Delta x} \right] \approx u_{tt}(\bar{x}, t).$$

Taking the limit as $\Delta x \to 0$, we have $\frac{T}{\rho} u_{xx}(x,t) = u_{tt}(x,t)$ which is the 1-dimensional wave equation. If this equation is written in the form $u_{tt} - \frac{T}{\rho} u_{xx} = 0$, it can be seen that this is a hyperbolic equation with $A = 1$, $C = -\frac{T}{\rho}$, $B = 0$, and therefore $B^2 - AC = \frac{T}{\rho} > 0$.

Note: if we write $\beta^2 = \frac{T}{\rho}$, T is a force having dimension $kg \cdot m/sec^2$, and the dimension of the density ρ is kg/m; therefore, the constant β^2 has dimension m^2/sec^2. This means that β can be thought of as a *velocity* with which a small disturbance moves along the string.

Initial Conditions: The wave equation is second order in t, so there must be two initial conditions specified. These will be given in the form of an initial position function $u(x, 0) = f(x)$ and an initial velocity function $u_t(x, 0) = g(x)$. Both f and g must be at least piecewise continuous on the interval $[0, L]$.

Boundary Conditions: We will first assume that the two ends of the string are fixed for all t; that is, $u(0, t) = u(L, t) = 0$ for all $t > 0$.

8.3.1 Solution of the Wave Equation by Separation of Variables.
Let $u(x, t) = X(x)T(t)$. Then, substituting into the equation $u_{tt} = \beta^2 u_{xx}$,

$$XT'' = \beta^2 X''T.$$

Dividing by $\beta^2 XT$:

$$\frac{XT''}{\beta^2 XT} = \frac{\beta^2 X''T}{\beta^2 XT} \Rightarrow \frac{T''}{\beta^2 T} = \frac{X''}{X}.$$

To have a function of t identical to a function of x implies that they must both equal the same constant. Therefore we can write

$$\frac{T''}{\beta^2 T} = \frac{X''}{X} = -\lambda,$$

and this results in the two ordinary differential equations $\begin{cases} X'' + \lambda X = 0 \\ T'' + \lambda \beta^2 T = 0. \end{cases}$

The Sturm-Liouville equation for X is exactly the same as in the case of the first heat equation, with the same homogeneous boundary conditions:

$$X'' + \lambda X = 0, \quad X(0) = X(L) = 0;$$

therefore, we know that the eigenvalues and corresponding eigenfunctions are $\lambda_n = \frac{n^2 \pi^2}{L^2}$ and $X_n(x) = c_n \sin\left(\frac{n\pi x}{L}\right)$, $n = 1, 2, \ldots$.

To find the corresponding function $T_n(t)$, note that the second-order equation

$$T_n'' + \frac{n^2 \pi^2 \beta^2}{L^2} T_n = 0$$

8.3 The 1-dimensional Wave Equation

has characteristic polynomial $r^2 + \left(\frac{n\pi\beta}{L}\right)^2 = 0$, with complex roots $r = \pm\frac{n\pi\beta}{L}\iota$. Thus the general solution is

$$T_n(t) = c_1 \cos\left(\frac{n\pi\beta}{L}t\right) + c_2 \sin\left(\frac{n\pi\beta}{L}t\right).$$

The general solution of the wave equation can now be written in the form

$$u(x,t) = \sum_{n=1}^{\infty} c_n X_n(x) T_n(t) = \sum_{n=1}^{\infty} \sin\left(\frac{n\pi x}{L}\right)\left[A_n \cos\left(\frac{n\pi\beta}{L}t\right) + B_n \sin\left(\frac{n\pi\beta}{L}t\right)\right],$$

with $A_n = c_n c_1$ and $B_n = c_n c_2$.

The first initial condition can be used to determine the constants A_n. When $t = 0$, $u(x,0) = \sum_{n=1}^{\infty} \sin\left(\frac{n\pi x}{L}\right) A_n$ can be seen to be a Fourier sine series for $f(x)$; therefore, $A_n = \frac{2}{L} \int_0^L f(x) \sin\left(\frac{n\pi x}{L}\right) dx$. Omitting the hard analysis involved in justifying term by term differentiation of the infinite series, the partial derivative of u with respect to t can be shown to be equal to

$$u_t(x,t) = \sum_{n=1}^{\infty} \sin\left(\frac{n\pi x}{L}\right)\left(\frac{n\pi\beta}{L}\right)\left[-A_n \sin\left(\frac{n\pi\beta}{L}t\right) + B_n \cos\left(\frac{n\pi\beta}{L}t\right)\right].$$

Therefore,

$$u_t(x,0) = \sum_{n=1}^{\infty} \sin\left(\frac{n\pi x}{L}\right)\left(\frac{n\pi\beta}{L}B_n\right) \equiv g(x)$$

is a Fourier sine series for $g(x)$. This means that $\frac{B_n n\pi\beta}{L} = \frac{2}{L} \int_0^L g(x) \sin\left(\frac{n\pi x}{L}\right) dx$ and $B_n = \frac{2}{n\pi\beta} \int_0^L g(x) \sin\left(\frac{n\pi x}{L}\right) dx$. The complete solution of the 1-dimensional wave equation can now be written as:

$$u(x,t) = \sum_{n=1}^{\infty} \sin\left(\frac{n\pi x}{L}\right)\left[A_n \cos\left(\frac{n\pi\beta}{L}t\right) + B_n \sin\left(\frac{n\pi\beta}{L}t\right)\right],$$

$$A_n = \frac{2}{L} \int_0^L f(x) \sin\left(\frac{n\pi x}{L}\right) dx, \qquad (8.12)$$

$$B_n = \frac{2}{n\pi\beta} \int_0^L g(x) \sin\left(\frac{n\pi x}{L}\right) dx.$$

Example 8.3.1. *Assume a guitar string 0.6m long is fastened at $x = 0$ and at $x = L = 0.6$. It is plucked in such a way that its initial position is given by*

$$f(x) = \begin{cases} \frac{1}{3}x & 0 < x < 0.3 \\ -\frac{1}{3}(x - 0.6) & 0.3 < x < 0.6 \end{cases}$$

Find the displacement $u(x,t)$ in the string for $t > 0$, $0 \le x \le 0.6$, as a function of $\beta^2 = T/\rho$.

Solution. From what we did above, the general solution is

$$u(x,t) = \sum_{n=1}^{\infty} \sin\left(\frac{n\pi x}{0.6}\right)\left[A_n \cos\left(\frac{n\pi\beta}{0.6}t\right) + B_n \sin\left(\frac{n\pi\beta}{0.6}t\right)\right].$$

Note that β determines the period of oscillation of u; that is, all of the functions $T_n(t)$ are periodic of period $P = \frac{2\pi}{\pi\beta/0.6} = \frac{1.2}{\beta}$. The larger β is, the shorter the period and the higher the frequency $\omega = 2\pi/P$. It is also clear that the value of β can be made larger by increasing the tension or decreasing the density of the string.

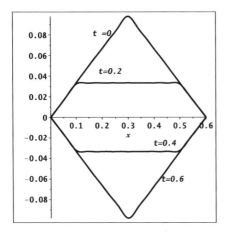
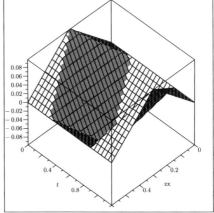

Figure 8.4. Position of the string at times $t = 0, 0.2, 0.4, 0.6$.

Figure 8.5. A 3-dimensional plot of the function $u(x, t)$ over 1 period.

In this example the coefficients B_n are all zero since the initial velocity $g(x)$ of the string is assumed to be zero. The values of the constants A_n are given by

$$A_n = \frac{2}{0.6} \int_0^{0.6} f(x) \sin\left(\frac{n\pi x}{0.6}\right) dx.$$

If we assume $\beta^2 = 1$, the string will oscillate with period equal 1.2 seconds. Figure 8.4 shows the position of the string at equally spaced time intervals over a half-period. In Figure 8.5 a 3-dimensional plot of $u(x, t)$ has also been made to show how the string oscillates over one full period. ■

In the next example we will assume a *small* amount of damping due to bending friction in the string, and see that the motion of the string looks more realistic in this case.

The Damped Wave Equation. Consider the equation

$$u_{tt} + 2cu_t = \beta^2 u_{xx}$$

where c is a small positive constant. The term $2cu_t$ represents a damping force proportional to the velocity u_t. Note that this is still a hyperbolic PDE since the second order derivatives have not changed. For simplicity, we will assume that the length of the string is $L = \pi$, and the constant $\beta^2 = 1$.

If we let $u(x, t) = X(x)T(t)$, then, with $\beta = 1$,

$$\frac{XT'' + 2cXT'}{XT} = \frac{X''T}{XT} \Rightarrow \frac{T'' + 2cT'}{T} = \frac{X''}{X} = -\lambda.$$

This gives us the two ordinary differential equations

$$X'' + \lambda X = 0, \quad T'' + 2cT' + \lambda T = 0.$$

8.3 The 1-dimensional Wave Equation

If we assume the two ends of the string are fixed, so $u(0,t) = u(\pi,t) = 0$ for all $t > 0$, then we again have the Sturm-Liouville problem

$$X'' + \lambda X = 0, \quad X(0) = X(\pi) = 0$$

with eigenvalues $\lambda_n = \dfrac{n^2\pi^2}{\pi^2} = n^2$ and corresponding eigenfunctions $X_n(x) = b_n \sin(nx)$. The equation for $T_n(t)$ is

$$T_n'' + 2cT_n' + n^2 T_n = 0,$$

with characteristic polynomial $r^2 + 2cr + n^2 = 0$. We will assume that $c < 1$, and then

$$r = \frac{-2c \pm \sqrt{4c^2 - 4n^2}}{2} = -c \pm \sqrt{c^2 - n^2}.$$

The discriminant $c^2 - n^2$ is negative for any positive integer $n \geq 1$, so the roots of the characteristic polynomial are complex and the general solution is

$$T_n(t) = a_n e^{-ct} \cos(\sqrt{n^2 - c^2}\, t) + b_n e^{-ct} \sin(\sqrt{n^2 - c^2}\, t).$$

This means that

$$u(x,t) = \sum_{n=1}^{\infty} \sin(nx) e^{-ct} \left[a_n \cos(\sqrt{n^2 - c^2}\, t) + b_n \sin(\sqrt{n^2 - c^2}\, t) \right].$$

If the initial conditions are $u(x,0) = f(x), \; u_t(x,0) = g(x)$, then

$$u(x,0) = \sum_{n=1}^{\infty} a_n \sin(nx) = f(x)$$

implies that $a_n = \dfrac{2}{\pi} \int_0^\pi f(x) \sin(nx) dx$. Again, omitting the details on term-by-term differentiation of the series, we can write the partial of u with respect to t as

$$u_t(x,t) = \sum_{n=1}^{\infty} \sin(nx)\left[-ce^{-ct}\left(a_n \cos(\sqrt{n^2-c^2}\,t) + b_n \sin(\sqrt{n^2-c^2}\,t)\right)\right]$$
$$+ \sin(nx)\left[e^{-ct}\sqrt{n^2-c^2}\left(-a_n \sin(\sqrt{n^2-c^2}\,t) + b_n \cos(\sqrt{n^2-c^2}\,t)\right)\right];$$

therefore,

$$u_t(x,0) = \sum_{n=1}^{\infty} \sin(nx)\left[-ca_n + \sqrt{n^2 - c^2}\, b_n\right] \equiv g(x).$$

This is a Fourier sine series for $g(x)$ with coefficients

$$-ca_n + \sqrt{n^2 - c^2}\, b_n = \frac{2}{\pi} \int_0^\pi g(x) \sin(nx) dx;$$

therefore,

$$b_n = \frac{1}{\sqrt{n^2 - c^2}} \left(ca_n + \frac{2}{\pi} \int_0^\pi g(x) \sin(nx) dx \right).$$

The solution of $u_{tt} + 2cu_t = u_{xx}$, $u(0,t) = u(\pi,t) = 0$, for all $t > 0$ with the given initial conditions is

$$u(x,t) = \sum_{n=1}^{\infty} \sin(nx)e^{-ct}\left[a_n \cos(\sqrt{n^2-c^2}\,t) + b_n \sin(\sqrt{n^2-c^2}\,t)\right], \quad (8.13)$$

$$a_n = \frac{2}{\pi}\int_0^\pi f(x)\sin(nx)dx, \quad b_n = \frac{1}{\sqrt{n^2-c^2}}\left(ca_n + \frac{2}{\pi}\int_0^\pi g(x)\sin(nx)dx\right)$$

Example 8.3.2. We will redo the "plucked" string problem from the previous subsection, letting

$$f(x) = \begin{cases} \dfrac{2}{\pi}x & 0 \leq x < \dfrac{\pi}{2} \\ -\dfrac{2}{\pi}(x-\pi) & \dfrac{\pi}{2} \leq x \leq \pi \end{cases}$$

and $g(x) \equiv 0$, and assuming a damping constant $c = 0.4$.

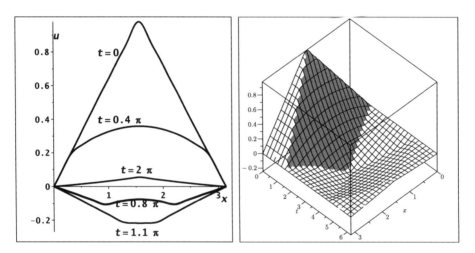

Figure 8.6. Position of the string at five different times.

Figure 8.7. A 3-dimensional plot of the function $u(x,t)$ over two oscillations.

A graph of $u(x,t)$ at $t = 0, 0.4\pi, 0.8\pi, 1.1\pi$, and 2π, and a 3-dimensional plot of $u(x,t)$ for $0 \leq t \leq 2\pi$ are shown in Figures 8.6 and 8.7.

It can be seen that with c equal to 0.4, the string moves down and back once as t goes from 0 to approximately 2π. However the solution is no longer a sum of exact harmonics; that is, the sines and cosines in successive terms do not have frequencies that are exact integral multiples of a fundamental frequency. In addition, the oscillations in the string damp out to zero as $t \to \infty$. □

The Wave Equation in Cylindrical Coordinates. Simple 3-dimensional vibration problems are often modelled by the equation

$$u_{tt} = b^2(u_{xx} + u_{yy} + u_{zz}) \equiv b^2\nabla^2 u,$$

8.3 The 1-dimensional Wave Equation

where ∇^2 is called the **laplacian operator**[1]. If the object being modelled is more easily described in terms of cylindrical coordinates, the laplacian itself can be converted into cylindrical coordinates. In one of the exercises at the end of this section you will be asked to show that the laplacian, expressed in cylindrical coordinates, has the form

$$\nabla^2 u = u_{xx} + u_{yy} + u_{zz} = U_{rr} + \frac{1}{r}U_r + \frac{1}{r^2}U_{\theta\theta} + U_{zz},$$

where $U(r, \theta, z) \equiv u(x(r, \theta), y(r, \theta), z) = u(r\sin(\theta), r\cos(\theta), z)$.

This formula should make it easy to see where the PDE $u_{tt} = u_{rr} + \frac{1}{r}u_r$, used to model the radially symmetric drumhead (at the end of Chapter 7), comes from. You should now be able to solve a problem of this type by separation of variables, and to find formulas for the coefficients a_n and b_n in terms of radially symmetric initial displacement and/or velocity functions. One of the exercises at the end of this section asks you to do this.

8.3.2 D'Alembert's Solution of the Wave Equation on an Infinite Interval.
To solve the wave equation $u_{tt} = \beta^2 u_{xx}$ on the infinite line $-\infty < x < \infty, t > 0$, it is easier to make the change of independent variables

$$r = x + \beta t, \quad s = x - \beta t,$$

and let $u(x, t) = w(r, s) = w(r(x, t), s(x, t))$.

To find u_{xx} and u_{tt} in terms of the new variables r and s, we will need the four derivatives

$$\frac{\partial r}{\partial x} = 1, \quad \frac{\partial s}{\partial x} = 1, \quad \frac{\partial r}{\partial t} = \beta, \quad \frac{\partial s}{\partial t} = -\beta.$$

Then, using the chain rule for differentiation,

$$u_t \equiv \frac{\partial u}{\partial t} = \frac{\partial w}{\partial r}\frac{\partial r}{\partial t} + \frac{\partial w}{\partial s}\frac{\partial s}{\partial t} = \beta(w_r - w_s)$$

and

$$u_{tt} = \frac{\partial}{\partial t}(\beta(w_r - w_s)) = \beta(w_{rr}\frac{\partial r}{\partial t} + w_{rs}\frac{\partial s}{\partial t} - w_{sr}\frac{\partial r}{\partial t} - w_{ss}\frac{\partial s}{\partial t})$$
$$= \beta(\beta w_{rr} - \beta w_{rs} - \beta w_{sr} + \beta w_{ss}) = \beta^2(w_{rr} - 2w_{rs} + w_{ss}).$$

Differentiating with respect to x,

$$u_x \equiv \frac{\partial u}{\partial x} = \frac{\partial w}{\partial r}\frac{\partial r}{\partial x} + \frac{\partial w}{\partial s}\frac{\partial s}{\partial x} = w_r + w_s$$

and

$$u_{xx} = \frac{\partial}{\partial x}(w_r + w_s) = w_{rr}\frac{\partial r}{\partial x} + w_{rs}\frac{\partial s}{\partial x} + w_{sr}\frac{\partial r}{\partial x} + w_{ss}\frac{\partial s}{\partial x} = w_{rr} + 2w_{rs} + w_{ss}.$$

If the formulas for u_{tt} and u_{xx} are substituted into the wave equation $u_{tt} = \beta^2 u_{xx}$,

$$\beta^2(w_{rr} - 2w_{rs} + w_{ss}) = \beta^2(w_{rr} + 2w_{rs} + w_{ss});$$

and cancelling like terms on each side gives $4\beta^2 w_{rs} = 0$. Since β is a nonzero constant, this implies that

$$w_{rs}(r, s) \equiv 0.$$

This is easily solved by integrating twice, first with respect to s:

$$w_r(r, s) = p(r),$$

---
[1] See p. 14 of *Partial Differential Equations* by W. A. Strauss, John Wiley & Sons, 1992.

where p is an arbitrary function of r (since for any such function $\frac{\partial p}{\partial s} = 0$). A second integration with respect to r gives:

$$w(r, s) = \int p(r) dr + G(s),$$

where G is an arbitrary function of s. Now we can write $w(r, s) = F(r) + G(s)$, and therefore

$$u(x, t) = F(x + \beta t) + G(x - \beta t)$$

where F and G can be any functions of a **single** variable. This is the **general solution** of the wave equation on the infinite line.

Now suppose we are given initial conditions $u(x, 0) = f(x)$, $u_t(x, 0) = g(x)$. The partial of u with respect to t is

$$u_t(x, t) = \frac{\partial}{\partial t}(F(x + \beta t) + G(x - \beta t)) = F'(x + \beta t) \cdot \beta + G'(x - \beta t) \cdot (-\beta).$$

Remember that F and G are functions of *one* variable. At $t = 0$ we have

$$u(x, 0) = F(x) + G(x) \equiv f(x). \tag{8.14}$$

The initial velocity condition gives

$$u_t(x, 0) = \beta F'(x) - \beta G'(x) = \beta \frac{d}{dx}(F(x) - G(x)) \equiv g(x).$$

Integrating the last equality,

$$F(x) - G(x) = \frac{1}{\beta} \int_0^x g(\tau) d\tau + C. \tag{8.15}$$

If we add equations (8.14) and (8.15) we get

$$2F(x) = f(x) + \frac{1}{\beta} \int_0^x g(\tau) d\tau + C \Rightarrow F(x) = \frac{1}{2}\left(f(x) + \frac{1}{\beta} \int_0^x g(\tau) d\tau + C\right).$$

Subtracting equations (8.14) and (8.15) gives

$$2G(x) = f(x) - \frac{1}{\beta} \int_0^x g(\tau) d\tau - C \Rightarrow G(x) = \frac{1}{2}\left(f(x) - \frac{1}{\beta} \int_0^x g(\tau) d\tau - C\right).$$

Using these formulas for $F(x)$ and $G(x)$, the function u can be written as

$$u(x, t) = F(x + \beta t) + G(x - \beta t)$$
$$= \frac{1}{2}\left(f(x + \beta t) + \frac{1}{\beta} \int_0^{x+\beta t} g(\tau) d\tau + C\right) + \frac{1}{2}\left(f(x - \beta t) - \frac{1}{\beta} \int_0^{x-\beta t} g(\tau) d\tau - C\right),$$

and the exact *analytic* solution of the wave equation is

$$u(x, t) = \frac{1}{2}(f(x + \beta t) + f(x - \beta t)) + \frac{1}{2\beta} \int_{x-\beta t}^{x+\beta t} g(\tau) d\tau. \tag{8.16}$$

8.3 The 1-dimensional Wave Equation

Example 8.3.3. *To study the behavior of a wave on an infinite line, we will assume the initial position function $u(x,0)$ is given by*

$$u(x,0) \equiv f(x) = \begin{cases} 1 & -1 \leq x \leq 1 \\ 0 & \text{otherwise} \end{cases}$$

and the initial velocity $g(x)$ is identically zero. Let the constant $\beta = 1$.

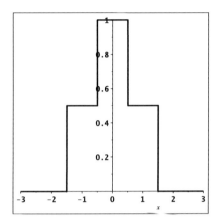
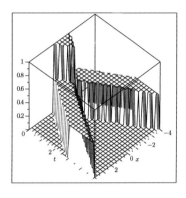

Figure 8.8. Graph of $u(x,t)$ at $t = 0.5$, and 3-dimensional plot of $u(x,t)$

In Figure 8.8, the position of the wave at time $t = 0.5$ is shown on the left, and the 3-dimensional plot shows the wave moving out along the infinite line. At any time $\bar{t}$ the function $u(x, \bar{t})$ is equal to $\frac{f(x+\bar{t})+f(x-\bar{t})}{2}$. This can be graphed by moving the graph of f $\bar{t}$ units to the right and $\bar{t}$ units to the left, adding the two functions, and dividing by two. When $t > 1$ the two graphs are completely separated, and the graph of u consists of two copies of $f(t)/2$ moving to the right and left with speed determined by β. ∎

Exercises 8.3. *The starred exercises require a computer.*

1. ⋆ Solve the wave equation $u_{tt} = u_{xx}$ for a piano string, 1 meter long, which is initially at equilibrium $u(x,0) = f(x) \equiv 0$ and is struck by hitting a piano key, giving the string an initial velocity of

$$u_t(x,0) = g(x) = \begin{cases} 10x^2 - 10x + 2.4 & \text{if } 0.4 \leq x \leq 0.6 \\ 0 & \text{otherwise.} \end{cases}$$

Assume $L = 1.0$, $u(0,t) = u(1,t) = 0$ for all $t > 0$.

2. ⋆ Solve the wave equation $u_{tt} = \beta^2 u_{xx}$ with initial displacement $u(x,0) = f(x) = 8x^4 - 16x^3 + 10x^2 - 2x$. The initial velocity $u_t(x,0) = g(x) \equiv 0$. Assume $L = 1.0$, $\beta = 1.0$. Plot $u(x,t)$ at several different times t.

3. Write a formula for the solution of the damped wave equation

$$u_{tt} + 2cu_t = \beta^2 u_{xx}$$

on an interval of length L. Assume the constant $c < \frac{\pi\beta}{L}$. The constants c, β, and L are otherwise arbitrary. Use the boundary conditions $u(0,t) = u(L,t) = 0$ for $t > 0$.

4. ★ Using the series solution derived in Exercise 3, find the solution of
$$u_{tt} + 0.3u_t = u_{xx}.$$
Use the initial functions $f(x)$ and $g(x)$ defined in Exercise 1. Plot $u(x,t)$ at several different times t.

5. Solve the equation $u_{tt} = \beta^2 u_{xx}$ on the infinite line, given the initial condition $u(x,0) = \sin(x)$, $u_t(x,0) \equiv 0$. <u>Describe</u> the behavior of $u(x,t)$ for $t > 0$. This is a model of a standing wave with no boundaries. How does the value of β affect the wave?

6. Find the analytic solution of
$$u_{tt} = \beta^2 u_{xx}, u(x,0) = f(x) = e^{-x^2}, u_t(x,0) = g(x) \equiv 0, -\infty < x < \infty.$$

7. Find the analytic solution of
$$u_{tt} = \beta^2 u_{xx}, u(x,0) = f(x) \equiv 0, u_t(x,0) = g(x) = \frac{1}{1+x^2}, -\infty < x < \infty.$$
Assume $\beta = \frac{1}{2}$, and sketch the solution at several values of t. What happens to the solution as $t \to \infty$?

8. In this problem you will be led through the conversion of the laplacian from rectangular to cylindrical coordinates $((x,y,z) \to (r\cos(\theta), r\sin(\theta), z))$.

    (a) Let
    $$r = \sqrt{x^2 + y^2}, \ \theta = \tan^{-1}\left(\frac{y}{x}\right), \ z = z,$$
    and show that
    $$\frac{\partial r}{\partial x} = \cos(\theta), \ \frac{\partial r}{\partial y} = \sin(\theta), \ \frac{\partial \theta}{\partial x} = \frac{-\sin(\theta)}{r}, \ \frac{\partial \theta}{\partial y} = \frac{\cos(\theta)}{r}.$$

    (b) Make the change of variables $u(x,y,z) \equiv U(r(x,y), \theta(x,y), z)$ and use the chain rule to write
    $$\frac{\partial u}{\partial x} = \frac{\partial U}{\partial r}\frac{\partial r}{\partial x} + \frac{\partial U}{\partial \theta}\frac{\partial \theta}{\partial x} = U_r \cos(\theta) - U_\theta \frac{\sin(\theta)}{r}.$$
    Use the chain rule once more to write
    $$u_{xx} = \frac{\partial}{\partial x}\left(U_r \cos(\theta) - U_\theta \frac{\sin(\theta)}{r}\right)$$
    $$= \frac{\partial}{\partial r}\left(U_r \cos(\theta) - U_\theta \frac{\sin(\theta)}{r}\right)\frac{\partial r}{\partial x} + \frac{\partial}{\partial \theta}\left(U_r \cos(\theta) - U_\theta \frac{\sin(\theta)}{r}\right)\frac{\partial \theta}{\partial x}.$$
    Work this out very carefully, remembering to use the product rule for differentiation. You should find (after eliminating two equal terms) that
    $$u_{xx} = U_{rr}\cos^2(\theta) - 2U_{r\theta}\frac{\sin(\theta)\cos(\theta)}{r} + U_{\theta\theta}\frac{\sin^2(\theta)}{r^2} + U_r \frac{\sin^2(\theta)}{r} + 2U_\theta \frac{\sin(\theta)\cos(\theta)}{r^2}.$$
    Assume that the function U is nice enough so that the mixed partial derivatives $U_{r\theta}$ and $U_{\theta r}$ are equal.

    (c) Use a completely analogous procedure to write u_{yy} in terms of r and θ.

    (d) Since x and y are both independent of z, the laplacian is obtained by adding the term u_{zz} to the sum of u_{xx} and u_{yy}. This should give you the final result
    $$u_{xx} + u_{yy} + u_{zz} = U_{rr} + \frac{1}{r}U_r + \frac{1}{r^2}U_{\theta\theta} + U_{zz}.$$

9. * A circular drumhead of radius 1 is initially at rest. It is struck in the center, giving it an initial velocity

$$g(r) = \begin{cases} -0.1 & 0 \leq r < 0.05 \\ 0 & 0.05 \leq r \leq 1 \end{cases}$$

along each radius. Assume the displacement $u(r, t)$ satisfies $u_{tt} = u_{rr} + \frac{1}{r}u_r$, $u(0, t)$ finite, $u(1, t) = 1$, $t \geq 0$, $u(r, 0) = f(r) \equiv 0$, $u_t(r, 0) = g(r)$, $0 \leq r \leq 1$. Find a series solution for $u(r, t)$ and plot the displacement at times $t = 0.1, 2.0, 3.9$. The Maple program at the end of Chapter 7 can be used by including the definition of $g(r)$, the formula for b_n, and the term $b_n \sin(z_n t)$ in the series.

8.4 Numerical Solution of Parabolic and Hyperbolic Equations

If it is not possible to solve a partial differential equation by separation of variables, or some other analytic method, the solution can often be approximated by numerical methods. The particular methods used depend on the type of the equation, and the methods described in this section are used for parabolic and hyperbolic PDEs. However, the equation does not need to be linear in the lower-order terms.

Numerical Approximations to Partial Derivatives. One way to solve partial differential equations numerically is by partitioning both the x-axis and the t-axis, and approximating the partial derivatives of the unknown function in terms of *differences* of function values at nearby points in the resulting 2-dimensional grid. The difference formulas that we will use are described below.

For a function f of a single variable x, we can approximate the first derivative by

$$f'(x) \approx \frac{f(x + \Delta x) - f(x)}{\Delta x}.$$

This is called a **forward difference approximation**.

If a function $f(x)$ is analytic at some point x, the Taylor series for f at x converges and has the form

$$f(x + \Delta x) = f(x) + f'(x)\Delta x + \frac{f''(x)}{2!}(\Delta x)^2 + \cdots + \frac{f^{(n)}(x)}{n!}(\Delta x)^n + \cdots.$$

Subtracting $f(x)$ from both sides and dividing by Δx,

$$\frac{f(x + \Delta x) - f(x)}{\Delta x} = f'(x) + \frac{f''(x)}{2!}\Delta x + \mathcal{O}((\Delta x)^2).$$

The term $\mathcal{O}((\Delta x)^n)$ is read as "on the order of $(\Delta x)^n$"; that is, it becomes approximately equal to a constant times $(\Delta x)^n$ as $\Delta x \to 0$. Using this notation, we can say that the forward difference approximation to $f'(x)$ has error $\mathcal{O}(\Delta x)$, and if Δx is small the error will be small.

A better approximation to $f'(x)$ can be obtained by using what is called a **central difference approximation**:

$$f'(x) \approx \frac{f(x + \Delta x) - f(x - \Delta x)}{2\Delta x}.$$

This is obtained by subtracting the Taylor series

$$f(x - \Delta x) = f(x) - f'(x)\Delta x + \frac{f''(x)}{2!}(\Delta x)^2 - \cdots + (-1)^n \frac{f^{(n)}(x)}{n!}(\Delta x)^n + \cdots$$

from the Taylor series for $f(x + \Delta x)$. This gives

$$f(x + \Delta x) - f(x - \Delta x) = 2f'(x)\Delta x + 2\frac{f'''(x)}{3!}(\Delta x)^3 + \dots,$$

and dividing by $2\Delta x$,

$$\frac{f(x + \Delta x) - f(x - \Delta x)}{2\Delta x} = f'(x) + \frac{f'''(x)}{3!}(\Delta x)^2 + \mathcal{O}((\Delta x)^4).$$

Solving this for $f'(x)$, we see that the central difference approximation has error $\mathcal{O}((\Delta x)^2)$. As Δx gets small, this gets small much faster than the error in the forward difference approximation.

The second derivative of a function of one variable can be approximated by

$$f''(x) \approx \frac{f(x + \Delta x) - 2f(x) + f(x - \Delta x)}{(\Delta x)^2}.$$

This is called the **central difference approximation to** $f''(x)$. It is obtained by adding the Taylor series for $f(x + \Delta x)$ and $f(x - \Delta x)$:

$$f(x + \Delta x) + f(x - \Delta x) = 2f(x) + 2\frac{f''(x)}{2}(\Delta x)^2 + 2\frac{f^{(4)}(x)}{24}(\Delta x)^4 + \cdots$$

from which we have

$$\frac{f(x + \Delta x) - 2f(x) + f(x - \Delta x)}{(\Delta x)^2} = f''(x) + \frac{f^{(4)}(x)}{12}(\Delta x)^2 + \mathcal{O}((\Delta x)^4),$$

so that the central difference approximation for $f''(x)$ can be seen to have an error $\mathcal{O}((\Delta x)^2)$.

The following example will give you an idea of how accurately forward and central differences can approximate derivatives,

Example 8.4.1. *Consider the function $f(x) = \ln(x)$. We know from calculus that for any $x > 0$, $f'(x) = \frac{1}{x}$ and $f''(x) = -\frac{1}{x^2}$; therefore, at $x = 2$ the exact values of these derivatives are $f'(2) = 0.5$ and $f''(2) = -0.25$. Using the forward difference approximation, with $\Delta x = 0.1$, the first derivative is approximated by*

$$f'(2) \approx \frac{f(2 + 0.1) - f(2)}{0.1} = \frac{\ln(2.1) - \ln(2)}{0.1} \approx 0.4879.$$

This is not a very good approximation to $f'(2) = 0.5000$. However, if we use $\Delta x = 0.001$, it becomes

$$f'(2) \approx \frac{f(2.001) - f(2)}{0.001} \approx 0.49988$$

which is much better. Using the central difference approximation, with $\Delta x = 0.1$,

$$f'(2) \approx \frac{f(2.1) - f(1.9)}{2(0.1)} \approx 0.500417$$

which is nearly as accurate as the forward difference approximation with $\Delta x = 0.001$. This means that a careful choice of difference formulas may be used to increase the accuracy in a computation.

8.4 Numerical Solution of Parabolic and Hyperbolic Equations

To examine the accuracy of the central difference approximation to the second derivative of $\ln(x)$ at $x = 2$, using the larger value $\Delta x = 0.1$,

$$f''(2) \approx \frac{\ln(2.1) - 2\ln(2) + \ln(1.9)}{(0.1)^2} \approx -0.25031.$$

This is very close to the exact value $-\frac{1}{4}$. ∎

One thing to be careful of when approximating derivatives numerically: if you take Δx too small, round-off error can quickly give meaningless results. Consider, for example, the computation of $\frac{f(2+\Delta x) - f(2-\Delta x)}{2\Delta x}$ when Δx is **very** small. Depending on the number of significant digits stored by the calculator or computer, if Δx is so small that $f(2+\Delta x)$ and $f(2-\Delta x)$ are the same to that many digits, you can end up with the meaningless value $f'(2) = 0$.

Numerical Solution of the Heat Equation. Consider the 1-dimensional heat equation $u_t = au_{xx}$, $u(x,0) = f(x)$ for $0 < x < L$. Make a 2-dimensional grid with x on the horizontal axis and t on the vertical axis, as shown below. If you want to find an approximation to $u(x,t)$ for $0 \le x \le L$ and $0 \le t \le T$, then let $\Delta x = \frac{L}{N}$ and $\Delta t = \frac{T}{M}$ for some large integers N and M. This partitions the two axes to give a rectangular grid. The points on the grid can be indexed by $0 \le i \le N$ along the x-axis and $0 \le j \le M$ along the vertical axis, where we can write $x_i = i \cdot \Delta x$ and $t_j = j \cdot \Delta t$.

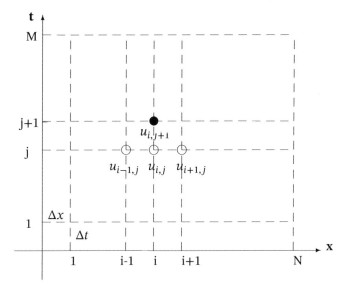

Using a forward difference approximation for $u_t(x,t)$ and a central difference approximation for $u_{xx}(x,t)$, the difference approximation to the partial differential equation $u_t = au_{xx}$ is

$$\frac{u(x, t+\Delta t) - u(x,t)}{\Delta t} = a\left(\frac{u(x+\Delta x, t) - 2u(x,t) + u(x-\Delta x, t)}{(\Delta x)^2}\right). \quad (8.17)$$

The reason for NOT using a central difference approximation to u_t is that we have only the initial condition $u(x,0) = f(x)$ with which to begin our calculation. The central difference formula requires knowing u at two previous values of t in order to compute u

at a third value of t. There is an alternative method, called the Crank-Nicolson scheme, which assumes that $\frac{u(x,t+\Delta t)-u(x,t)}{\Delta t}$ represents a central difference centered at time $t + \frac{\Delta t}{2}$, and writes the right-hand side of the equation as an average of u_{xx} at times t and $t+\Delta t$. It is complicated by the fact that at each time step it is necessary to solve a system of linear equations for u at each value of x.

From equation (8.17), we obtain the following approximate formula for values of u at time $t + \Delta t$ entirely in terms of values of u at the previous time t:

$$u(x, t + \Delta t) = u(x, t) + \frac{a\Delta t}{(\Delta x)^2} (u(x + \Delta x, t) - 2u(x, t) + u(x - \Delta x, t)). \quad (8.18)$$

If we let $U_{i,j}$ denote the approximation to $u(x_i, t_j) \equiv u(i \cdot \Delta x, j \cdot \Delta t)$, equation (8.18) can be written in the form

$$U_{i,j+1} = U_{i,j} + C(U_{i+1,j} - 2U_{i,j} + U_{i-1,j}), \quad (8.19)$$

where C is the constant $\frac{a\Delta t}{(\Delta x)^2}$. There is a known condition on C, namely that $C \leq \frac{1}{2}$, in order for this method to be *stable*. If $C > \frac{1}{2}$, we will see that round-off errors become magnified as t increases, and the numerical solution will show large meaningless oscillations.

Since the initial function $f(x)$ determines the value of $u(x, 0)$ for any x, the values of $U_{i,0} = f(i \cdot \Delta x)$ can be stored, for $i = 0, 1, \ldots, N$. Before using equation (8.19) to find the values of $U_{i,j+1}$ for any integer $j \geq 0$, the boundary conditions on u must be used to assign values to $U_{0,j}$ and $U_{N,j}$. Note that this means we are able to solve the heat equation with arbitrary (possibly nonconstant) boundary conditions $u(0, t) = T_1(t)$ and $u(L, t) = T_2(t)$. This is not possible to do when solving the equation by separation of variables. Do you see why?

Example 8.4.2. *Solve the heat equation $u_t = 0.5 u_{xx}$, with $u(x, 0) = f(x) = 2x(10 - x)$, holding the boundary temperatures constant at $u(0, t) = 60$ and $u(10, t) = -20$. Use the method of separation of variables, and compare the results to a numerical solution.*

Solution. This problem can be solved by separation of variables, and using the formulas derived in Subsection 8.2.2 it can be shown that the solution is

$$u(x, t) = 60 - 8x + \sum_{n=1}^{\infty} b_n \sin\left(\frac{n\pi x}{10}\right) e^{-\frac{n^2 \pi^2 t}{200}},$$

where

$$b_n = \frac{2}{10} \int_0^{10} (2x(10 - x) - (60 - 8x)) \sin\left(\frac{n\pi x}{10}\right) dx.$$

With 50 terms used in the series, Figure 8.9 shows the temperature in the rod at four different times. Even with 50 terms in the Fourier series, the graph at $t = 0$ is very rough. This is caused by the Gibbs phenomenon and accentuated due to the fact that the initial temperature does not satisfy the boundary conditions.

To solve the problem numerically, N was chosen to be 40, making $\Delta x = 0.25$. To compare the values obtained from the two methods at time $t = 2.0$, three numerical computations were run, using $M = 80, 40,$ and 20. The value of $C = a\Delta t/(\Delta x)^2 = (0.5)(\frac{2}{M})/(0.25)^2$ is less than $\frac{1}{2}$ in the first two cases and greater than $\frac{1}{2}$ when $M = 20$. In the table below the values of the 50-term series solution at $t = 2$, $x = 0, 2, \ldots, 10$

8.4 Numerical Solution of Parabolic and Hyperbolic Equations

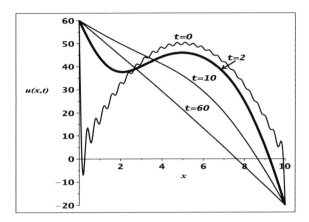

Figure 8.9. Temperature at times $t = 0, 2, 10$ and 60

are shown in the left column. The corresponding numerically computed values are given in the next three columns. With $M = 20$ it is clear that the solution has become unstable. Notice that the values of the temperature at $t = 2$ agree to at least 3 significant digits when $M = 80$.

x	Series Soln.	Numerical Solutions		
		$M = 80$ $C = 0.2$	$M = 40$ $C = 0.4$	$M = 20$ $C = 0.8$
0	60	60	60	60
2	37.666	37.580	37.456	-611750
4	44.282	44.272	44.236	-5373
6	43.912	43.914	43.925	-4841
8	25.081	25.108	25.143	196440
10	-20	-20	-20	-20

∎

Numerical Solution of the Wave Equation. The wave equation can also be solved numerically. Consider the 1-dimensional wave equation $u_{tt} = \beta^2 u_{xx}$, $u(x, 0) = f(x)$, $u_t(x, 0) = g(x)$ for $0 < x < L$, with arbitrary boundary conditions $u(0, t) = T_1(t)$ and $u(L, t) = T_2(t)$. Using a central difference approximation for both $u_{tt}(x, t)$ and $u_{xx}(x, t)$, the partial differential equation is approximated by a difference equation of the form

$$\frac{u(x, t + \Delta t) - 2u(x, t) + u(x, t - \Delta t)}{(\Delta t)^2} = \beta^2 \left(\frac{u(x + \Delta x, t) - 2u(x, t) + u(x - \Delta x, t)}{(\Delta x)^2} \right), \quad (8.20)$$

where β^2 is the positive constant equal to the tension divided by the density.

Notice that in this case both second-order differences are central differences, thus accurate to $\mathcal{O}((\Delta t)^2)$. If a grid is set up exactly like the one for the heat equation, and if we use indices i and j on the x and t axes, (8.20) can be solved for $U_{i,j+1} \approx u(x, t + \Delta t)$

to get

$$U_{i,j+1} = 2U_{i,j} - U_{i,j-1} + C(U_{i+1,j} - 2U_{i,j} + U_{i-1,j}), \qquad (8.21)$$

where C is the constant $\beta^2 \frac{(\Delta t)^2}{(\Delta x)^2}$. For stability of the method it is necessary for the value of C to be less than 1. Note that, unlike the heat equation, $U_{i,j+1}$ depends on values at two previous times t_j and t_{j-1}. This means that in order to get started we need to not only store values of $u(x, 0)$ for all x, but also need to be able to get an approximate value of $u(x, -\Delta t)$ at each value of x. But unlike the heat equation, we are given the time derivative $u_t(x, 0) = g(x)$ at every value of x. The function g can be used to get approximate values for $u(x, -\Delta t)$ by writing

$$g(x) = u_t(x, 0) \approx \frac{u(x, \Delta t) - u(x, -\Delta t)}{2\Delta t} \implies u(x, -\Delta t) \approx u(x, \Delta t) - 2\Delta t g(x).$$

This is also a central difference, so our approximations are all good to $\mathcal{O}((\Delta t)^2)$.

The first step is to store values of the initial function $f(x)$ in $U_{i,0}$. This can be done by a simple loop setting $U_{i,0} = f(i\Delta x)$, for i from 1 to $N-1$. Then the second step uses (8.21) with $u(x, -\Delta t)$ replaced by the approximation $u(x, \Delta t) - 2\Delta t g(x)$. This can be written in the form

$$U_{i,1} = U_{i,0} + \Delta t g(i\Delta x) + 0.5C(U_{i+1,0} - 2U_{i,0} + U_{i-1,0}).$$

For steps $j = 2, 3, \ldots, M$ equation (8.21) can be used exactly as written. At each time step $j = 0, \ldots, M$ it is necessary to first set the boundary values $U_{0,j} = T_1(j \cdot \Delta t)$ and $U_{N,j} = T_2(j \cdot \Delta t)$.

Example 8.4.3. *Solve the problem $u_{tt} = 0.25 u_{xx}$ on the interval $[0, L]$, $L = \pi$, with*

$$u(x, 0) = 0.2 \sin(x), \ u_t(x, 0) = 0,$$

and boundary conditions $u(0, t) = T_1(t) = 0$, $u(\pi, t) = T_2(t) = 0.1 \sin(t)$ for $0 \le t \le 4$. Draw a graph of the function $u(x, t)$ at time $t = 4$.

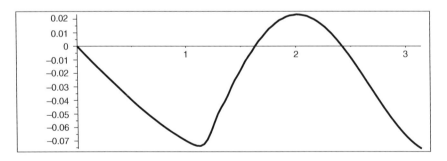

Figure 8.10. Position of the rope at time $t = 4.0$

This might represent the behavior of a rope under tension, where the right-hand end is periodically being moved up and down a small amount. Note that because of the non-constant boundary conditions this problem cannot be solved by separation of variables.

Solution. Be careful to choose values for Δt and Δx so that $\Delta t < \frac{\Delta x}{\beta}$. If we let $N = 100$ and $M = 200$, and run the program out to $T = 10$, we will have $\Delta t = T/M = 0.05$

8.4 Numerical Solution of Parabolic and Hyperbolic Equations

and $\Delta x = L/N \approx 0.0314$. The value of β^2 is 0.25, so $\frac{\Delta x}{\beta} \approx 0.0628$ and the condition is satisfied. Figure 8.10 shows the position of the rope at time $t = 4$. ■

Example 8.4.4. *To demonstrate the accuracy of a numerical solution of the wave equation, this example will compute the numerical solution of the damped wave equation*

$$u_{tt} + 2cu_t = u_{xx}, u(0,t) = u(\pi,t) = 0, c = 0.4. \tag{8.22}$$

We will assume the same initial conditions as in Example 8.3.2 on page 304:

$$u(x,0) = f(x) = \begin{cases} \frac{2}{\pi}x & 0 \leq x < \frac{\pi}{2} \\ -\frac{2}{\pi}(x-\pi) & \frac{\pi}{2} \leq x \leq \pi \end{cases}$$

and $u_t(x,0) = g(x) \equiv 0$. The series solution of this problem was shown to be

$$u(x,t) = \sum_{n=1}^{\infty} \sin(nx)e^{-ct} \left[a_n \cos(\sqrt{n^2 - c^2}\, t) + b_n \sin(\sqrt{n^2 - c^2}\, t) \right] \tag{8.23}$$

$$a_n = \frac{2}{\pi} \int_0^{\pi} f(x)\sin(nx)dx, \quad b_n = \frac{1}{\sqrt{n^2 - c^2}} \left(ca_n + \frac{2}{\pi} \int_0^{\pi} g(x)\sin(nx)dx \right). \tag{8.24}$$

The PDE (8.22) can be approximated in terms of three central differences as

$$\frac{u(x,t+\Delta t) - 2u(x,t) + u(x,t-\Delta t)}{(\Delta t)^2} + 2c \left(\frac{u(x,t+\Delta t) - u(x,t-\Delta t)}{2\Delta t} \right)$$

$$= \frac{u(x+\Delta x, t) - 2u(x,t) + u(x-\Delta x, t)}{(\Delta x)^2}.$$

By combining similar terms and then solving for $u(x, t+\Delta t)$, we get a difference equation of the form

$$U_{i,j+1} = \frac{1}{1 + c\Delta t} \left[2\left(1 - \left(\frac{\Delta t}{\Delta x}\right)^2\right)U_{i,j} + (c\Delta t - 1)U_{i,j-i} + \left(\frac{\Delta t}{\Delta x}\right)^2 (U_{i+1,j} + U_{i-1,j}) \right]. \tag{8.25}$$

The initial function $f(x)$ can be used to compute $U_{i,0}$ for $i = 0$ to $i = N$. For each $j \geq 0$ the values of $U_{0,j}$ and $U_{N,j}$ need to be set to zero before computing the interior points $i = 1$ to $i = N-1$. To compute values of $U_{i,j+1}$ for $j = 0$, the condition $u_t(x,0) = 0$ must be used. This can be accomplished by setting the central difference $\frac{U_{i,1} - U_{i,-1}}{2\Delta t} = 0$, or equivalently $U_{i,-1} = U_{i,1}$ for $1 \leq i \leq N-1$. For $j = 0$ equation (8.25) then becomes

$$U_{i,1} = U_{i,0} + \frac{1}{2}\left(\frac{\Delta t}{\Delta x}\right)^2 (U_{i+1,0} - 2U_{i,0} + U_{i-1,0})$$

for $1 \leq i \leq N-1$.

The function $u(x,t)$ computed by the series method was graphed in Figure 8.6 on page 304. It shows a rather interesting shape at time $t = 0.8\pi$, so this will be the time chosen to compare the two methods. A numerical Maple routine for the PDE was written with $N = 200$, $\Delta x = \frac{\pi}{N}$, and $\Delta t = \pi \cdot 10^{-4}$. The graph in Figure 8.11 shows the series solution (solid line) together with the numerical approximation (dotted) at time 0.8π. They appear to be essentially the same. The table below compares values of $u(x, 0.8\pi)$ at $x = 0, 0.1\pi, \ldots, 0.9\pi, \pi$ for the two different methods. It can be seen that there is essentially an agreement in the first two significant digits. This could be improved by making Δt smaller.

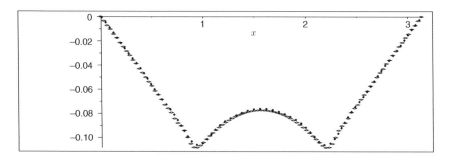

Figure 8.11. Comparison of analytic and numerical solutions at $t = 0.8\pi$

x	numerical computation	series computation
0	0	0
0.1π	-0.035290	-0.035990
0.2π	-0.071471	-0.072984
0.3π	-0.106144	-0.108465
0.4π	-0.084426	-0.085881
0.5π	-0.076065	-0.077504
0.6π	-0.084428	-0.085878
0.7π	-0.106144	-0.108462
0.8π	-0.071479	-0.072981
0.9π	-0.035295	-0.035989
1.0π	7.6×10^{-7}	0

■

Exercises 8.4. *The starred exercises require a computer.*

1. For the function $f(x) = e^x$, approximate $f'(0)$ by the forward difference formula. Try several values of Δx. How small must Δx be to have the derivative exact to 4 decimal places? Repeat this for $f'(0)$ approximated by the central difference formula, and for $f''(0)$ approximated by the central difference formula.

2. Redo Exercise 1 using $f(x) = e^x$, but approximating the three derivatives at $x = 4$. Does the size of Δx required to obtain a 4 decimal place accuracy appear to depend on the size of the derivative? Explain.

The next three problems refer to a simplified version of the diffusive logistic equation described in Section 7.1:

$$\frac{\partial P}{\partial t} = 0.1 \frac{\partial^2 P}{\partial x^2} + P(1 - P), \quad 0 < x < 1, \ t > 0. \tag{8.26}$$

3. Write a difference approximation to (8.26), and solve it for $P(x, t + \Delta t)$ in terms of values of P at time t. Use a forward difference for $\frac{\partial P}{\partial t}$ and a central difference for $\frac{\partial^2 P}{\partial x^2}$.

4. Assume a constant initial condition $P(x, 0) = C, C > 0$. If the boundary conditions are $P_x(0, t) = P_x(1, t) = 0$ for $t > 0$, show that the term representing $\frac{\partial^2 P}{\partial x^2}$ in the

8.4 Numerical Solution of Parabolic and Hyperbolic Equations

difference equation will always equal zero. What constant value would you expect P to approach as $t \to \infty$? Think carefully, using your difference equation to explain your answer.

5. ★ *If the boundary conditions are $P(0, t) = P(1, t) = 0$ for $t > 0$, what do you think will happen to $P(x, t)$ as $t \to \infty$?*

The Maple program below can be used to compute the solution of (8.26):

```
> N := 20; delx := 0.05; delt := 0.01; M := 1000;
> B := 0.1*delt/delx^2; C := 0.5;
> for i from 0 to N do P[i, 0] := C; od;
> for j from 0 to M-1 do
    .....
    P[i, j+1] := P[i, j]+B*(P[i+1, j]-2.0*P[i, j]
        +P[i-1, j])+delt*P[i, j]*(1.0-P[i, j]); od; od:
```

To simulate zero boundary conditions, replace "....." with

```
P[0,j]:=0; P[N,j]:=0; for i from 1 to N-1 do
```

To simulate the conditions $P_x(0, t) = P_x(1, t) = 0$ for $t > 0$, replace "....." with

```
P[-1,j]:=P[1,j]; P[N+1,j]:=P[N-1,j]; for i from 0 to N do
```

To draw a graph of $P(x, M\Delta t)$, use the statements

```
> L:=[ ]: for i from 0 to N do L:=[op(L), [i*delx,P[i,M]]]; od:
> with(plots): pointplot(L);
```

6. ★ *A 15 ft insulated water pipe is open on both ends. The left end is in a heated basement with constant temperature $55°F$. The right end is $1'$ outside the wall where the daily temperature is $u(15, t) = g(t) = 20 + 25 \sin\left(\frac{2\pi t}{24}\right)$, t in hours. Assume that the temperature $u(x, t)$ of the water in the pipe, initially at $55°$ throughout, satisfies the heat equation $u_t = 0.2 u_{xx}$ for $t > 0$. Solve the PDE for $0 \le t \le 240$. Draw graphs of $u(11, t), u(12, t)$, and $u(13, t)$ over the 240-hour period. In about how many days would you expect the water in the pipe inside the basement to freeze. Explain in your own terms.*

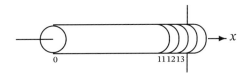

7. ★ *Write a computer program to solve the wave equation numerically. Choose some* **nonconstant** *boundary functions $T_1(t)$ and $T_2(t)$ and run the program with parameters and initial function to simulate a situation of your choice. Describe what happens in terms of the problem you are simulating.*

8.5 Laplace's Equation

Our third example of a second-order linear PDE in two variables is **Laplace's equation**, which has the form

$$u_{xx} + u_{yy} = 0. \tag{8.27}$$

If the 2-dimensional heat equation is written in the form

$$u_{tt} = a(u_{xx} + u_{yy}),$$

and the temperature $u(x, t)$ is no longer changing with time, then this equation reduces to (8.27); therefore, Laplace's equation is often referred to as the steady-state heat equation.

It can be easily checked that equation (8.27) is an **elliptic** partial differential equation by noting that $A = C = 1$ and $B = 0$ imply $B^2 - AC = -1 < 0$.

As an example of a physical situation where this equation arises, consider the temperature $u(x, y)$ in a thin rectangular metal plate which is insulated on the top and bottom so that heat cannot flow in the z-direction. If the temperatures on all four edges of the rectangle are specified appropriately as functions of x or y and *kept fixed*, then given any initial temperature distribution the solution of the heat equation in the interior of the rectangular plate will approach its steady state value; that is, it will approach the solution of (8.27).

Solution of Laplace's Equation by Separation of Variables. The method of separation of variables that we applied to the heat equation and the wave equation can also be used to solve equation (8.27) if it is assumed that three sides of the rectangle are held at temperature $0°F$. The temperature on the fourth side can be specified by an arbitrary piecewise continuous function. Since Laplace's equation is linear and homogeneous, we can find four different series solutions, each one satisfying a nonzero condition on a different side, and add them together to get a solution which satisfies arbitrary conditions around the entire boundary of the rectangle. We will assume first that the boundary conditions are as shown in the figure below. If $u(x, y) = X(x)Y(y)$

8.5 Laplace's Equation

is substituted into (8.27), and the result is divided by XY, then
$$\frac{X''Y}{XY} + \frac{XY''}{XY} = 0 \Rightarrow \frac{X''}{X} = -\frac{Y''}{Y} = -\lambda.$$
The two ordinary differential equations in x and y are
$$X''(x) + \lambda X(x) = 0, \quad Y''(y) - \lambda Y(y) = 0.$$
The boundary conditions $u(0, y) = X(0)Y(y) = 0$ and $u(a, y) = X(a)Y(y) = 0$ for all y in the interval $[0, b]$ imply that $X(0) = X(a) = 0$. These boundary conditions give us the same Sturm-Liouville problem for $X'' + \lambda X = 0$ that we have solved twice before. The eigenvalues will be $\lambda_n = \frac{n^2\pi^2}{a^2}$ and the corresponding eigenfunctions are $X_n(x) = c_n \sin\left(\frac{n\pi x}{a}\right)$. The equation for Y_n, with $\lambda_n = \frac{n^2\pi^2}{a^2}$, becomes $Y_n'' - \frac{n^2\pi^2}{a^2} Y_n = 0$, which has solution
$$Y_n(y) = a_n \cosh\left(\frac{n\pi y}{a}\right) + b_n \sinh\left(\frac{n\pi y}{a}\right).$$
The series solution, which will be called $u_1(x, y)$, can be written as
$$u_1(x, y) = \sum_{n=1}^{\infty} \sin\left(\frac{n\pi x}{a}\right)\left[A_n \cosh\left(\frac{n\pi y}{a}\right) + B_n \sinh\left(\frac{n\pi y}{a}\right)\right],$$
and the coefficients A_n and B_n must be chosen to satisfy the remaining two conditions $u_1(x, b) = 0$, $u_1(x, 0) = f(x)$, for $0 \leq x \leq a$. The condition
$$u_1(x, 0) = f(x) = \sum_{n=1}^{\infty} A_n \sin\left(\frac{n\pi x}{a}\right)$$
implies that the A_n are the coefficients in the Fourier sine series for $f(x)$; therefore, $A_n = \frac{2}{a}\int_0^a f(x) \sin\left(\frac{n\pi x}{a}\right) dx$. The other condition implies that
$$u_1(x, b) = \sum_{n=1}^{\infty} \sin\left(\frac{n\pi x}{a}\right)\left[A_n \cosh\left(\frac{n\pi b}{a}\right) + B_n \sinh\left(\frac{n\pi b}{a}\right)\right] \equiv 0,$$
and therefore, for each $n = 1, 2, \ldots$, we must have $A_n \cosh\left(\frac{n\pi b}{a}\right) + B_n \sinh\left(\frac{n\pi b}{a}\right) = 0$. This means that
$$B_n = -A_n \frac{\cosh\left(\frac{n\pi b}{a}\right)}{\sinh\left(\frac{n\pi b}{a}\right)} = -A_n \coth\left(\frac{n\pi b}{a}\right),$$
and the solution u_1 can be written as
$$u_1(x, y) = \sum_{n=1}^{\infty} \sin\left(\frac{n\pi x}{a}\right) A_{n1}\left[\cosh\left(\frac{n\pi y}{a}\right) - \coth\left(\frac{n\pi b}{a}\right) \sinh\left(\frac{n\pi y}{a}\right)\right],$$
$$A_{n1} = \frac{2}{a}\int_0^a f(x) \sin\left(\frac{n\pi x}{a}\right) dx.$$
The other three cases correspond to boundary conditions specified as shown below.

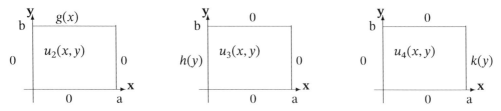

Check that the corresponding solutions are:

$$u_2(x,y) = \sum_{n=1}^{\infty} \sin\left(\frac{n\pi x}{a}\right)[B_{n2} \sinh\left(\frac{n\pi y}{a}\right)], \quad B_{n2} = \frac{\frac{2}{a}\int_0^a g(x)\sin\left(\frac{n\pi x}{a}\right)dx}{\sinh\left(\frac{n\pi b}{a}\right)}, \quad (8.28)$$

$$u_3(x,y) = \sum_{n=1}^{\infty} \sin\left(\frac{n\pi y}{b}\right) A_{n3}[\cosh\left(\frac{n\pi x}{b}\right) - \coth\left(\frac{n\pi a}{b}\right)\sinh\left(\frac{n\pi x}{b}\right)], \quad (8.29)$$

$$A_{n3} = \frac{2}{b}\int_0^b h(y)\sin\left(\frac{n\pi y}{b}\right)dy, \quad (8.30)$$

$$u_4(x,y) = \sum_{n=1}^{\infty} \sin\left(\frac{n\pi y}{b}\right)[B_{n4}\sinh\left(\frac{n\pi x}{b}\right)], \quad B_{n4} = \frac{\frac{2}{b}\int_0^b k(y)\sin\left(\frac{n\pi y}{b}\right)dy}{\sinh\left(\frac{n\pi a}{b}\right)}. \quad (8.31)$$

Example 8.5.1. *Consider the rectangle $R = \{0 \le x \le 15, 0 \le y \le 10\}$ with temperatures along the boundary given by $u(x,0) = f(x) = 0.7x(15-x), u(x,10) = g(x) \equiv 0, u(0,y) = h(y) = 20\sin(\frac{\pi y}{5})$, and $u(15,y) = k(y) = y(10-y)$.*

Using a Maple program, the series for $u(x,y) = u_1(x,y) + u_2(x,y) + u_3(x,y) + u_4(x,y)$, with 20 terms in each series was plotted as a 3-dimensional surface above the rectangle R. A contour plot showing where the temperature values are $-10°, 0°, 5°, 10°, 15°, 20°$ and $30°$ is also shown below (see Figures 8.12 and 8.13). In this example, the boundary functions were chosen so that $u = 0$ at all four corners. This guaranteed a continuous solution everywhere inside the rectangle. ∎

Numerical Solution of Laplace's Equation. If the rectangle R is partitioned along the x and y axes, by letting $\Delta x = a/N$ and $\Delta y = b/M$ for integers N and M, the central difference formula for u_{xx} and u_{yy} can be used to write the following approximation to equation (8.27):

$$\frac{u(x+\Delta x, y) - 2u(x,y) + u(x-\Delta x, y)}{\Delta x^2}$$
$$+ \frac{u(x, y+\Delta y) - 2u(x,y) + u(x, y-\Delta y)}{\Delta y^2} = 0.$$

If Δx and Δy can be chosen to be equal, then letting $\Delta x = \Delta y = h$, the equation can be multiplied by h^2 on both sides, resulting in

$$u(x+h,y) - 2u(x,y) + u(x-h,y) + u(x,y+h) - 2u(x,y) + u(x,y-h) = 0.$$

This can be solved for $u(x,y)$ in the form

$$u(x,y) = \frac{u(x+h,y) + u(x-h,y) + u(x,y+h) + u(x,y-h)}{4}. \quad (8.32)$$

8.5 Laplace's Equation

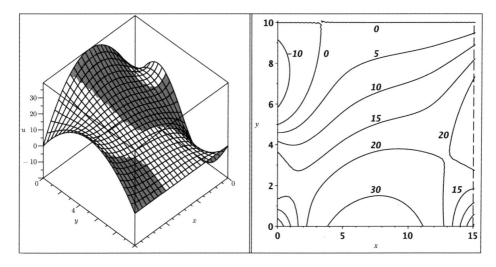

Figure 8.12. Three-dimensional plot of the steady-state temperature.

Figure 8.13. Contour plot of the steady-state temperature.

Note that this says that the temperature at each grid point in the interior of R is the average of the temperatures at the four nearest grid points. If you go on in mathematics to take a course in complex variables, you will find that this property is a characteristic of any "harmonic" function; that is, a function that satisfies Laplace's equation.

If all of the boundary values are given, (8.32) produces a system of linear equations for the unknown temperatures in the grid.

Example 8.5.2. *We will numerically approximate the temperatures that were calculated by the series solution in Example 8.5.1. With $a = 15$ and $b = 10$, we can let $\Delta x = \Delta y = h = 5$ and use the grid shown below. There are only two unknown temperatures to be computed, labelled T_1 and T_2. They should be approximations to $u(5, 5)$ and $u(10, 5)$, respectively. The boundary temperatures were calculated from the formulas used for $f(x), g(x), h(y),$ and $k(y)$ from Example 8.5.1.*

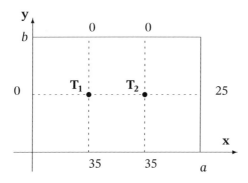

Using (8.32), the two linear equations for T_1 and T_2 are:

$$4T_1 = 0 + T_2 + 35 + 0, \quad 4T_2 = 0 + 25 + 35 + T_1.$$

Written in the form

$$4T_1 - T_2 = 35$$
$$-T_1 + 4T_2 = 60$$

the equations can be solved to give $T_1 = 13.3333$ and $T_2 = 18.3333$. These compare to the values $T_1 = u(5, 5) \approx 12.061$ and $T_2 = u(10, 5) \approx 16.340$ obtained from the series solution in Example 8.5.1. ∎

It is clear that a much better numerical approximation would result if the step size h is decreased. If we take $h = 2.5$, which is one half of the original h, the number of unknown temperatures inside the rectangle increases to 15 (Check it!). Similarly, the number of linear equations in the system increases to 15. The refined grid is shown in Figure 8.14, with temperatures on the boundary again computed from the four functions $f, g, h,$ and k.

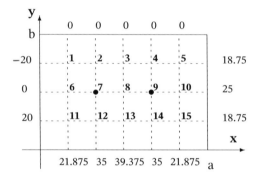

Figure 8.14. Grid with $h = 2.5$

The first three equations, solved for $u_1, u_2,$ and u_3, are

$$u_1 = \frac{1}{4}(0 + u_2 + u_6 - 20) \rightarrow 4u_1 - u_2 - u_6 = -20 \qquad (8.33)$$

$$u_2 = \frac{1}{4}(0 + u_3 + u_7 + u_1) \rightarrow -u_1 + 4u_2 - u_3 - u_7 = 0 \qquad (8.34)$$

$$u_3 = \frac{1}{4}(0 + u_4 + u_8 + u_2) \rightarrow -u_2 + 4u_3 - u_4 - u_8 = 0 \qquad (8.35)$$

If you put the coefficients of the u_i, $1 \le i \le 15$, into a 15×15 matrix, you will see a definite pattern. Exercise 5. asks you to describe the form of the matrix.

Computer methods for solving large systems of linear equations have been around for a long time, and they are very easy to apply. This is a topic that is covered in both linear algebra and numerical analysis courses. Because Laplace's equation arises in so many areas of physics, engineering and applied mathematics, much work has been done over the past seventy years (since large computers became readily available) to find faster and more accurate methods for solving it. The method of finite elements, which treats problems on nonrectangular regions, has been developed completely over this period, and even today is an active area of research.

Exercises 8.5. *The starred exercises require a computer.*

8.5 Laplace's Equation

1. *Derive the series solution for $u(x, y)$ where u satisfies Laplace's equation inside the rectangle R, and the boundary conditions are:*

   $u(0, y) = u(a, y) = 0$, $0 \le y \le b$; $u(x, 0) = 0$, $u(x, b) \equiv g(x)$, $0 \le x \le a$.

   Your answer should look like $u_2(x, y)$ in equation (8.28).

2. *Use the numerical method for solving Laplace's equation to find approximations to T_1, T_2, and T_3 in the L-shaped region shown here. The boundary temperatures are all given in the diagram.*

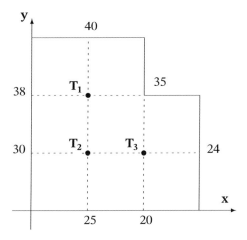

   Note: The method of separation of variables does not work for a nonrectangular region of this type.

3. ★ *Use a computer program like Maple to set up and compute the numerical solution to Example 8.5.2, using $h = 2.5$. The temperatures given by the series in Example 8.5.1 (with 20 terms) are:*

x	2.5	5.0	7.5	10.0	12.5
$y = 7.5$	-1.0737	4.7192	7.0714	8.6165	11.176
$y = 5.0$	6.8162	12.059	15.055	16.337	18.086
$y = 2.5$	16.458	21.849	25.008	24.026	20.392

   These can be used to check the accuracy of your answer.

4. *If Δx and Δy are halved again, so that $h = 1.25$, how many unknown temperatures must be computed in the numerical solution in Example 8.5.2?*

5. *If the system of linear equations in the previous two exercises is written in matrix form as*

   $$A\vec{u} = \vec{b},$$

   where $\vec{u}$ is the column vector of unknown temperatures ordered by going across each row of the rectangle in succession, the matrix A has a very distinctive form. Describe it.

8.6 Student Project: Harvested Diffusive Logistic Equation

In a very recent paper [H. Chen, H. Xing, X. He, *Bifurcation and stability of solutions to a logistic equation with harvesting*, Math. Methods in the Appl. Sciences, 2014], the authors consider a 1-dimensional diffusive logistic equation with harvesting,

$$\frac{\partial u}{\partial t} = \frac{\partial^2 u}{\partial x^2} + a(x)u - b(x)u^2 - \lambda h(x), \tag{8.36}$$

for a population $u(x,t)$ satisfying Neuman boundary conditions $u'(0,t) = u'(1,t) = 0$, for $t > 0$. Notice that (8.36) is a parabolic PDE, but it is neither linear nor homogeneous (if $\lambda \neq 0$), and it can have nonconstant coefficients.

Using advanced methods in the theory of PDEs, the authors prove that under certain conditions on the nonconstant functions $a(x)$, $b(x)$, and $h(x)$ this partial differential equation has exactly two positive steady-state solutions (that is, solutions of (8.36) with $\frac{\partial u}{\partial t} \equiv 0$) for all λ small enough. The upper solution is stable and the lower one is unstable. There is a unique value λ^* of the harvesting parameter λ at which these two solutions coalesce, and for $\lambda > \lambda^*$ no positive steady-state solutions exist.

It is assumed that the functions $b(x)$ and $h(x)$ are positive on $[0, 1]$. To prove the results stated above, it is necessary that the growth function $a(x)$ be strictly bounded above by $\pi^2/4$ on $[0, 1]$. In addition, the smallest eigenvalue $\mu = \mu_1$ of the Sturm-Liouville problem

$$v'' + (a(x) + \mu)v = 0, v'(0) = v'(1) = 0 \tag{8.37}$$

must be negative. Under these conditons on $a(x)$, however, it may be possible to have a negative growth rate in some portions of the x-interval.

If we let the equation model the growth of an aquatic population on a 1-dimensional stretch across a lake, the Neuman boundary conditions model a situation where the population cannot flow across the edges of the lake. The function $a(x)$ determines the growth rate of the population at points across the lake, the carrying capacity at point x is modelled by $a(x)/b(x)$ when λ is zero, and $h(x)$ describes the harvesting pattern across the lake.

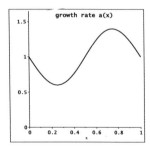
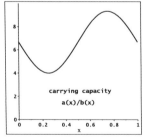
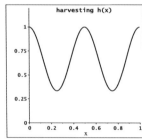

Figure 8.15. The three functions $a(x)$, $a(x)/b(x)$, and $h(x)$

8.6 Student Project: Harvested Diffusive Logistic Equation

Example. *The three functions*

$$a(x) = 1 - 0.4\sin(2\pi x) \tag{8.38}$$

$$b(x) \equiv 0.15 \tag{8.39}$$

$$h(x) = \frac{2 + \cos(4\pi x)}{3.0} \tag{8.40}$$

are chosen arbitrarily, and graphs of the growth rate $a(x)$, the carrying capacity of the lake $a(x)/b(x)$, and the harvesting distribution across the lake $h(x)$ are shown in Figure 8.15.

To check that the smallest eigenvalue $\mu = \mu_1$ of (8.37) satisfies $\mu_1 < 0$, solve (8.37), over a range of values for μ, using *initial conditions* $v(0) = 1, v'(0) = 0$. It is then only necessary to check that the smallest value of μ, for which the solution of (8.37) satisfies $v'(1) = 0$, is negative. Every scalar multiple of this function will be a solution of the SL-problem, and hence $c\{v(x)\}$ will be the family of eigenfunctions for the eigenvalue μ_1. It is well known that the eigenfunction corresponding to the smallest eigenvalue of a Sturm-Liouville problem has constant sign on the interval $0 < x < 1$, and this can be used to check that you have found the smallest eigenvalue. Using the coefficient functions given by equations (8.38)-(8.40), the minimum eigenvalue is found to be $\mu_1 \approx -1.006075$. The corresponding eigenfunction $v(x)$ can then be obtained by solving (8.37) with $\mu = \mu_1$ and initial conditions $v(0) = 1, v'(0) = 0$.

It was also shown in the paper that the bifurcation value λ^* of the parameter λ is bounded by

$$\lambda^* < \frac{\mu_1^2}{4b_0} \frac{\int_0^1 v(x)dx}{\int_0^1 h(x)v(x)dx}$$

where μ_1 is the smallest eigenvalue, defined above, b_0 is the minimum of the function $b(x)$ on $0 \leq x \leq 1$, and $v(x)$ is an eigenfunction corresponding to the eigenvalue μ_1. Notice that it does not matter which eigenfunction $cv(x)$ is used, since the constant multiplying v cancels out in the formula for λ^*. In this example, the condition implies that $\lambda^* < 2.53901$. Check this, using your own computer software.

For values of λ strictly less than λ^*, the theorem implies that there are two solutions w_1 and w_2 of the ODE

$$w''(x) + (a(x) - b(x)w(x))w(x) - \lambda h(x) = 0, \tag{8.41}$$

satisfying the boundary conditions $w'(0) = 0$ and $w'(1) = 0$. Note that this is a non-linear boundary-value problem, and you will have to let $w'(0) = 0$ and iterate on $w(0)$ until you find the two values which produce solutions with $w'(1) = 0$. The solutions with these two values of $w(0)$ can be seen to be the steady-state solutions of the PDE (8.36); that is, they are solutions when the time derivative u_t is equal to zero. With λ arbitrarily chosen to be 0.9, the two solutions were found to have initial values approximately equal to 0.636305 and 5.86033. Let w_1 denote the upper (stable) solution with $w_1(0) = 5.86033$ and w_2 denote the lower (unstable) solution with $w_2(0) = 0.636305$. Graphs of these two solutions, found numerically, are shown in Figure 8.16.

According to the paper, any solution of the PDE (8.36), starting with $u(x, 0) = f(x)$ strictly above $w_2(x)$ should converge to the stable steady-state $w_1(x)$. Any solution

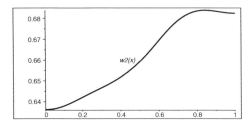 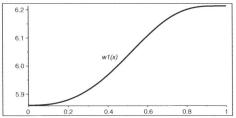

Figure 8.16. The lower and upper solutions of (8.41)

starting from an initial function $u(x, 0) = g(x)$ strictly below $w_2(x)$ will become negative after some time $t > T$. Note that this is very similar to the results for the harvested logistic ODE, for which the population converges to a stable limit *value* for small enough harvesting and large enough initial population, but goes extinct for all initial populations if the harvesting is greater than the *critical harvesting rate* $\lambda = \lambda^*$. In the case of the PDE (8.36), try to use the graph of the growth function $a(x)$, and the distribution of the harvesting across the lake, to explain the shape of the steady-state solution to which the population converges.

The numerical method we used to solve the heat equation in Section 8.4 can be used to solve the nonlinear PDE (8.36). A Maple program for solving the equation numerically is included in Exercises 8.6. The graphs in Figure 8.17 show the results for $\lambda = 0.9$. In the left-hand graph, the initial function was $u(x, 0) \equiv 6.25$, and for several time steps it can be seen that as t increases, $u(x, t)$ is approaching the stable steady state $w_1(x)$ from above. In the middle graph, a constant initial function $u(x, 0) = 3.0$ between w_1 and w_2 led to solutions which increased to the steady state $w1$ as t increased. In the right-hand graph the initial function was $u(x, 0) \equiv 0.4$, below the unstable steady state solution, and the solutions can be seen to be decreasing as t increases. When the solution goes through zero at a value of x, it is assumed that the population has become extinct at that point.

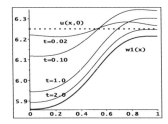 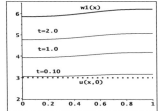 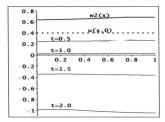

Figure 8.17. In the left and middle figures, the initial functions $u(x, 0)$ are $f_1(x) \equiv 6.25$ and $f_2(x) \equiv 3.0$, and the solutions can be seen to be approaching w_1 as t increases. In the right-hand figure the initial function is below the unstable steady state solution w_2 and solutions can be seen to be tending to $-\infty$ with increasing t.

Exercises 8.6. *The starred exercises require a computer.*

8.6 Student Project: Harvested Diffusive Logistic Equation

1. *Show that the partial differential equation*
$$u_t = u_{xx} + a(x)u - b(x)u^2 - \lambda h(x)$$
   cannot be solved by separation of variables unless both $b(x) \equiv 0$ and $h(x) \equiv 0$.

2. *Write a difference equation for the PDE in Exercise 1 and solve it for $u(x, t + \Delta t)$ in terms of values of u at time t.*

3. ★ *Using the parameters given in (8.38)-(8.40), increase the value of the harvesting parameter to $\lambda = 2.0$. Find the initial value Z_1 of the stable equilibrium solution of the PDE. Draw a graph of $w_1(x)$. How does increasing the harvesting change the equilibrium solution?*

4. ★ *Show that the function $a(x) = 0.7 + 0.8\sin(5.0x)$ satisfies both properties required by the theorem to guarantee that the PDE will have an upper and lower equilibrium solution if the harvesting parameter λ is small enough. Sketch a graph of $a(x)$ and describe what is happening at the right end of the lake.*

5. ★ *Using the parameters $b(x) = 0.15$, $h(x) = x$, $\lambda = 1.0$, and the growth function $a(x) = 0.7 + 0.8\sin(5.0x)$ from Exercise 4., solve the PDE with constant initial functions $u(x, 0) \equiv C$. Use the Maple program below, or write your own PDE solver. Run the program out to time $t = 1.0$, with enough different constant initial functions to be able to estimate the values of Z_1 and Z_2 (the initial values of the stable and unstable equilibrium solutions). Remember that solutions starting above the stable solution will decrease and those below will increase - if they are not below the unstable solution.*

Maple program to solve the PDE

```
> a:=x-> 0.7 + 0.8*sin(5.0*x):  b:=x->0.15:
> h:=x-> x:   lam:=1.0:
> N := 40:  delx:=1.0/N:
> M := 5000:  delt:=0.0002:  B := delt/delx^2:
####  store the initial function u(x,0)
> for i from 0 to N do u[i, 0] := C; od;    #### try different constants C
####  solve the PDE from time t=0 to t=M*delt
> for j from 0 to M-1 do
    u[-1, j] := u[1, j]; u[N+1, j] := u[N-1, j];
    for i from 0 to N do
      u[i, j+1] := u[i, j]+B*(u[i+1, j]-2*u[i, j]+u[i-1,j])
         +delt*(a(i*delx)*u[i, j]-b(i*delx)*u[i, j]^2-h(i*delx));
    od; od:
> L:=[]: for i from 0 to N do L:=[op(L),[i*delx,u[i,M]]]: od:
> with(plots): pointplot(L);
```

6. ★ *Use the method given in Section 8.6 to find the upper and lower equilibrium solutions of the PDE, with the parameters used in Exercise 5.. Remember that the upper and lower equilibrium solutions of the PDE are the two solutions of the boundary-value problem*
$$z'' + (a(x) - b(x)z)z - \lambda h(x) = 0, \quad z'(0) = z'(1) = 0. \tag{8.42}$$

How well do your estimates in Exercise 5. agree with these values?

7. ★ *Choose your own parameter functions a, b, and h and show that they satisfy the conditions of the theorem. Find the bound λ^* on the critical harvesting value λ. Let $\lambda = \frac{1}{2}\lambda^*$ and use the Maple program to find the **smallest** constant initial function $u(x,0) \equiv C$ for which the solution $u(x,t)$ tends to the stable equilibrium solution $z_1(x)$ as $t \to \infty$. Does $u(x,0) \equiv C$ lie entirely above or entirely below the unstable equilibrium solution? Explain.*

For your further enjoyment. *Remember the spruce-budworm equation, an ODE that was described back on page 28 in Section 2.2. A group of students, working with two faculty members at the University of Hartford, have recently published a very interesting paper[2] in which they study a version of the spruce-budworm equation with a spatial diffusion term added. The journal in which their paper appears, describes itself as a journal which encourages "high quality mathematical research involving students from all academic levels." The first ten pages of their paper should be readable and of interest to anyone who has worked on the problems in this section. It might hopefully inspire you to do a bit of mathematical research of your own.*

---

[2] H. Al-Khalil, C. Brennan, R. Decker, A. Demirkaya, and J. Nagode, Numerical existence and stability of solutions to the spruce budworm model, *Involve: A Journal of Mathematics, Vol. 10,* 2017, pp. 857–879.

Appendix

Appendix A

Answers to Odd-numbered Exercises

Section 1.1

1. first-order; independent variable t; dependent variable y; no parameters

3. first-order; independent variable t; dependent variable P; parameters r, k, β, and α

5. third-order; independent variable t; dependent variable x; parameter ω

7. second-order; independent variable t; dependent variable θ; no parameters

9. $2x'' + 6x' + 4x = 2(4e^{-2t}) + 6(-2e^{-2t}) + 4(e^{-2t}) = (8 - 12 + 4)e^{-2t} \equiv 0$.
   The solution is defined for all t.

11. $t^2 x'' + 3tx' + x = t^2(2t^{-3}) + 3t(-t^{-2}) + t^{-1} = (2 - 3 + 1)t^{-1} \equiv 0$.
    The solution is defined for $t \neq 0$.

13. $P' = rCe^{rt} = r(Ce^{rt}) = rP$. The solution is defined for all t.

15. $x' = \frac{1}{2}(t^2 + 4t + 1)^{-\frac{1}{2}}(2t + 4) = \frac{t+2}{\sqrt{t^2+4t+1}} \equiv \frac{t+2}{x}$.
    The solution exists if $t^2 + 4t + 1 \geq 0$, i.e., if $t < -2 - \sqrt{3}$ or $t > -2 + \sqrt{3}$.

17. $t^2 y'' + ty' + (t^2 - \frac{1}{4})y = t^2 \left(\sin(t)(-t^{-\frac{1}{2}} + \frac{3}{4}t^{-\frac{5}{2}}) - t^{-\frac{3}{2}}\cos(t) \right)$
    $+ t\left(\sin(t)(-\frac{1}{2}t^{-\frac{3}{2}}) + \cos(t)(t^{-\frac{1}{2}}) \right) + (t^2 - \frac{1}{4})\left(\sin(t)(t^{-\frac{1}{2}}) \right) \equiv 0$.
    The solution exists for all $t > 0$ and also for all $t < 0$.

331

19. $x' = e^t(-2\sin(2t)) + e^t\cos(2t) - \frac{1}{2}e^t 2\cos(2t) - \frac{1}{2}e^t\sin(2t) = -\frac{5}{2}e^t\sin(2t)$.

$y' = e^t(-2\sin(2t)) + e^t\cos(2t) + 2e^t 2\cos(2t) + 2e^t\sin(2t) = 5e^t\cos(2t)$.

$x - y = -\frac{5}{2}e^t\sin(2t) \equiv x'$.

$4x + y = 5e^t\cos(2t) \equiv y'$.

The solution exists for all t.

21. $x' = 15e^{5t} - 2e^{-2t} \equiv x + 3y$.

$y' = 20e^{5t} + 2e^{-2t} \equiv 4x + 2y$.

The solution exists for all t.

Section 1.2

1. $x' = Ce^t \equiv x - 2 = (2 + Ce^t) - 2$.

3. $y' = Ce^t - 1 \equiv y + t = (Ce^t - 1 - t) + t$.

5. $x' = 2Ce^{2t} \equiv 2x$, for any C. $x(0) = Ce^{2(0)} = C = -1 \Rightarrow x(t) = -e^{2t}$.

   The solution exists for all t.

7. $y' = (\sec(t + C))^2 = 1 + (\tan(t + C))^2 \equiv 1 + y^2$. $y(0) = 1 = \tan(C) \Rightarrow C = \frac{\pi}{4}$.

   The solution $y(t) = \tan(t + \frac{\pi}{4})$ exists for $-\frac{3\pi}{4} < t < \frac{\pi}{4}$.

9. With $x = C_1\sin(t) + C_2\cos(t) + t^2 - 2$, $x'' = -C_1\sin(t) - C_2\cos(t) + 2$, so $x'' + x = t^2$.

   $x(0) = C_1(0) + C_2(1) - 2 = C_2 - 2 = 0 \Rightarrow C_2 = 2$.

   $x'(0) = C_1\cos(0) - C_2\sin(0) = C_1 = 1 \Rightarrow C_1 = 1$.

   The solution $x(t) = \sin(t) + 2\cos(t) + t^2 - 2$ exists for all t.

11. (a) With $x(t) = \sqrt{\frac{1}{C-2t}} \equiv (C - 2t)^{-\frac{1}{2}}$,

    $x'(t) = -\frac{1}{2}(C - 2t)^{-\frac{3}{2}}(-2) = (C - 2t)^{-\frac{3}{2}} = (x(t))^3$.

    (b) $x(t) \equiv 0$ is a solution, but is not equal to $\sqrt{\frac{1}{C-2t}}$ for any value of C.

    (c) $x(0) = \sqrt{\frac{1}{C-0}} = 1$ implies that $C = 1$ and $x(t) = \sqrt{\frac{1}{1-2t}}$.

    (d) The solution $x(t)$ has a vertical asymptote at $t = \frac{1}{2}$.

13. The general solution is $y(t) = \frac{1}{2} + Ce^{-2t}$, and $y(0) = \frac{1}{2} + C$; therefore, the five curves you need to plot are $y(t) = \frac{1}{2} - \frac{3}{2}e^{-2t}, y(t) = \frac{1}{2} - \frac{1}{2}e^{-2t}, \ldots, y(t) = \frac{1}{2} + \frac{5}{2}e^{-2t}$.
This is a one-parameter family of curves which all approach $\frac{1}{2}$ as $t \to \infty$.

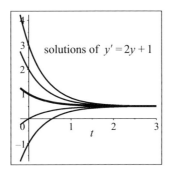

Section 1.3

- **PHYSICS:**

  (a) $v(0) = \frac{mg}{k} + Ce^0 = 9.8 + C = 0 \Rightarrow C = -9.8$
  $v(5) = 9.8 - 9.8e^{-5} \approx 9.734 \text{m/sec} \approx 21.8 \text{mph}$

  (b) Terminal velocity $\frac{mg}{k} = 5\text{mph} \Rightarrow \frac{mg}{k} = \frac{5}{2.237} \approx 2.235 \text{m/sec}$; therefore, $\frac{k}{m} = \frac{g}{mg/k} \approx \frac{9.8}{2.235} \approx 4.4$. Note that terminal velocity is independent of initial velocity.

- **MATHEMATICS:**

  (a) Using implicit differentiation on $x^2 + 3y^2 = C$ gives $2x + 6yy' = 0$, and solving for y', $y' = -\frac{2x}{6y}$, so $y' = -\frac{x}{3y}$.

  (c) See the figure.

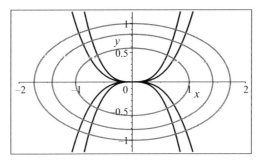

- **ENGINEERING:**

  (a) Using the solution given in the text, $i(t) = Ce^{-3t} + \frac{10}{1+9}(3\sin(t) - \cos(t))$. As $t \to \infty$, the exponential term goes to zero, so the steady state solution is $i_{ss}(t) = 3\sin(t) - \cos(t)$.

- **ECOLOGY:**

  (a) Using the solution in the text, $P(t) = \frac{N}{1-Ce^{-rt}} = \frac{1000}{1-Ce^{-0.5t}}$.
  $P(0) = 20 \Rightarrow 1 - C = \frac{1000}{20} = 50$, and $C = -49$.
  After 7 days, $P(7) = \frac{1000}{1+49e^{(-0.5)(7)}} \approx 403$ ants.

- **NEUROLOGY:**

    (a) The curve with the largest K grows fastest around $z = 0$.

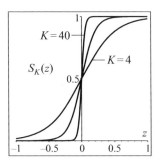

Section 2.1

1. Not separable

3. Separable

5. $\frac{dx}{dt} = (\frac{1}{t})x$

    $\int \frac{dx}{x} = \int \frac{1}{t} dt$

    $\ln |x| = \ln |t| + C$

    $|x| = e^{\ln |t| + C} = e^C |t|$; therefore, $x = At$ where $A = \pm e^C$

7. $\frac{dx}{dt} = (1)(x + 5)$

    $\int \frac{dx}{x+5} = \int 1 dt$

    $\ln |x + 5| = t + C$

    $|x + 5| = e^{t+C} = e^t e^C$; therefore, $x = Ae^t - 5$ where $A = \pm e^C$

9. $\frac{dx}{dt} = (\cos(t))(x)$

    $\int \frac{dx}{x} = \int \cos(t) dt$

    $\ln |x| = \sin(t) + C$

    $x(t) = Ae^{\sin(t)}$ where $A = \pm e^C$

11. $\frac{dx}{dt} = (1 + 2t + 3t^2)(\frac{1}{x})$

    $\int x dx = \int (1 + 2t + 3t^2) dt$

    $\frac{x^2}{2} = t + t^2 + t^3 + C$

    $x(t) = \pm \sqrt{2t + 2t^2 + 2t^3 + 2C}$

Section 2.1

13. $\frac{dx}{dt} = t(1 + x^2)$

    $\int \frac{dx}{1+x^2} = \int t\,dt$

    $\arctan(x) = t^2/2 + C$

    $x(t) = \tan(t^2/2 + C)$

15. $\frac{dy}{dt} = y + 1$

    $\int \frac{dy}{y+1} = \int 1\,dt$

    $\ln|y + 1| = t + C$ therefore, $y = Ae^t - 1$ where $A = \pm e^C$

    $y(0) = A - 1 = 2, A = 3$

    $y = 3e^t - 1$

17. $\frac{dx}{dt} = (\cos(t))(x), \quad x(0) = 1$

    General solution (see Exercise 9):

    $x(t) = Ae^{\sin(t)}$

    $x(0) = Ae^{\sin(0)} = A = 1$

    $x(t) = e^{\sin(t)}$

19. $\frac{dx}{dt} = (t + 1)(\cos(x))^2, \; x(0) = 1$

    $\int \sec^2(x)\,dx = \int (t+1)\,dt$

    $\tan(x) = t^2/2 + t + C$

    $x(t) = \arctan(t^2/2 + t + C)$

    $x(0) = \arctan(C) = 1 \Rightarrow C = \tan(1)$

    $x(t) = \arctan(t^2/2 + t + \tan(1))$

21. $T(t) = 90 - \alpha e^{-kt}$

    $T(0) = 90 - \alpha = 40 \Rightarrow \alpha = 50$

    $T(5) = 90 - 50e^{-5k} = 50 \Rightarrow k \approx 0.04463$

    $T(20) = 90 - 50e^{-0.04463(20)} \approx 69.52^0$

    As $t \Rightarrow \infty, T \Rightarrow 90^0$

23. $\frac{dy}{dx} = -\frac{x}{3y}$

    $\int 3y\,dy = \int -x\,dx$

    $\frac{3y^2}{2} = -\frac{x^2}{2} + C$

    $x^2 + 3y^2 = 2C = K$ (ellipses)

25. (b) The point $(1.5, 151.3)$ is well below the curve. The 1930s depression and World War II could be factors.

    (c) $P(2.1) \approx 292.7$

    (d) $P(3) \approx 328.7$ seems reasonable

    (e) If you use $t = 0$ for 1900 and $t = 1$ for 1950, then $ce^0 = 76.2$ and $ce^r = 151.3$, so $c = 76.2$ and $r \approx 0.686$. This gives a value $p(4) \approx 1185$ which seems unreasonably large.

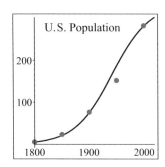

Section 2.2

1. Slope field for $x' = x + \frac{t}{2}$

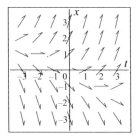

3. Slope field for $x' = x(1 - \frac{x}{2})$

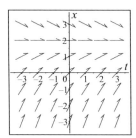

5. Slope field for $x' = \frac{t}{x}$

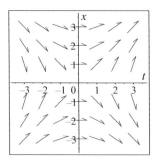

Section 2.3

7. Slope field for $x' = x^3 + t$ (using isoclines)

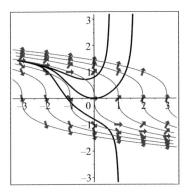

9. Slope field for $x' = t^2 - x$ (using isoclines)

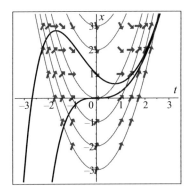

15. Slope field for $x' = 1 + \frac{t}{2}$

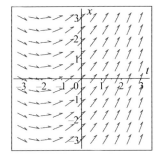

Section 2.3

1. Both separable and linear

3. Separable, not linear

5. Separable (autonomous), but not linear

7. $x' + 2x = e^{-2t}\sin(t)$ is linear
$\mu(t) = e^{\int p(t)dt} = e^{\int 2dt} = e^{2t}$

$e^{2t}(x' + 2x) = e^{2t}(e^{-2t}\sin(t)) = \sin(t)$

$\frac{d}{dt}(e^{2t}x) = \sin(t)$

$e^{2t}x = \int \sin(t)dt = -\cos(t) + C$

$x(t) = e^{-2t}(-\cos(t) + C) \Rightarrow x(t)$ oscillates and tends to 0 as $t \to \infty$.

9. $x' - \frac{2}{t}x = \frac{1}{t} + 1$

$\mu(t) = e^{\int p(t)dt} = e^{\int(-\frac{2}{t})dt} = t^{-2}$

$t^{-2}(x' - \frac{2}{t}x) = \frac{d}{dt}(t^{-2}x) = t^{-2}(\frac{1}{t} + 1) = t^{-3} + t^{-2}$

$t^{-2}x = \int(t^{-3} + t^{-2}) = \frac{t^{-2}}{-2} + \frac{t^{-1}}{-1} + C$

$x(t) = -\frac{1}{2} - t + Ct^2 \Rightarrow \lim_{t \to \infty} x(t) = \pm\infty$ depending on the value of the constant C.

11. $x' + x = e^{2t}$, $x(0) = 1$

$\mu(t) = e^{\int p(t)dt} = e^{\int 1 dt} = e^t$

$e^t(x' + x) = \frac{d}{dt}(e^t x) = e^t(e^{2t}) = e^{3t}$

$e^t x = \int e^{3t}dt = \frac{e^{3t}}{3} + C$

$x(t) = e^{-t}(\frac{e^{3t}}{3} + C) = \frac{e^{2t}}{3} + Ce^{-t}$

$x(0) = \frac{1}{3} + C = 1 \Rightarrow C = \frac{2}{3}$

13. If $t \neq 0$, $x' + \frac{1}{t}x = 3t - 1$, $x(1) = 0$

$\mu(t) = e^{\int p(t)dt} = e^{\int(\frac{1}{t})dt} = e^{\ln(t)} = t$

$t(x' + \frac{1}{t}x) = \frac{d}{dt}(tx) = t(3t - 1) = 3t^2 - t$

$tx = \int(3t^2 - t)dt = t^3 - \frac{t^2}{2} + C$

$x(t) = t^2 - \frac{t}{2} + \frac{C}{t}$

$x(1) = 1 - \frac{1}{2} + C = 0 \Rightarrow C = -\frac{1}{2}$

15. (a) It is linear if $p = 0$ or 1.
 (b) $v' = -\frac{k}{m}v + g \Rightarrow v' + 0.6v = 9.8$
 $\frac{d}{dt}(e^{0.6t}v) = e^{0.6t}(9.8)$
 $e^{0.6t}v = 9.8\left(\frac{e^{0.6t}}{0.6}\right) + C$
 $v(t) = \frac{9.8}{0.6} + Ce^{-0.6t} \approx 16.33 + Ce^{-0.6t}$
 (c) As $t \to \infty$, $v(t) \to 16.33$ meters/sec ≈ 36.5 mph
 (d) The terminal velocity does not depend on initial velocity since the term involving the constant C goes to 0 as $t \to \infty$.

17. Let $V(t) = 150 - t/2$

    $x' = (1.5)(3) - 2x(t)/V(t)$

    $x' + 4x/(300 - t) = 4.5, \ x(0) = 20$

    Solution $x(t) = 1.5(300 - t) - \dfrac{430(300-t)^4}{300^4}$

    Solving $x(t)/V(t) = 1$ gives $t \approx 33.92$ min.

    $V(33.92) \approx 133$ gal. left in the tank

Section 2.4

1. Both $f(t, x) = \dfrac{x^2}{1+t^2}$ and $\dfrac{\partial f}{\partial x} = \dfrac{2x}{1+t^2}$ are continuous for all values of x and t; therefore, the equation has a unique solution through every initial point (t_0, x_0). Solutions cannot intersect anywhere in the plane.

3. $f(t, x) = x^2 - t$ and $\dfrac{\partial f}{\partial x} = 2x$ are both continuous for all t and x; therefore there is a unique solution through any initial point. Solution curves can never intersect.

5. Since $\dfrac{\partial f}{\partial x} = \dfrac{\partial}{\partial x}(x^{2/3}) = \dfrac{2}{3}x^{-1/3}$ is not defined when $x = 0$, the theorem does not apply when $x = 0$.

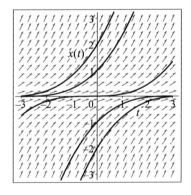

Let $\dfrac{dx}{dt} = x^{2/3}$; then $\int (x^{-2/3})dx = \int dt \Rightarrow 3x^{1/3} = t + C \Rightarrow x(t) = \left(\dfrac{t+C}{3}\right)^3$.

$x(t) = \left(\dfrac{t}{3}\right)^3$ is one solution through $(0,0)$ but $x(t) \equiv 0$ is also a solution through that point. What about other points where $x = 0$?

Section 2.5

1. Yes, $\dfrac{\partial}{\partial y}(x + y) = 1 = \dfrac{\partial}{\partial x}(x)$.

3. Yes, $\dfrac{\partial}{\partial y}(\sin(y)) = \cos(y) = \dfrac{\partial}{\partial x}(x \cos(y))$.

5. Yes, $\dfrac{\partial}{\partial y}(2xy) = 2x = \dfrac{\partial}{\partial x}(x^2 + y^2)$.

7. $\frac{\partial F}{\partial x} = x + y \Rightarrow F = \frac{x^2}{2} + xy + Q(y)$

   $\frac{\partial F}{\partial y} = x + Q'(y) = x + 1 \Rightarrow Q'(y) = 1$

   $Q(y) = y \Rightarrow F = \frac{x^2}{2} + xy + y$

   Solving $\frac{x^2}{2} + xy + y = C$ for y gives $y = \frac{C}{x+1} - \frac{x^2}{2(x+1)}$.

9. $\frac{\partial F}{\partial x} = y + \sin(y) \Rightarrow F = xy + x\sin(y) + Q(y)$

   $\frac{\partial F}{\partial y} = x + x\cos(y) + Q'(y) = 1 + x + x\cos(y)$

   $Q'(y) = 1 \Rightarrow Q(y) = y$

   $xy + x\sin(y) + y = C$ (implicit soln.)

11. $\frac{\partial}{\partial y}(1 + xy) = x = \frac{\partial}{\partial x}(\frac{1}{2}x^2)$

    $\frac{\partial F}{\partial x} = 1 + xy \Rightarrow F = x + \frac{x^2}{2}y + Q(y)$

    $\frac{\partial F}{\partial y} = \frac{x^2}{2} + Q'(y) \Rightarrow Q'(y) = 0$

    $x + \frac{x^2}{2}y = C \Rightarrow y = \frac{2}{x^2}(C - x)$

    $y(1) = 2(C - 1) = 1 \Rightarrow C = \frac{3}{2}$

    $y(x) = \frac{2}{x^2}(\frac{3}{2} - x)$

13. $\frac{\partial}{\partial y}(\sin(y)) = \cos(y) = \frac{\partial}{\partial x}(x\cos(y))$

    $\frac{\partial F}{\partial x} = \sin(y) \Rightarrow F = x\sin(y) + Q(y)$

    $\frac{\partial F}{\partial y} = x\cos(y) + Q'(y) \Rightarrow Q'(y) = 0$

    $x\sin(y) = C \Rightarrow y = \arcsin(\frac{C}{x})$

    $y(1) = \arcsin(C) = \frac{\pi}{2} \Rightarrow C = 1$

    $y(x) = \arcsin(\frac{1}{x})$

15. $P' = 2P - \frac{1}{2}P^2, N = 2, v \equiv 1/P$

    $v' = -2v + \frac{1}{2}, \mu = e^{\int 2dt} = e^{2t}$

    $v = \frac{1}{4} + Ce^{-2t}$

    $P = \frac{1}{v} = \frac{1}{\frac{1}{4}+Ce^{-2t}} = \frac{4}{1+4Ce^{-2t}}$

17. $y' = -y + e^t y^2, N = 2, v \equiv 1/y$

    $v' - v = -e^t, \mu = e^{-t}$

    $v = -te^t + Ce^t, v(0) = C = 1$

    $v(t) = (1 - t)e^t \Rightarrow y(t) = \frac{e^{-t}}{1-t}$

19. $y' = -y + ty^3$, $N = 3$, $v \equiv 1/y^2$

$v' - 2v = -2t \Rightarrow v = t + \frac{1}{2} + Ce^{2t}$

$y = \frac{1}{\sqrt{v}} = \pm\sqrt{\frac{1}{t + \frac{1}{2} + Ce^{2t}}}$

21. Since $y(t) = \left(\frac{2}{3}e^t + Ce^{4t}\right)^{-1/2}$, the solution will tend to zero if $C > 0$ and will have a vertical asymptote if $C < 0$; therefore $C = 0$, so $y(0) = \sqrt{\frac{3}{2}} \approx 1.2247$. The special solution is $y(t) = \sqrt{\frac{3}{2}}e^{-t/2}$.

Section 2.6

1. (a)

t_j	x_j	x'_j	$x_j + \frac{1}{8}x'_j$
0	1.0	−1.0	0.8750
0.125	0.8750	−0.7500	0.781250
0.250	0.781250	−0.531250	0.714844
0.375	0.714844	−0.339844	0.672364
0.500	0.672364	−0.172364	0.650818
0.625	0.650818	−0.025818	0.647591
0.750	0.647591	0.102409	0.660392
0.875	0.660392	0.214608	0.687218
1.0	0.687218		

(b)

Δt	$x(1)$	error in $x(1)$
0.5	0.50000	0.235759
0.25	0.632813	0.102946
0.125	0.687218	0.048541

(c) As Δt decreases by half, the error decreases by about one half, as expected.

3. Using the Maple instructions

```
soln := dsolve({x'(t)=t-x(t),x(0)=1},
type=numeric,method=rkf45, abserr=0.5e-7);
ans := soln(2.0);,
```

The answer $[t = 2.0, x(t) = 1.270670239]$ is accurate to 6 decimal places.

5. (a)

t_j	x_j	x'_j	$x_j + 0.2x'_j$
0	1.0	1.00	1.200
0.2	1.2	1.44	1.488
0.4	1.488	2.214144	1.930829
0.6	1.930829	3.728100	2.676449
0.8	2.676449	7.163378	4.109124
1.0	4.109124	16.884903	7.486105
1.2	7.486105		

(b) $\frac{dx}{dt} = x^2 \Rightarrow \int x^{-2} dx = \int 1 dt$

$-\frac{1}{x} = t + C \Rightarrow x(t) = -\frac{1}{t+C}$

$x(0) = -\frac{1}{C} = 1 \Rightarrow C = -1$

$x(t) = \frac{1}{1-t}$, so x has a vertical asymptote at $t = 1$. In part (a) the numerical solution skips right over this asymptote, because of the large step size.

7. Using Maple, in both cases it printed the message "Error, (in soln) cannot evaluate the solution further right of .99999999, probably a singularity"

9. $x' + 0.2x = e^{-0.2t} \sin(t)$, $x(0) = -1$

$e^{0.2t}(x' + 0.2x) = \sin(t)$

$e^{0.2t} x = \int \sin(t) dt = -\cos(t) + C$

$x(t) = e^{-0.2t}(C - \cos(t))$

$x(0) = C - 1 = -1 \Rightarrow C = 0$

$x(t) = -e^{-0.2t} \cos(t) \Rightarrow x(5) = -0.1043535$

11. $tx' = 2x + t^2$, $x(1) = 1$

if $t \neq 0$, $x' - \frac{2}{t}x = t \Rightarrow \mu = e^{\int(-\frac{2}{t})dt} = t^{-2}$

$t^{-2} x' - 2t^{-3} x = \frac{d}{dt}(t^{-2}x) = t^{-1}$

$t^{-2} x = \ln|t| + C \Rightarrow x(t) = t^2 \ln|t| + Ct^2$

$x(1) = C = 1 \Rightarrow x(t) = t^2(1 + \ln|t|)$

$x(5) = 65.23595$

13. $y' = -y + e^t y^2$, $y(0) = 1$

Let $v(t) = 1/y(t)$

$v' - v = -e^t \Rightarrow e^{-t} v' - e^{-t} v = -1$

$\frac{d}{dt}(e^{-t} v) = -1 \Rightarrow e^{-t} v = -t + C$

$v(t) = -te^t + Ce^t$ and $v(0) = 1/y(0) = 1 \Rightarrow C = 1$

$y(t) = 1/v(t) = \frac{1}{e^t(1-t)}$

This function has a vertical asymptote at $t = 1$, so the solution does not exist at $t = 5$.

Section 2.7

1. Phase line for $x' = x(1 - x/4)$. See figure on the left below

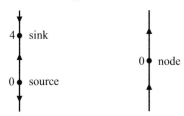

3. Phase line for $x' = x^2$. The only equilibrium is $x = 0$. See figure on the right above.

5. Phase line for $\frac{dP}{dt} = P(1-\frac{P}{5}) - \frac{0.7P^2}{(0.05)^2+P^2}$. Equilibria are at $P = 0, 0.83795, 4.1584$, and 0.0035872 (this one is hard to see)

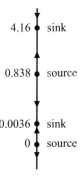

7. Phase line for $x' = rx(1 - x/N)$

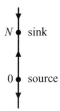

If $x(0) > 0$ the population must tend to the sink at $x = N$ as $t \to \infty$.

9. If $v' = g - \frac{k}{m}(v)^p$, q, k, m, p all positive

$g - \frac{k}{m}(v)^p = 0 \Rightarrow v = \left(\frac{mg}{k}\right)^{\frac{1}{p}}$. The single equilibrium is a sink. Given any initial value $v(0)$, $v(t)$ will approach the sink as $t \to \infty$.

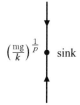

11. $x' = f(x, h) = 0.5x(4 - x) - h \equiv 2x(1 - \frac{x}{4}) - h$

   (a) If $h = 0$, the carrying capacity is 4 (4000 fish).

   (b) If $h = 1$, $f(x, 1) = -0.5x^2 + 2x - 1$ has roots $2 \pm \sqrt{2}$.
   The phase line has a sink and a source

   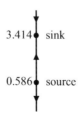

   (c) If $h = 1$, $x(0) = 0.5$ (500 fish), the population ultimately goes extinct.

   (d)

   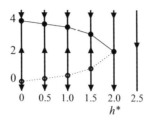

   (e) The bifurcation value of h is $h^* = \frac{rN}{4} = 2.0$.

Section 3.1

1. Linear, homogeneous

3. Linear, nonhomogeneous

5. Nonlinear

7. The solution will be continuous for all t (constant functions are continuous everywhere).

9. $-\frac{\pi}{2} < t < \frac{\pi}{2}$ because $\sec(t) = \frac{1}{\cos(t)}$ is continuous on $-\frac{\pi}{2} < t < \frac{\pi}{2}$ and $\frac{1}{2+t}$ is continuous on $-2 < t < \infty$.

11. Divide by $(t^2 - 1)$; the coefficient $\frac{t}{t^2-1}$ is continuous on $-1 < t < 1$, so the solution is continuous on $(-1, 1)$

13. $x_1'' + x_1 = -\cos(t) + \cos(t) = 0$

   $x_2'' + x_2 = -\sin(t) + \sin(t) = 0$

   $W(x_1, x_2) = x_1 x_2' - x_2 x_1' = (\cos t)(\cos t) - (\sin t)(-\sin t) = \cos^2 t + \sin^2 t = 1$, so x_1 and x_2 form a fundamental solution set on the entire t-axis

15. $t^2 x_1'' - 2t x_1' + 2x_1 = t^2(0) - 2t(1) + 2(t) = 0$
$t^2 x_2'' - 2t x_2' + 2x_2 = t^2(2) - 2t(2t) + 2(t^2) = 0$
$W(x_1, x_2) = x_1 x_2' - x_2 x_1' = t(2t) - t^2(1) = t^2 > 0$ on $t > 0$, so x_1 and x_2 form a fundamental solution set on $(-\infty, 0)$ and also on $(0, \infty)$

17. (a) $x'' + x \equiv x'' + (0)x' + (1)x = 0$, so $p(t) \equiv 0$. We need to show that $\frac{dW}{dt} = -p(t)W \equiv 0$. In Exercise 13 the Wronskian $W = 1$, so $\frac{dW}{dt} = \frac{d}{dt}(1) \equiv 0$

(c) In standard form the equation is $x'' - \frac{2}{t}x' + \frac{2}{t^2}x = 0$, so $p(t) = -\frac{2}{t}$. We need to show that $\frac{dW}{dt} = -p(t)W = \frac{2}{t}W$. In Exercise 17 the Wronskian $W = t^2$ so $\frac{dW}{dt} = 2t$ and $-p(t)W = -(-\frac{2}{t})t^2 = 2t$

Section 3.2

1. $P(r) = r^2 + 5r + 6 = (r+3)(r+2) \Rightarrow r = -2, -3$
$x(t) = C_1 e^{-2t} + C_2 e^{-3t}$

3. $P(r) = r^2 + 6r + 9 = (r+3)^2 \Rightarrow$ double root of -3
$x(t) = C_1 e^{-3t} + C_2 t e^{-3t}$

5. $P(r) = r^2 + 4r + 5 = 0 \Rightarrow r = \frac{-4 + \sqrt{16-20}}{2} = -2 \pm \frac{\sqrt{-4}}{2} = -2 \pm \iota$
$x(t) = C_1 e^{-2t} \cos(t) + C_2 e^{-2t} \sin(t)$

7. $P(r) = r^3 + r^2 + 4r + 4 = 0$
$P(-1) = -1 + 1 - 4 + 4 = 0 \Rightarrow -1$ is a root.
Dividing by $(r+1)$, $P(r) = (r+1)(r^2 + 4) \Rightarrow$ roots are $-1, \pm 2\iota$.
$x(t) = c_1 e^{-t} + A_1 \cos(2t) + B_1 \sin(2t)$

9. $r^2 + 4r + 4 = (r+2)^2 \Rightarrow r = -2$ is a double root.
$x(t) = C_1 e^{-2t} + C_2 t e^{-2t} \Rightarrow x(0) = C_1 = 2$
$x'(t) = -2 C_1 e^{-2t} + C_2 e^{-2t} - 2 C_2 t e^{-2t}$
$x'(0) = -2 C_1 + C_2 = -1 \Rightarrow C_2 = 3$
$x(t) = 2 e^{-2t} + 3t e^{-2t}$. See graph.

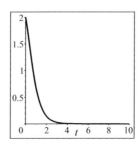

11. $r^2 + 2r + 10 = 0 \Rightarrow r = -1 \pm 3\iota$

$x(t) = e^{-t}(C_1 \cos(3t) + C_2 \sin(3t))$

$x'(t) = -e^{-t}(C_1 \cos(3t) + C_2 \sin(3t)) + e^{-t}(-3C_1 \sin(3t) + 3C_2 \cos(3t))$

$x(0) = C_1 = 1$

$x'(0) = -C_1 + 3C_2 = 0 \Rightarrow C_2 = \frac{1}{3}$

$x(t) = e^{-t}(\cos(3t) + \frac{1}{3}\sin(3t)) \equiv re^{-t}\sin(3t + \phi)$

$r = \sqrt{C_1^2 + C_2^2} = \frac{\sqrt{10}}{3}$

$\tan(\phi) = \frac{C_1}{C_2} = 3$

$x(t) = \frac{\sqrt{10}}{3} e^{-t} \sin(3t + \tan^{-1}(3))$

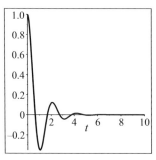

13. We need to make $R(\cos(bt)\cos(\theta) - \sin(bt)\sin(\theta))$ identical to $C_1 \sin(bt) + C_2 \cos(bt)$. Letting $C_1 = -R\sin(\theta)$ and $C_2 = R\cos(\theta)$, we need $R = \sqrt{C_1^2 + C_2^2}$ and $\tan(\theta) = -C_1/C_2$; that is, $\theta = \tan^{-1}(-C_1/C_2)$ with π added to θ if $C_2 < 0$.

15. This is the same as problem 11. So write
$$x(t) = e^{-t}(\cos(3t) + \frac{1}{3}\sin(3t))$$
as
$$\frac{\sqrt{10}}{3} e^{-t} \cos(3t - \tan^{-1}(1/3)).$$
The graph should be identical to the graph in #11.

Section 3.3

1. underdamped, unforced

3. underdamped, forced

5. overdamped, forced

7. system is undamped

$r^2 + 10 = 0$, with roots $r = \pm\sqrt{10}\iota$

$x(t) = C_1 \cos(\sqrt{10}t) + C_2 \sin(\sqrt{10}t)$

$x'(t) = -\sqrt{10}C_1 \sin(\sqrt{10}t) + \sqrt{10}C_2 \cos(\sqrt{10}t)$
$x(0) = C_1 = 2, x'(0) = \sqrt{10}C_2 = 0$
$x(t) = 2\cos(\sqrt{10}t)$

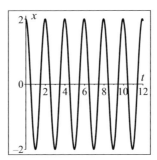

9. system is underdamped
$r^2 + 2r + 5 = 0$, with roots $-1 \pm 2\iota$
$x(t) = e^{-t}(C_1 \sin(2t) + C_2 \cos(2t))$
$C_1 = C_2 = -1 \rightarrow x(t) = -e^{-t}(\sin(2t) + \cos(2t))$
$R = \sqrt{C_1^2 + C_2^2} = \sqrt{2}$,
$\theta = \tan^{-1}(1) + \pi = \frac{5}{4}\pi$
$x(t) = \sqrt{2}e^{-t}\sin(2t + \frac{5}{4}\pi)$

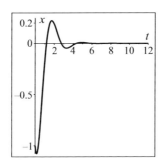

11. Find the largest k so that the solution of
$$1000x'' + 500x' + kx = 0, \quad x(0) = 2.5, \quad x'(0) = 0$$
stays above $x = -0.3$.
Write it as $x'' + \frac{1}{2}x' + Kx = 0$ where $K = k/1000$.
$r^2 + \frac{1}{2}r + K$ has roots $r = -\frac{1}{4} \pm \sqrt{1/16 - K}$, and they can be complex, so $r = -1/4 \pm \sqrt{K - 1/16}\iota$.
$$x(t) = e^{-t/4}(C_1 \cos(\sqrt{K - 1/16}\,t) + C_2 \sin(\sqrt{K - 1/16}\,t)).$$
Using the initial conditions, $C_1 = 2.5$ and $C_2 = \frac{2.5}{4\sqrt{K-1/16}}$.
The value of x will be a minimum at one half period past $t = 0$.
Setting $x\left(\frac{\pi}{\sqrt{K-1/16}}\right) = -0.3$ gives $K \approx 0.1997$; therefore, $k \approx 200$.

Section 3.4

1. $r^2 + 3r + 2 = (r+1)(r+2) \Rightarrow r = -1, -2$
   $x_h(t) = C_1 e^{-t} + C_2 e^{-2t}$ so let $x_p(t) = Ae^{2t}$.

3. $r^2 + 4 = 0 \Rightarrow r = 0 \pm 2i$
   $x_h(t) = C_1 \cos(2t) + C_2 \sin(2t)$ so $x_p(t) = t(A\cos(2t) + B\sin(2t))$.

5. $r^2 + 1 = 0 \Rightarrow r = 0 \pm i$
   $x_h(t) = C_1 \cos(t) + C_2 \sin(t)$ so $x_p(t) = A + Bt + Ct^2 + Dt^3$.

7. $x_h = C_1 e^{-t} + C_2 e^{-2t}$ so $x_p = Ae^{2t}$
   $x_p'' + 3x_p' + 2x_p = (4Ae^{2t}) + 3(2Ae^{2t}) + 2(Ae^{2t}) = 12Ae^{2t} \equiv 6e^{2t} \Rightarrow A = \frac{1}{2}$
   $x_p(t) = \frac{1}{2}e^{2t}$

9. $x_h = C_1 e^{-2t} + C_2 t e^{-2t}$
   $x_p = A + Bt + Ct^2 \Rightarrow x_p'' + 4x_p' + 4x_p$
   $= (2C) + 4(B + 2Ct) + 4(A + Bt + Ct^2)$
   $= (2C + 4B + 4A) + t(8C + 4B) + t^2(4C) \equiv 2 + (0)t + t^2$
   $C = \frac{1}{4}, \ B = -\frac{1}{2}, \ A = \frac{7}{8}$
   $x_p(t) = \frac{7}{8} - \frac{1}{2}t + \frac{1}{4}t^2$

11. $x_h = C_1 \sin(t) + C_2 \cos(t)$
    $x_p = A \sin(2t) + B \cos(2t)$
    $x_p'' + x_p = (-4A \sin(2t) - 4B \cos(2t)) + (A \sin(2t) + B \cos(2t))$
    $= -3A \sin(2t) - 3B \cos(2t) \equiv \cos(2t) \Rightarrow A = 0, \ B = -\frac{1}{3}$
    $x(t) = C_1 \sin(t) + C_2 \cos(t) - \frac{1}{3}\cos(2t)$

13. $x_h = C_1 e^{-t} + C_2 e^{-4t}$
    $x_p = x_a + x_b = (A + Bt) + (Ce^t)$
    $x_a'' + 5x_a' + 4x_a = 0 + 5(B) + 4(A + Bt) \equiv t$
    $5B + 4A = 0, 4B = 1 \Rightarrow B = \frac{1}{4}, \ A = -\frac{5}{16}$
    $x_b'' + 5x_b' + 4x_b = 10Ce^t \equiv e^t \Rightarrow C = \frac{1}{10}$
    $x(t) = C_1 e^{-t} + C_2 e^{-4t} - \frac{5}{16} + \frac{1}{4}t + \frac{1}{10}e^t$

15. $x_h = C_1 e^t + C_2 e^{4t}$, so $x_p = A + Bt$
    $x_p'' - 5x_p' + 4x_p = (-5B + 4A) + 4Bt \Rightarrow B = \frac{1}{4}, \ A = \frac{5}{16}$
    $x(t) = C_1 e^t + C_2 e^{4t} + \frac{5}{16} + \frac{1}{4}t$
    Solving $x(0) = 1, \ x'(0) = 0$ gives $C_1 = 1, \ C_2 = -\frac{5}{16}$.
    $x(t) = e^t - \frac{5}{16}e^{4t} + \frac{5}{16} + \frac{1}{4}t$

Section 3.4

17. $x_h = C_1 \cos(t) + C_2 \sin(t)$, so $x_p = Ae^{-t}$
$x_p'' + x_p = 2Ae^{-t} \equiv 4e^{-t} \Rightarrow A = 2$
$x(t) = C_1 \cos(t) + C_2 \sin(t) + 2e^{-t}$
Setting $x(0) = 1$, $x'(0) = 0$ gives $C_1 = -1$, $C_2 = 2$.
$x(t) = -\cos(t) + 2\sin(t) + 2e^{-t}$

19. $r^2 + 4r + 5 = 0 \Rightarrow r = -2 \pm \iota$
$x_h = e^{-2t}(C_1 \cos(t) + C_2 \sin(t))$
$x_p = A\cos(t) + B\sin(t)$
$x_p'' + 4x_p' + 5x_p = (5A + 4B - A)\cos(t) + (5B - 4A - B)\sin(t) \equiv \cos(t)$
Solving $4A + 4B = 1$, $4B - 4A = 0$ gives $A = B = \frac{1}{8}$.
$x(t) = e^{-2t}(C_1 \cos(t) + C_2 \sin(t)) + \frac{1}{8}(\cos(t) + \sin(t))$
$x(0) = x'(0) = 0 \Rightarrow C_1 = -\frac{1}{8}, C_2 = -\frac{3}{8}$
$x(t) = e^{-2t}(-\frac{1}{8}\cos(t) - \frac{3}{8}\sin(t)) + \frac{1}{8}(\cos(t) + \sin(t))$

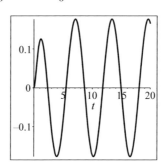

As $t \to \infty$, the exponential function e^{-2t} drives the homogeneous solution to zero, and $x(t)$ approaches the particular solution $x_p(t)$.

The following problems are on the variation of parameters.

21. $W(x_1, x_2) = e^{-t}(-3e^{-3t}) - (-e^{-t})e^{-3t} = -2e^{-4t}$
$v_1 = -\int \frac{x_2 f}{W} = -\int \frac{(e^{-3t})(2e^t)}{(-2e^{-4t})} dt = \int e^{2t} dt = \frac{e^{2t}}{2}$
$v_2 = \int \frac{x_1 f}{W} = \int \frac{(e^{-t})(2e^t)}{(-2e^{-4t})} dt = -\frac{e^{4t}}{4}$
$x_p = x_1 v_1 + x_2 v_2 = \frac{e^t}{4}$

23. $W(x_1, x_2) = e^{-t}(0) - (-e^{-t})(1) = e^{-t}$
$v_1 = -\int \frac{x_2 f}{W} = -\int \frac{e^t}{1+e^{-t}} dt = -e^t + \ln(e^t + 1)$
$v_2 = \int \frac{x_1 f}{W} = \int \frac{(e^{-t})(e^t)}{1+e^{-t}} dt = \ln(e^t + 1)$
$x_p = x_1 v_1 + x_2 v_2 = (e^{-t} + 1)\ln(e^t + 1) - 1$
As $t \Rightarrow \infty$, $x_p(t)$ is asymptotic to the line $x = t - 1$.
If $f(t) \equiv 1$, the particular solution is $x_p(t) = t$.

25. For the homogeneous equation, $r^2 + (-2-1)r + 2 = r^2 - 3r + 2$ has roots $r = 1, 2$.
Let $x_1 = t$, $x_2 = t^2$, $f(t) = \frac{3}{t}$.
$W(x_1, x_2) = t(2t) - t^2(1) = t^2$
$v_1 = -\int \frac{x_2 f}{W} = -\int \frac{3}{t} dt = -3\ln(t)$
$v_2 = \int \frac{x_1 f}{W} = \int \frac{3}{t^2} dt = -\frac{3}{t}$
$x_p = x_1 v_1 + x_2 v_2 = -3t\ln(t) - 3t$

27. $r^2 + (0-1)r - 2 = (r-2)(r+1) \Rightarrow$
$x_h = C_1 t^{-1} + C_2 t^2$
Let $x_1 = \frac{1}{t}$, $x_2 = t^2$, $f(t) = \frac{2}{t}$.
$W(x_1, x_2) = \frac{1}{t}(2t) - t^2(-\frac{1}{t^2}) = 3$
$v_1 = -\int \frac{x_2 f}{W} = -\frac{2}{3}\int t\, dt = -\frac{t^2}{3}$
$v_2 = \int \frac{x_1 f}{W} = \frac{2}{3}\int \frac{1}{t^2} dt = -\frac{2}{3t}$
$x_p = x_1 v_1 + x_2 v_2 = -\frac{t}{3} - \frac{2t}{3} = -t$

Section 3.5

1. $x'' + 4x = 1 + e^{-t}$, $x(0) = 0$, $x'(0) = 1$
$x_h(t) = C_1 \cos(2t) + C_2 \sin(2t)$
$x_p(t) = \frac{1}{4} + \frac{1}{5} e^{-t}$
Using the initial conditions, $x(t) = -0.45 \cos(2t) + 0.6 \sin(2t) + \frac{1}{4} + \frac{1}{5} e^{-t}$.

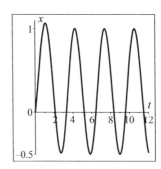

The solution does not approach a constant as $t \to \infty$ because there is no damping to make the homogeneous solution die out.

3. Let $x_p = A\sin(t) + B\cos(t)$
$x_p'' + x_p' + Kx_p = (-A - B + KA)\sin(t) + (-B + A + KB)\cos(t) \equiv (1)\sin(t)$
Solving $(K-1)A - B = 1$, $A + (K-1)B = 0$, $A = \frac{K-1}{(K-1)^2+1}$, $B = \frac{-1}{(K-1)^2+1}$.
Writing $A\sin(t) + B\cos(t) = R\sin(t+\phi)$, $R = \sqrt{\frac{(K-1)^2+1}{((K-1)^2+1)^2}} = \sqrt{\frac{1}{(K-1)^2+1}}$.

To make $R < 0.5$, $K > 1 + \sqrt{3} \approx 2.732$.

A graph of the solution, with $K = 2.732$ is shown below.

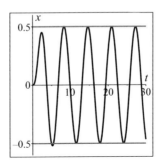

5. $I'' + 2I' + 5I = 6\cos(3t)$, $I(0) = 10$, $I'(0) = 0$

$I_h = e^{-t}(C_1 \cos(2t) + C_2 \sin(2t))$

$I_p = A\cos(3t) + B\sin(3t) \Rightarrow (-6A - 4B) = 0$, $(-4A + 6B) = 6$

$I(t) = I_h - \frac{6}{13}\cos(3t) + \frac{9}{13}\sin(3t)$

Using the initial conditions, $C_1 = \frac{136}{13}$, $C_2 = \frac{109}{26}$.

A graph of $I(t)$ is shown below, with the time $4T = 4$ marked.

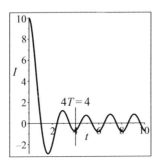

7. If the function $f(t)$ is of the form $C_1 e^{bt}(\sin(at))$ or $C_2 e^{bt}(\cos(at))$ or a linear combination of these, the form of x_p can be $Ae^{bt}(\sin(at)) + Be^{bt}(\cos(at))$, since any derivative of this function can be seen to have the same form.

For the function $f(t) = 4e^{-0.2t}\sin(t)$, let $x_p = Ae^{-0.2t}\sin(t) + Be^{-0.2t}\cos(t)$; then

$x_p' = (-0.2A - B)e^{-0.2t}\sin(t) + (A - 0.2B)e^{-0.2t}\cos(t)$,

$x_p'' = (-0.96A + 0.4B)e^{-0.2t}\sin(t) + (-0.4A - 0.96B)e^{-0.2t}\cos(t)$,

$x_p'' + 5x_p' + 6x_p = (4.04A - 4.6B)e^{-0.2t}\sin(t) + (4.6A + 4.04B)e^{-0.2t}\cos(t)$

and solving $4.04A - 4.6B = 4$, $4.6A + 4.04B = 0$ gives $A \approx 0.4311$, $B \approx -0.4909$.

Section 3.6

Cauchy-Euler problems

1. (i) $P(r) = r^2 + (5-1)r + 3 = r^2 + 4r + 3 = (r+1)(r+3)$
   $x(t) = C_1 t^{-1} + C_2 t^{-3} \to 0$ as $t \to \infty$ and is infinite at $t = 0$. Corresponds to Graph B.
   (ii) $P(r) = r^2 + (1-1)r + 9 = r^2 + 9$ has roots $0 \pm 3\imath$.
   $x(t) = C_1 t^0 \cos(3\ln(t)) + C_2 t^0 \sin(3\ln(t))$
   As $t \to 0$ x oscillates infinitely often. Corresponds to Graph C.
   (iii) $r^2 + (-3-1)r + 3 = r^2 - 4r + 3 = (r-1)(r-3)$
   $x(t) = C_1 t + C_2 t^3 \to 0$ as $t \to 0$. Corresponds to Graph A.

3. $t^2 x'' - 2x = 2t$
   $r^2 + (0-1)r - 2 = r^2 - r - 2 \to r = -1, 2$
   $x_1 = t^{-1}, \ x_2 = t^2, \ f = 2t/t^2 = 2/t$
   $W = t^{-1}(2t) - (-t^{-2})t^2 = 3$
   $v_1 = -\int (x_2 f/W) dt = -t^2/3$
   $v_2 = \int (x_1 f/W) dt = -\frac{2}{3} t^{-1}$
   $x_p = x_1 v_1 + x_2 v_2 = t^{-1}(-t^2/3) + t^2(-\frac{2}{3} t^{-1}) = -t$

5. $x'' - \frac{2}{t} x' + \frac{2}{t^2} x = \frac{3}{t}$
   $r^2 - 3r + 2 = 0 \to r = 1, 2$
   $x_1 = t, \ x_2 = t^2, \ f = 3/t$
   $W = t(2t) - t^2(1) = t^2$
   $v_1 = -\int (3/t) dt = -3 \ln(t)$
   $v_2 = \int (3/t^2) dt = -3/t$
   $x_p = x_1 v_1 + x_2 v_2 = -3t \ln(t) - 3t$
   $x(t) = C_1 t + C_2 t^2 - 3t(1 + \ln(t)), x(1) = C_1 + C_2 - 3$
   $x'(t) = C_1 + 2C_2 t - 6 - 3\ln(t), x'(1) = C_1 + 2C_2 - 6$
   Setting $x(1) = 1$, $x'(1) = 0$ gives $C_1 = C_2 = 2$.
   $x(t) = 2t + 2t^2 - 3t(1 + \ln(t))$. The graph is below.

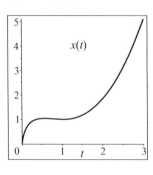

Section 3.6

7. $x'' + \frac{1}{4t^2}x = \sqrt{t}$, $x(1) = x'(1) = 0$

$4r^2 - 4r + 1 = 0 \Rightarrow r = \frac{1}{2}$ is a double root.

$x_1 = t^{\frac{1}{2}}$, $x_2 = t^{\frac{1}{2}}\ln(t)$, $f = t^{\frac{1}{2}}$

$W(t^{\frac{1}{2}}, t^{\frac{1}{2}}\ln(t)) = 1$

$v_1 = -\int (t\ln(t))dt = -\frac{t^2}{2}\ln(t) + \frac{t^2}{4}$

$v_2 = \int (t)dt = \frac{t^2}{2}$

$x_p = v_1 x_1 + v_2 x_2 = \frac{t^{5/2}}{4}$

$x(t) = C_1 t^{\frac{1}{2}} + C_2 t^{\frac{1}{2}}\ln(t) + \frac{t^{5/2}}{4}$

Setting $x(1) = x'(1) = 0$ gives $C_1 = -1/4$, $C_2 = -1/2$.

$x(t) = -\frac{1}{4}t^{\frac{1}{2}} - \frac{1}{2}t^{\frac{1}{2}}\ln(t) + \frac{t^{5/2}}{4}$

L'Hôpital's rule can be used to show that $\lim_{t \to 0} x(t) = 0$. The graph is below.

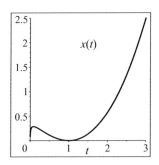

9. $x'(t) = te^{-\frac{t^2}{2}}$, $x''(t) = e^{-\frac{t^2}{2}}(t^2 - 1)$ $\quad x(0) = e^0 = 1$ and $x'(0) = 0$

$x'' + tx' + x = e^{-\frac{t^2}{2}}(t^2 - 1) + t(-te^{-\frac{t^2}{2}}) + e^{-\frac{t^2}{2}} = e^{-\frac{t^2}{2}}(t^2 - 1 - t^2 + 1) \equiv 0$

11. Divide the equation by t^2 to get $x'' + \frac{1}{t}x' + \left(\frac{t^2 - N^2}{t^2}\right)x = 0$.

(i) $p(t) = \frac{1}{t}, q(t) = \left(\frac{t^2 - N^2}{t^2}\right)$, so $p_0 = \lim_{t \to 0} tp(t) = 1$ and $q_0 = \lim_{t \to 0} t^2 q(t) = -N^2$.

The indicial equation is $r^2 + (p_0 - 1)r + q_0 = r^2 - N^2 = 0$ with roots $r = \pm N$.

(ii) For $N = 1$ there is one series solution of the form $x(t) = t\sum_{n=0}^{\infty} a_n t^n$. Differentiate the series to get x' and x''. Putting these into the equation gives:

$$\sum_{n=1}^{\infty} n(n+1)a_n t^{n+1} + \sum_{n=0}^{\infty} (n+1)a_n t^{n+1} + \sum_{n=0}^{\infty} a_n t^{n+3} - \sum_{n=0}^{\infty} a_n t^{n+1} = 0.$$

Letting $n + 3 = m + 1$ in the third sum and $n = m$ in the other three we have

$$\sum_{m=1}^{\infty} m(m+1)a_m t^{m+1} + \sum_{m=0}^{\infty} (m+1)a_m t^{m+1} + \sum_{m=2}^{\infty} a_{m-2} t^{m+1} - \sum_{m=0}^{\infty} a_m t^{m+1} = 0.$$

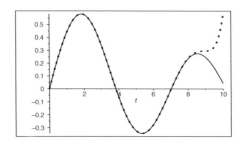

When $m = 0$ the terms in a_0 cancel, so a_0 is arbitrary. With $m = 1$, we get $3a_1 = 0$, so a_1 must be zero. For any $m \geq 2$, the recursion relation gives $a_m = -\frac{a_{m-2}}{m(m+2)}$. Therefore all a_m with m odd are zero, and for even m we have
$$a_2 = -\frac{a_0}{2\cdot 4}, a_4 = -\frac{a_2}{4\cdot 6} = \frac{a_0}{2\cdot 4\cdot 4\cdot 6}, \cdots.$$
Looking at this very carefully, it can be seen that $a_{2m} = \frac{(-1)^m a_0}{2^{2m} m!(m+1)!}$. Letting $a_0 = 1/2$, gives the formula for the **Bessel function of order 1 of the first kind**:
$$J_1(t) = \sum_{m=0}^{\infty} \frac{(-1)^m (t/2)^{2m+1}}{m!(m+1)!}.$$

(iii) The graph shows plots of $BesselJ(1.0, t)$ and eleven terms in the series (dotted graph) on the interval $0 \leq t \leq 10$. Just before $t = 10$ the round-off in computing the series becomes evident.

Section 3.7

1. $x'' = F(x, x') = -x - x'(1 - x^2)$
$\vec{T}(1.5, 2) = 2\vec{i} + (-1.5 - 2(1 - 1.5^2))\vec{j} = 2\vec{i} + \vec{j}$
$\vec{T}(-1, 1) = \vec{i} + (1 - (1 - 1))\vec{j} = \vec{i} + \vec{j}$
$\vec{T}(-2, -2) = -2\vec{i} - 4\vec{j}$

3. $x'' = F(x, x') = -x - 3x' - x^3$
$\vec{T}(-2, 3) = 3\vec{i} + \vec{j} \qquad \vec{T}(0, 1) = \vec{i} - 3\vec{j} \qquad \vec{T}(2, -3) = -3\vec{i} - \vec{j}$

5. It appears that with initial velocity $x'(0) = 2.5$ the position of the mass increases to about $0.7m$ and then goes monotonically to zero. With initial velocity $x'(0) = 4$ the position increases to about $1.4m$ before returning to zero.

7. Phase plane for $x'' + 5x' + 6x = 0$

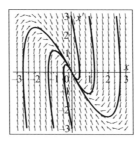

9. Phase plane for $x'' + x' + 5x = 0$

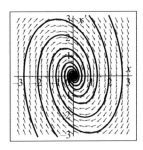

Section 4.1

1. Let $x_1 \equiv x, x_2 \equiv x'$
$$x'_1 = x_2$$
$$x'_2 = \sin(t) - 2x_1 - 5x_2$$

   The system is linear, nonhomogeneous, dimension 2.

3. Let $x_1 \equiv x, x_2 \equiv x', x_3 \equiv x''$
$$x'_1 = x_2$$
$$x'_2 = x_3$$
$$x'_3 = 2 + t - 3x_1 - 4x_3$$

   The system is linear, nonhomogeneous, dimension 3.

5. The system is already in the form (5.1). It is a homogeneous linear system of dimension 2.

7. Let $x_1 \equiv x, x_2 \equiv y, x_3 \equiv y', x_4 \equiv y''$
$$x'_1 = x_1 \cdot x_2$$
$$x'_2 = x_3$$
$$x'_3 = x_4$$
$$x'_4 = x_1 + x_4 + x_2^2$$

   The system is nonlinear of dimension 4.

9. The system equations are
$$x' = 5x - 7y$$
$$y' = 3x - 2y$$

   The system is linear, homogeneous, dimension 2.

11. (a) The system is not linear; there are several products of dependent variables, such as the term $ax(t)z(t)$ in the first equation.

    (b) The growth of the prey population is modeled by the equation for $z'(t)$. The predator X affects the growth of Z negatively due to the term $-kz(t)x(t)$.

(c) If the prey dies out, that is $z(t) \to 0$, then the equation modeling the growth of X becomes $x'(t) = -rx(t)$, and X will decrease exponentially to zero as $t \to \infty$.

(d) If both predators disappear, so $x(t) = y(t) = 0$, then the equation for the growth of Z becomes a logistic equation $z'(t) = qz(t)(1 - z(t))$, and Z will tend to its carrying capacity $z(t) = 1$.

Section 4.2

1. $\mathbf{A} + \mathbf{B} = \begin{pmatrix} 1 & 3 \\ -2 & 7 \end{pmatrix}$

3. $\mathbf{C}(\mathbf{A} + \mathbf{B})$ is not defined since $\mathbf{C}$ has dimension 2×3 and $\mathbf{A} + \mathbf{B}$ has dimension 2×2.

5. $\mathbf{ABC} = \begin{pmatrix} -21 & -29 & 5 \\ -18 & -17 & 20 \end{pmatrix}$

7. $\mathbf{AB} = \begin{pmatrix} -1 & 7 \\ 7 & 6 \end{pmatrix}$, $\mathbf{BA} = \begin{pmatrix} -8 & 7 \\ -7 & 13 \end{pmatrix}$

9. $\text{tr}(\mathbf{B}) = -1 + 3 = 2$ $\det \mathbf{B} = (-1) \cdot (3) - (1) \cdot (2) = -5$

11. $\mathbf{A}^{-1} = \frac{1}{8+3} \begin{pmatrix} 4 & -1 \\ 3 & 2 \end{pmatrix} = \begin{pmatrix} \frac{4}{11} & \frac{-1}{11} \\ \frac{3}{11} & \frac{2}{11} \end{pmatrix}$

$\mathbf{B}^{-1} = \frac{1}{-3-2} \begin{pmatrix} 3 & -2 \\ -1 & -1 \end{pmatrix} = \begin{pmatrix} \frac{-3}{5} & \frac{2}{5} \\ \frac{1}{5} & \frac{1}{5} \end{pmatrix}$

$\mathbf{D}^{-1} = \frac{1}{0+3} \begin{pmatrix} -4 & -1 \\ 3 & 0 \end{pmatrix} = \begin{pmatrix} \frac{-4}{3} & \frac{-1}{3} \\ 1 & 0 \end{pmatrix}$

13. In matrix form the system is $\begin{pmatrix} -1 & 2 \\ 1 & 3 \end{pmatrix} \begin{pmatrix} x \\ y \end{pmatrix} = \begin{pmatrix} 0 \\ 5 \end{pmatrix}$. The inverse of the matrix of coefficients is $\begin{pmatrix} \frac{-3}{5} & \frac{2}{5} \\ \frac{1}{5} & \frac{1}{5} \end{pmatrix}$, and the solution is $\begin{pmatrix} x \\ y \end{pmatrix} = \begin{pmatrix} \frac{-3}{5} & \frac{2}{5} \\ \frac{1}{5} & \frac{1}{5} \end{pmatrix} \begin{pmatrix} 0 \\ 5 \end{pmatrix} = \begin{pmatrix} 2 \\ 1 \end{pmatrix}$.

15. $\begin{pmatrix} x_1 \\ x_2 \end{pmatrix}' = \begin{pmatrix} 0 & 1 \\ -5/2 & -2 \end{pmatrix} \begin{pmatrix} x_1 \\ x_2 \end{pmatrix} + \begin{pmatrix} 0 \\ e^{2t}/2 \end{pmatrix}$

17. $\begin{pmatrix} x_1 \\ x_2 \\ x_3 \\ x_4 \end{pmatrix}' = \begin{pmatrix} 0 & 1 & 0 & 0 \\ 0 & 0 & 1 & 0 \\ 0 & 0 & 0 & 1 \\ 1 & -1 & 3 & -2 \end{pmatrix} \begin{pmatrix} x_1 \\ x_2 \\ x_3 \\ x_4 \end{pmatrix}$

Section 4.3

1. Eigenpairs $\left\{-1,\begin{pmatrix}2\\1\end{pmatrix}\right\},\left\{-4,\begin{pmatrix}1\\-1\end{pmatrix}\right\}$

3. Single eigenpair $\left\{-2,\begin{pmatrix}1\\1\end{pmatrix}\right\}$

5. Complex eigenpairs $\left\{1+3\iota,\begin{pmatrix}1\\i\end{pmatrix}\right\},\left\{1-3\iota,\begin{pmatrix}1\\-i\end{pmatrix}\right\}$

7. Eigenpairs for **G** are $\left(-4,\begin{pmatrix}0\\0\\1\end{pmatrix}\right),\left(2,\begin{pmatrix}0\\1\\1\end{pmatrix}\right)$ and $\left(1,\begin{pmatrix}1\\1\\2\end{pmatrix}\right)$

9. (a) $\dfrac{0.45750}{3}\begin{pmatrix}3\\3\\5\end{pmatrix}=\begin{pmatrix}0.45750\\0.45750\\0.76250\end{pmatrix}$

   $(-0.40825-0.40825i)\begin{pmatrix}1\\-1+\iota\\0\end{pmatrix}=\begin{pmatrix}-0.40825-0.40825\iota\\0.81650\\0\end{pmatrix}$

   (b) $(0.45750)^2+(0.45750)^2+(0.76249)^2=1.000004\approx 1$
   $(-0.40825)^2+(-0.40825)^2+(0.81650)^2=1.000008\approx 1$

10. If $\vec{X}=e^{rt}\vec{U}$ is a solution of $\vec{X}'=\mathbf{A}\vec{X}$, then we know that $\mathbf{A}\vec{U}=r\vec{U}$. Now let $\vec{Y}=e^{rt}(k\vec{U})$ (note that $e^{rt}k$ is a scalar). Then

    $$\frac{d}{dt}\vec{Y}=kre^{rt}\vec{U}=ke^{rt}(r\vec{U})=ke^{rt}(\mathbf{A}\vec{U})=\mathbf{A}(ke^{rt}\vec{U})=\mathbf{A}e^{rt}(k\vec{U})=\mathbf{A}\vec{Y}$$

    using the properties of scalar and matrix multiplication.

Section 4.4

1. $\vec{x}(t)=c_1 e^{2t}\begin{pmatrix}1\\0\end{pmatrix}+c_2 e^{-3t}\begin{pmatrix}1\\-5\end{pmatrix}$ $\quad\begin{cases}x(t)=c_1 e^{2t}+c_2 e^{-3t}\\y(t)=-5c_2 e^{-3t}\end{cases}$

3. Using the eigenpair $\left(2\iota,\begin{pmatrix}-2-2\iota\\1\end{pmatrix}\right)$,

   $\vec{x}(t)=c_1\left(\cos(2t)\begin{pmatrix}-2\\1\end{pmatrix}-\sin(2t)\begin{pmatrix}-2\\0\end{pmatrix}\right)+c_2\left(\sin(2t)\begin{pmatrix}-2\\1\end{pmatrix}+\cos(2t)\begin{pmatrix}-2\\0\end{pmatrix}\right)$

   $\begin{cases}x(t)=c_1(-2\cos(2t)+2\sin(2t))+c_2(-2\sin(2t)-2\cos(2t))\\y(t)=c_1\cos(2t)+c_2\sin(2t)\end{cases}$

   Your answer may look quite different if you use a different eigenpair, but if initial conditions are given they will both give the same answer.

5. $\vec{x}(t)=c_1 e^{-2t}\begin{pmatrix}1\\1\end{pmatrix}+c_2 e^{-2t}\left(t\begin{pmatrix}1\\1\end{pmatrix}+\begin{pmatrix}1\\\frac{3}{2}\end{pmatrix}\right)$

   $\vec{x}(0)=c_1\begin{pmatrix}1\\1\end{pmatrix}+c_2\begin{pmatrix}1\\\frac{3}{2}\end{pmatrix}=\begin{pmatrix}2\\3\end{pmatrix}$

The solution of this linear system is $c_1 = 0, c_2 = 2$; therefore,

$$\vec{x}(t) = \begin{pmatrix} x(t) \\ y(t) \end{pmatrix} = 2e^{-2t}\left(t\begin{pmatrix} 1 \\ 1 \end{pmatrix} + \begin{pmatrix} 1 \\ \frac{3}{2} \end{pmatrix}\right).$$

7. $x(t) = C_1 e^t - C_2 e^{-t}, y(t) = C_1 e^t + C_2 e^{-t}$

9. $x(t) = C_1 e^t + C_2 e^{2t}, y(t) = C_1 e^t$

11. $x(t) = e^t \sin t, y(t) = e^t \cos t$

13. The equation for $x(t)$ is separable:
$$\frac{dx}{dt} = -x \Rightarrow \int \frac{dx}{x} = -\int dt \Rightarrow \ln(x) = -t + C \Rightarrow x = e^C e^{-t}$$
$x(0) = 1 \Rightarrow e^C = 1 \Rightarrow x(t) = e^{-t}$
$\frac{dy}{dt} = e^{-t}, \; y(0) = -1 \Rightarrow y(t) = -e^{-t}$

Answers to the Mixing Problem

15. Setting $x(t) = y(t)$ gives the value $t \approx 21.3$ minutes when the two amounts are equal.

17. The volume in Tank A is $V_A(t) = 50 + t$.
The volume in Tank B is $V_B(t) = 40 - t$.
The system of equations becomes
$$x'(t) = \frac{y(t)}{V_B} - \frac{2x(t)}{V_A}$$
$$y'(t) = \frac{2x(t)}{V_A} - \frac{3y(t)}{V_B}.$$
The equations are defined only for $t < 40$. At time $t = 40$ Tank B becomes empty.

19. (a) The equation for x' contains an added input term: $x' = -3\frac{x}{50} + \frac{y}{40} + 2p$
so the system is
$$\begin{pmatrix} x \\ y \end{pmatrix}' = \begin{pmatrix} -0.06 & 0.025 \\ 0.06 & -0.075 \end{pmatrix}\begin{pmatrix} x \\ y \end{pmatrix} + \begin{pmatrix} 2p \\ 0 \end{pmatrix}$$

(b) To find the particular solution, let $\vec{x}_p = \begin{pmatrix} a \\ b \end{pmatrix}$, then
$$\vec{x}'_p = \begin{pmatrix} 0 \\ 0 \end{pmatrix} = \mathbf{A}\begin{pmatrix} a \\ b \end{pmatrix} + \begin{pmatrix} 2p \\ 0 \end{pmatrix}$$
$$\begin{pmatrix} a \\ b \end{pmatrix} = \mathbf{A}^{-1}\begin{pmatrix} -2p \\ 0 \end{pmatrix} = \begin{pmatrix} -25 & -\frac{25}{3} \\ -20 & -20 \end{pmatrix}\begin{pmatrix} -2p \\ 0 \end{pmatrix} = \begin{pmatrix} 50p \\ 40p \end{pmatrix}$$

(c) With $p = 0.1$ and $t = 0$,
$$\vec{x}(0) = \begin{pmatrix} 0.61625 & -0.47001 \\ 0.78755 & 0.88266 \end{pmatrix}\begin{pmatrix} c_1 \\ c_2 \end{pmatrix} + \begin{pmatrix} 5 \\ 4 \end{pmatrix} = \begin{pmatrix} 5 \\ 1 \end{pmatrix} \Rightarrow$$
$$\begin{pmatrix} c_1 \\ c_2 \end{pmatrix} = \begin{pmatrix} 0.61625 & -0.47001 \\ 0.78755 & 0.88266 \end{pmatrix}^{-1}\begin{pmatrix} 0 \\ -3 \end{pmatrix} \approx \begin{pmatrix} -1.5425 \\ -2.0225 \end{pmatrix}$$

Section 4.5

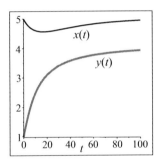

(d) A graph of x and y is shown above.
(e) In this case the concentration in both tanks must approach 0.1 pounds per gallon as $t \to \infty$; therefore, $x(t) \to 5$ and $y(t) \to 4$ as $t \to \infty$.

Section 4.5

1. $\mathbf{A} = \begin{pmatrix} 2 & 0 \\ 0 & -4 \end{pmatrix}$ is diagonal, so $e^{\mathbf{A}t} = \begin{pmatrix} e^{2t} & 0 \\ 0 & e^{-4t} \end{pmatrix}$

3. $\mathbf{A} = \mathbf{D} + \mathbf{N} = \begin{pmatrix} -1 & 0 \\ 0 & -1 \end{pmatrix} + \begin{pmatrix} 0 & 2 \\ 0 & 0 \end{pmatrix}$

   $\mathbf{N}^2 = \mathbf{0}$, so $e^{\mathbf{N}t} = \mathbf{I} + \mathbf{N}t = \begin{pmatrix} 1 & 2t \\ 0 & 1 \end{pmatrix}$

   $\mathbf{D}$ diagonal $\Rightarrow e^{\mathbf{D}t} = \begin{pmatrix} e^{-t} & 0 \\ 0 & e^{-t} \end{pmatrix}$

   $\mathbf{DN} = \mathbf{ND}$, so
   $$e^{\mathbf{A}t} = e^{\mathbf{D}t}e^{\mathbf{N}t} = \begin{pmatrix} e^{-t} & 2te^{-t} \\ 0 & e^{-t} \end{pmatrix}$$

5. $\mathbf{A}^2 = \begin{pmatrix} 1 & 0 \\ 1 & 0 \end{pmatrix}\begin{pmatrix} 1 & 0 \\ 1 & 0 \end{pmatrix} = \begin{pmatrix} 1 & 0 \\ 1 & 0 \end{pmatrix} = \mathbf{A}$

   So $\mathbf{A}^3 = \mathbf{A}^4 = \cdots = \mathbf{A}$.

   $e^{\mathbf{A}t} = \mathbf{I} + \mathbf{A}\left(t + \dfrac{t^2}{2!} + \dfrac{t^3}{3!} + \cdots\right) = \begin{pmatrix} 1 & 0 \\ 0 & 1 \end{pmatrix} + \begin{pmatrix} 1 & 0 \\ 1 & 0 \end{pmatrix}(e^t - 1) = \begin{pmatrix} e^t & 0 \\ e^t - 1 & 1 \end{pmatrix}$

7. $\vec{\mathbf{x}} = \begin{pmatrix} e^{2t} & 0 \\ 0 & e^{-4t} \end{pmatrix}\begin{pmatrix} 1 \\ -1 \end{pmatrix} = \begin{pmatrix} e^{2t} \\ -e^{-4t} \end{pmatrix}$

   $x' = 2e^{2t} = 2x$, $\quad y' = 4e^{-4t} = -4y$

9. $\vec{\mathbf{x}} = \begin{pmatrix} e^{-t} & 2te^{-t} \\ 0 & e^{-t} \end{pmatrix}\begin{pmatrix} 1 \\ -1 \end{pmatrix} = \begin{pmatrix} e^{-t}(1-2t) \\ -e^{-t} \end{pmatrix}$

   $x' = \dfrac{d}{dt}\left(e^{-t}(1-2t)\right) = -3e^{-t} + 2te^{-t} = -x + 2y$, $\quad y' = \dfrac{d}{dt}(-e^{-t}) = e^{-t} = -y$

11. $\vec{\mathbf{x}} = \begin{pmatrix} e^t & 0 \\ e^t - 1 & 1 \end{pmatrix}\begin{pmatrix} 1 \\ -1 \end{pmatrix} = \begin{pmatrix} e^t \\ e^t - 2 \end{pmatrix}$

    $x' = e^t = x$, $\quad y' = \dfrac{d}{dt}(e^t - 2) = e^t = x$

13. $\mathbf{A}^2 = \begin{pmatrix} 0 & 1 \\ 1 & 0 \end{pmatrix}\begin{pmatrix} 0 & 1 \\ 1 & 0 \end{pmatrix} = \begin{pmatrix} 1 & 0 \\ 0 & 1 \end{pmatrix} = \mathbf{I}$

so $\mathbf{A}^3 = \mathbf{A}, \mathbf{A}^4 = \mathbf{I}, \cdots$

$$e^{\mathbf{A}t} = \begin{pmatrix} 1 & 0 \\ 0 & 1 \end{pmatrix} + \begin{pmatrix} 0 & 1 \\ 1 & 0 \end{pmatrix}t + \begin{pmatrix} 1 & 0 \\ 0 & 1 \end{pmatrix}\frac{t^2}{2} + \cdots$$

$$= \begin{pmatrix} 1 + \frac{t^2}{2!} + \frac{t^4}{4!} \cdots & t + \frac{t^3}{3!} + \cdots \\ t + \frac{t^3}{3!} + \cdots & 1 + \frac{t^2}{2!} + \frac{t^4}{4!} \cdots \end{pmatrix}$$

$$= \begin{pmatrix} \cosh(t) & \sinh(t) \\ \sinh(t) & \cosh(t) \end{pmatrix}$$

$$\vec{x}(t) = \int_0^t e^{\mathbf{A}(t-s)} \begin{pmatrix} 2 \\ 0 \end{pmatrix} ds + e^{\mathbf{A}t}\begin{pmatrix} 1 \\ 1 \end{pmatrix}$$

$$= \int_0^t \begin{pmatrix} \cosh(t-s) & \sinh(t-s) \\ \sinh(t-s) & \cosh(t-s) \end{pmatrix}\begin{pmatrix} 2 \\ 0 \end{pmatrix} ds + \begin{pmatrix} \cosh(t) & \sinh(t) \\ \sinh(t) & \cosh(t) \end{pmatrix}\begin{pmatrix} 1 \\ 1 \end{pmatrix}$$

$$= \int_0^t \begin{pmatrix} 2\cosh(t-s) \\ 2\sinh(t-s) \end{pmatrix} ds + \begin{pmatrix} \cosh(t) + \sinh(t) \\ \sinh(t) + \cosh(t) \end{pmatrix}$$

$$= \begin{pmatrix} -2\sinh(0) + 2\sinh(t) \\ -2\cosh(0) + 2\cosh(t) \end{pmatrix} + \begin{pmatrix} \cosh(t) + \sinh(t) \\ \sinh(t) + \cosh(t) \end{pmatrix}$$

$x(t) = 3\sinh(t) + \cosh(t)$

$y(t) = -2 + 3\cosh(t) + \sinh(t)$

Section 5.1

1. There is only one equilibrium point at $(5, 4)$.

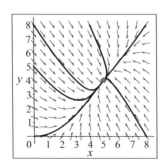

3. The equilibrium points are $(0,0), (1,0)$, and $(\frac{1}{2}, 1)$.

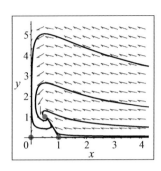

5. The only equilibrium point is $(0,0)$. Every solution tends to a limit cycle as $t \to \infty$.

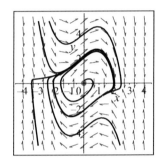

Section 5.2

1. $tr(\mathbf{A}) = -1$, $\det(\mathbf{A}) = 2$. $(0,0)$ is a spiral sink.

3. $tr(\mathbf{A}) = -3$, $\det(\mathbf{A}) = -2$. $(0,0)$ is a saddle point.

5. $tr(\mathbf{A}) = 0$, $\det(\mathbf{A}) = 2$. $(0,0)$ is a center.

7. Eigenpairs: $\left(-\frac{1}{2} \pm \frac{\sqrt{7}}{2}i, \begin{pmatrix} 0.530 \pm 0.468i \\ -0.707 \end{pmatrix}\right)$.   Spiral sink at $(0,0)$.

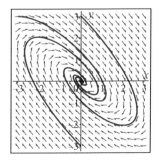

9. Eigenpairs: $\left(0.562, \begin{pmatrix} 0.402 \\ 0.916 \end{pmatrix}\right), \left(-3.562, \begin{pmatrix} -0.977 \\ -0.214 \end{pmatrix}\right)$ Saddle point at $(0,0)$.

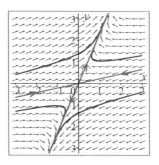

11. Eigenpairs: $\left(\pm\sqrt{2}i, \begin{pmatrix} -1 \pm (-\frac{\sqrt{2}}{2}i) \\ 1 \end{pmatrix}\right)$. Center at $(0,0)$.

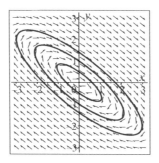

13. Equation is overdamped. $\begin{pmatrix} x \\ y \end{pmatrix}' = \begin{pmatrix} 0 & 1 \\ -2 & -4 \end{pmatrix} \begin{pmatrix} x \\ y \end{pmatrix}$

    $tr(\mathbf{A}) = -4$, $\det(\mathbf{A}) = 2$. $(0,0)$ is a sink.

15. Equation is underdamped. $\begin{pmatrix} x \\ y \end{pmatrix}' = \begin{pmatrix} 0 & 1 \\ -\frac{1}{3} & -\frac{2}{3} \end{pmatrix} \begin{pmatrix} x \\ y \end{pmatrix}$

    $tr(\mathbf{A}) = -\frac{2}{3}$, $\det(\mathbf{A}) = \frac{1}{3}$. $(0,0)$ is a spiral sink.

17. Equation is overdamped. $\begin{pmatrix} x \\ y \end{pmatrix}' = \begin{pmatrix} 0 & 1 \\ -4 & -5 \end{pmatrix} \begin{pmatrix} x \\ y \end{pmatrix}$

    $tr(\mathbf{A}) = -5$, $\det(\mathbf{A}) = 4$. $(0,0)$ is a sink.

> A spring-mass system is
> underdamped $\Leftrightarrow$ $(0,0)$ is a spiral sink
> overdamped $\Leftrightarrow$ $(0,0)$ is a sink
> critically damped $\Leftrightarrow$ $(tr(\mathbf{A}), \det(\mathbf{A}))$ lies on the curve $\det = \frac{tr^2}{4}$

19. $\begin{pmatrix} x \\ y \end{pmatrix}' = \begin{pmatrix} 0 & 1 \\ -\frac{k}{m} & -\frac{b}{m} \end{pmatrix} \begin{pmatrix} x \\ y \end{pmatrix}$

The characteristic polynomial of the matrix is

$$\det\begin{pmatrix} 0-r & 1 \\ -\frac{k}{m} & -\frac{b}{m}-r \end{pmatrix} = \frac{1}{m}(mr^2 + br + k).$$

Since $m \neq 0$, this has the same roots as the characteristic polynomial for the second-order equation.

In both cases, these roots determine the exponential terms in the solution.

21. (a) We must have $\text{tr}(\mathbf{A}) = a + d = 0$ and $\det(\mathbf{A}) = ad - bc = 0$; that is, $d = -a$ and $bc = -a^2$. If $a \neq 0$, then neither b nor c can be 0, and the matrix must have the form given in (5.4). If $a = 0$, the other 3 forms correspond to $b = c = 0$; $c = 0, b \neq 0$; and $b = 0, c \neq 0$.

(b) If $\mathbf{A} = \begin{pmatrix} 0 & 0 \\ 0 & 0 \end{pmatrix}$, then $\vec{\mathbf{x}}' = \mathbf{A}\vec{\mathbf{x}} \Rightarrow \begin{cases} x' = 0 \\ y' = 0. \end{cases}$

This says that every point in the phase plane is an equilibrium point.

If $\mathbf{A} = \begin{pmatrix} 0 & b \\ 0 & 0 \end{pmatrix}$, then $\begin{cases} x' = by \\ y' = 0 \end{cases} \Rightarrow \begin{cases} y(t) = y_0 \\ x(t) = x_0 + by_0 t. \end{cases}$

Every point on the x-axis ($y_0 = 0$) is an equilibrium point. Trajectories through (x_0, y_0), where $y_0 \neq 0$, are horizontal lines, moving right if $by_0 > 0$ and left if $by_0 < 0$.

If $\mathbf{A} = \begin{pmatrix} 0 & 0 \\ c & 0 \end{pmatrix}$, then $\begin{cases} x' = 0 \\ y' = cx \end{cases} \Rightarrow \begin{cases} x(t) = x_0 \\ y(t) = y_0 + cx_0 t. \end{cases}$

Every point on the y-axis ($x_0 = 0$) is an equilibrium point. Trajectories through (x_0, y_0), where $x_0 \neq 0$, are vertical lines, moving up if $cx_0 > 0$ and down if $cx_0 < 0$.

(c) An eigenvector corresponding to eigenvalue 0 must satisfy

$$\begin{pmatrix} a & -b \\ \frac{a^2}{b} & -a \end{pmatrix}\begin{pmatrix} u \\ v \end{pmatrix} = \begin{pmatrix} au - bv \\ \frac{a^2}{b}u - av \end{pmatrix} = \begin{pmatrix} 0 \\ 0. \end{pmatrix}$$

Every vector that satisfies this equation must have $au = bv$, and hence be of the form $K\begin{pmatrix} b \\ a \end{pmatrix}$ for some constant K.

(d) $\mathbf{A}\vec{\mathbf{u}}^* = \begin{pmatrix} a & -b \\ \frac{a^2}{b} & -a \end{pmatrix}\begin{pmatrix} 0 \\ -1 \end{pmatrix} = \begin{pmatrix} b \\ a \end{pmatrix}$; therefore, the general solution of the system is

$$\vec{\mathbf{x}}(t) = c_1 e^{0t}\vec{\mathbf{u}}_1 + c_2 e^{0t}(t\vec{\mathbf{u}}_1 + \vec{\mathbf{u}}^*) = c_1\begin{pmatrix} b \\ a \end{pmatrix} + c_2\left(t\begin{pmatrix} b \\ a \end{pmatrix} + \begin{pmatrix} 0 \\ -1 \end{pmatrix}\right)$$

$$= \begin{pmatrix} b(c_1 + c_2 t) \\ a(c_1 + c_2 t) - c_2 \end{pmatrix} = \begin{pmatrix} x(t) \\ y(t) \end{pmatrix}.$$

(e) From the solution obtained in (d), we see that $y(t) = \frac{a}{b}x(t) - c_2$.

(f) If $y = \frac{a}{b}x$, then $\vec{\mathbf{x}}'(t) = \begin{pmatrix} a & -b \\ \frac{a^2}{b} & -a \end{pmatrix}\begin{pmatrix} x \\ \frac{a}{b}x \end{pmatrix} = \begin{pmatrix} 0 \\ 0 \end{pmatrix}.$

(g) Assume first that the trajectory lies above the line $y = \frac{a}{b}x$; then, at any point on the trajectory, $y = \frac{a}{b}x + \varepsilon$, $\varepsilon > 0$. The tangent vector at the point is

$$\vec{x}'(t) = \begin{pmatrix} a & -b \\ \frac{a^2}{b} & -a \end{pmatrix} \begin{pmatrix} x \\ \frac{a}{b}x + \varepsilon \end{pmatrix} = \begin{pmatrix} -b\varepsilon \\ -a\varepsilon \end{pmatrix}.$$

If $b > 0$, $x' = -b\varepsilon < 0$, so the trajectory is moving left. If $b < 0$ it is moving right. The case where the trajectory lies below the line can be treated similarly.

(h) Phase plane for $\vec{x}' = \begin{pmatrix} 1 & -2 \\ \frac{1}{2} & -1 \end{pmatrix} \vec{x}$

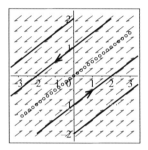

The four phase planes below show what happens if the entries in the second row of the matrix are changed by small amounts. In each case, the values of c and d are chosen to make the point $(\text{tr}(\mathbf{A}), \det(\mathbf{A}))$ lie just inside the appropriate region ($\varepsilon = 0.35$).

sink at $(0, 0)$

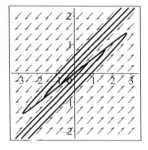

The entries in the second row of the matrix were changed to $c = \frac{1}{2} + \frac{\varepsilon}{2} + \frac{\varepsilon^2}{16}$, $d = -1 - \varepsilon$.

spiral sink at $(0, 0)$

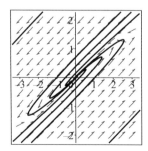

Section 5.3

The entries in the second row of the matrix were changed to $c = \frac{1}{2} + \frac{\varepsilon}{2} + \frac{\varepsilon^2}{4}$, $d = -1 - \varepsilon$.

center at $(0,0)$

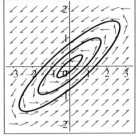

The entries in the second row of the matrix were changed to $c = \frac{1}{2} + \frac{\varepsilon}{2}$, $d = -1$.

saddle point at $(0,0)$

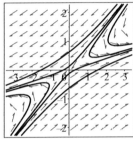

The entries in the second row of the matrix were changed to $c = \frac{1}{2} - \frac{\varepsilon}{2}$, $d = -1$.

Section 5.3

1. Using the Jacobian matrix $J(x,y) = \begin{pmatrix} 1 - 2x - y & -x \\ -\frac{1}{2}y & \frac{3}{4} - 2y - \frac{1}{2}x \end{pmatrix}$, the type of the equilibrium points is:

   $(0,0)$, source

   $(1,0)$ and $(0, \frac{3}{4})$, saddles

   $(\frac{1}{2}, \frac{1}{2})$, sink

   A phase plane is shown below:

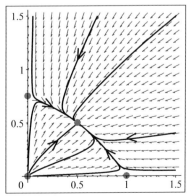

3. The equilibrium points (g, w) are:

   $(0, 0)$, source

   $(1, 0)$ and $(0.373270, 0.723398)$, saddles

   $(0, 1)$ and $(0.651049, 0.455001)$, sinks

   A phase plane is shown below:

   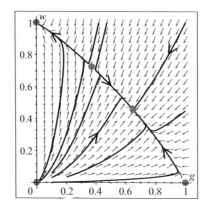

5. At an equilibrium solution the three functions

   $$F(x, y, z) = x(-0.2 + 0.5z)$$
   $$G(x, y, z) = y(-0.1 - 0.05x + 0.4z)$$
   $$H(x, y, z) = z(0.4 - 0.4z - 0.2x - 0.2y)$$

   must all be equal to zero; therefore, to have all three variables x, y, and z unequal to zero, they must satisfy the linear system

   $$\begin{pmatrix} 0 & 0 & 0.5 \\ -0.05 & 0 & 0.4 \\ 0.2 & 0.2 & 0.4 \end{pmatrix} \begin{pmatrix} x \\ y \\ z \end{pmatrix} = \begin{pmatrix} 0.2 \\ 0.1 \\ 0.4 \end{pmatrix}$$

   with solution $x = 1.2$, $y = 0$, $z = 0.4$. Since $y = 0$, the species Y is extinct; i.e., there is no equilibrium point where all three species coexist.

   The other equilibrium points $(0, 0, 0)$, $(0, 0, 1)$, $(0, 1.5, 0.25)$ can be found by assuming that one or more of the populations is zero. If $z = 0$, the prey population is extinct, and both x and y can be seen to decrease to zero as $t \to \infty$.

   The graph shows time plots of $x(t)$, $y(t)$, and $z(t)$.

   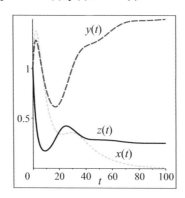

A 3 × 3 Jacobian matrix can be computed for this system by taking partial derivatives of F, G, and H with respect to each of the three independent variables. The eigenvalues of the Jacobian matrix will be found to all have negative real parts at one of the equilibrium solutions, and, as in the case of a 2-dimensional system, this will mean that this equilibrium is a sink. All solutions starting close enough to it will converge to it as $t \to \infty$.

7. The plots of $x(t)$ are shown for $\mu = 0.5, 2$, and 7, and in each case an estimate of the period is given. It is clear that the period increases monotonically with μ.

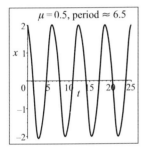
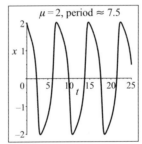
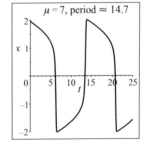

Section 5.4

1. Write $f(x, y) = x(-ax^2 + (1 + a)x - 1 - \varepsilon y) = 0$, $g(x, y) = y(x - 0.5) = 0$. $g(x, y) = 0 \Rightarrow$ either $y = 0$ or $x = 0.5$. If $y = 0$, then $x = 0, 1$, or $\frac{1}{a}$, and $(0, 0)$, $(\frac{1}{a}, 0)$, and $(1, 0)$ are equilibria. If $y \neq 0$, then $x = 0.5$ and $f(x, y) = 0 \Rightarrow y = \frac{a-2}{4\varepsilon}$. Therefore, $(\frac{1}{2}, \frac{a-2}{4\varepsilon})$ is an equilibrium.

(b) The Jacobian $J(x, y)$ is $\begin{pmatrix} -3ax^2 + 2(1 + a)x - 1 - \varepsilon y & -\varepsilon x \\ y & x - \frac{1}{2} \end{pmatrix}$.

This can be used to show that $(0, 0)$ is a sink for any value of a, $(1, 0)$ and $(\frac{1}{a}, 0)$ are saddles if $a > 2$.

$J(\frac{1}{2}, \frac{a-2}{4\varepsilon}) = \begin{pmatrix} \frac{1}{2} & -\frac{\varepsilon}{2} \\ \frac{a-2}{4\varepsilon} & 0 \end{pmatrix}$, with $\text{tr}(J) = \frac{1}{2}$ and $\det(J) = \frac{a-2}{8}$.

Setting $\det(J) = (\text{tr}(J))^2/4$, this point will be a source if $a < 2.5$ and a spiral source if $a > 2.5$. The bifurcation value is $a^* = 2.5$.

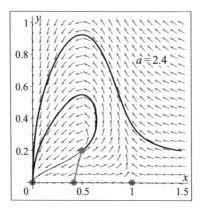

The only effect ε has is on the size of $\bar{y}$ in the fourth equilibrium point. It does not affect the type of any of the four equilibrium points.

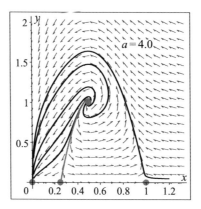

With $\varepsilon = \frac{1}{2}$, phase portraits are shown for $a = 2.4 < a^*$ and $a = 4 > a^*$. The stable manifold of the saddle point at $(\frac{1}{a}, 0)$ has been drawn in both graphs. It can be seen to separate solutions having different behavior as $t \to \infty$.

3. It is easy to see that $(0,0)$, $(0,3)$, and $(N,0)$ are equilibrium points.

The Jacobian $J(x,y) = \begin{pmatrix} 0.2 - 0.4\frac{x}{N} + 0.1y & 0.1x \\ 0.05y & 0.6 - 0.4y + 0.05x \end{pmatrix}$ can be used to show that for any $N > 0$, $(0,0)$ is a source, and $(0,3)$ and $(N,0)$ are saddles.

To be a coexistent state, $(\bar{x}, \bar{y})$ must satisfy

$$\begin{cases} 0.2\left(1 - \dfrac{\bar{x}}{N}\right) + 0.1\bar{y} = 0 \\ 0.6\left(1 - \dfrac{\bar{y}}{3}\right) + 0.05\bar{x} = 0. \end{cases} \quad (*)$$

Using Cramer's Rule, these have a solution:

$$\bar{x} = \frac{0.1}{0.04/N - 0.005}, \quad \bar{y} = \frac{0.01 + 0.12/N}{0.04/N - 0.005}.$$

Using (∗), the Jacobian at $(\bar{x}, \bar{y})$ can be simplified to

$$J(\bar{x}, \bar{y}) = \begin{pmatrix} -0.2\frac{\bar{x}}{N} & 0.1\bar{x} \\ 0.05\bar{y} & -0.2\bar{y} \end{pmatrix}.$$

$\text{tr}(J) = -0.2\left(\frac{\bar{x}}{N} + \bar{y}\right) < 0$ and $\det(J) = 0.04\frac{\bar{x}}{N}\bar{y} - 0.005\bar{x}\bar{y} > 0$ if $N < 8$.

Therefore $(\bar{x}, \bar{y})$ is a sink if $N < 8$.

At $N = 8$ the point $(\bar{x}, \bar{y})$ goes to ∞. Using the Poincaré-Bendixson Theorem, what must happen to any solution starting in the positive quadrant?

Phase planes are shown for $N = 4$ and $N = 10$.

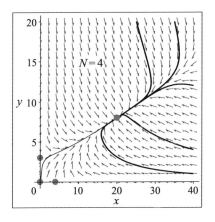
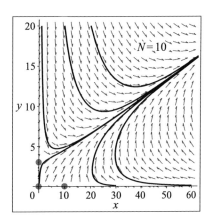

Section 6.1

1. $\dfrac{1}{s-3}$

3. $\mathcal{L}(2 + \cos(t)) = \dfrac{2}{s} + \dfrac{s}{s^2+1} = \dfrac{3s^2+2}{s(s^2+1)}$

5. $\mathcal{L}(2e^t - 3\cos(t)) = \dfrac{2}{s-1} - \dfrac{3s}{s^2+1} = \dfrac{-s^2+3s+2}{(s-1)(s^2+1)}$

7. $\mathcal{L}(\sin(3t + 2)) = \mathcal{L}(\sin(3t)\cos(2) + \cos(3t)\sin(2)) = \dfrac{3\cos(2)+s\sin(2)}{s^2+9}$

9. $L(\sin(bt)) = e^{-st}\left(\dfrac{-s\sin(bt)-b\cos(bt)}{(-s)^2+b^2}\bigg|_0^\infty\right) = 0 - \left(\dfrac{-s\sin(0)-b\cos(0)}{s^2+b^2}\right) = \dfrac{b}{s^2+b^2}$

11. Using integration by parts,

$L(t) = \int_0^\infty e^{-st} t\, dt = \dfrac{te^{-st}}{-s}\bigg|_0^\infty - \int_0^\infty \dfrac{e^{-st}}{-s} dt = 0 + \dfrac{1}{s}\int_0^\infty e^{-st} dt = \dfrac{1}{s}\dfrac{e^{-st}}{-s}\bigg|_0^\infty = \dfrac{1}{s^2}$

13. $\mathcal{L}(f'''(t)) = L((f''(t))') = sL(f''(t)) - f''(0)$
    $= s(s^2 F(s) - sf(0) - f'(0)) - f''(0) \equiv s^3 F(s) - s^2 f(0) - sf'(0) - f''(0)$

15. The graph of $g(t)$ is shown on the left below.
    It is piecewise continuous.

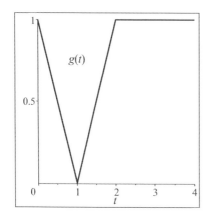
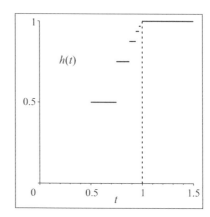

17. The graph of $h(t)$ is shown on the right above.

    It is not piecewise continuous since it has an infinite number of jumps between 0 and 1.

Section 6.2

1. $f(t) = 2e^{-3t}$

3. $f(t) = 2 + 3e^{-5t}$

5. $f(t) = \mathcal{L}^{-1}\left(\frac{2s}{s^2+9} + \frac{3}{s^2+9}\right) = 2\cos(3t) + \sin(3t)$

7. $f(t) = 2e^{-t} - e^{-2t}$

9. $f(t) = \frac{4}{3}e^{-t} - \frac{3}{2}e^{-2t} + \frac{1}{6}e^{-4t}$

11. $\mathcal{L}(x' + 2x) = \mathcal{L}(e^{3t})$

    $sX(s) - x(0) + 2X(s) = \frac{1}{s-3}$

    $(s+2)X(s) = 5 + \frac{1}{s-3} = \frac{5(s-3)+1}{s-3}$

    $X(s) = \frac{5s-14}{(s-3)(s+2)}$

13. $\mathcal{L}(x'' + 4x' + 2x) = \mathcal{L}(\cos(3t))$

    $(s^2 X(s) - sx(0) - x'(0)) + 4(sX(s) - x(0)) + 2X(s) = \frac{s}{s^2+9}$

    $(s^2 + 4s + 2)X(s) = s - 1 + 4 + \frac{s}{s^2+9} = \frac{(s+3)(s^2+9)+s}{s^2+9}$

    $X(s) = \frac{s^3+3s^2+10s+27}{(s^2+9)(s^2+4s+2)}$

15. $2(sX(s) - x(0)) + 5X(s) = \frac{1}{s+1}$

    $(2s+5)X(s) = 2 + \frac{1}{s+1} = \frac{2(s+1)+1}{s+1}$

    $X(s) = \frac{2s+3}{(s+1)(2s+5)} = \frac{1/3}{s+1} + \frac{4/3}{2s+5} \Rightarrow x(t) = \frac{1}{3}e^{-t} + \frac{2}{3}e^{-\frac{5}{2}t}$

Section 6.3

17. $s^2 X(s) + 3sX(s) + 2X(s) = \dfrac{4}{s+3}$

    $(s^2 + 3s + 2)X(s) = \dfrac{4}{s+3}$

    $X(s) = \dfrac{4}{(s+1)(s+2)(s+3)} = \dfrac{2}{s+1} - \dfrac{4}{s+2} + \dfrac{2}{s+3}$

    $\Rightarrow x(t) = 2e^{-t} - 4e^{-2t} + 2e^{-3t}$

Section 6.3

1. $\mathcal{L}(e^{-t}\cos(2t)) = \dfrac{s+1}{(s+1)^2 + 2^2} = \dfrac{s+1}{s^2+2s+5}$

3. $\mathcal{L}(t\sin(5t)) = -\dfrac{d}{ds}\left(\dfrac{5}{s^2+25}\right) = \dfrac{10s}{(s^2+25)^2}$

5. $\dfrac{2}{s^3} - \dfrac{6}{s^2} + \dfrac{4}{s}$

7. $\mathcal{L}(te^{-t}\sin(2t)) = \dfrac{2(2)(s+1)}{((s+1)^2+4)^2} = \dfrac{4s+4}{(s^2+2s+5)^2}$

9. $\mathcal{L}(t^{n+1}) = \mathcal{L}(t \cdot t^n) = -\dfrac{d}{ds}(\mathcal{L}(t^n)) = -\dfrac{d}{ds}\left(\dfrac{n!}{s^{n+1}}\right) = -n!\dfrac{d}{ds}(s^{-n-1}) = \dfrac{(n+1)!}{s^{n+2}}$

11. $f(t) = te^{2t} + 3t$

13. $\mathcal{L}^{-1}\left(\dfrac{s+3}{(s-2)^2}\right) = \mathcal{L}^{-1}\left(\dfrac{A}{s-2} + \dfrac{B}{(s-2)^2}\right) = \mathcal{L}^{-1}\left(\dfrac{1}{s-2} + \dfrac{5}{(s-2)^2}\right) = e^{2t} + 5te^{2t}$

15. $\mathcal{L}^{-1}\left(\dfrac{s+5}{(s+3)(s^2+1)}\right) = \mathcal{L}^{-1}\left(\dfrac{1/5}{s+3} + \dfrac{-\frac{1}{5}s + \frac{8}{5}}{s^2+1}\right) = \dfrac{1}{5}e^{-3t} - \dfrac{1}{5}\cos(t) + \dfrac{8}{5}\sin(t)$

17. $\mathcal{L}^{-1}\left(\dfrac{s+3}{(s+2)^2+2^2}\right) = e^{-2t}\cos(2t) + \dfrac{1}{2}e^{-2t}\sin(2t)$

19. $(s^2 X(s) - s + 1) + 4(sX(s) - 1) + 8X(s) = 0$

    $\rightarrow (s^2 + 4s + 8)X(s) = s + 3$

    $\rightarrow X(s) = \dfrac{s+3}{s^2+4s+8}$

    Using the answer to Exercise 17, $x(t) = e^{-2t}\cos(2t) + \dfrac{1}{2}e^{-2t}\sin(2t)$

21. $x(t) = \dfrac{1}{2} + \dfrac{1}{6}e^{-t}\sin(3t) - \dfrac{1}{2}e^{-t}\cos(3t)$

23. $sX(s) - x(0) = -2X(s) + 2Y(s) + \dfrac{1}{s}$

    $sY(s) - y(0) = X(s) - 3Y(s) + \dfrac{1}{s^2+1}$

    With $\vec{x} = \begin{pmatrix} X(s) \\ Y(s) \end{pmatrix}, \mathbf{A} = \begin{pmatrix} -2 & 2 \\ 1 & -3 \end{pmatrix}$,

    $\vec{b} = \begin{pmatrix} 1 + \frac{1}{s} \\ \frac{1}{s^2+1} \end{pmatrix}, (s\mathbf{I} - \mathbf{A})\vec{x} = \vec{b}$

    $\vec{x} = \dfrac{1}{s^2+5s+4}\begin{pmatrix} s+3 & 2 \\ 1 & s+2 \end{pmatrix}\begin{pmatrix} \frac{1+s}{s} \\ \frac{1}{s^2+1} \end{pmatrix}$

$$X(s) = \frac{s^4+4s^3+4s^2+6s+3}{s(s+1)(s+4)(s^2+1)}$$

$$Y(s) = \frac{s^3+2s^2+3s+1}{s(s+1)(s+4)(s^2+1)}$$

$$x(t) = \frac{3}{4} + \frac{1}{3}e^{-t} + \frac{43}{204}e^{-4t} - \frac{5}{17}\cos(t) + \frac{3}{17}\sin(t)$$

$$y(t) = \frac{1}{4} + \frac{1}{6}e^{-t} - \frac{43}{204}e^{-4t} - \frac{7}{34}\cos(t) + \frac{11}{34}\sin(t)$$

Section 6.4

1. $f(t) = 2 - 3u(t-1) + 2u(t-2) - u(t-5)$

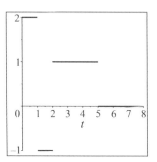

3. $f(t) = \cos(t)u(t) - (1+\cos(t))u(t-\pi) + (\sin(t)+1)u(t-2\pi) - \sin(t)u(t-3\pi)$

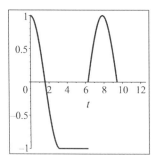

5. $f(t) = tu(t) - u(t-1) - u(t-2) - (t-2)u(t-3)$

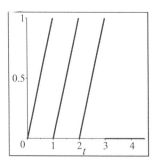

7. $F(s) = e^{-2s}\left(\frac{2}{s^3} + \frac{4}{s^2} + \frac{5}{s}\right)$

9. $F(s) = \frac{s}{s^2+1} + e^{-\pi s}\left(-\frac{1}{s} - \frac{s}{s^2+1}\right) + e^{-2\pi s}\left(\frac{1}{s} + \frac{1}{s^2+1}\right) - e^{-3\pi s}\left(\frac{1}{s^2+1}\right)$

Section 6.4

11. $f(t) = u(t-3)\cos(t-3)$

13. $f(t) = \frac{1}{2}\sin(2t) + u(t-1)\cos(2(t-1))$

15. $x(t) = e^{-2t} + u(t-3)(1 - e^{6-2t})$

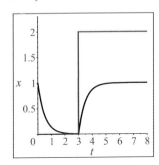

17. $x(t) = te^{-2t} + u(t-2)\left(\frac{1}{4} + \left[\frac{3}{4} - \frac{t}{2}\right]e^{4-2t}\right)$

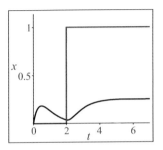

19. (a) $h(t) = \sin(t) + u(t-\pi)(-\sin(t) - \sin(t)) + u(t-2\pi)(\sin(t) - (-\sin(t))) + \cdots$
$= \sin(t) - 2u(t-\pi)\sin(t) + 2u(t-2\pi)\sin(t) - 2u(t-3\pi)\sin(t) + \cdots$
$\equiv \sin(t) + 2u(t-\pi)\sin(t-\pi) + 2u(t-2\pi)\sin(t-2\pi)$
$\quad + 2u(t-3\pi)\sin(t-3\pi) + \cdots$

(b) $\mathcal{L}(h(t)) = H(s) = \frac{1}{s^2+1} + 2e^{-\pi s}\frac{1}{s^2+1} + 2e^{-2\pi s}\frac{1}{s^2+1} + \cdots$

(c) $sI(s) + I(s) = H(s)$
$I(s) = \frac{1}{s+1}\left(\frac{1}{s^2+1} + 2\sum_{k=1}^{\infty}\frac{e^{-k\pi s}}{s^2+1}\right) = \frac{1}{(s+1)(s^2+1)} + 2\sum_{k=1}^{\infty}\frac{e^{-k\pi s}}{(s+1)(s^2+1)}$
Using partial fractions, $f(t) = \mathcal{L}^{-1}\left(\frac{1}{(s+1)(s^2+1)}\right) = \frac{1}{2}\left(e^{-t} + \sin(t) - \cos(t)\right)$
$i(t) = f(t) + 2\sum_{k=1}^{\infty}u(t-k\pi)f(t-k\pi)$
A graph of $i(t)$ is shown below.

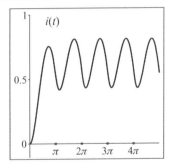

Section 6.5

1. $F(s) = \left(\frac{1}{s+2}\right)\left(\frac{1}{s+2}\right) = P(s)Q(s)$
   $p(t) = q(t) = e^{-2t}$
   $f(t) = p(t) \star q(t) = \int_0^t e^{-2\tau}e^{-2(t-\tau)}d\tau = te^{-2t}$

3. $F(s) = \left(\frac{1}{s^2}\right)\left(\frac{1}{s+5}\right) = P(s)Q(s)$
   $p(t) = t, \quad q(t) = e^{-5t}$
   $f(t) = p(t) \star q(t) = \int_0^t \tau e^{-5(t-\tau)}d\tau = \frac{t}{5} - \frac{1}{25} + \frac{e^{-5t}}{25}$

5. $F(s) = \left(\frac{1}{s^2+1}\right)\left(\frac{1}{s^2+1}\right) = P(s)Q(s)$
   $p(t) = q(t) = \sin(t)$
   $f(t) = p(t) \star q(t) = \int_0^t \sin(\tau)\sin(t-\tau)d\tau = \frac{1}{2}\sin(t) - \frac{1}{2}t\cos(t)$
   You may have to use the formulas for $\sin(2t)$ and $\cos(2t)$ to simplify your answer.

7. $x(t) = \left(e^{-t} - e^{-2t}\right) + u(t-2)\left(e^{-(t-2)} - e^{-2(t-2)}\right)$

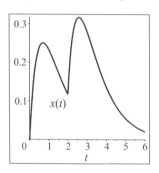

9. (a) $h(t) = \mathcal{L}^{-1}\left(\frac{1}{s^2+4s+4}\right) = \mathcal{L}^{-1}\left(\frac{1}{(s+2)^2}\right) = te^{-2t}$
   (b) $x(t) = (1 + 2e^{-t}) \star te^{-2t}$
   $= \int_0^t \tau e^{-2\tau}\left(1 + 2e^{-(t-\tau)}\right)d\tau$
   $= -\frac{5}{2}te^{-2t} - \frac{9}{4}e^{-2t} + 2e^{-t} + \frac{1}{4}$
   (c) The steady state solution is $x_p(t) = \lim_{t\to\infty} x(t) = \frac{1}{4}$

Section 7.1

1. $\frac{XT'}{aXT} = \frac{aX''T + bX'T}{aXT} \to \frac{T'}{aT} = \frac{X''}{X} + \frac{b}{a}\frac{X'}{X} = -\lambda$
   $\to \begin{cases} T' + a\lambda T = 0 \\ X'' + \frac{b}{a}X' + \lambda X = 0 \end{cases}$

3. $\frac{XT''}{bXT} = \frac{bX''T}{bXT} + \frac{xXT}{bXT} \to \frac{T''}{bT} = \frac{X''}{X} + \frac{1}{b}x = -\lambda$
   $\to \begin{cases} T'' + b\lambda T = 0 \\ X'' + \frac{1}{b}xX + \lambda X = 0 \end{cases}$

5. $\dfrac{XT'}{aXT} = \dfrac{aX''T}{aXT} + \dfrac{bXT}{aXT} + \dfrac{cX^2T^2}{aXT} \rightarrow \dfrac{T'}{aT} = \dfrac{X''}{X} + \dfrac{b}{a} + \dfrac{c}{a}XT$

In this case it is impossible to separate the variables x and t.

7. The three cases are $\lambda = 0, \lambda < 0, \lambda > 0$.

If $\lambda = 0$ the characteristic polynomial $r^2 = 0$ has a double root $r = 0$ and $X(x) = a + bx$. In this case $X(0) = a = 0$ and $X(1) = 0 + b \cdot 1 = 0 \rightarrow b = 0$; therefore, there are no nonzero solutions.

If $\lambda < 0$, say $\lambda = -K^2$, the characteristic polynomial $r^2 - K^2 = 0$ has distinct real roots $r = \pm K$ and $X(x) = ae^{Kx} + be^{-Kx}$. In this case $X(0) = a+b = 0 \longrightarrow b = -a$, and $X(1) = ae^K - ae^{-K} = 0$. The only solution is $K = 0$, so there are no nonzero solutions.

If $\lambda > 0$, say $\lambda = K^2$, the characteristic polynomial $r^2 + K^2 = 0$ has complex conjugate roots $r = \pm Ki$. The solution is $X(x) = a\sin(Kx) + b\cos(Kx)$. In this case, $X(0) = b = 0$ and $X(1) = a\sin(K) = 0$. This implies that the problem has a nonzero solution $C\sin(Kx)$ for any $K = n\pi$ for any positive integer n.

The boundary-value problem has a nonzero solution if, and only if, $\lambda = K^2 = n^2\pi^2$ for some integer $n \geq 1$.

Section 7.2

1. (a) even
   (b) odd
   (c) neither
   (d) even
   (e) even
   (f) odd
   (g) even
   (h) odd

3. Let the solution of the ODE be $X(x) = A\cosh(Kx) + B\sinh(Kx)$. Then $X(0) = A\cosh(0) + B\sinh(0) = A \cdot 1 + B \cdot 0 = A = 0$ and $X(1) = B\sinh(K) = 1 \rightarrow B = \dfrac{1}{\sinh(K)}$. The solution of the boundary-value problem for any nonzero K is $X(x) = \dfrac{\sinh(Kx)}{\sinh(K)}$.

5. If $m \neq n$, $\int_{-L}^{L} \cos\left(\dfrac{n\pi t}{L}\right)\cos\left(\dfrac{m\pi t}{L}\right) = 2\dfrac{1}{2}\int_0^L \left[\cos\left(\dfrac{(n-m)\pi t}{L}\right) + \cos\left(\dfrac{(n+m)\pi t}{L}\right)\right]dt =$
$\left[\dfrac{L}{(n-m)\pi}\sin\left(\dfrac{(n-m)\pi t}{L}\right) + \dfrac{L}{(n+m)\pi}\sin\left(\dfrac{(n+m)\pi t}{L}\right)\right]\Big|_0^L = 0$.

7. $\int_{-L}^{L}\left(\cos\left(\dfrac{n\pi t}{L}\right)\right)^2 dt = 2\dfrac{1}{2}\int_0^L\left(1+\cos\left(\dfrac{2n\pi t}{L}\right)\right)dt = L + \int_0^L \cos\left(\dfrac{2n\pi t}{L}\right)dt = L$.

9. The Fourier coefficients are $a_0 = 2$ and for $n > 0, a_n = 0$.

$$b_n = \int_{-1}^{1}(1+t)\sin(n\pi t)dt = \dfrac{-2(-1)^n}{n\pi}.$$

The graph shows the periodic function and the Fourier approximation.

376 Appendix A Answers to Odd-numbered Exercises

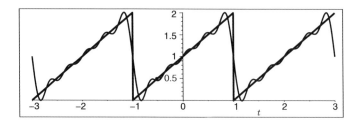

11. The function $\sin(t)$ is odd, so $a_n = 0$ for all n, and

$$b_n = \int_{-1}^{1} (\sin(t))\sin(n\pi t)dt = \frac{(1+n\pi)\sin(n\pi - 1) + (1 - n\pi)\sin(n\pi + 1)}{(n\pi + 1)(n\pi - 1)}.$$

The graph shows the periodic function and the Fourier approximation.

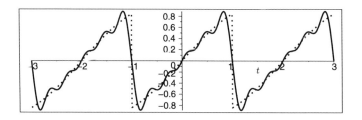

13. Let $n = 1$ and $m = 3$. Then $\int_{-1}^{1} t^n t^m dt = \int_{-1}^{1} t^4 dt = \frac{t^5}{5}\big|_{-1}^{1} = \frac{2}{5} \neq 0$ so these functions do NOT form an orthogonal set on $[-1, 1]$.

15. Using the integral formula $c_n = \frac{\int_{-1}^{1} e^t P_n(t)dt}{\int_{-1}^{1} (P_n(t))^2 dt}$, approximations to the coefficients are:

$c_0 \approx 1.175201, c_1 \approx 1.103638, c_2 \approx 0.357814, c_3 \approx 0.070456, c_4 \approx 0.009965.$

The graph of $\sum_{n=0}^{4} c_n P_n(t)$ lies almost exactly on the graph of e^t.

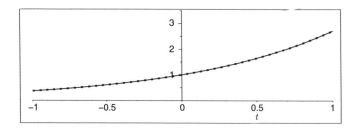

17. The coefficients are $a_0 = \frac{2}{\pi} \int_0^{\pi} \sin(t) dt = \frac{4}{\pi}$,

$$a_n = \frac{2}{\pi} \int_0^{\pi} \sin(t)\cos(nt) dt = \frac{2(1 + (-1)^n)}{\pi(1 - n^2)} = \frac{4}{\pi(1 - n^2)} \text{ if } n \text{ even, } 0 \text{ if } n \text{ odd.}$$

The function $F(t) = \frac{2}{\pi} + \frac{4}{\pi}\sum_{k=1}^{4} \frac{\cos(2kt)}{1-(2k)^2}$ is plotted below. The cosine series con-

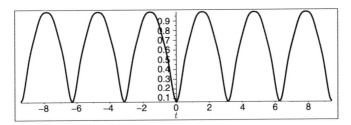

verges to the continuous function $|\sin(t)|$ on $(-\infty, \infty)$.

19. If N is odd, the Fourier series for the function
$$f(t) = \begin{cases} -1 & \text{if } -L \leq t < 0 \\ 1 & \text{if } 0 \leq t \leq L \end{cases}$$
is $F(t) = \frac{4}{\pi}\left[\sin(\pi t) + \frac{\sin(3\pi t)}{3} + \cdots + \frac{\sin(N\pi t)}{N}\right]$. The function $f(t)$ has a jump of magnitude $T = 2$ at $t^* = 0$. To compute the overshoot, use $F\left(\frac{1}{N+1/2}\right) - f\left(\frac{1}{N+1/2}\right)$. The overshoot for $N = 3$ is ≈ 0.17960, for $N = 11$ is ≈ 0.17945, and for $N = 21$ is ≈ 0.17913; therefore, the magnitude of the overshoot remains nearly constant at $9\%T = 0.18$. See the graphs below.

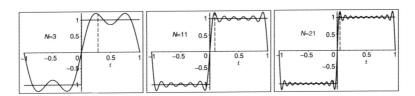

Section 7.3

1. The eigenvalues are $\lambda_n = n^2$ and eigenfunctions $\phi_n(t) = \sin(nt)$.

3. The eigenvalues are $\lambda_n = n^2$ and the eigenfunctions are $\phi_n(t) = \cos(nt)$.

5. To make $e^{cx}(aX'' + bX' + \lambda X)$ have the form of a Sturm-Liouville equation, we need $\frac{d}{dx}(ae^{cx}) = be^{cx}$; therefore $ac = b$ and $c = b/a$.

7. It seems to fit slightly better.

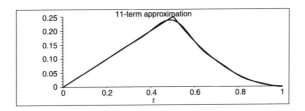

Section 8.1

1.
$$f_x = 2e^{2x+3y}, \quad f_y = 3e^{2x+3y}$$
$$f_{xx} = 4e^{2x+3y}, \quad f_{xy} = f_{yx} = 6e^{2x+3y}, \quad f_{yy} = 9e^{2x+3y}$$

3.
$$h_x = 4\cos(4x-3y), \quad h_y = -3\cos(4x-3y)$$
$$h_{xx} = -16\sin(4x-3y), \quad h_{xy} = h_{yx} = 12\sin(4x-3y), \quad h_{yy} = -9\sin(4x-3y)$$

5.
$$u_t = -Ke^{-Kt}\sin(ax), \quad u_x = ae^{-Kt}\cos(ax)$$
$$u_{xx} = -a^2 e^{-Kt}\sin(ax) \equiv -Ke^{-Kt}\sin(ax)$$

if $K = a^2$.

7. $u_{tt} - 4u_{xx} + 2u_t = 0 \to A = 1, B = 0, C = -4$; therefore,
$$AC - B^2 = -4 < 0 \to$$
hyperbolic.

9. $u_{xx} + bu_x + u - u_t = 0 \to A = 1, B = C = 0$; therefore,
$$AC - B^2 = 0 \to$$
parabolic.

11. If the second and third matrices are multiplied first, then
$$(x \ y)\begin{pmatrix} A & B \\ B & C \end{pmatrix}\begin{pmatrix} x \\ y \end{pmatrix}$$
$$= (x \ y)\begin{pmatrix} Ax + By \\ Bx + Cy \end{pmatrix} = x(Ax + By) + y(Bx + Cy) = Ax^2 + 2Bxy + Cy^2.$$

13. To show that $\mathbf{MU} = \mathbf{UD}$, write $\mathbf{MU} = \mathbf{M}\begin{pmatrix} \vec{v}_1 & \vec{v}_2 \\ \downarrow & \downarrow \end{pmatrix} = \begin{pmatrix} \mathbf{M}\vec{v}_1 & \mathbf{M}\vec{v}_2 \\ \downarrow & \downarrow \end{pmatrix}$
$$= \begin{pmatrix} \lambda_1 v_{11} & \lambda_2 v_{21} \\ \lambda_1 v_{12} & \lambda_2 v_{22} \end{pmatrix} \text{ and}$$
$$\mathbf{UD} = \begin{pmatrix} \vec{v}_1 & \vec{v}_2 \\ \downarrow & \downarrow \end{pmatrix}\begin{pmatrix} \lambda_1 & 0 \\ 0 & \lambda_2 \end{pmatrix} = \begin{pmatrix} \lambda_1 v_{11} & \lambda_2 v_{21} \\ \lambda_1 v_{12} & \lambda_2 v_{22} \end{pmatrix}.$$

15. If the change of independent variables $\begin{pmatrix} x \\ y \end{pmatrix} = \mathbf{U}\begin{pmatrix} r \\ s \end{pmatrix}$ is made, then $(x \ y)$
$$= \begin{pmatrix} x \\ y \end{pmatrix}^T = \left(\mathbf{U}\begin{pmatrix} r \\ s \end{pmatrix}\right)^T = \begin{pmatrix} r \\ s \end{pmatrix}^T \mathbf{U}^T = (r \ s)\mathbf{U}^T.$$

Now the quadratic form $Q = (x \ y)\begin{pmatrix} A & B \\ B & C \end{pmatrix}\begin{pmatrix} x \\ y \end{pmatrix} = ((r \ s)\mathbf{U}^T)\mathbf{M}$
$$\left(\mathbf{U}\begin{pmatrix} r \\ s \end{pmatrix}\right) = (r \ s)\begin{pmatrix} \lambda_1 & 0 \\ 0 & \lambda_2 \end{pmatrix}\begin{pmatrix} r \\ s \end{pmatrix} = \lambda_1 r^2 + \lambda_2 s^2.$$

17. The eigenpairs of the matrix $\mathbf{M} = \begin{pmatrix} 1 & -3 \\ -3 & 1 \end{pmatrix}$ are $\left(4, \begin{pmatrix} 1 \\ -1 \end{pmatrix}\right)$ and $\left(-2, \begin{pmatrix} 1 \\ 1 \end{pmatrix}\right)$.

Let $\begin{pmatrix} x \\ y \end{pmatrix} = \begin{pmatrix} 1 & 1 \\ -1 & 1 \end{pmatrix} \begin{pmatrix} r \\ s \end{pmatrix}$ so $x = r + s$, $y = -r + s$. Then

$$Q(x, y) = (r + s)^2 - 6(r + s)(-r + s) + (-r + s)^2 = 8r^2 - 4s^2.$$

The quadratic form Q is hyperbolic.

Section 8.2

1. Using equations 8.6, with 20 terms in the series,

$$u(x,t) = \sum_{n=1}^{\infty} b_n \sin\left(\frac{n\pi x}{2}\right) e^{\left(-\frac{n^2\pi^2 t}{4}\right)}, \quad b_n = \frac{2}{2}\int_0^2 20x \sin\left(\frac{n\pi x}{2}\right) dx.$$

The graph shows $u(x, t)$ at times $t = 0, 0.05, 0.2, \ldots, 1.0$.

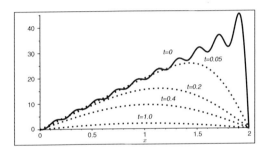

3. If the left end is insulated and the right end is held at $0°$, the equation for $u(x, t)$ is

$$u(x,t) = \sum_{n=0}^{\infty} c_n \cos\left(\frac{(2n+1)\pi x}{2L}\right) e^{-a\left(\frac{(2n+1)\pi}{2L}\right)^2 t}; \quad b_n = \frac{\int_0^2 20x \cos\left(\frac{(2n+1)\pi x}{2L}\right) dx}{\int_0^2 \cos^2\left(\frac{(2n+1)\pi x}{2L}\right) dx}.$$

The integral in the denominator of b_n is equal to one for $n > 0$.

The graph below shows $u(x, t)$ at times $t = 0, 0.05, 0.2, \ldots, 4.0$. Notice that the temperature builds up at the insulated end before decreasing to zero.

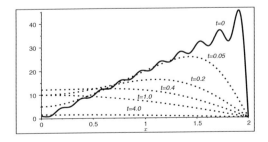

5. The graph shows the temperature at the insulated end for $0 < t < 50$. The temperature at $x = 10$ rises until all of the temperatures to the left are smaller, and then it drops to 0.

380 Appendix A Answers to Odd-numbered Exercises

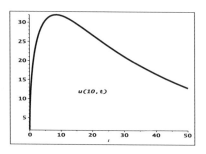

7. If the PDE is written in the form $KP_{xx} - P_t + rP(1-P/N) = 0$, the only second-order term is P_{xx}, and $AP_{xx} + 2BP_{xt} + CP_{tt} = KP_{xx} + 0P_{xt} + 0P_{tt}$ implies that $AC - B^2 = 0$; therefore, it is parabolic.

The equation is not linear because of the term $-rP^2/N$.

Section 8.3

1. A graph of the initial velocity function is shown below. The displacement spreads

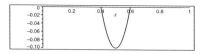

out, returns to zero, and then goes up and spreads out again. The right-hand graph below shows the position of the string over a full period $0 \le t \le 2.0$.

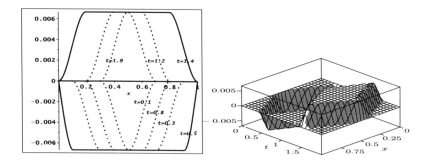

3. Assuming $c < \frac{\pi\beta}{L}$, the solution of $u_{tt} + 2cu_t = \beta^2 u_{xx}$, $u(x,0) = f(x)$, $u_t(x,0) = g(x)$, $u(0,t) = u(L,t) = 0, t \ge 0$ is

$$u(x,t) = \sum_{n=1}^{\infty} \sin\left(\frac{n\pi x}{L}\right) e^{-ct} \left[a_n \cos\left(t\sqrt{\left(\frac{n\pi\beta}{L}\right)^2 - c^2}\right) + b_n \sin\left(t\sqrt{\left(\frac{n\pi\beta}{L}\right)^2 - c^2}\right) \right]$$

$$a_n = \frac{2}{L} \int_0^L f(x) \sin\left(\frac{n\pi x}{L}\right) dx, \quad b_n = \frac{ca_n + \frac{2}{L}\int_0^L g(x)\sin\left(\frac{n\pi x}{L}\right) dx}{\sqrt{\left(\frac{n\pi\beta}{L}\right)^2 - c^2}}$$

Section 8.3

5.
$$u(x,t) = \frac{1}{2}(\sin(x+\beta t) + \sin(x-\beta t))$$
$$= \frac{1}{2}(\sin(x)\cos(\beta t) + \cos(x)\sin(\beta t) + \sin(x)\cos(\beta t) - \cos(x)\sin(\beta t)) = \sin(x)\cos(\beta t).$$

At any fixed value of x the displacement is a cosine wave with amplitude $\sin(x)$. The period of the wave is $\frac{2\pi}{\beta}$. The wave has fixed points at $x = 0, \pm\pi, \pm 2\pi, \ldots$.

7. The graph on the left shows the initial velocity $g(x) = \frac{1}{1+x^2}$. With $\beta = \frac{1}{2}$, the graph on the right shows the solution $u(x,t) = \tan^{-1}(x+t/2) - \tan^{-1}(x-t/2)$ at several different times.

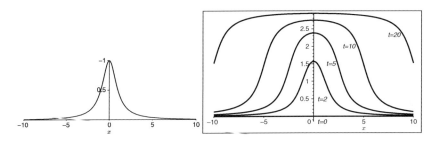

9. Using the formula
$$u(r,t) = \sum_{1}^{\infty} J_0(z_n r)(a_n \cos(z_n t) + b_n \sin(z_n t))$$

from Chapter 7, the velocity is
$$\frac{\partial u}{\partial t} = \sum_{1}^{\infty} J_0(z_n r) z_n(-a_n \sin(z_n t) + b_n \cos(z_n t)).$$

With $f(r) \equiv 0$, the a_n are all zero, and therefore
$$u_t(r,t) = \sum_{1}^{\infty} b_n z_n J_0(z_n r) = g(r).$$

Using the orthogonality of the functions $J_0(z_n r)$, with weight function r, the coefficients are
$$b_n = \frac{\int_0^1 rg(r)J_0(z_n r)dr}{z_n \int_0^1 r(J_0(z_n r))^2 dr}.$$

The graphs below show the circular drumhead at times $t = 0.1, 2.0, 3.9$.

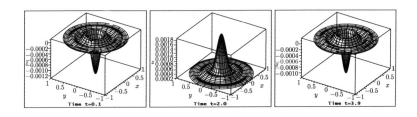

Section 8.4

1.

Δx	$f'(0) \approx \dfrac{e^{\Delta x}-1.0}{\Delta x}$	$f'(0) \approx \dfrac{e^{\Delta x}-e^{-\Delta x}}{2\Delta x}$	$f''(0) \approx \dfrac{e^{\Delta x}-2.0+e^{-\Delta x}}{(\Delta x)^2}$
0.10	1.051709	1.001668	1.000834
0.01	1.005017	1.000017	1.000007
0.001	1.000500	1.000000	0.999800
0.0001	1.000050	1.000000	1.000000

The first-order forward derivative requires $\Delta x \leq 0.0001$, and the second-order derivatives require $\Delta x \leq 0.01$.

3. The difference equation is

$$P(x,t+\Delta t) = P(x,t) + \left(\frac{0.1\Delta t}{(\Delta x)^2}\right)(P(x+\Delta x,t) - 2P(x,t) + P(x-\Delta x,t))$$

$$+\Delta t \left(P(x,t) - P^2(x,t)\right).$$

5. To simulate the boundary conditions, $u(0,k)$ and $u(N,k)$ are set to zero at each time step $k \geq 0$. As $t \to \infty$, $u(x,t)$ approaches the other equilibrium solution $u(x,t) \equiv 0$.

Section 8.5

1. The two ODEs are $Y'' - \lambda Y = 0$ and $X'' + \lambda X = 0$, $X(0) = X(a) = 0$. The Sturm-Liouville equation has eigenvalues $\lambda_n = \dfrac{n^2\pi^2}{a^2}$ and eigenfunctions $X_n(x) = \sin\left(\dfrac{n\pi x}{a}\right)$. The equation in Y has solutions $Y_n(y) = A_n \cosh\left(\dfrac{n\pi y}{a}\right) + B_n \sinh\left(\dfrac{n\pi y}{a}\right)$. The general solution of the PDE is

$$u(x,y) = \sum_{n=1}^{\infty} \sin\left(\frac{n\pi x}{a}\right)\left[A_n \cosh\left(\frac{n\pi y}{a}\right) + B_n \sinh\left(\frac{n\pi y}{a}\right)\right].$$

When $y = 0$, $u(x,0) = \sum_{n=1}^{\infty} A_n \sin\left(\dfrac{n\pi x}{a}\right) \equiv 0$ implies $A_n = 0$ for $n \geq 1$. When $y = b$, $u(x,b) = \sum_{n=1}^{\infty} \left(B_n \sinh\left(\dfrac{n\pi b}{a}\right)\right) \sin\left(\dfrac{n\pi x}{a}\right) = g(x)$. Therefore,

$$B_n = \frac{2}{\sinh\left(\frac{n\pi b}{a}\right)} \int_0^a g(x) \sin\left(\frac{n\pi x}{a}\right) dx.$$

3. Labelling the temperatures as shown in Figure 8.14, the system of equations can be written in matrix form as $\mathbf{A}\begin{pmatrix} T_1 \\ T_2 \\ T_3 \\ T_4 \\ T_5 \\ T_6 \\ T_7 \\ T_8 \\ T_9 \\ T_{10} \\ T_{11} \\ T_{12} \\ T_{13} \\ T_{14} \\ T_{15} \end{pmatrix} = \begin{pmatrix} -20.0 \\ 0 \\ 0 \\ 0 \\ 18.75 \\ 0 \\ 0 \\ 0 \\ 0 \\ 25.0 \\ 41.875 \\ 35.0 \\ 39.375 \\ 35.0 \\ 40.625 \end{pmatrix}$, where the form of the matrix $\mathbf{A}$ is described in the answer to Exercise 5. The solution of this linear system is given in the table below.

x	2.5	5.0	7.5	10.0	12.5
$y = 7.5$	-2.1472	4.3673	7.2031	8.9227	11.5713
$y = 5.0$	7.0440	12.4134	15.5224	16.9165	18.6123
$y = 2.5$	17.9097	22.7199	25.5565	24.6086	20.9615

These are similar to the series solution, but to obtain better agreement, Δx and Δy would need to be quite a bit smaller.

5. For the 15-variable system, the matrix $\mathbf{A}$ is a "block" matrix, consisting of three different kinds of 5×5 blocks:

$$\mathbf{0} = \begin{pmatrix} 0 & 0 & 0 & 0 & 0 \\ 0 & 0 & 0 & 0 & 0 \\ 0 & 0 & 0 & 0 & 0 \\ 0 & 0 & 0 & 0 & 0 \\ 0 & 0 & 0 & 0 & 0 \end{pmatrix}, \quad -\mathbf{I} = \begin{pmatrix} -1 & 0 & 0 & 0 & 0 \\ 0 & -1 & 0 & 0 & 0 \\ 0 & 0 & -1 & 0 & 0 \\ 0 & 0 & 0 & -1 & 0 \\ 0 & 0 & 0 & 0 & -1 \end{pmatrix}, \quad \mathbf{R} = \begin{pmatrix} 4 & -1 & 0 & 0 & 0 \\ -1 & 4 & -1 & 0 & 0 \\ 0 & -1 & 4 & -1 & 0 \\ 0 & 0 & -1 & 4 & -1 \\ 0 & 0 & 0 & -1 & 4 \end{pmatrix}.$$

The form of the 15×15 matrix $\mathbf{A}$ is

$$\mathbf{A} = \begin{pmatrix} \mathbf{R} & -\mathbf{I} & \mathbf{0} \\ -\mathbf{I} & \mathbf{R} & -\mathbf{I} \\ \mathbf{0} & -\mathbf{I} & \mathbf{R} \end{pmatrix}.$$

Section 8.6

1. Let $u(x,t) = X(x)T(t)$. Then

$$\frac{XT'}{XT} = \frac{X''T + a(x)XT + b(x)X^2T^2 - \lambda h(x)}{XT} \rightarrow$$

$$\frac{T'}{T} = \frac{X''}{X} + a(x) + b(x)XT - \frac{\lambda h(x)}{XT}.$$

To make the right-hand side a function depending only on x, the last two terms must vanish.

3. With $\lambda = 2.0$, the stable solution of the boundary-value problem

$$w'' + (a(x) - b(x)w)w - \lambda h(x) = 0, \quad w'(0) = 0, \quad w'(1) = 0$$

has initial value $w_1(0) = Z_1 \approx 4.5$. The graph below compares the stable equilibrium for $\lambda = 2.0$ (with constant 1.16 added to it) to that with $\lambda = 0.9$. The increased harvesting appears to decrease the population across the lake by approximately 1.16, but also tends to decrease it slightly more in regions where the harvesting is largest.

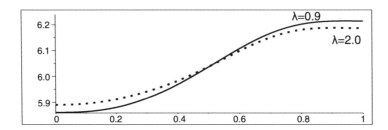

5. If the constant initial function is close to the stable equilibrium solution, $u(x, t)$ will tend to it quickly as t increases. With $u(x, 0) \equiv 4.95$, the solution of the PDE converged to a solution with $u(x, 0) \approx 5.2$. Near the unstable equilibrium, solutions starting with a constant initial function tend to stay close to the equilibrium before going either up or down. The solution with initial function $u(x, 0) \equiv 0.658$ stayed close to a solution with $u(x, 0) \approx 0.73$ for t from 0 to 2.0.

Appendix B

Derivative and Integral Formulas

The following formulas are the ones that you need to have committed to memory. Having to use a calculator to do these simple derivatives and integrals makes work in differential equations much too tedious.

Appendix B Derivative and Integral Formulas

Differentiation Formulas

$\frac{d}{dx}(C) = 0$, C a constant

$\frac{d}{dx}(Cf(x)) = Cf'(x)$,
where $f'(x) \equiv \frac{d}{dx}(f(x))$

$\frac{d}{dx}(f(x) \pm g(x)) = f'(x) \pm g'(x)$

$\frac{d}{dx}(f(x) \cdot g(x)) = f(x)g'(x) + f'(x)g(x)$

$\frac{d}{dx}\left(\frac{f(x)}{g(x)}\right) = \frac{g(x)f'(x) - g'(x)f(x)}{(g(x))^2}$

$\frac{d}{dx}(x^n) = nx^{n-1}$ and $\frac{d}{dx}(x) = 1$

$\frac{d}{dx}(\sin(x)) = \cos(x)$

$\frac{d}{dx}(\cos(x)) = -\sin(x)$

$\frac{d}{dx}(e^x) = e^x$

$\frac{d}{dx}(\ln(x)) = \frac{1}{x}$

$\frac{d}{dx}(\tan^{-1}(x)) = \frac{1}{1+x^2}$

$\frac{d}{dx}(f(g(x))) = f'(g(x))g'(x)$, chain rule

Integration Formulas

$\int a\, dx = ax + C$

$\int (c_1 f(x) + c_2 g(x))\, dx$
$= c_1 \int f(x)\, dx + c_2 \int g(x)\, dx$

$\int x^n\, dx = \frac{x^{n+1}}{n+1} + C$, if $n \neq -1$

$\int \frac{1}{x}\, dx = \ln|x| + C$

$\int e^x\, dx = e^x + C$

$\int \sin(x)\, dx = -\cos(x) + C$

$\int \cos(x)\, dx = \sin(x) + C$

$\int \frac{1}{1+x^2}\, dx = \tan^{-1}(x) + C$

$\int \frac{1}{a+bx}\, dx = \frac{\ln|a+bx|}{b} + C$

$\int u\, dv = uv - \int v\, du$,
Integration by Parts

A very useful formula:

$\int e^{ax}\, dx = \frac{e^{ax}}{a} + C$

Appendix C

Cofactor Method for Determinants

The formula for the determinant of a 2 × 2 matrix is

$$\det\begin{pmatrix} a & b \\ c & d \end{pmatrix} = ad - bc,$$

and it is simple to compute. To find the determinant of an $n \times n$ matrix with $n > 2$, there are various methods, all of which are fairly complicated and involve a lot of arithmetic. The method of cofactors, described here, reduces the determinant of an $n \times n$ matrix to a sum of n determinants of size $(n-1) \times (n-1)$. For relatively small values of n, this can sometimes be useful, especially if a large percentage of the elements in the matrix are zero.

It is first necessary to give two definitions.

Definition C.1. *For an $n \times n$ matrix $\mathbf{A}$, the **minor** $\mathbf{M}_{ij}$ of the element a_{ij} in the ith row and jth column is the matrix obtained from $\mathbf{A}$ by deleting its ith row and jth column.*

Example C.0.1. *If* $\mathbf{A} = \begin{pmatrix} a_{11} & a_{12} & a_{13} \\ a_{21} & a_{22} & a_{23} \\ a_{31} & a_{32} & a_{33} \end{pmatrix} = \begin{pmatrix} 1 & 2 & 3 \\ 4 & 5 & 6 \\ 7 & 8 & 9 \end{pmatrix}$, *then the minors of a_{12} and a_{31} are* $\mathbf{M}_{12} = \begin{pmatrix} 4 & 6 \\ 7 & 9 \end{pmatrix}$ *and* $\mathbf{M}_{31} = \begin{pmatrix} 2 & 3 \\ 5 & 6 \end{pmatrix}$. □

Definition C.2. *The $(\mathbf{i}, \mathbf{j})$-**cofactor** of an element a_{ij} in the matrix $\mathbf{A}$ is denoted by A_{ij} and is defined as*

$$A_{ij} = (-1)^{i+j} \det \mathbf{M}_{ij} \quad \text{(The cofactor is a scalar, not a matrix.)}$$

In the above example, the cofactors A_{12} and A_{31} are

$$A_{12} = (-1)^{1+2} \det(\mathbf{M}_{12}) = (-1)^3 \det\begin{pmatrix} 4 & 6 \\ 7 & 9 \end{pmatrix} = -(36 - 42) = 6$$

and
$$A_{31} = (-1)^{3+1} \det(\mathbf{M}_{31}) = (-1)^4 \det\begin{pmatrix} 2 & 3 \\ 5 & 6 \end{pmatrix} = 12 - 15 = -3.$$

The determinant of a square matrix can now be defined as a weighted sum of its cofactors.

Theorem C.1. *For any $n \times n$ matrix $\mathbf{A} = (a_{ij})$, the determinant of $\mathbf{A}$ can be written as*
$$\det \mathbf{A} = \sum_{j=1}^{n} a_{ij} A_{ij},$$
where i can be any row index, $1 \leq i \leq n$. It can also be written
$$\det \mathbf{A} = \sum_{i=1}^{n} a_{ij} A_{ij},$$
where j can be any column index.

Example C.0.2. *Find the determinant of the matrix $\mathbf{A}$ in the previous example, first by expanding in cofactors along the second row of $\mathbf{A}$ and then by cofactors of the third column.*

Using row 2,
$$\det \mathbf{A} = a_{21}A_{21} + a_{22}A_{22} + a_{23}A_{23}$$
$$= 4(-1)^{2+1}\det\begin{pmatrix} 2 & 3 \\ 8 & 9 \end{pmatrix} + 5(-1)^{2+2}\det\begin{pmatrix} 1 & 3 \\ 7 & 9 \end{pmatrix} + 6(-1)^{2+3}\det\begin{pmatrix} 1 & 2 \\ 7 & 8 \end{pmatrix}$$
$$= -4(18-24) + 5(9-21) - 6(8-14) = 24 - 60 + 36 = 0.$$

Using column 3,
$$\det \mathbf{A} = a_{13}A_{13} + a_{23}A_{23} + a_{33}A_{33}$$
$$= 3(-1)^{1+3}\det\begin{pmatrix} 4 & 5 \\ 7 & 8 \end{pmatrix} + 6(-1)^{2+3}\det\begin{pmatrix} 1 & 2 \\ 7 & 8 \end{pmatrix} + 9(-1)^{3+3}\det\begin{pmatrix} 1 & 2 \\ 4 & 5 \end{pmatrix}$$
$$= 3(32-35) - 6(8-14) + 9(5-8) = -9 + 36 - 27 = 0.$$

Check that you get the same value for the determinant by using any other row or column. ∎

In Chapter 4, the claim is made that the characteristic polynomial $P(\lambda) = \det(\mathbf{A} - \lambda \mathbf{I})$ of an $n \times n$ matrix $\mathbf{A}$ is a polynomial of degree n in λ. This is easy to see when the method of cofactors is used to evaluate the determinant. When $n = 2$,
$$P_2(\lambda) = \det\begin{pmatrix} a - \lambda & b \\ c & d - \lambda \end{pmatrix} = (a-\lambda)(d-\lambda) - bc = \lambda^2 - (a+d)\lambda + (ad-bc).$$

For $n = 3$, if we expand the determinant
$$P_3(\lambda) = \det\begin{pmatrix} a_{11} - \lambda & a_{12} & a_{13} \\ a_{21} & a_{22} - \lambda & a_{23} \\ a_{31} & a_{32} & a_{33} - \lambda \end{pmatrix}$$

by cofactors of the first row, we are essentially adding together three polynomials of degree ≤ 2 multiplied either by a constant or the term $a_{11} - \lambda$; therefore, $P_3(\lambda) = -\lambda^3 +$ a polynomial of degree ≤ 2. Induction on n then implies that for any positive integer $n \geq 2$, $P_n(\lambda) = (-\lambda)^n +$ a polynomial of degree $\leq n - 1$; that is, P_n is a polynomial of degree n in λ.

Appendix D

Cramer's Rule for Solving Systems of Linear Equations

Given a system of n linear equations in n unknowns, written in matrix form as

$$\mathbf{A}\vec{\mathbf{x}} = \begin{pmatrix} a_{11} & a_{12} & \cdots & a_{1n} \\ a_{21} & a_{22} & \cdots & a_{2n} \\ & & \ddots & \\ a_{n1} & a_{n2} & \cdots & a_{nn} \end{pmatrix} \begin{pmatrix} x_1 \\ x_2 \\ \vdots \\ x_n \end{pmatrix} = \begin{pmatrix} b_1 \\ b_2 \\ \vdots \\ b_n \end{pmatrix} = \vec{\mathbf{b}},$$

it is possible to obtain the solution $\vec{\mathbf{x}}$ in terms of the determinant of $\mathbf{A}$ and determinants of matrices formed from $\mathbf{A}$ and the right-hand side vector $\vec{\mathbf{b}}$. The following theorem is usually proved in a course in linear algebra.

Theorem D.1. (*Cramer's Theorem*) *Given the system* $\mathbf{A}\vec{\mathbf{x}} = \vec{\mathbf{b}}$ *of n equations in n unknowns* $x_1, x_2, \ldots, x_n$, *if* $\det \mathbf{A} \neq 0$, *the solution can be written as*

$$x_j = \frac{\det(\mathbf{M}_j)}{\det(\mathbf{A})}, \quad j = 1, 2, \ldots, n,$$

where $\mathbf{M_j}$ *is the matrix* $\mathbf{A}$ *with the jth column replaced by the column vector* $\vec{\mathbf{b}}$.

Example D.0.1. *Solve the system of equations*

$$\begin{array}{rcl} x_1 + 3x_2 - x_3 & = & 6 \\ 4x_1 - x_2 & = & 7 \\ -x_1 + x_2 + 5x_3 & = & -6 \end{array} \equiv \begin{pmatrix} 1 & 3 & -1 \\ 4 & -1 & 0 \\ -1 & 1 & 5 \end{pmatrix} \vec{\mathbf{x}} = \vec{\mathbf{b}} = \begin{pmatrix} 6 \\ 7 \\ -6 \end{pmatrix}.$$

We first use the method of cofactors (see Appendix C) to find the determinant of $\mathbf{A}$.

Expanding about the third column,

$$\det \mathbf{A} = \det \begin{pmatrix} 1 & 3 & -1 \\ 4 & -1 & 0 \\ -1 & 1 & 5 \end{pmatrix}$$

$$= (-1)(-1)^{1+3} \det \begin{pmatrix} 4 & -1 \\ -1 & 1 \end{pmatrix} + 0 + 5(-1)^{3+3} \det \begin{pmatrix} 1 & 3 \\ 4 & -1 \end{pmatrix}$$

$$= -(4-1) + 5(-1-12) = -68 \neq 0.$$

Since $\det \mathbf{A} \neq 0$, *we can now write*

$$x_1 = \frac{\det \begin{pmatrix} 6 & 3 & -1 \\ 7 & -1 & 0 \\ -6 & 1 & 5 \end{pmatrix}}{\det \mathbf{A}} = \frac{((-1)(7-6) + 5(-6-21))}{-68} = \frac{-136}{-68} = 2$$

$$x_2 = \frac{\det \begin{pmatrix} 1 & 6 & -1 \\ 4 & 7 & 0 \\ -1 & -6 & 5 \end{pmatrix}}{\det \mathbf{A}} = \frac{((-1)(-24+7) + 5(7-24))}{-68} = \frac{-68}{-68} = 1$$

$$x_3 = \frac{\det \begin{pmatrix} 1 & 3 & 6 \\ 4 & -1 & 7 \\ -1 & 1 & -6 \end{pmatrix}}{\det \mathbf{A}} = \frac{((1)(6-7) - 3(-24+7) + 6(4-1))}{-68} = \frac{68}{-68} = -1.$$

∎

In solving for x_3, the determinant in the numerator was evaluated using cofactors of the first row.

You should check that the values $x_1 = 2$, $x_2 = 1$, and $x_3 = -1$ satisfy the equations.

Appendix E

The Wronskian

A concise way of determining whether a linear combination of two solutions of the second-order linear homogeneous equation

$$x'' + p(t)x' + q(t)x = 0 \tag{E.1}$$

is a general solution of the equation is provided by a function called the Wronskian. It will be assumed in what follows that the coefficient functions $p(t)$ and $q(t)$ are continuous for all values of t. If this is not the case, it can be shown that the same arguments can be used if we simply restrict t to an interval in which p and q are continuous.

Suppose we have two solutions x_1 and x_2 of (E.1), and we form the linear combination $x = C_1 x_1 + C_2 x_2$. We know from Lemma 3.1 that x is a solution of (E.1) for any constants C_1 and C_2. We want to know if this linear combination gives us all possible solutions. Theorem 3.2 shows that to have all possible solutions we need to be able to find values for C_1 and C_2 such that $x = C_1 x_1 + C_2 x_2$ can be made to satisfy arbitrary initial conditions at any initial time $t = t_0$; that is, we need to find conditions on x_1 and x_2 such that the system of simultaneous linear equations

$$\begin{aligned} x(t_0) &= C_1 x_1(t_0) + C_2 x_2(t_0) = x_0 \\ x'(t_0) &= C_1 x_1'(t_0) + C_2 x_2'(t_0) = v_0 \end{aligned} \tag{E.2}$$

has a solution C_1, C_2 for arbitrary values of x_0, v_0, and t_0. This is just a system of two linear equations in the two unknowns C_1 and C_2, and a well-known theorem in linear algebra states that a unique solution exists for any values of x_0 and v_0 if, and only if, the determinant of the coefficients of C_1 and C_2 is unequal to zero; that is, if, and only if,

$$\begin{vmatrix} x_1(t_0) & x_2(t_0) \\ x_1'(t_0) & x_2'(t_0) \end{vmatrix} = x_1(t_0)x_2'(t_0) - x_1'(t_0)x_2(t_0) \neq 0 \text{ for any } t_0.$$

Definition E.1. *If $x_1(t)$ and $x_2(t)$ are solutions of the second-order linear homogeneous equation* (E.1), *the determinant*

$$W(x_1, x_2)(t) \equiv \det\begin{pmatrix} x_1(t) & x_2(t) \\ x_1'(t) & x_2'(t) \end{pmatrix} \equiv x_1(t)x_2'(t) - x_1'(t)x_2(t)$$

*is called the **Wronskian** of the functions x_1 and x_2.*

The following theorem shows that it is only necessary to check the value of the Wronskian at a single value of t.

Theorem E.1. *If x_1 and x_2 are two solutions of the equation $x'' + px' + qx = 0$, then $W(x_1, x_2)$ is either zero for all t or unequal to zero for all t.*

Proof. The Wronskian is a function of t, and the proof consists of showing that W satisfies the first-order differential equation $\dfrac{dW}{dt} = -pW$, where $p(t)$ is the coefficient of x' in (E.1). To see this, use the product rule for differentiation to write

$$\frac{d}{dt}W(x_1, x_2) = \frac{d}{dt}(x_1 x_2' - x_1' x_2) = x_1 x_2'' + x_1' x_2' - x_1' x_2' - x_1'' x_2$$
$$= x_1 x_2'' - x_1'' x_2 = x_1(-px_2' - qx_2) - (-px_1' - qx_1)x_2$$
$$= -p(x_1 x_2' - x_1' x_2) = -pW(x_1, x_2),$$

where the hypothesis that x_1 and x_2 satisfy (E.1) has been used to write $x_1'' = -px_1' - qx_1$ and $x_2'' = -px_2' - qx_2$. The resulting first-order differential equation $W' = -pW$ has solution $W(t) = W_0 e^{-\int_{t_0}^{t} p(s)ds}$, where W_0 is the value of $W(x_1(t), x_2(t))$ at $t = t_0$. Therefore, if $W_0 = 0$, the Wronskian is zero for all values of t, and if $W_0 \neq 0$, W is an exponential function that is unequal to zero for all values of t for which it is defined. $\square$

At this point we have proved the following theorem.

Theorem E.2. *If x_1 and x_2 are two solutions of the homogeneous equation (E.1), and their Wronskian $x_1 x_2' - x_1' x_2$ is not equal to zero at some value $t = t_0$, then $x = C_1 x_1 + C_2 x_2$ is a general solution of (E.1).*

Definition E.2. *A pair of solutions x_1 and x_2 of (E.1) is called a **fundamental solution set** of (E.1) if $W(x_1(t), x_2(t))$ is unequal to 0 at some value of t.*

The following lemma shows that for a second-order linear equation, an equivalent condition for a pair of solutions to be a fundamental solution set is that they not be constant multiples of each other.

Lemma E.1. *Given two nonzero solutions x_1, x_2 of (E.1), their Wronskian $W(x_1, x_2)$ is identically equal to zero if, and only if, $x_2 = Kx_1$ for some constant K.*

Proof. If $x_2 = Kx_1$, then

$$W(x_1, x_2) = W(x_1, Kx_1) = x_1(Kx_1)' - (x_1)'Kx_1 = K(x_1 x_1' - x_1' x_1) \equiv 0.$$

Conversely, suppose $W(x_1, x_2) \equiv 0$. Since we are assuming that x_1 is not the zero function, we can find a value of t, say t_0, where $x_1(t_0) \neq 0$, and since the solution x_1 is a continuous function it will be unequal to zero in an entire interval I around t_0. Using the quotient rule for differentiation,

$$\frac{d}{dt}\left(\frac{x_2}{x_1}\right) = \frac{x_1 x_2' - x_1' x_2}{x_1^2} = \frac{W(x_1, x_2)}{x_1^2} \equiv 0$$

on the interval I, and this implies that x_2/x_1 is a constant K in the interval I; that is, $x_2(t) \equiv Kx_1(t)$. Since $x_2(t_0) = Kx_1(t_0)$, $x_2'(t_0) = Kx_1'(t_0)$, the two solutions x_1 and $x_2 = Kx_1$ both satisfy the same initial conditions at t_0. By the Existence and Uniqueness Theorem they must be identical functions for all t for which they are defined. $\square$

Appendix F

Table of Laplace Transforms

formula	function $f(t)$	Laplace transform $F(s) = L[f(t)]$
1	1	$\frac{1}{s}, s > 0$
2	e^{at}	$\frac{1}{s-a}, s > a$
3	$\cos bt$	$\frac{s}{s^2+b^2}, s > 0$
4	$\sin bt$	$\frac{b}{s^2+b^2}, s > 0$
5	t^n	$\frac{n!}{s^{n+1}}, s > 0$
6	$af(t) + bg(t)$	$aF(s) + bG(s)$
7	$f'(t)$	$sF(s) - f(0)$
8	$f''(t)$	$s^2 F(s) - sf(0) - f'(0)$
9	$e^{at} f(t)$	$F(s - a)$
10	$t^n f(t)$	$(-1)^n F^{(n)}(s)$
11	$e^{at} \cos bt$	$\frac{s-a}{(s-a)^2+b^2}$
12	$e^{at} \sin bt$	$\frac{b}{(s-a)^2+b^2}$

Continued on next page

formula	function $f(t)$	Laplace transform $F(s) = L[f(t)]$
13	$t \cos bt$	$\dfrac{s^2-b^2}{(s^2+b^2)^2}$
14	$t \sin bt$	$\dfrac{2bs}{(s^2+b^2)^2}$
15	$t^n e^{at}$	$\dfrac{n!}{(s-a)^{n+1}}$
16	$u(t-c)$	$\dfrac{1}{s}e^{-cs}$
17	$\delta(t-c)$	e^{-cs}
18	$u(t-c)f(t-c)$	$e^{-cs}F(s)$
19	$f(t) \star g(t)$	$F(s) \cdot G(s) \equiv \mathcal{L}(f(t)) \cdot \mathcal{L}(g(t))$

Appendix G

Review of Partial Derivatives

In this appendix we review the definitions and notation for first- and second-order partial derivatives of functions of two independent variables. These are the derivatives that appear in the second-order partial differential equations in Chapter 8.

For a function y depending on a single independent variable t, the derivative $y'(t)$ can be thought of as the instantaneous rate of change of y at time t. This derivative, when it exists, can be calculated as follows:

$$y'(t) = \frac{d}{dt}(y(t)) = \lim_{\Delta t \to 0}\left(\frac{y(t+\Delta t) - y(t)}{\Delta t}\right).$$

If a function u is a function of more than one variable, say $u = u(x, y)$, then we need to use partial derivatives to measure the rate of change of u with respect to each of its independent variables. For example, the instantaneous rate of change of u with respect to x is called the **partial derivative of u with respect to x** and is defined by

$$u_x(x, y) \equiv \frac{\partial}{\partial x}(u(x, y)) = \lim_{\Delta x \to 0}\left(\frac{u(x+\Delta x, y) - u(x, y)}{\Delta x}\right).$$

This is just the derivative of u with respect to x, assuming that y is held fixed.

Similarly, the **partial derivative of u with respect to y** is given by

$$u_y(x, y) \equiv \frac{\partial}{\partial y}(u(x, y)) = \lim_{\Delta y \to 0}\left(\frac{u(x, y+\Delta y) - u(x, y)}{\Delta y}\right).$$

It will also be necessary to use second-order partial derivatives, and these are defined in the same way. For example,

$$u_{xx}(x, y) \equiv \frac{\partial}{\partial x}\left(\frac{\partial u}{\partial x}\right) \equiv \frac{\partial^2 u}{\partial x^2} = \lim_{\Delta x \to 0}\left(\frac{u_x(x+\Delta x, y) - u_x(x, y)}{\Delta x}\right).$$

The other second-order partial derivatives are $u_{yy}(x, y) \equiv \frac{\partial}{\partial y}\left(\frac{\partial u}{\partial y}\right) \equiv \frac{\partial^2 u}{\partial y^2}$ and the mixed derivatives u_{xy} and u_{yx}. For the functions we will be considering, the two mixed partial derivatives are equal and can be denoted by either $u_{xy} = \frac{\partial}{\partial x}\left(\frac{\partial u}{\partial y}\right)$ or $u_{yx} =$

$\frac{\partial}{\partial y}\left(\frac{\partial u}{\partial x}\right)$. Look back at your calculus text to see what the conditions are on the function $u(x, y)$ to ensure that the two mixed partial derivatives *are* equal.

Example G.0.1. *For the function* $u(x, y) = \sin(4x)\cos(3y)$, *find all first and second partial derivatives, and show that u satisfies the partial differential equation* $u_{xx} = \frac{16}{9}u_{yy}$.

Solution. The six derivatives are

$$u_x = 4\cos(4x)\cos(3y), \quad u_{xx} = -16\sin(4x)\cos(3y), \quad u_{yx} = -12\cos(4x)\sin(3y),$$
$$u_y = -3\sin(4x)\sin(3y), \quad u_{yy} = -9\sin(4x)\cos(3y), \quad u_{xy} = -12\cos(4x)\sin(3y).$$

To show that u satisfies the given equation,

$$u_{xx} = -16\sin(4x)\cos(3y) = \frac{16}{9}(-9\sin(4x)\cos(3y)) = \frac{16}{9}u_{yy}. \quad \blacksquare$$

Index

absolute error, 61
acceleration, 1, 12
air resistance, 12
ambient temperature, 25
amplitude, 91
analytic function, 128
analytic solution, 4

beats, 119
Bernoulli equation, 56
   linear, 56
   method of solution, 56
Bessel equation, 81
   $J_N(t)$, 135
bifurcation, 33
   Andronov-Hopf, 218, 220
   competing species model, 211
   definition, 74
   pitchfork, 75
   predator-prey model, 209
   spring-mass model, 208
   two-dimensional system, 208
bifurcation diagram, 75
bifurcation value, 74
boundary-value problem, 8

Cauchy-Euler equation, 114
   characteristic polynomial, 126
   definition, 125
   general solution, 126
   solution by variation of parameters, 114
chain rule, 4, 60, 126
change of variable, 126
characteristic polynomial
   $\det(\mathbf{A} - r\mathbf{I})$, 160
   Cauchy-Euler equation, 126
   complex roots, 91
   for $a_n x^{(n)} + \cdots + a_0 x = 0$, 93
      degree= n, 388
   for $ax'' + bx' + cx = 0$, 88
      discriminant, 89
   multiplicity of roots, 93
circular drumhead problem, 279
   Maple program, 283
   series solution, 283
column vector, 150

computer algebra system(CAS), 11
conservation of mass, 41
convolution integral
   definition, 252
   Laplace transform, 252
cosh(t), 181, 267
Cramer's rule
   solution of n equations in n unknowns, 389
critical harvesting rate, 77
critical point, 185
cubic polynomial
   roots, 55

damped pendulum, 206
delta function, 253
derivative, 1
   central difference approximation, 309
   forward difference approximation, 309
   of a linear combination, 83
   of a sum, 83
   rate of change, 1
determinant, 154
   2×2 matrix, 154, 387
   method of cofactors, 154, 388
differential equation, 2
   analytic solution, 4, 113
   first-order, 3
      autonomous, 20, 71
      Bernoulli equation, 56
      definition, 19
      exact, 51
      linear, 36
      separable, 20
   general solution, 7, 9
      algebraic formula, 9
      geometric description, 30
   linear, 9
   nonlinear, 9, 10, 12
   numerical solution, 4
   order, 3
   ordinary, 2
   partial, 2
   second-order, 3, 8
      autonomous, 136
      definition, 81
      linear, 81

linear homogeneous, 81
   phase plane for $x'' = F(x, x')$, 139
system of first order, 145, 146
   converting to a system, 146
   solution, 5
differentiation formulas, 385
diffusion coefficient, 262
direction field, 184
discriminant, 89
double spring-mass system
   definition, 148
   equilibrium position, 148
   first-order system, 148
   force equations, 148
   in matrix form, 156
   initial conditions, 148
   natural frequency, 156
dynamical systems, 16, 219

ecology, 14
ecosystem, 15
   ant population, 18
   deer population, 32
   spruce-budworm population, 77
   Western grasslands model, 206
eigenpair, 155, 159
eigenvalue, 154, 159
   algebraic multiplicity, 164
   solution of $\det(\mathbf{A} - r\mathbf{I}) = 0$, 160
eigenvector, 155, 159
   normalized to unit length, 164
   solution $\vec{\mathbf{u}}$ of $(\mathbf{A} - r\mathbf{I})\vec{\mathbf{u}} = \vec{\mathbf{0}}$, 160
electrical circuit, 13
   inductance, 13
   resistance, 13
   RL-circuit, 13, 17
      steady-state solution, 14
   RLC-circuit
      example, 124
      resonance, 122
      time constant, 123
      transient, 123
elementary functions, 29, 132
equilibrium point, 185
   hyperbolic, 199, 202
equilibrium solution, 71
   center, 190
   node, 73
   nonlinear system
      type, 200
   saddle point, 190
   semi-stable, 73
   sink, 73, 189
   source, 73, 189
   spiral sink, 191
   spiral source, 191
   stable, 73
   two-dimensional system, 185
   unstable, 73
equivalence sign, 4

error function $\mathrm{erf}(t)$, 132
Euler's method, 60
   second-order autonomous equation, 136
even function, 266
exact differential equation, 51
   definition, 52
   integrating factor, 55
   method of solution, 53
   test for exactness, 52
existence and uniqueness, 44
   first-order equation $x' = f(t, x)$, 44
   for $x' = -t/x$, 50
   second-order equation
      $x'' = F(x, x')$, 136
      for $x'' + p(t)x' + q(t)x = f(t)$, 84
   two-dimensional system, 185
explicit solution, 21
exponential function, 6
exponential growth, 15
exponential order, 222

fixed point, 185
forced spring-mass equation, 114
Fourier cosine series, 271
Fourier series
   trigonometric Fourier series, 268
      coefficients, 268
      convergence, 270
Fourier sine series, 271
friction
   air resistance, 12
fundamental solution set, 85
   definition, 392

gain function, 122
   using Laplace transforms, 257
geometric series
   partial sum, 248
Gibbs phenomenon, 272, 274
graphical methods, 28
gravitational acceleration, 12

Hartman-Grobman Theorem, 199, 200, 202
heat equation
   boundary conditions, 291
      nonhomogeneous, 296
   derivation, 289
   initial conditions, 291
   insulated ends, 294
Heaviside function, 239
   using the SIGN function, 242
Heun's method, 64
hyperbolic equilibrium, 199
   definition, 202

identity, 4
   equivalence sign, 4
identity matrix, 154
implicit solution, 21
improper integral, 223
improved Euler method, 64

Index

algorithm, 65
  order, 64
impulse response, 256
indefinite integral, 20, 21
induction proof, 238
initial condition, 7, 8
initial-value problem, 7, 8
integral transform, 221
integrating factor, 38
integration formulas, 385
inverse Laplace transform, 227
inverse matrix, 157
  for $s\mathbf{I} - \mathbf{A}$, 237
  solving linear equations, 157
irreducible quadratic, 233
isocline, 31

Jacobian matrix, 201

Laplace transform
  complete table, 393
  definition, 223
  inverse
    definition, 227
    linear operator, 227
  linear operator, 224
  of $f'(t)$, 225
  solution of a differential equation, 229
Laplace's equation, 2, 3
  solution on a rectangle, 318
laplacian, 290, 305
  in cylindrical coordinates, 305
Legendre equation, 81
Legendre polynomials, 273
limit cycle, 204, 205, 220
  stable, 217
linear differential equation
  autonomous
    impulse response, 256
  first-order
    definition, 36
    general solution, 38
    homogeneous, 36
    homogeneous solution, 37
    integrating factor, 38
    standard form, 36
  fundamental solution set, 85
  second-order
    associated homogeneous equation, 82
    general solution $x_h + x_p$, 83
    homogeneous solution x_h, 82
    particular solution x_p, 83
    series solution, 127
linear system
  nonhomogeneous
    solution by Laplace transforms, 236
    solution by undetermined coefficients, 175
    solution using matrix exponential, 178
Lipschitz condition, 45
local truncation error, 63
logistic growth equation, 14, 24
  analytic solution, 15, 24
    interval of existence, 47
  as a Bernoulli equation, 56
  carrying capacity, 15, 24
  existence and uniqueness, 46
  intrinsic growth rate, 15, 24
  shape of solution curves, 15
  slope field, 30
  with harvesting, 32
    bifurcation diagram, 76
    critical harvesting rate, 77
    deer population, 32
    equilibrium points, 75
    phase line, 76

Maclaurin series, 128
Maple instructions
  DEplot, 35
    scene, 184
  Digits, 68
  dsolve, 11, 70
  Eigenvectors(A), 163
  invlaplace, 231
  laplace, 231
  MatrixExponential, 181
  op, 70
  sol, 70
  with(DEtools), 35
  with(inttrans), 231
  with(LinearAlgebra), 163
Mathematica instructions
  Dsolve, 11
  Eigensystem[A], 164
  First, 71
  InverseLaplaceTransform, 232
  LaplaceTransform, 231
  NDSolve, 71
  ScaleFunction, 35
  VectorFieldPlot, 35
matrix
  cofactor A_{ij}, 387
  minor $\mathbf{M}_{ij}$, 387
  notation for, 155
  square
    main diagonal, 153
    multiplicative identity, 154
matrix algebra, 150
  basic operations
    addition, 150
    algebraic properties, 152
    matrix multiplication, 151
    scalar multiplication, 151
matrix
  derivative, 153
  derivative of a product, 179
  eigenvalue, 154
  equality, 150
  nilpotent, 178
  self-adjoint, 282
  size, 150

square, 150
zero, 150
matrix(definition), 150
multiplicative inverse, 157
matrix exponential
   definition, 176
   diagonal matrix, 176
   fundamental matrix, 177
   properties, 177
method of cofactors, 162
mixing problem
   three-compartment
      matrix exponential solution, 181
mixing problem
   concentration, 42
   flow rates, 42
   one-compartment, 41
      blood flow, 247
      example, 41
      piecewise continuous input, 244
      variable flow rates, 42, 44
   pollutant, 41
   three-compartment, 180
   two-compartment, 169
      analytic solution, 173
      long-term behavior, 170

nerve cell activity, 15
   neuron, 78
   oscillatory solution, 16
   periodic input, 16
   response function, 15
      $S(z)$, 215
      $S_a(z)$, 79
   single neuron equation, 78
      bifurcation diagram, 80
      oscillatory input, 80
      phase line, 79
   threshold level, 15, 79
   Wilson-Cowan system, 215
neurology
   response function
      arctan, 18
      S(z), 18
      tanh, 18
   single-neuron model, 15
neuron, 78
Newton's Law of Cooling, 25
   ambient temperature, 25
   explicit solution, 25
Newton's second law, 1, 141, 148
node, 73
nonlinear system
   equilibrium solution
      type, 200
      linearization about an equilibrium, 201
nullcline for x', 199
nullcline for y', 199
numerical methods, 59
   nth-order numerical method, 64

danger in use, 70
error of order n $\mathcal{O}((\Delta t)^n)$, 63
error propogation, 63
Euler's method, 60
   algorithm, 61
   algorithm for $x'' = F(x, x')$, 136
   first-order method global truncation error
      $\mathcal{O}(\Delta t)$, 63
   Heun's method, 64
   improved Euler method, 64
      truncation error, 64
   local truncation error, 63
   round-off error, 62, 68
   Runge-Kutta method
      order, 66
   significant digits, 68
   truncation error, 63

odd function, 266
one-parameter family, 12
   ellipses, 13
   parabolas, 13
ordinary differential equation, 2
orthogonal curves, 13, 17, 27
orthogonal set of functions, 266

parameter
   bifurcation value, 33, 74
   definition, 3
   experimentally determined, 26
parametric space curve, 138
partial derivative, 3
   equality of mixed partials, 52
partial differential equation, 2
   classification, 286
   definition, 261
   diffusive logistic equation, 316
      with harvesting, 324
   linear, 285
      homogeneous, 286
   second-order, 285
   steady-state solution, 325
partial fraction expansion, 24, 228
   `expand`, 228
particular solution, 7
pendulum, 141
   damped pendulum model, 206
pendulum equation, 141
   damped
      phase plane, 142
   undamped
      phase plane, 142
phase angle, 91
phase line
   construction, 71
phase plane
   definition, 139
   two-dimensional linear system
      both eigenvalues zero, 196
      center, 190
      direction of rotation, 191

Index

equal eigenvalues, 194
improper node, 195
saddle point, 190
sink, 189
source, 189
spiral sink, 191
spiral source, 191
zero eigenvalue, 193
two-dimensional system, 184
vector field, 139
phase portrait, 140, 184
piecewise continuous function, 221
piecewise linear approximation, 61, 138
pitchfork bifurcation, 75
Poincaré-Bendixson Theorem, 204
population growth, 1, 23
competing species, 206
Lotka-Volterra model, 198
with parameter, 211
cooperating species model, 214
exponential growth, 1, 24
predator-prey model, 207
group defense behavior, 218
nonlogistic growth, 213
with one parameter, 209
predator-prey system, 184
spruce-budworm equation, 77
spruce-budworm equation, 28
two-predator, one prey, 149, 207
U. S. census data, 27
power series
change of index, 129
derivative, 128
recursion relation, 130
sums of, 129
pure resonance, 118
graph, 118

quadratic form, 287

ramp function, 243
RC-circuit
piecewise continuous input, 250
reaction-diffusion equation, 262
recursion formula, 128
recursion relation, 130
resonant frequence, 122
response function, 15
$erf(t)$, 132
$S(z)$, 15
derivative, 79
RL-circuit, 44
phase line, 78
steady-state current, 78
RLC-circuit, 204
example, 124
power surge, 254
resonance, 122
van der Pol equation, 204
round-off error, 62
row vector, 150

Runge-Kutta method, 66
algorithm, 66

sawtooth function, 226, 247
example, 249
scalar, 150
semi-stable equilibrium, 73
separable equation, 20
method of solution, 21
separatrix, 190, 203
series solutions, 127
sign function, 242
single neuron equation, 78
sinh(t), 181, 267
sink, 73
skydiver, 12
as a Bernoulli equation, 59
phase line, 78
problem, 17
solution of the equation, 44
terminal velocity, 12, 17, 59
slope field, 12, 28
autonomous equation, 30, 48, 71
definition, 29
for $x' = -t/x$, 50
for $x' = x^2 - t$, 67
separating solution, 67
symmetry, 58
using a rectangular grid, 29
using isoclines, 31
for $x' = x^2 - t$, 31
slow-release drug, 247
solution
constant solution, 23, 47
for a Bernoulli equation, 57
equilibrium solution, 71
explicit, 21
general solution, 7, 8, 23
of $a_n x^{(n)} + \cdots + a_0 x = 0$, 93
of $ax'' + bx' + cx = 0$, 89
implicit, 21, 52
long-term behavior, 25, 40
monotonic, 72
n-parameter family, 6
one-parameter family, 7, 10, 14, 20, 21
cubics, 17
ellipses, 13
parabolas, 13
particular solution, 7, 22
two-parameter family, 8, 9, 83
solution of $\vec{x}' = \mathbf{A}\vec{x}$
constant 2×2 matrix $\mathbf{A}$
complex eigenvalues, 166
distinct real eigenvalues, 166
equal eigenvalues, 167
source, 73
space curve
parametric, 138
space-filling curves, 46
special function

Bessel function $J_N(t)$, 135
error function $erf(t)$, 132
spring-mass equation, 1
  beats
    graph, 120
  damped
    gain function, 122
    resonant frequency, 122
  damped system
    resonance, 121
  damping coefficient, 2
  forced
    discontinuous driving function, 221
    steady-state solution, 114
    transient solution, 114
  hard spring, 143
  periodic forcing, 115
  pure resonance, 117
    graph, 118
  soft spring, 143
  spring constant, 2
  undamped
    beats, 119
    pure resonance, 118
  variable damping, 129
spring-mass system
  undamped
    beats, 117
    natural frequency, 117
    resonance, 117
spruce-budworm equation, 77
spruce-budworm equation, 28
square matrix
  determinant, 154
  trace, 154
stable equilibrium, 73
stable manifold, 203
Stable Manifold Theorem, 203
stationary point, 185
steady-state solution, 114
Sturm-Liouville boundary-value problem, 274
  eigenfunction, 275
    orthogonality of eigenfunctions, 275
  eigenvalue, 275
  homogeneous unmixed boundary conditions, 275
  weight function, 275
system of first-order equations
  linear system, 147
    dimension, 147
    homogeneous, 147
    matrix form, 155
  two-dimensional
    phase plane, 184

tangent line approximation, 60
tangent vector, 184, 188
  sketching, 142
  to a space curve, 139
Taylor series, 60, 128, 176

error formula, 60
  for $f(x, y)$, 201
terminal velocity, 44
time constant, 123
time plot, 138
total differential, 51
trace, 154
trace-determinant plane, 192
trajectory, 28, 140, 186
transient, 123
transient solution, 114
trigonometric formula
  $A\sin(bt) + B\cos(bt)$ as a cosine, 94
  $A\sin(bt) + B\cos(bt)$ as a sine, 91
  for $\sin(A) - \sin(B)$, 120
truncation error, 63
tuning fork, 119

U. S. census data, 27
undetermined coefficients, 102
  form of x_p
    if x_p is a homogeneous solution, 107
    if x_p sum of functions, 108
unit impulse function
  definition, 253
  Laplace transform, 253
unit step function
  definition, 239
  Laplace transform, 242
unstable equilibrium, 73

van der Pol equation, 2, 81, 204
  limit cycle, 205
variation of parameters, 109
vector
  dot product, 151
  linear dependence, 153
  linear independence, 153
  linearly independent
    determinant test, 154
  notation for, 155
vector field, 184
velocity, 1
vertical asymptote, 10, 11, 25, 30, 40, 45, 48, 58, 72, 136

wave equation
  D'Alembert's solution, 305
  derivation, 299
  initial conditions, 300
  with damping, 302
Wilson-Cowan system, 215
  equilibrium solutions, 216
  with parameter, 217
word size, 62
Wronskian, 85
  definition, 391
  differential equation for, 392
  zeros of, 85, 392